Techniques for Pollination Biologists

Techniques
for Pollination Biologists

CAROL ANN KEARNS
Environmental, Population, and Organismic Biology
University of Colorado and
Mountain Research Station, University of Colorado

DAVID WILLIAM INOUYE
Departments of Zoology and of Botany
University of Maryland
and Rocky Mountain Biological Laboratory

UNIVERSITY PRESS OF COLORADO

Copyright © 1993 by the University Press of Colorado
P.O. Box 849
Niwot, Colorado 80544

The University Press of Colorado is a cooperative publishing enterprise supported, in part, by Adams State College, Colorado State University, Fort Lewis College, Mesa State College, Metropolitan State College of Denver, University of Colorado, University of Northern Colorado, University of Southern Colorado, and Western State College.

Library of Congress Cataloging-in-Publication Data
Kearns, Carol Ann, 1955–
    Techniques for pollination biologists / Carol Ann Kearns, David William Inouye.
    p. cm.
    Includes bibliographical references and index.
    ISBN 0-87081-279-3 (alk. paper). — ISBN 0-87081-281-5 (pbk., alk. paper)
    1. Pollination — Research — Technique. I. Inouye, David William, 1950– . II. Title.
QK926.K425 1993
582'.01662 — dc20   93-14994
CIP

The paper used in this publication meets the minimum requirements of the American National Standard for Information Sciences—Permanence of Paper for Printed Library Materials. ANSI Z39.48–1984

∞

10 9 8 7 6 5 4 3 2

Contents

Contents

Contents

Contents

Contents

Contents

Contents

Contents

Contents

Preface

The stimulus for this book came from the awareness that there has been no single source for much of the practical information needed for experimental field studies of pollination biology. This is in part a consequence of the diversity of fields that contribute to the discipline. There have been handbooks of techniques for laboratory studies for at least 125 years (e.g., Frey's 1867 *Handbuch der Histologie und Histochemie des Menschen*), and there are some recent books describing general ecological techniques, but we perceived a need for a similar handbook for pollination studies. Having met several pollination biologists who had notebooks with copies of various stain recipes, microscope techniques, and so on, we have tried to incorporate all of this and other information from the literature into a single book. If we have neglected techniques that you use, please send them to us for incorporation in a possible future edition. In writing this book C. A. K. was primarily responsible for Chapters 4, 5, and 6, and Appendices 1 and 5, and D. W. I. for Chapters 1, 2, 3, 7, 8, and 9, and Appendices 2, 3, and 4.

We have been lucky to work at the Rocky Mountain Biological Laboratory, where many pollination biologists gather each summer and where one can often find answers to questions about how best to carry out some technique or experiment. For those not so fortunate, we hope that this book will provide some answers, and even raise some questions.

David W. Inouye
Department of Zoology
University of Maryland
College Park, MD 20742
e-mail:
di5@umail.umd.edu

Carol A. Kearns
Environmental, Population, and
Organismic Biology
University of Colorado
Boulder, CO 80309
e-mail:
kearns@colorado.edu

Acknowledgments

Many pollination biologists have contributed to this book through their publications, responses to our requests for additional information or tricks of the trade, or reviews of earlier versions of chapters. Several researchers who also work at the Rocky Mountain Biological Laboratory were particularly helpful, including James Thomson, Nick Waser, Mary Price, Lisa Rigney, Randy Mitchell, Mark Schlessman, Ward Watt, Carol Boggs, and Mitch Cruzan. We also thank the other people who commented on the manuscript, including Mike Zimmerman, Olle Pellmyr, Heidi Dobson, Jeff Mitton, Sara Hiebert, Joe Patt, Doug Gill, Carl Elliger, Dave Carr, Michelle Dudash, Charlie Fenster, Allison Snow, Ed Southwick, Kermit Ritland, and Gary Bernard. Sue Short supplied much of the information in Appendix 1. Sarah Corbett provided names and addresses of some sources for equipment and supplies in England. A number of colleagues shared copies of unpublished data or figures, unpublished manuscripts, computer programs, manuscripts in press, or answered other questions for us. They include Jeff Mitton, Helen Young, Peter Wetherwax, Robert Cruden, Robert Wyatt, Jens Olesen, Rod Peakall, Lynn Carpenter, Stephen Chaplin, Norman Gary, Bastiaan Meeuse, Joe Patt, Brian Inouye, Sharon Kinsman, Alberto Búrquez, Jim Cane, James Cresswell, Barbara Thomson, Peter Turchin, Dan Livingstone, Alex Motten, and Stephen Murphy. Many others also provided reprints, suggested sources of equipment, or answered our questions in letters, calls, and electronic mail. Sigma Chemical Company's technical service representatives were helpful in locating chemicals for Appendix 3. Earthwatch and its Research Corps helped to fund the fieldwork that stimulated our interest in pollination biology.

The University of Colorado's Mountain Research Station, the Rocky Mountain Biological Laboratory, and the University of Maryland's Departments of Zoology and Botany were also supportive of the effort required to produce this manuscript. The interlibrary loan department of McKeldin Library at the University of Maryland assisted greatly by obtaining references.

Permission to use quotations or figures was granted by the following:

American Journal of Botany (Figure 3–2)
Analytical Biochemistry (Figure 5–1)
Animal Behavior Society (Petersen 1990, in Appendix 4)
Blackwell Scientific Publications (Figures 7–14, 7–24)
Canadian Journal of Zoology (Figures 7–19B, 7–20, 7–28)
Condor (Figure 7–17)
CRC Publishing (Tables 5–2, 8–1)
Ecological Society of America (Figures 4–2, 7–9)
Entomological Society of America (Figure 7–27)
C. Galen (poem, Chapter 5)
L. Goldwasser (Figure 7–16)
International Bee Research Association (Figure 7–10)
Journal of Chromatography (Figure 5–2)
Journal of the Lepidopterists' Society (Figure 7–15)
National Research Council of Canada (Figure 7–13)
New York Botanical Garden and Fryxell (Figure 6–1)
N. Newfield (Figure 7–6)
Phytochemical Bulletin (Table 5–3)
Springer Verlag (Figures 7–18, 7–22, 7–23, 7–26)
University of Nijmegen, Department of Botany (Figure 7–19A)
University of Texas Press (Table 5–5)

Techniques for Pollination Biologists

1. Introduction

Sexual reproduction in angiosperms has three sequential stages: pollination, fertilization, and seed maturation (Lyons et al. 1989). The first of these stages is primarily the concern of field biologists, while the second is more suitably studied in the laboratory, because pollen grains and ovules are not easily studied in the field. This handbook is designed primarily for these first two areas of study.

Pollination biology draws from many biological and some chemical or physical disciplines. Taxonomy is represented in the identification of both plants and their pollinators; sensory physiology in the description of pollinators' senses of smell and color; physiology in work on flight temperatures of pollinators or of nectar secretion and digestion; morphology in the description of flowers or pollinator mouthparts; population and quantitative genetics in the study of floral traits, plant population structure, and breeding systems; animal behavior in the study of pollinator movements and responses to flowers; and meteorology in the study of wind pollination and how weather affects flowering and the activity of pollinators. Throughout, elements of botany and zoology are intermingled, as are fieldwork and laboratory studies. Thus the list of what should be included in a study of pollination biology can become extensive.

Experimental studies of pollination biology actually started at least two centuries ago. For example, in 1736 James Logan published "Experiments Concerning the Impregnation of the Seed of Plants" (Baker 1983; for a comprehensive history of the field of pollination biology, see this reference and Proctor and Yeo 1973). Clements and Long (1923) review the early history of experimental pollination, which they attribute to the French biologist F. Plateau. His first pollination paper, on artificial flowers, appeared in 1877 and his last in 1910. The twenty papers he published during this period describe his work on the significance of color and odor in attraction of pollinators and his studies of the foraging behavior of pollinators. He and his contemporaries conducted experiments on the senses of pollinators (for example, by removing antennae and looking at color choices), manipulated floral rewards by adding honey, and altered floral displays by removing parts of flowers and covering parts of inflorescences.

Although all of this early phase in anthecology took place in Europe, Clements and Long conducted a wide variety of manipulative experiments in the Colorado Rocky Mountains seventy years ago. During the 1950s and 1960s most studies of pollination were ecological and descriptive in nature. Since then, accompanying the development of evolutionary ecology, experimental and manipulative studies have once again become more common. Despite this renaissance, there has been no one source to turn to for information about techniques used in pollination studies.

The equipment available for use by pollination biologists has changed significantly in the past few decades or even years. Previously, pollen counts could only be made painstakingly with a microscope; now they can be done with a particle counter. New insights into the biology of pollen tubes have been made possible with the development of fluorescence microscopy and appropriate staining techniques. Nectar concentration can be measured easily in the field with hand-held refractometers. And microclimate in or around flowers can be measured with electronic instead of mechanical instruments. Laboratory techniques such as high-performance liquid chromatography and nuclear magnetic resonance spectroscopy permit identification of floral odor components, and electrophoresis or DNA fingerprinting allow genotyping of individual plants. We can probably look forward to continued progress in this technological vein, including the ability to manipulate flowering and flower development with the knowledge gained recently about the genes controlling the development and growth of flowers in plants. These techniques can already be used to generate mutations in flower color and floral morphology (e.g., Coen and Meyerowitz 1991, Luo et al. 1991). Although pollination biologists don't seem to have taken advantage of these capabilities yet for experimental studies, they probably will soon.

Pollination is studied for a variety of reasons. Because plant-pollinator interactions can provide some of the best examples of coevolution, evolutionary biologists may choose to study their intricacies. Flowers and pollinators have also provided the basis for studies of phenotypic selection and reproductive success (e.g., Campbell 1989, 1991a). Ecologists seeking systems that can be easily manipulated may turn to flower-visiting insects because of their accessibility and visibility and because data are often easy to collect. In many cases both plants and flower visitors are good candidates for studies of community ecology (e.g., Inouye 1978). Behavioral biologists may find pollinators well suited for studies of optimality theory. Population geneticists may also find pollination systems conducive to answering questions about the movement of genes in a population, by using morphological or electrophoretic genetic markers to study gene flow or neighborhood size. Pollination studies can also provide insight into a plant population's genetic structure, sexual selection, and sex allocation, or facilitate studies of the cost of reproduction in plants (e.g., Primack and Hall 1990). Thus pollination ecology can have significance for conservation biology (e.g., how to manage a small population of an inbred species that is self-incompatible, or for hand-pollination of a species that has lost its native pollinator), restoration ecology (e.g., whether a plant can be reestablished in the absence of a specialist pollinator), and perhaps even sustainable agriculture. Therefore pollination techniques are important for a variety of studies in which pollination per se isn't the endpoint.

There are also important economic reasons for studying pollination. Pollinators are essential for many flower and fruit crops (e.g., Free 1970a). The demand for pollinators appears to be growing in the United States (Torchio 1990), both for introduced bees such as honeybees and managed populations of native bees (Torchio 1987). Bumblebees are now important pollinators of greenhouse plants (Eijnde 1990). In Europe one commercial company (Koppert) rears more than 10,000 bumblebee colonies per year for greenhouse pollination of tomatoes; a similar business (Bees-under-Glass Pollination Services) in Canada expects to rear about 5,000 colonies per year (Sanford 1991). Corbet (1991) reviewed a variety of applied pollination studies aimed at managing the interaction among plants, insects, and humans. Kevan et al. (1990) discuss the changes in agricultural pollination in North America that may be necessitated by the spread of Africanized honeybees, and pathogenic mites that affect honeybees. They suggest that there will be an increased reliance on native or wild pollinators, about which there is still much to be learned.

There has been a lamentable lack of interaction between field biologists who study pollination and plant breeders working with crop plants. Although not all techniques developed by the latter group will be useful for ecological studies, certain techniques could well be adopted. You may find it instructive to look at some of the many books on applied plant breeding (e.g., Lawrence 1968, Janick and Moore 1975, Sedgley and Griffin 1989). Some of the products of plant breeding experiments may also be useful for ecological studies. For example, varieties with different petal sizes or numbers could be used for studies on pollinator behavior. Studies of the degree of pollen dispersal by insects from genetically engineered crops is an area of current interest (e.g., Umbeck et al. 1991) that may help to further interaction between field biologists and plant breeders.

Despite the diversity of reasons for and the varied perspectives used in pollination studies it is possible to outline the sorts of information generally sought. Baker and Hurd (1968) made a list of information that one might include in a pollination study:

1. identification of plants and flower visitors;
2. attractive devices;
3. breeding systems and floral behavior;
4. collection and study of flower visitors throughout the distribution range of the plants;
5. detailed observations on behavior of flower visitors;
6. study of pollinators for daily or seasonal cycles or periodicities;
7. collection of flower visitors from other plant species in the habitat of the study plant;
8. analysis of pollen carried by flower visitors;

To this list we would add:

9. description of the phenology of flowering and of floral parts;
10. studies of pollen carryover;
11. studies of the behavior and physiology of pollinators;
12. micrometeorological studies;
13. chemical analyses of nectar and pollen;
14. morphological measurements of both flowers and flower visitors;
15. studies of inflorescence structure;
16. comparisons of different visitors to a flower species.

We have tried to include in this book descriptions of techniques necessary for all of these components of a pollination study. The contents probably reflect our own research experience in northern hemisphere temperate regions, but many of the techniques should be universally applicable. The contents also reflect the bias in the literature and our own work toward insects as pollinators, but we have tried to include information of relevance for both avian and mammalian systems, as well as abiotic pollination. Some of the methods we describe call for stains, chemicals, or somewhat specialized equipment. We have listed sources and representative prices for many of these in Appendix 2 (Sources of Equipment and Supplies) and Appendix 3 (Chemicals and Stains). We have not, however, covered in much detail many of the histological techniques that are important parts of some laboratory studies of pollination (Shivanna and Rangaswamy [1992] covers a variety of laboratory techniques for studying pollen). In most sections we have described a variety of techniques that can be used. The one that is most appropriate for you will depend on what equipment you have access to, your budget, and the characteristics of the system you are studying.

We have not had the opportunity to try all of these techniques and cannot confirm that all of them will work as reported in the literature. We have tried to indicate any techniques or chemicals that are potentially hazardous and urge you to use caution and common sense in employing them.

Although studies of pollination can be simple and elegant, there are also a number of potential pitfalls that may confound either the execution or interpretation of an experimental study. We have tried to anticipate some of these problems and at the end of each chapter present a list of things to keep in mind when planning a study. An additional important point to bear in mind is the (often neglected) distinction between animals seen visiting a flower (flower visitors) and those that effect pollination (pollinators). In the case of nectar robbers (Inouye 1980a), which bite holes in corollas to extract nectar or pollen illegitimately, the distinction is often obvious. But in the case of nectar theft, in which a flower visitor extracts nectar without contacting the reproductive parts of a flower and therefore doesn't pollinate it, the distinction may not be obvious. To confirm that a flower visitor is also a pollinator, you need to make the following observations:

1. Pollen is transferred from the visitor to the stigma.
2. Pollen is transferred between flowers on a plant or among plants.
3. If there is the potential that the pollen may not be viable (e.g., if it has been in contact with an ant; Peakall et al. 1991), you may also want to confirm fertilization or seed production.

We hope that you won't find these lists of caveats and factors to consider too intimidating. They should not distract from the fact that studies of pollination biology can be a lot of fun, intellectually challenging, and rewarding (if not financially, at least intellectually!). Many natural pollination systems are also aesthetically pleasing to work with, with beautiful flowers and flower visitors as subjects. Although we describe a number of techniques that require access to supplies and equipment, remember that "the primary technique of pollination ecology . . . is the same today as in Sprengel's or Darwin's days: consistent observation of what really happens in nature, in the original, natural habitat of the plant under investigation" (Faegri and van der Pijl 1979). This kind of careful observation, combined with manipulative experiments, will provide answers to almost any question in anthecology.

Although they do not emphasize techniques, a number of pollination books may prove useful as sources of ideas, background about study organisms, photographs, drawings, descriptions of marvelously complex pollination mechanisms, and so on. These include Darwin (1876, 1877), Müller (1883), Kerner von Marilaun (1902), Knuth (1906–1909), Grant and Grant (1965), Proctor and Yeo (1973), A. J. Richards (1978), Faegri and van der Pijl (1979), Bentley and Elias (1983), Jones and Little (1983), Real (1983), Meeuse and Morris (1984), Barth (1985), and Willemstein (1987). To keep up with all of the literature pertinent to pollination biology is a daunting task. The bibliography for this book includes 1,223 references, encompassing 223 different journals or series, as well as many books. If you want to narrow your reading list, here is a list of the 23 most commonly cited journals in the bibliography (and the number of times cited):

American Journal of Botany (102)
Oecologia (96)
Ecology (66)
Evolution (54)
American Naturalist (26)
Biotropica (21)
Science (21)
Journal of the Kansas Entomological Society (19)
Oikos (19)
Canadian Journal of Botany (17)
Nature (18)
Australian Journal of Ecology (17)
American Midland Naturalist (15)
Environmental Entomology (15)
Annals of the Entomological Society of America (14)
Canadian Journal of Zoology (14)
Journal of Apicultural Research (13)
Condor (13)
Stain Technology (13)
Plant Systematics and Evolution (12)
New Phytologist (11)
Ecological Entomology (10)
Functional Ecology (10)

The temporal distribution of the references cited in the bibliography (Figure 1–1) indicates that pollination biology is an active

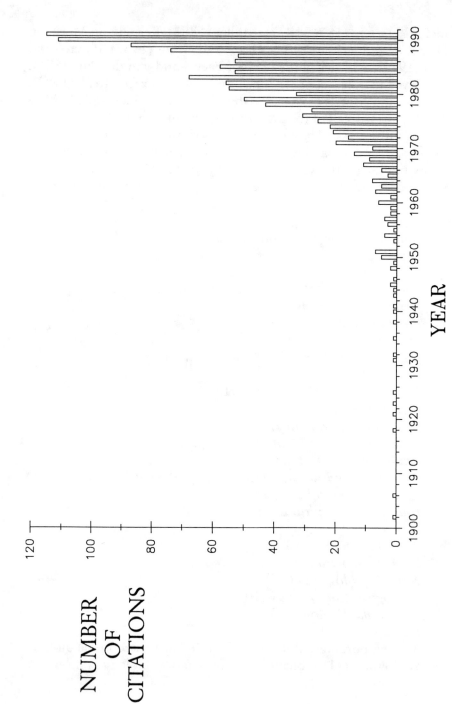

Figure 1–1. The temporal distribution of references cited in the bibliography.

field, with new papers and new techniques appearing every year for the past several decades. We hope that this book will help to continue the trend.

2. Plants

The techniques described in this chapter are appropriate for use on whole plants. Techniques specific to flowers, pollen, or nectar are described in the following chapters.

A. Collecting and Preserving Plants

Systematics is still a very dynamic field of study. Thus the name of a species you are studying now, or even the family to which it is assigned, may differ in time, creating some confusion for future researchers interested in your work. To avoid ambiguity about the species you are studying you should collect voucher specimens and deposit them in a permanent reference collection, such as an herbarium at a museum, academic institution, or field station.

For descriptions of how to press plants correctly for herbarium specimens, see references such as Radford et al. (1974). Wagner (1991) suggests a rule of thumb for collecting: one should not collect any specimen unless there are at least twenty individuals present, and one should not collect more than one out of twenty plants (the "1-in-20" rule). This is probably a wise criterion for avoiding overcollecting, but Pavlovic et al. (1992) argue that an even more conservative approach is needed to incorporate the concept of population size. They suggest a limit of 1% of a population if it is 100–500 plants, 5% if it is over 500 plants, and only a portion of a plant if the population comprises fewer than 100 individuals. Collecting may not be justified at all in smaller populations, and they suggest photography for documentation in such cases. They also suggest a criterion of avoiding more than 5% loss of seed production in a population.

B. Marking or Tagging Plants

Field research involving plants may require a trade-off between marking individual plants so you can find them again easily and making them conspicuous to herbivores, collectors, or vandals. Study sites located in areas protected from most human intrusion are particularly attractive though not always available. Many field stations offer access to protected study sites (for a list and descriptions of habitats and facilities, write to the Organization of Biological Field Stations for a copy of their directory of field stations; see Appendix 2).

An obvious way of marking plants involves the use of vinyl or polyethylene flagging (see Appendix 2), which you can wrap around branches, stems, or trunks and label with a permanent marker. Vinyl is more colorfast, but polyethylene remains pliable in cold temperatures. J. Thomson (*personal communication*) has found that fluorescent colors fade faster than other colors. Flagging especially designed for work in extremely low temperatures is also available. Flagging is easily removed, however, and doesn't usually last more than a single field season or year, so it may not be suitable for some uses. Long exposure to solar radiation makes it brittle, and if it is windy, pieces may break off. If you use flagging to mark plants or plots early in the growing season, you may also find it difficult to relocate later in the season if it is hidden by vegetation. If you have to use inconspicuous markers, they can be made from wire twist-ties wrapped around the stems of herbaceous plants, or from embroidery thread (e.g., Zimmerman 1981, 1987).

Wire stake flags, available in different lengths and with different colors and sizes of polyethylene or vinyl flags, also work well for marking the location of herbaceous plants within a field season and have the advantage of not requiring anything to be tied to the plant itself. These flags may also be useful for supporting stems or, if multiple flags are used per plant, for supporting net bags over plants. They do not, however, work well for long-term marking because the flags may break off and the wire stakes are easily knocked over or pulled out.

More permanent markers required for long-term studies can be purchased or made. A large variety of sizes and shapes of aluminum, copper, brass, plastic, or stainless steel tags are available commercially, some with individual numbers on them. We have found that the softer aluminum tags that can be engraved by writing with a pen do not last long (i.e., more than a year) when placed on the ground to mark herbaceous plants. An alternative that works well is tags cut from aluminum beverage cans (tags cut from steel cans rust and quickly become illegible). Use metal-cutting shears or sturdy scissors to cut the top and bottom off and cut down the side to form a sheet. Flatten the sheet by pulling it over the edge of a table, cut it into tags of the size desired, and then write on a tag by pressing hard with a ballpoint pen. We have tags like these that are still legible after 18 years in the field. The tag can be punched with a hole punch, or with a nail, and then pinned to the ground with the nail. We have found that 16-penny nails (approximately 8 cm long) work well. If you are working in a site where gophers or other fossorial rodents occur or where other soil disturbance results in the burying of tags, use a metal detector to locate buried tags (you may also find a variety of other interesting metallic objects!). If you need to use thicker metal that can't be embossed with a pen, get a set of stamping dies at a hardware store for marking it.

C. Preventing Visitation

For some studies it may be necessary to prevent animals from visiting flowers. Examples might include studies of breeding systems to determine whether plants are self-pollinating or not, manipulations of nectar levels in studies of foraging, or studies of nectar production. Comparisons of the significance of nocturnal versus diurnal pollination (e.g., Bertin and Willson 1980, Eguiarte and Búrquez 1988, Goldingay et al. 1991, Jennersten and Morse 1991) might also require regulating access to flowers. Pellmyr (1989) used a sequential bagging experiment to determine the time of pollination in *Trollius europaeus* (Ranunculaceae), which has long-lived flowers, by exposing flowers to pollinators for up to 9 days after flowering started. Several simple techniques facilitate this kind of manipulation.

Plants that are either grown from seed or transplanted into pollinator-free greenhouses or growth chambers may be sufficient to answer some questions, such as whether they can produce seeds in the absence of pollinators. The controlled conditions afforded by a growth chamber may also facilitate careful studies of the environmental variables that affect nectar or seed production. However, if the plants are wind-pollinated, it may be necessary to bag them even if they are grown in greenhouses. Antonovics and Schmitt (1986) used glassine bags (30 x 10 cm) to enclose grass inflorescences. The bags were held on a cane and sealed at the base with a plug of cotton by tying a string around the bag.

If plants must be left in the field, there are also several ways to prevent visits by potential pollinators. Probably the first means of doing this was with glass chimneys, which have the disadvantage of being relatively heavy and fragile. Hocking (1968) used truncated 30° cones of acetate sheet, 30 cm high, with the top and three ventilation holes (all 4 cm in diameter) made low down on the sides covered with nylon netting (Figure 2–1A). These also have the advantage of nesting for easy storage or transport. Kevan (1972a; see Section D.2.) used a similar method. Another simple technique is to cover plants or inflorescences with some sort of fabric. For example, large patches of flowers can be covered temporarily with lightweight gauze, cheesecloth, or nylon mesh (e.g., mosquito netting, bridal veil, or floating row cover made for covering crop plants). Nylon or other synthetic fabrics have some advantage in areas where the fabric may become wet, as they may shed water more readily. In some situations it may be possible simply to lay pieces of fabric on top of the vegetation to deter most insect visitors. If the flowers are large enough, and of the right shape, you may be able to keep the petals from opening by putting a rubber band around them; this method can also be used after a controlled pollination (Lawrence 1968).

Birds or bats can be deterred with chicken wire. Chicken wire will admit insects, which may facilitate studies of the relative significance of different groups of visitors. For example, Vanstone and Paton (1988) used nylon bird netting (1-cm mesh; available at garden shops, for preventing birds from harvesting fruits or berries) to exclude birds but not insects. Waser (1978, 1979) and Ford (1979) covered plants with chicken wire or mist nets to exclude birds and butterflies while allowing access by other insects. Pellmyr (1989) used cages with a 5-mm-square mesh to enable flies but not

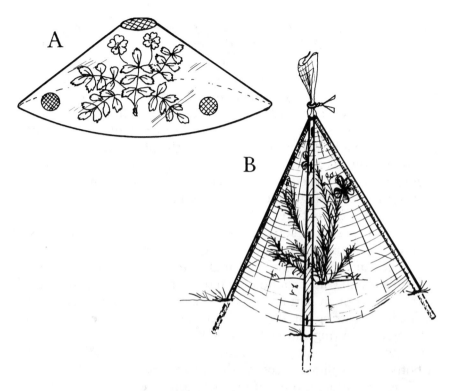

Figure 2–1. Hocking's (1968) design for acetate cones for covering plants (A). Design by Douglas and Sun, using stakes and mosquito netting (B).

bumblebees to visit flowers. Large screen flight cages can also be used to enclose whole plants or populations of plants.

Individual inflorescences, stalks, or plants can be covered with net bags. The bags can be made from a variety of materials, but the most commonly used fabric is bridal veil, a nylon mesh material available at most fabric stores. If you are concerned with excluding very small insects and can't find bridal veil with a fine mesh, try using multiple layers of fabric (e.g., Barrett and Helenurm 1987) or look for mosquito netting in a hardware store or camping store. Cheesecloth is another possibility, but it does not last as well and gets heavier when wet. If you're good with a sewing machine you can add a drawstring around the base of the bag or, alternatively, just tie a string around the bag once it's in place on the plant. Those without access to sewing machines (or the sewing-impaired) may find it easier to use twist-ties, which are also easily removed to gain access to the plant. We have found tomato cages, which are conical, commercially available wire structures about 1 m high used to support tomato plants, very useful for supporting net bags over plants. If a bag gets wet and is unsupported, it may come into contact with the plant inside and even cause it to fall over and break. Wooden, fiberglass, or metal stakes can also serve the same purpose of supporting a bag so it doesn't contact the plant inside.

Kelly Douglas and I-Fang Sun (R. Cruden, *personal communication*) have developed a flexible cage, basically a tripod covered with mosquito netting, that can be used to cover plants and is easy to transport (Figure 2–1B). The mosquito netting is stapled to just one of the three stakes, starting about 10 cm from the base (approximately the distance that the stake will be inserted into the ground). The three stakes are driven at an angle into the soil around the plant to form a rough cone around it, and the net is then wrapped around the stakes. Enough netting should extend above the stakes so you can tie it with string (or wire or plastic tie), and the other end of the loose side of netting can be fixed to the ground with soil or staked with a wire stake. The cage can then be opened from either the top or bottom to facilitate access to the flowers inside.

This technique of preventing visitation by enclosing plants can be used to study the relative significance of diurnal versus nocturnal pollinators for plants that are visited by both. For example, in a study of a white *Ipomopsis* (Polemoniaceae) species, we found that syrphid flies were visiting flowers during the day, as were nectar-robbing bumblebees, while white-lined sphinx moths were visiting the flowers at night (Inouye, *personal observation*). By covering some stalks during the day and others at night, we could study the seed set produced by each group of pollinators. Bagging flowers during the day also prevented any effects of nectar robbing by bumblebees on moth visitation. Goldingay et al. (1991) used the same technique to compare nocturnal and diurnal pollination of mammal-pollinated flowers in Australia.

Pollination bags made for crop plants can also be used for field studies. Some bags are constructed of brown paper, some are treated with wax to be more waterproof, some have clear windows, and some are made of nonwoven polyolefin that is semitransparent. Although they are typically designed for use with crop plants such

16

as corn, tobacco, or sunflowers, or for use on coniferous trees, commercially available bags can also be used on a variety of flowers. D. Gill (*personal communication*) used Delnet (polyolefin) bags to prevent visitation to pink lady-slipper orchids by supporting the bags with pieces of sticks from the forest floor. Another alternative might be bags designed for soil samples, made of polyethylene, unbleached muslin, or tyvek (see Appendix 2, pollination bags).

A disadvantage of net covers, pollination bags, or covers for individual flowers such as gelatin capsules or dialysis tubing (described in the next chapter) is that they alter the environment of the flowers or plant inside. They will provide shading, which might affect the phenology of flowering, patterns of resource allocation, the production of nectar, or seed set, through effects on the temperature, solar radiation, and relative humidity of the flowers or plants. In many cases this will not influence what is being studied, but in other cases these effects may have to be measured, controlled for, or just ignored. Corbet and Delfosse (1984) found that irradiance to bagged plants was reduced by 14% to 36% (depending on the type of material used for the bag; bridal veil or white mesh); Pleasants' (1983) observation that nectar production was reduced on shady days suggests that bagging may affect nectar production in at least some species. Temperatures and relative humidities inside the bags were sometimes higher and sometimes lower than those outside the bags, differing by up to 1°C and 8% relative humidity. Cumming and Righter (1948) suggested trimming the needles off of bagged branches of conifers to facilitate access to the reproductive parts and reduce transpiration inside some types of bags, but this might also adversely affect development of seeds.

Cruden et al. (1983) measured temperature effects of several bagging materials; brown paper bags, white pollination bags, and glassine bags raised temperatures 10°–15°C compared to unbagged controls. Mosquito netting did not raise the temperature as much but could actually lower it through evaporative cooling if it was wet. Southwick et al. (1981) and Southwick (1983) reported that temperatures inside fine nylon mesh bags (28 openings/cm) were about 1°–2°C different from ambient temperatures. Corbet and Willmer (1981) found that at air temperatures of 26.5°–29°C, the temperature inside three permeable muslin bags covering inflorescences in the sun was 3.5°, 5°, and 6°C above ambient; temperatures in bags in the shade were only 0.5°C above ambient. They also found that relative humidity was altered by the bags. At ambient relative humidities ranging from 69% to 97% the relative humidity inside two bags in full sun was 4% to 5.5% lower; in the shade humidity in seven bags averaged 2.3±0.94% higher (range -0.5% to 7.5%). Corbet and Willmer caution that microclimatic effects of less permeable bags can be more severe.

Wyatt et al. (1992) investigated the influence of four different materials used for bagging flowers on nectar production in milkweeds (*Asclepias;* Asclepiadaceae): bridal veil (mesh size 10 x 10 threads/cm), pellon (a soft, white fabric with irregular mesh size), brown kraft paper, and clear polyethylene plastic (18 μm thick). They measured temperature and relative humidity inside the bags every three hours from dawn to dusk. The plastic bags caused the greatest variation in temperature and humidity from control values; temperatures were up to 5°C higher, and relative humidity remained consistently high when other treatments showed dramatic declines. Inflorescences in plastic bags also appeared to have reduced floral lifespans. Pellon and paper bags increased temperature and humidity relative to controls, whereas bridal veil bags did not raise either significantly. The volume and concentration of nectar differed most strongly in inflorescences enclosed in the plastic bags, which produced notably greater volumes of significantly more dilute nectar, probably as a consequence of reduced evaporation. The other types of bags produced small and inconsistent differences in nectar volume and concentration, with the smallest differences produced by the bridal veil bags.

If you suspect that thrips may be effecting some pollination, and might influence results of exclusion experiments, try using an insecticide. Cruden et al. (1990) used Malathion in a situation like this and found that although a single application was not sufficient to eliminate the thrips, applications repeated every other day were. Baker and Cruden (1991) found that thrips and aphids played a significant role (up to half of the fruit set of flowers not visited by larger pollinators) in self-pollination of *Ranunculus sceleratus* (Ranunculaceae) and *Potentilla rivalis* (Rosaceae). They also used Malathion (50% Acme Malathion spray; 3 teaspoons to 1 gallon of water) to remove aphids and thrips. A weekly spray with a synthetic pyrethoid, Resmethrin, or biweekly spraying with a mixture of Dycarb and Talstar, Nufilm, or soapy water were not as effective at removing aphids and thrips. These observations argue for inclusion of a combination insecticide + cage treatment in experimental protocols to determine the significance of autonomous self-pollination versus insect self-pollination by these small insects (Baker and Cruden 1991). Beware of potential effects of the insecticide on plants. Vaughton and Ramsey (1991) found that the insecticide they used (not specified) caused flowers on some inflorescences to open prematurely.

Ornduff (1975a) found that plants transplanted into planter boxes and put on the roof of a building were still visited by flies but not by the small halictid bees that visited flowers in the field. Similar sorts of spatial manipulations might facilitate presentation of plants to particular groups of pollinators.

Another reason for preventing access by insects to flowers is to prevent damage to flowers by ants. Galen (1985) put Tanglefoot on the base of stems of her plants to keep ants from damaging flowers. Some plants may react adversely to Tanglefoot, however. J. Mohan (*personal communication*) found that *Campsis radicans* (Bignoniaceae) stems were killed by its application. This could be prevented by putting a band of plastic or tape (for example, a section of plastic tubing split lengthwise) around the stem before applying the Tanglefoot.

D. Abiotic Pollination

Studies of abiotic pollination are relatively uncommon compared to studies of zoophilous species, and there are not as many techniques developed (or perhaps required) for them. Some of the techniques described in other chapters are applicable to abiotic pollination.

1. Hydrophily

There are several documented cases of water pollination, including examples of pollination on the surface (ephydrophily) and in water (hyphydrophily). Proctor and Yeo (1973) and Faegri and van der Pijl (1979) review many of the known cases. Flowers and pollen grains of hydrophilous species may be quite different from their terrestrial counterparts (Ducker and Knox 1976, Guo et al. 1990, Cox 1991). Studies of hydrophily so far have been descriptive rather than experimental in nature, so there do not appear to be many techniques specific to this kind of pollination that we can report. Cox (1988, 1991) and Cox et al. (1990) describe some methods, such as use of diving gear for making observations.

2. Anemophily

If you are uncertain about the mode of pollination of your study species, you should check for the occurrence of wind pollination. If you suspect both zoophily and anemophily are important, then you may wish to determine the relative contributions of insects and wind to pollination. To do this you can either bag flowers or look for evidence of pollen being carried by the wind. The reason for bagging flowers is to exclude insects while still allowing access to airborne pollen; seed set from this treatment can then be compared to seed set when both sources of pollen are excluded. Kevan (1972a) made insect excluders from clear acetate sheets (1/32 inch thick), which he rolled into a truncate cone with a base diameter of 9 inches and top diameter of 2 inches. The cones were 12 inches high and completely covered the high arctic flowers he was studying. He made three 2-inch holes about 4 inches from the base and covered these and the top with fine mesh to serve as vents while still allowing pollen to enter (that some plants set seed was an indication that it did). Argus (1974) used a plastic bottle to build an insect excluder by cutting it to make a frame that fit around a branch and then covering it with insect netting. To verify that pollen would enter the container he placed a glycerin-jelly slide (see Chapter 4, Section C) inside each excluder. Pellmyr (1986a) used two kinds of bags to determine the possible occurrence of anemophily, self-pollination, or apomixis. To test for wind pollination, he excluded insects with double layers of 1-mm mesh insect net before flowering began. To test for self-pollination or apomixis he used bags made from white nylon stockings (with a much finer mesh). Arroyo and Squeo (1987) also used two types of bags, of flexible plastic netting (to exclude insects but not pollen) and cotton fabric (to exclude insects and airborne pollen).

Sacchi and Price (1988) also used insect exclusion as a way to test the relative contribution of insects and wind to pollination. Their treatments included control flowers open to both wind and insects, insect exclosures made from nylon bags with 0.9-mm mesh, and tests for apomictic seed production that used white paper bags to exclude both insects and wind. They tested for pollen flow through the mesh bags by using pollen samplers. A pair of slides coated with silicone grease was set out near a male plant, and one slide was covered with a mesh bag. There was no significant difference in the number of pollen grains deposited on the slides after 4–6 days, indicating that the bags did not interfere with windborne pollen. Sacchi and Price also tested for a treatment effect of the bags on seed production by hand-pollinating some flowers that were enclosed in bags; this treatment resulted in "abundant seed set."

To determine whether pollen is being carried by the wind, you can put out slides coated with silicone grease at different distances from a potential pollen source, and then look for pollen on the slides. A more quantitative method is use of a volumetric pollen sampler, which will sample a known volume of air (e.g., Sacchi and Price 1988). For more information on both of these techniques see Chapter 4, Section C.

Sacchi and Price (1988) measured the amount of pollen reaching female *Salix* (Salicaceae) inflorescences by putting out simulated catkins. They cut pieces of glass tubing that were similar in size to real catkins, coated them with silicone grease, and placed them adjacent to pistillate catkins on branches of their study plants. They collected these later and counted the number of pollen grains with a microscope. This technique showed them that although abundant willow pollen becomes airborne, relatively few grains reached the artificial catkins. A problem of this technique is that aerodynamics may play a major role in wind pollination (e.g., Niklas 1985a), and the flow of air over a glass tube is probably quite different from that around a real catkin.

Photography of plants and pollen in a wind tunnel can be a useful tool for studies of the biomechanics of wind-pollinated flowers. Use of a strobe light produces a series of dots indicating the trajectory of a pollen grain. Tiny helium-filled bubbles can also be used for this purpose. For methods and examples of this technique, see Niklas 1984, 1985a, 1985b, 1985c; Niklas and Buchmann 1988.

For a theoretical treatment of pollen dispersal through wind pollination, including factors such as wind speed, wind turbulence, amount of pollen released, and terminal velocity of pollen grains, see Levin and Kerster (1974) and Niklas (1985a).

E. Flowering Phenology

"Flowering phenology" refers to the seasonal timing of flowering. It is of significance for both ecological and evolutionary reasons, because flowers are important food resources in ecological time and provide a mechanism for reproductive isolation or speciation over evolutionary time. For example, Schmitt (1983a) found that individual variation in first flowering date was related to reproductive success in *Linanthus androsaceus* (Polemoniaceae), and English-Loeb and Karban (1992) found that timing of flowering by clones of *Erigeron glaucus* (Asteraceae) had consequences for both plant reproductive success and herbivory.

Community-level studies of flowering phenology can address each time scale, as the temporal pattern of resource presentation to flower visitors is important both in ecological time (e.g., as a way of maintaining pollinators in a community throughout the growing season) and from an evolutionary perspective (e.g., which factors determine the sequence and overlap of flowering). It has also been studied from a mechanistic approach, to determine what the developmental paths are leading to flower production, or to elucidate the cues, either intrinsic or environmental, that are responsible for the timing of flowering.

Studies of flowering phenology can consist simply of records of the first date of flowering, but such qualitative studies are not as useful as quantitative ones. They may suffice for an arboretum list of when visitors can expect to see certain species in bloom, or for comparisons of variation in flowering over latitudinal (e.g., Reader 1983) or altitudinal gradients. However, the additional information that can be gained from quantitative studies that generate "flowering curves" (i.e., number of flowers or inflorescences open plotted against census date) will often make the extra effort well worthwhile.

Quantitative studies of flowering phenology are relatively simple to conduct; they only involve counting numbers of flowers. They differ, however, in terms of scale and frequency of counting. At one extreme, a study of the phenology of anther dehiscence might involve only a few flowers and observations made hourly, whereas at the other extreme the observations might be made monthly and involve thousands of inflorescences. The frequency of observations, and the scale of the study, will have to be determined by the nature of the study and the availability of observers' time. David W. Inouye (DWI) found, for example, that in montane Colorado and Australia, a study of flowering phenology with observations made every other day in 2 x 2–m plots requires only 1–2 hours each census date for about 25 plots. If censuses are made less frequently than this, one risks missing the flowering of species that are only in bloom (at least in a given plot) for a day or two. However, in some cases (e.g., Heideman 1989) even monthly surveys may provide useful data.

Plots for phenological studies can be marked with short lengths of steel, such as rebar (concrete reinforcing rod, available at hardware stores or steel suppliers), pounded into the ground far enough to prevent them from being pulled out or bent over easily. Plastic surveyor's caps placed on top may make them easier to relocate and prevent injury to grazing animals that might brush up against them. Cotton string placed around the stakes makes a good way of delineating the outlines of the plots. In a 2 x 2–m plot one can reach into any part of the plot to lift up flowers or inflorescences for closer inspection, without having to step into the plot. A rectangular plot would also work well for investigators with shorter arms.

J. Thomson (*personal communication*) has used a single stake to indicate the center of a circular plot. A lightweight metal or wooden bar with a pivot hole in one end is placed over the stake and swung around the stake. The flowers can be counted as the bar passes over them. Different radii can be used for counting rare and common plants, and the single stake is less conspicuous than a plot with more stakes and string.

If a plot has an extremely large number of small flowers that would take an inordinate time to count, it may be practical to subdivide it and count only a part of it (e.g., Rotenberry 1990). Or to keep track of which flowers you have counted, subdivide the plot with string. Hand-held tally meters, which are available with five or more individual counters, may also facilitate counting. If data are collected at the level of individual inflorescences, such a counter can be used to keep track of the frequency distribution of inflorescences with different numbers of flowers open.

Familiarity with the plots and with species found in them is important for collection of phenological data (Inouye, *personal observation*). It helps to have a search image for each species of flower, as some of them may be quite small or cryptic, and this may only be developed after an initial field season. Access to a good flora with color pictures of flowers can be a big advantage when working in an unfamiliar area for the first time. It is also important to be consistent in use of criteria for counting or not counting a flower or inflorescence as being in bloom. For example, close examination of composite capitulae may be necessary to determine whether there are still florets open, which is probably a better criterion than the more easily observed presence or absence of ray florets.

These methods will have to be modified for use on other growth forms, such as trees, or other habitats, such as forests. If flowers are not easily reached for counting, then methods such as using binoculars (e.g., Heideman 1989) to count inflorescences or other sampling units may have to be substituted for counts of individual flowers. If the flowers are too small to count with binoculars, or not visible in the canopy, another method is to collect flowers as they fall. For example, House (1989) used litter traps constructed of squares of cloth suspended from four posts 80 cm above the ground. She measured the rate and duration of flower production in rain forest trees by counting flowers as they fell from the canopy, in units of flower trapped per m^2 crown shadow per unit time.

In studies of trees or clonal plants, it might also be important to consider the potential for genetic differentiation in flowering among branches or ramets of the same genet (Whitham and Slobodchikoff 1981, Gill 1986). The sampling units can also be adjusted to reflect the perspective of the flower visitors, which may make distinctions at the level of individual flowers (if they are large enough), capitulae, stems, inflorescences, and so on (Rotenberry 1990).

For a long-term study with multiple observers, it is important that the methods be consistent and carefully documented. One possible way to do this is to videotape a representative census, with a running commentary on how decisions are being made. A videotape is also a good way to document the location of plots a researcher may want to locate in the future in order to duplicate an earlier study. Alternatively, a photographic record accompanied by a list of compass directions and distances between plots could be constructed.

If data from the plots are stored using a microcomputer spreadsheet program, a variety of statistics can be calculated easily (e.g., length of flowering period, maximum number of flowers in bloom during the season, number of species in bloom on a given date, diversity indices, etc.), and sections of the updated spreadsheet can be printed to produce data sheets for recording additional data (see Figure 2–2).

Long-term data on flowering phenology can provide insight into the temporal and quantitative variation of resource availability for pollinators and, by comparison with weather data, into the environmental factors that may be responsible for this variation. For example, Inouye and McGuire (1991) described the variation in flowering of a long-lived herbaceous perennial wildflower from the Rocky Mountains over an 18-year period and reported a significant correlation between numbers of flowers in a given summer and the amount of snowfall during the previous winter (Figure 2–3). Inouye et al. (1991) described a negative correlation between the availability of four wildflower species visited by Broad-tailed Hummingbirds (*Selasphorus platycercus*) and the number of birds banded at the Rocky Mountain Biological Laboratory over an 11-year period. These kinds of studies are also of interest given current forecasts of climate change, as the statistical relationships that may be discovered between environmental variables and flowering may permit prediction of the consequences of climate change for flowering and for pollinators.

If a knowledge of resource availability for flower visitors is the goal of a study, then the data on flower availability that are derived from a study of phenology can be combined with data on nectar, oil, or pollen production to estimate resource availability.

Phenology can also be studied on another scale, that of inflorescences or even individual flowers. For some studies it may be

Erythronium Meadow # 1 - 1991

Date	5/30	6/1	6/3	6/5	6/7	6/9	6/11	6/13	6/15	6/17	6/19	6/21	6/23	6/25	6/27	6/29	7/1	7/4
Claytonia lanceolata	177	53	81	67	40	97	37	31	2	3								
Thlaspi		14	62	86	161	271	400	446	365	537	353	243	177	97	17	5	3	
montanum		7	20	25	42	46	58	62	86	82	71	56	46	29	9	2	2	
Erythronium				9	18	23	8											
grandiflorum				9	16	17	7											
Ranunculus inamoenus							1	1	1	2	3	1					1	
Valeriana acutiloba							2	5	5	6	6	7	7	5	5	5	6	5
Draba											1	6	20	29	45	62	74	70
aurea											1	2	6	6	8	10	13	14
Sedum rosea													1	8	20	25	39	32
Pedicularis															1	5	8	14
bracteosa															1	1	2	2
Hydrophyllum															3	2	3	8
fendleri															1	1	1	1

Figure 2–2. Example of part of a data sheet summarizing flowering phenology from a 2 x 2–m plot near the Rocky Mountain Biological Laboratory. Each column contains data for a single census date. The rows indicate the number of flowers (if there is a single row for a species) or the number of flowers and the number of plants (the second row if there are two rows for a species).

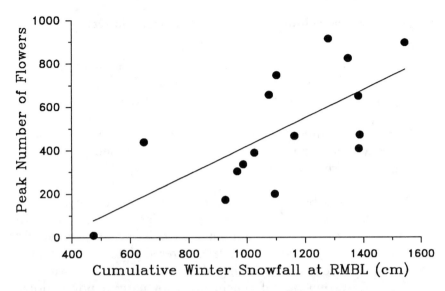

Figure 2–3. The relationship between snowpack and numbers of flowers produced by *Delphinium nelsonii* (Ranunculaceae). Data from a study site near the Rocky Mountain Biological Laboratory.

important to know the timing of pollen presentation, nectar production, or stigma receptivity (methods for determining these variables are addressed elsewhere in this book). Devlin and Stephenson (1985) separated floral development of a protandrous flower species into six phases, which might prove useful as categories for other studies as well:

1. immature staminate = flowers recently open but not yielding pollen
2. mature staminate = pollen present
3. old staminate = no pollen; style elongating
4. immature pistillate = stigma exserted but receptive surface not exposed
5. mature pistillate = receptive surface open
6. old pistillate = flower wilting

Lindsey (1982) used the following categories for scoring phenology of hermaphroditic and staminate flowers of two species of Apiaceae:

Hermaphroditic Flowers		Staminate Flowers
Tight bud	1	Tight bud
Stigmas evident	2	Cracking, filaments extending
Stigmas receptive, floral cracking	3	Full anthesis
Stigmas past, full anthesis	4	Stamens dropping
Styles withered, past anthesis, fruit developing	5	All stamens shed

It is also possible to manipulate phenology for experimental studies. If your study species flowers in response to an environmental cue, it may be possible to alter the cue to change flowering time. For example, Augspurger (1980) watered plants that normally flower in response to the beginning of the rainy season and was able to trigger flowering out of synchrony with natural populations. Waser (1979) took advantage of latitudinal differences in flowering time and transplanted flowering plants from one latitude to another, and also (1978) used altitudinal differences in flowering to create an artificial overlap in flowering times of two species. It is also possible to manipulate photoperiod or temperature to alter flowering times. Construction of a plastic "greenhouse" around a 2 x 2–m plot can advance flowering by up to a couple of weeks (Inouye, *personal observation*). If the environmental cue initiates floral bud development, and that development takes more than one season, the manipulation will be a long-term one. For example, flower stalks and buds of *Frasera speciosa* (Gentianaceae) are preformed three years in advance of their appearance above ground (Inouye 1986), and in order to initiate flowering out of synchrony with control populations (which flower about every 3–5 years; Taylor and Inouye 1985) plants must be watered during the summer three years before you want to see flowering plants.

Phenological data collected for an entire community, or at least for a guild (e.g., all hummingbird-pollinated plants in a community), could potentially provide clues about the factors that shape the flowering curves. For instance, if there is competition for pollinators, one might expect significant temporal displacement of the flowering curves for sequentially flowering species (Waser 1978). A variety of techniques have been suggested for examining phenological data for such temporal heterogeneity. We will not discuss these different techniques, but the following references may be useful if you wish to investigate them: Poole and Rathcke 1979, Estabrook et al. 1982, Gleeson 1982, Waser 1983, Fleming and Partridge 1984, Kochmer and Handel 1986, Murray et al. 1987, Rathcke 1988, Pleasants 1990, Rotenberry 1990. For a discussion of some of the problems of using detrended correspondence analysis, see Beals (1984) and Wartenberg et al. (1987).

F. Seeds

1. Seed Predation

If you need to count seeds or fruits or germinate seeds produced as a result of your pollination experiments, predispersal losses to seed predators can be a serious problem. Failure to consider this potential source of seed (or fruit) loss can make it difficult to count seeds, or lead to substantial underestimates of fruit or seed production. For example, Inouye and Taylor (1979) found levels of seed predation that sometimes exceeded 90%, Louda (1982) found 44% to 73% flower and seed predation, and Andersen (1989) reported destruction of 95% of seeds. One side effect of bagging flowers for pollination treatments may be the exclusion of seed predators. An alternative method is the use of insecticides. For instance, Vaughton (1990) found that the proportion of inflorescences that produced fruits and the number of seeds produced per infructescence both increased by 45% when she excluded insects from *Banksia* (Proteaceae) inflorescences with insecticide. She used endosulfan, a nonsystemic, contact insecticide (60–80 mL of a 0.3% solution per inflorescence). The insecticide was applied during the bud stage but not during the flowering period to reduce any adverse effects of the insecticide on pollen viability, anthesis rates, or the behavior of pollinators. Beware of other potential effects of insecticides on plants: Vaughton and Ramsey (1991) found that the insecticide they used (not specified) caused flowers on some inflorescences to open prematurely.

If you do find substantial predispersal seed predation, don't neglect the possibility that the same species of insect can serve as both pollinator and seed predator, perhaps even in a mutualistic relationship with the plant. Pettersson (1991) describes a system (*Silene* [Caryophyllaceae] and *Hadena* moths) in which the noctuid moth serves as both. Other relationships Pettersson lists include *Yucca* (Agavaceae) and *Tegeticula* moths (Addicott 1986), *Ficus* and agaonid wasps (Wiebes 1979), *Davilla* and curculionid beetles (Gottsberger 1977), *Nuphar lutea* (Nymphaeaceae) and *Donacia* beetles (Schneider and Moore 1977), and *Trollius* (Ranunculaceae) and *Chiastocheta* flies (Pellmyr 1989).

2. Seed Viability

For many studies the production of seeds is sufficient as an indication of successful pollination. In some cases, however, it may be necessary or of interest to confirm the viability of these seeds. The definitive test is germination, but this is not always as simple as it might appear. Many plants have seeds with dormancy mechanisms, and the seeds may require scarification, vernalization, or other treatments before they will germinate. Detailed consideration of these techniques is beyond the scope of this book. An alternative to actual germination is a chemical test; Freeland (1976) describes these techniques. The most widely used test is the tetrazolium test, which is based on the visual reduction (development of red staining) of 2,3,5-triphenyl-2H-tetrazolium chloride (Moore 1962, Hatton 1989); this vital stain is also used for evaluating pollen quality (Heslop-Harrison et al. 1984). For example, Sobrevila (1989) used this technique to assay seed viability. Reduction of the stain is indicative of the activity of dehydrogenase enzymes that cause reduction activity in living tissue (Roberts 1972). Tetrazolium testing is used widely in agriculture to test dormant seeds, rate seed lot vigor, and supplement germination tests (Grabe 1970). The Association of Official Seed Analysts, in their publication *Tetrazolium Testing Handbook for Agricultural Seeds* (Grabe 1970) presents generalized, standardized procedures.

The following procedure is based on guidelines from Grabe (1970) for testing seeds.

Materials:
seeds
staining dishes (watch glasses for small seeds, petri dishes
  for larger ones)
razor blade, dissecting knife or pin for piercing seeds
stereo microscope, or magnifier for larger seeds
dropper
blotting paper
lactophenol and a dispensing bottle for applying it

Procedure:
1. Test at least 200 randomly selected seeds. Divide the lot into several replicates of fewer than 100 seeds.
2. Prepare tetrazolium solution. Dissolve 1 g of 2,3,5-triphenyl-2H-tetrazolium chloride (TTC) in 100 mL of distilled water to make a 1.0% stock solution. This solution is used for intact seeds. Further dilution to 0.5%, 0.2%, or 0.1% is common. The 0.1% solution is generally used on seeds that have been bisected through the embryo. The pH of the solution usually falls between 6 and 8. More acidic solutions will not stain properly, and when this occurs the salt must be dissolved in a phosphate buffer instead of water.
   Phosphate buffer:
   Solution A: Dissolve 9.078 g KH_2PO_4 in 1,000 ml H_2O
   Solution B: Dissolve 11.876 g $Na_2HPO_4 \cdot H_2$ in 100 ml H_2O
   Mix 400 ml of solution A with 600 ml of solution B.
   Dissolve 10 g of TTC in the buffer for a 1.0% tetrazolium solution with a neutral pH.
3. Prepare lactophenol solution, with 20 parts lactic acid:20 parts phenol:40 parts glycerin:20 parts water. NOTE: **Lactophenol is toxic.** Avoid inhalation or skin contact. This solution is used to clear grass seeds so that the embryo is visible through the lemma and palea after tetrazolium staining.
4. Tests can be performed at temperatures between 20° and 45°C. Within this range, reaction speed approximately doubles from 20° to 30°C and again from 30° to 40°C.

5. Soften seeds by placing them between moist blotter paper or paper towels overnight, or in a beaker of warm water for 3–4 hours. Seeds should be soft enough to bisect easily. (For some species this step can be omitted.) Don't let seeds swell too rapidly, or the seed coat may burst and tissue damage may occur.

6. Seeds with tough seed coats will not imbibe stain well. You may want to (1) bisect seeds through the embryo (corn, grass seeds; (2) pierce the seed coat (small grass seeds); or (3) remove the seed coat with a razor, needle, or forceps (dicots with impermeable seed coats). After cutting or piercing a seed, place it in testing solution immediately so that it doesn't dehydrate.

7. Place seeds in staining dishes and cover them completely with tetrazolium solution, allowing enough excess for large seeds to imbibe some of the fluid. Staining time will vary. Moore (1962) states that overnight staining at 20°C or 4–6 hours at 30°C is appropriate for most crop seeds. Insufficient staining will make results difficult to interpret whereas excessive staining will cause tissue damage.

8. After staining, use the dropper to remove excess solution. If you can not evaluate seeds immediately, replace the solution with water to prevent dehydration.

9. For grasses, apply a few drops of lactophenol to a dish of about 100 grass seeds after blotting any remaining tetrazolium solution. Clearing takes about 10–30 minutes.

10. Examine seeds under magnification to evaluate staining patterns. Knowledge of the areas of rapid, early cell division will aid in evaluations. In grasses these include tips of radicals and seminal roots and the base of plumules. In dicots these are usually the radical and plumule.

In general, tetrazolium tests are simpler and less time consuming than seed germination tests, but they may not provide identical results. You may wish to confirm the validity of the tetrazolium tests by conducting germination tests, too. If both tests are conducted properly, sampling error may still result in a 3% to 5% difference (Grabe 1970). Drawbacks with the tetrazolium test are that it "can be slow and tedious, particularly with small seeds" (Hutchings

1986). Hutchings also warns that staining patterns differ among species, and that even nonviable seeds may be stained.

G. Suggestions for Planning Studies

1. Effects of plant or inflorescence size on visitation. If the plants you are studying vary significantly in height, size, or even just inflorescence size, beware of the potential effects of this variation on visitation. Visitation has been shown to increase with plant height (Hainsworth et al. 1984, Larson and Larson 1990) or decrease with plant size (Andersson 1988). Visitation rates have also been shown to vary with inflorescence size (e.g., Willson and Bertin 1979, Thomson 1988). Firmage and Cole (1988) found that reproductive success (fruits produced or pollinia removed) increased with inflorescence size. The relationships between visitation and pollination may not coincide, for example in self-incompatible species in which there is limited pollen carryover (e.g., Andersson 1988). There is also an increased potential for geitonogamy as inflorescence size increases. Pleasants and Zimmerman (1990) found that visits per inflorescence, flowers visited per visit, and visits per flower all increased with increasing inflorescence size in two species of montane wildflowers. Dudash (1991) found that large plants contributed disproportionately through both female and male function to the production of offspring. Flowers of large plants produced more pollen per flower and had more pollen grains deposited on their stigmas than flowers of small plants, and seed production per fruit was also greater for fruits of large plants.

2. Effect of stress on maternal plants. Stressed wild radish plants selectively abort seeds sired by particular pollen donors (Marshall and Ellstrand 1988). Stress can also have effects on pollen viability (Young and Stanton 1990a).

3. Density effects. Density of flowering plants could affect the foraging behavior and profitability of foraging by pollinators, and hence their effects on pollination. Some studies have found correlations between flower density and visitation rates or seed set (e.g., Thomson 1981a, Cibula and Zimmerman 1984, Schmitt et al. 1987, Allison 1990), but others have not found any (e.g., Roubik et al. 1982, Schmitt 1983b). Density can also affect phenology and the production of cleistogamous versus chasmogamous flowers (Schmitt et al. 1987), and outcrossing rate (e.g., Ellstrand et al. 1978). Position effects, such as whether plants are growing at the edge versus the interior of a patch, can also be important.

4. Distance effects. Some studies have shown that seed production can be influenced by the distance between the pollen source and the recipient. This effect has been manifested by differences among treatments in germination and survivorship (Waser and Price 1983, 1989, 1991a) or by a reduction in fruit set and numbers of seeds per fruit (Redmond et al. 1989). Such an effect could be controlled for by using pollen from only one donor plant, or from plants from only one distance (either fixed or varying slightly, such as the nearest neighbor). Using a single donor plant might cause potential problems, too, if there is variation among individuals in success as a pollen donor (e.g., Marshall 1991). Fenster (1991a) used information on pollen- and seed-dispersal distances to collect pollen for crosses in another experiment. He took pollen from plants at distances corresponding to distances in terms of neighborhood units of self, within a genetic neighborhood, between adjacent neighborhoods, between far neighborhoods in the same subpopulation, and between neighborhoods in different subpopulations of increasing distance (three different distances).

5. Competition for pollinators. The presence of plants competing for pollinators can affect pollination of your study species. Beattie (1976) found that proximity and composition of competing flowering species affected pollination of *Viola*, and Campbell (1985) and Kohn and Waser (1985) demonstrated experimentally the consequences of competition for pollination. You may need to control for this kind of effect by choosing study areas judiciously or by manipulations.

3. Flowers

A. Marking or Tagging Flowers

Marking of individual flowers is often required for pollination studies, and there are a variety of possible techniques. The best technique for use with a particular flower species will depend on its morphology, longevity, and the nature of the study. If the study involves natural pollinators, then care should be taken not to use a technique that might alter the attractiveness of the flower by changing its visual or odor cues (for example, drawing in hummingbirds by using red flags or tags). Inconspicuous tags can be made by tying different colors of sewing or embroidery thread on the pedicels of individual flowers (e.g., Pleasants 1980, McGuire and Armbruster 1991) or whole umbels (Thomson and Barrett 1981). It is also possible to write directly on flowers or pedicels with different colors of felt-tip markers with indelible ink (e.g., Thomson and Barrett 1981, Morse 1987) or paint (Zimmerman and Pyke 1988a). Different colors of paint can be used to signify visits by different species of pollinators or variation in their behavior (e.g., pollen and/or nectar collection, vs. nectar robbing) (Kendall and Smith 1975, 1976). Some colors will fade, so if you need long-term marks, try the colors before using them (J. Thomson, *personal communication,* reports that reds are especially prone to fading, but that blues and blacks are usually good). Sarah Corbet (*personal communication*) found that if she marked the calyces of flowers with fluorescent highlighter pens, she could find marked flowers in the dark with an ultraviolet light.

If you need to associate more information with a flower or plant than can be provided by the methods described above, jewelry tags (small rectangles of stiff paper with threads tied through a hole at one end, available at office supply stores) may prove a satisfactory alternative. These tags will also reflect the visible components of light from ultraviolet lamps at night, making them relatively easy to find. J. Thomson (*personal communication*) recommends gummed labels, wrapped around the stem or pedicel and overlapping back on themselves. Use pencil or indelible ink to write on the tags if they may get wet from dew or precipitation. Or you may be able to print directly on them with a printer after generating the text on a computer. This kind of label is available in fluorescent colors, will last a whole field season in temperate climates, can be cut to the desired size, and can be quickly applied and easily read (as they stand out like a flag). It is also possible to add a second label to the first, indicating, for example, that the flower has been pollinated. If you need to harvest fruits, and the labels are applied to a pedicel, you can just cut the pedicel with the label attached. It is also easy to keep these labels organized by marking them while they are still on their backing sheet. You can make labels for your whole study, and then apply them.

In tropical areas or other habitats where cotton or even polyester threads may not be satisfactory, thin copper wire may be a viable substitute. Wire that is covered with a variety of colors of insulation, such as is used for telephone wire in large cables, when each wire must still be individually recognizable, also works well (Y. Linhart, *personal communication*). Goldblatt and Bernhardt (1990) used different colors of wire to mark different treatments in an experimental pollination study. Corbet and Delfosse (1984) used 1–2 mm lengths of colored plastic drinking straws, slit lengthwise, as tags for individual flowers by putting them around the pedicels.

Colored tape has also been used to mark flowers. Waterproof tape is available in a variety of colors, and small pieces can be wrapped around the pedicels of individual flowers. Put pieces of the tape on a sheet of glass, cut pieces of the desired size with a razor blade, and mark them with a pen. Cane (1991) makes durable tags (which have lasted at least four field seasons) from plastic paper clips and embossed labels. The label is prepared with a relatively inexpensive Dymo label maker, and a small hole is punched in one end with a point punch. Slitting the lower corner of the plastic clip with a chisel creates an opening for the label and for placing the clip around the plant stalk. Tags can be color coded to indicate different treatments.

Paint can also be used to mark flowers. Acrylic paints seem to do well and are easy to work with because the brush can be cleaned with water after use, but enamel paints such as model paints can also be used. You may have to experiment to find out what type of paint works with a particular flower species. For example, C. Galen (*personal communication*) found that enamel paints worked well on *Polemonium* (Polemoniaceae) but not on *Silene* (Caryophyllaceae). We have used different colors of Liquid Paper but found that it sometimes flaked off the sepals after a week or two. If you are painting sepals to mark flowers, make sure that they are not deciduous, or you may lose your marks. C. Galen (*personal communication*) found that permanent markers (e.g., Sharpie markers) worked well for writing on *Ranunculus* (Ranunculaceae) stems.

B. Preventing Visitation

For small flowers, or even parts of umbels, we have found dialysis tubing a convenient way to isolate flowers (e.g., Schemske et al. 1978, Schemske 1980a, Bierzychudek 1987). Cut the tubing to the desired length, and tie it at either end with sewing thread. Although it comes dry, you will have to soak it first before you can handle it to cover flowers. A short roll of dialysis tubing in a container of water is easy to take into the field. Sausage casing is a traditional bagging material for some tree crops (Madden and Malstrom 1975). Pleasants and Chaplin (1983) and Zimmerman (1988a) used halves of large gelatin capsules (available from pharmacists) to cover flowers to prevent visitation, in order to measure nectar production. The problems they encountered were that some capsules were blown off during a storm, and they can melt in rain (M. Zimmerman, *personal communication*). Lynda Delph (*personal communication*) uses small plastic centrifuge tubes, which have attached lids, to prevent visits to cushion plants in the alpine tundra. Put a pin through the cap and into the cushion in order to anchor the tube.

Covering the pistil is another technique for preventing pollination. An advantage of this technique is that it will prevent autogamy but still leave open the possibility of controlled hand-pollination. It also leaves the anthers intact and minimizes the potential environmental effects of bagging whole flowers. Cruzan (1989) placed hollow dried grass straws over the gynoecia of glacier lily (*Erythronium grandiflorum;* Liliaceae) flowers to prevent pollen deposition on stigmas. Paper or plastic drinking straws also serve well as female condoms, but you may have to experiment to find the right size (try plastic cocktail straws or coffee stirrers for smaller flowers; you can slit the base of the straw with a razor blade to permit some degree of expansion). Thomson, Shivanna, et al. (1989) covered entire small flowers with friction-fitting caps made from plastic drinking straws. To make a cap, split a piece of straw lengthwise and then form it into the desired diameter with a thin ring sliced from plastic tubing. This technique has the advantage of allowing the construction of different sizes of caps for different sizes of flowers (by using different size lengths of straw and different sizes of plastic rings), but is much more work than finding a straw of the appropriate diameter. Because it is only constrained by the ring in one spot, the split straw permits considerable adjustment.

Richardson and Stephenson (1991) covered styles of *Campanula americana* (Campanulaceae) with glassine photographic paper (used for storing negatives) taped shut with adhesive tape. They found that this paper allows gas exchange, and stigmas became receptive and were successfully hand-pollinated. An advantage of this covering was that they could monitor the onset of receptivity (i.e., opening of stigmatic lobes) by shining a flashlight through the glassine sleeve. Another technique for protecting just parts of flowers is to use cotton to obstruct access by pollinators to nectaries. Clements and Long (1923) used this technique on a variety of flowers.

Bagging flowers is likely to have the same sorts of effects on microclimate that bagging whole plants does, with possible consequences for phenology, anthesis, nectar production, and dehiscence. See Chapter 2, Section C.

Gill et al. (1982) used Colgate toothpaste to exclude insects but not hummingbirds from the tropical passionflower *Passiflora vitifolia* (Passifloraceae). They smeared toothpaste on the vine on either side of the flower stalk, and on top of a layer of waterproof duct tape wrapped around the base of the corolla. Ants would not walk across the toothpaste barrier, and the odor repelled stingless bees (*Trigona*). To exclude hummingbirds but not bees or ants, they sealed the dorsal entrance to the nectar chambers with tape. Because the odoriferous compounds in many brands of toothpastes are terpenes, it might be worth trying components of mint or wintergreen. As a sense of smell has been demonstrated in one species of hummingbird (Goldsmith and Goldsmith 1982), it seems that this effect depends on a differential response by birds and bees, and not on an inability of birds to detect the odor.

If the object of preventing visitation is simply to exclude the possibility of seed set, for example, for manipulations in a study of the cost of reproduction, it may be sufficient to remove the stigma of a flower. However, it may be important to remove the style as well, as pollen deposited on the cut style can germinate in some cases and still effect pollination (e.g., in *Erythronium grandiflorum;* Lisa Rigney, *personal communication*).

C. Morphological Measurements

Floral morphology is important for taxonomy and can be ecologically important as well. Corolla morphology can serve as an isolating mechanism, with corolla length or shape excluding flower visitors with the wrong length of proboscis or bill shape. Floral morphology can also result in placement of different species of pollen on different parts of a common pollinator's body. Thus it may be important for pollination studies to quantify a variety of morphological measurements of plants, inflorescences, or flowers. These measurements can also provide insights into the nature of selective pressures that have generated floral morphologies (e.g., Murray et al. 1987, Fenster 1991b). Floral color (e.g., Stanton 1987a, 1987b) or scent (although not strictly speaking a morphological character, we include it in this category) can serve as isolating mechanisms, or as important components of pollinator attraction.

1. Floral Measurements and Intercorrelations
Floral measurements such as corolla length have occasionally been made for taxonomic studies, but data are not widely available.

So if you are interested in morphological aspects of plant-pollinator interactions, you will probably have to make your own measurements. Unless you're fortunate (or smart) enough to be working with very large flowers, you will probably find it easier to make these measurements with help of a magnifying glass or dissecting microscope. We have used table-mounted magnifiers with built-in lights and found these easy to work with, but alternatives also include jeweler's head-mounted magnifying lenses (e.g., Optivisor). A fine pair of forceps may help in manipulating the flowers if they are small; locking forceps that can hold the flower without finger pressure may also help. The measurements are probably best made with a vernier caliper, which permits measurements as small as 0.02 mm. The fine points on the calipers make it easy to align them with a particular point on the flower, and if they are made of stainless steel they require little care in the field. J. Thomson (*personal communication*) finds digital calipers easier to work with and suggests that they require less maintenance than vernier calipers in dusty areas. It is also possible to use a caliper connected to a computer, to bypass data entry (see Appendix 2).

If you are working with large flowers, another alternative to measuring them in the field is to dissect floral parts, photocopy them, and later measure or digitize them. Computerized image analysis systems might also facilitate some measurements.

If you are making more than one morphological measurement on flowers, look for correlations among them. Campbell (1989) and Campbell et al. (1991) found that several floral traits of *Ipomopsis aggregata* (Polemoniaceae) were highly correlated with each other, even though there was substantial variation in the characters. For example, the correlation between the proportion of time spent in the pistillate floral phase and the distance the stigma was exserted beyond the corolla tube was $r = 0.95$. There were also significant ($P<0.05$) correlations between mean corolla length and mean stigma exsertion, and mean corolla length and the mean distance from the base of the calyx to the distal end of the highest (or lowest) anther. If the correlation is high enough, you may be able to omit measurements of highly correlated characters in some analyses.

There have been very few mechanistic studies of pollinator-mediated selection so far (Campbell et al. 1991) (see Chapter 6, Section N). The techniques described below can be used to provide the morphological data necessary for such studies of phenotypic selection.

a. Corolla tubes

For studies of resource partitioning by flower visitors, effective corolla length may prove to be the most significant floral dimension. Although it may seem as though this would be a simple measurement to make, in fact it is often difficult to tell where to make the measurement on the distal end of the flower, as some flower visitors will push harder than others or have different head sizes, which permit them to extract nectar from different points in the tube. Two different methods have been developed to provide functional measurements that are more informative than measurements of the whole corolla tube. Plowright (1987) attempted to avoid this problem by observing the fit between a bumblebee head (mounted on an insect pin) and flowers of different species to measure an "effective corolla depth." The insect pin was pushed through the head so it emerged between the mandibles. The pin was inserted into the corolla until it touched the nectar, and then the head was slid down the pin until it was impeded by the corolla. The length of the pin still protruding from the mandibles was then used as a measurement of effective corolla depth. Although this worked well for flowers with medium and long corolla tubes, it was less effective for open flowers.

Barrow and Pickard (1985) used a similar technique. They took the largest and smallest bumblebee workers they could find at their study site, killed them, and dried them with their heads outstretched in line with the longitudinal axis of the body and their wings folded. A probe handle was created for each bee by pushing a needle through it longitudinally, from the anus, into the abdomen. These "bees on a stick" were pushed in turn into flowers, and the distance from the distal end of the labrum to the bottom of the corolla tube or spur was recorded. Although it might be possible to determine this by backlighting the flower, they obtained visual access by cutting small holes in the corollae with a scalpel. These measurements indicated a maximum and minimum "exclusive corolla length" for a given plant species. They made field observations of bees feeding on the flowers to determine how far into the corolla bees would push. They compared their data with measurements for the same species made with a different method by Prŷs-Jones (1982). Prŷs-Jones measured the distance between observed nectar levels and the place on the corolla where it increased substantially in diameter. The distances determined by Prŷs-Jones were 0.74 mm (23.4%) longer than those measured with the bee probes, but this difference was not statistically significant; the greatest discrepancies in measurements were on flowers with wide, tapering corollas.

If it is not possible to use one of these functional methods, try to pick an easily repeatable measurement. For example, for legumes Harder and Cruzan (1990) measured the length of the pistil from its base to the point where the style bent sharply upward.

Brink (1980) and Brink and deWet (1980) pressed flowers in the field, and then determined the distance to the nectary later by putting individual flowers into folded strips of clear acetate film, projecting them with a photographic enlarger, bending a fine copper wire to conform with the distance representing the depth of the nectary, and then measuring the wire. Steiner and Whitehead (1990) used a similar technique for fresh *Diascia* (Scrophulariaceae) flowers, which have a curved spur, by inserting a wire into the corolla to straighten out the spur, and then measuring the wire. In some cases this method may have to be adjusted because it does not take into consideration how far a flower visitor might insert its head into the corolla.

Campbell et al. (1991) demonstrated that corolla width can also be an important morphological characteristic, influencing male function during pollination. This effect was not mediated by visitation rate but by the amount of pollen exported per visit. They measured corolla width at the opening of the tube, using an average of eight flowers per plant to obtain a mean for each plant.

The data presented in Table 9–1 (Chapter 9), showing means and standard deviations for some measurements of corolla length and other floral characters, may prove useful in determining the number of measurements you will need to estimate a mean with a particular degree of accuracy.

b. Other floral measurements

For studies of pollen transfer or pollinator behavior other floral measurements may also be necessary. Both intra- and interspecific measurements may be important. Intraspecific variation is known to influence components of male and female fitness (e.g., Beare and Perkins 1982, Stanton et al. 1991), and interspecific differences can provide insight into differences in pollinator faunas and resource partitioning. These measurements might include petal size, corolla diameter, stigma lobe length, distance from anthers to stigma, exsertion of stigma or anthers from the corolla mouth, or insertion point of filaments in the corolla tube (e.g., Galen et al. 1987, Galen 1989, Campbell 1989, Stanton et al. 1991). Some of these measurements may be particularly important for studies of heteromorphic flowers (e.g., Barrett 1985). The same tools described above for measurements of corolla length will probably suffice for these other measurements as well. For some studies it may also be important to count floral parts such as anthers. Philbrick and Bogle (1988) found substantial variation in anther number in *Podostemum ceratophyllum* (Podostemaceae); it varied from two to seven anthers per flower in one population.

Campbell (1989) found that the distance that stigmas are exerted from corollas of *Ipomopsis aggregata* can play a prominent role in pollen transfer. Corolla width was also significantly correlated with male fitness in one year of her study. Both Campbell (1989) and Galen (1989) found that flower size components were highly correlated. Galen (1989) used principal components analysis to reduce a set of morphological variables to independent axes of floral variation, which accounted for 77% of the variation of four traits she measured. However, Morse and Fritz (1985) found few correlations among characters at the within-clone level for *Asclepias syriaca* (Asclepiadaceae). Be aware that there can also be significant differences in morphological measurements among individuals or clones. For example, Morse and Fritz (1985) found "strong among-clone differences" in ten characters of pollinaria, anthers, and alar fissures measured from the flowers of seven clones of common milkweed (*Asclepias syriaca*).

Another aspect of floral morphology that is probably important for pollinators is the overall size, which determines the distance at which a flower can be seen. Dafni (1991) determined the planar projection of flowers by measuring them manually with a portable digital planimeter (Planix 7P by Tamaya; A. Dafni, *personal communication*). He drew the floral image at a scale of 1:1 from the angle of approach used by pollinators; flowers were fixed to a table, a clear plate (apparently glass or Plexiglas) was placed just above the flower, and its outline was traced onto a plastic sheet (e.g., overhead projector sheet). In a study of 13 species of Labiateae, he found a positive relationship between flower planar projection and the reward per flower (nectar volume or joules per flower). Dafni suggests making a distinction between the "en-face" planar projection seen by an approaching pollinator and the profile projection. Floral size may also be correlated with other characteristics. Harder et al. (1985) found that nectar and pollen production were positively correlated with flower size in *Erythronium americanum* (Liliaceae), and Dafni (1991) found significant relationships between planar projection and both flower tube length and flower longevity among 13 species of Labiatae.

c. Flower orientation

Some species of flowers are heliophilic, tracking the diurnal pattern of movement of the sun. This behavior, sometimes observed in bowl-shaped flowers that may function as solar heaters, can result in warming of the flower to temperatures significantly above that of the ambient air (Kevan 1975). The increased temperature is apparently attractive to some insects, which bask in the flowers, and probably serves to hasten physiological processes such as the growth of pollen tubes and maturation of ovules or seeds. A compass can be used to provide information about both the sun's position and flower orientation. Reese and Barrows (1980) determined the sun's position by aligning a lensatic compass with the shadow cast by a perpendicular nail on a horizontal plane, and the direction a flower was facing by aligning its longitudinal axis with the compass. Smith (1975) prevented heliotropic flowers from moving by tying them to stakes, to study the consequences of heliotropism for insect pollination.

Stanton and Galen (1989) developed a simple heliotropometer for measuring the angular deviation of flowers from the sun. To make one, mark a flat cardboard disk (50 mm radius) with concentric rings at 1-mm intervals. Insert a 40-mm-long pointer (e.g., a dissecting needle) perpendicular to the center of the disk. When the disk is aligned so that its surface is parallel to the orientation of the flower, the angular deviation from the sun can be calculated using the trigonometric relationship: sun angle = arctan (length of shadow in mm/40). If you find that measurements of angles greater than 51.3° are required, you may want to make a larger disk in order to measure longer shadows. Stanton and Galen (1989) just used a truncated distribution of measurements and analyzed data with nonparametric statistics, with six even classes of solar orientation determined from the cosine of the angle of solar deviation (the sixth class was angles greater than 51.3°). This technique does not provide information about the compass orientation of the flowers, but that could be determined with the help of a compass. The magnitude of diurnal movement during a day can be established by measurements made with a protractor, compass, and level to measure the angle made by each flower's central axis with respect to a horizontal line directed toward some compass point (e.g., east) (Stanton and Galen 1989).

It may also be valuable to measure orientation of flowers relative to vertical and horizontal axes, for studies of the evolution of bilateral symmetry, or as it relates to accessibility to pollinators. Schlessman (1986a) used a pocket level to measure the difference between the horizontal and the orientation of a plane defined by petal lobes (essentially perpendicular to the corolla tube).

2. Inflorescence Structure

The structure of an inflorescence is likely to influence factors such as the foraging efficiency of flower visitors, the degree of outcrossing, and other aspects of plant reproductive success. For example, Schemske (1980b) found that reproductive success in a pollinator-limited orchid varied significantly with inflorescence size in an orchid, and Wyatt (1982) reviewed data on how inflorescence architecture (flower number, arrangement, and phenology) affects pollination and fruit set. For these reasons it may be important to describe inflorescence structure, either by drawing it in a figure, or using some generally understood terminology. Although there is apparently no universally accepted and endorsed terminology yet, Weberling's (1989) book on the morphology of inflorescences presents a possible (although rather complex) vocabulary for this task.

Lindsey and Bell (1985) described one way to analyze inflorescence structure of umbels, by measuring umbel and umbellet diameter, tabulating the number of umbellets per umbel and flowers per umbellet, and then calculating umbel density using the formula:

$$\text{Density} = (\text{radius of umbellet})^2 \cdot \frac{\text{number of umbellets}}{(\text{radius of umbel})^2}$$

This sort of quantitative index should prove useful for comparisons of umbels within and between species. In other cases measurements of the length of an inflorescence and the number of flowers on it may suffice for a description.

It may also be important to describe the structure of inflorescences for studies of foraging behavior. In a study of optimal foraging of bumblebees, Pyke (1979) measured the distance between flowers on an inflorescence, and the angular distribution of the flowers on the stalk. He used the same measurements from a hummingbird-pollinated plant to create artificial inflorescences with different degrees of similarity to natural inflorescences (disposable syringe needles inserted into a styrofoam rod; see Chapter 7, Section K.3 on artificial flowers; Pyke 1981a).

It is also possible to manipulate natural inflorescences in order to investigate aspects of both plant reproductive biology and flower-visitor behavior. The simplest type of manipulation is cutting off flowers, but combining flowers may also be simple to accomplish to generate larger inflorescences. Willson and Rathcke (1974) used both methods to alter inflorescence size of *Asclepias syriaca*. Jones and Buchmann (1974) turned flowers 90° and 180° from the original axis to investigate the responses of bee pollinators; they also manipulated ultraviolet reflectance patterns of individual flowers.

3. Energy Content

The energy content of flowers can be subdivided into a structural component (floral tissue), nectar content, and energy devoted to pollen. Harder and Barrett (1992) measured the structural energy in *Pontederia* (Pontederiaceae) flowers by drying flowers from which nectar had been removed with filter paper wicks, and then subjecting the samples to microbomb calorimetry (Phillipson 1964). See Chapter 4, Section N, for information on how to measure the energy content of pollen samples, and Chapter 5, Section B.3, for how to measure the energy content of nectar. Although these approaches imply that biomass or carbon are the best currencies with which to measure the cost of flowers, another approach is to consider mineral nutrients as the most appropriate measure of allocation. Ashman and Baker (1992) measured concentrations of nitrogen and phosphorus in sepals, petals, gynoecium tissue, and pollen and found that pollen had the highest requirement for phosphorus. They also found seasonal variation in allocation patterns. They used a micro-Kjeldahl technique for nitrogen analysis and an ascorbic acid technique to measure phosphorus.

D. Floral Fragrance

Floral fragrances are potentially important not only as attractants for pollinators but also as isolating mechanisms if they are selectively attractive to different species (e.g., Dobson 1987). Williams (1983) has a nice review of the role of floral fragrances as cues in animal behavior, with particular reference to orchids. Not surprisingly, floral odors often have complex compositions, including the major chemical groups terpenoids, benzenoids, and aliphatics. Examples of these components include sesquiterpene and other alcohols, aldehydes, keto compounds, many monoterpenes (e.g., geraniol, limonene, cineole, camphor, menthol), simple aromatics (e.g., vanillin, methyl salicylate, eugenol), aminoid compounds (nitrogen-containing compounds such as indole, skatole), fatty acid derivatives, and methyl esters (Bergström et al. 1980, Williams 1983, Dobson et al. 1990). If the flowers you are studying are long-lived, be aware that floral odor may change over the flowering period (e.g., Patt, French, et al. 1991). Daily variation and variation among individual plants can also be important. Patt, Rhoades, and Corkill (1988) found that the quantity of floral emissions from *Platanthera stricta* (Orchidaceae) varied widely among individual plants (e.g., a range of 0.02 to 26.26 µg/hour/inflorescence during the day when collected on charcoal). Although this is a greater range than has been reported in other studies, it suggests that any study of floral odor should include a large number of plants to look for such variation. They also found significant nocturnal/diurnal differences in *Platanthera stricta;* the charcoal-trapped emissions ranged from 1.20 to 1.26 µg/hour/inflorescence. Altenburger and Matile (1990) also found diurnal variations in fragrance.

This kind of variation is one of the problems facing researchers who wish to characterize floral fragrances through quantitative and qualitative studies. Although there is general consensus among flower fragrance researchers about which general methods are most accurate and reliable, individuals and laboratories differ in details that are appropriate to the kind of flower material they are studying (H. Dobson, *personal communication*). These differences can lead to very different results and conclusions regarding the flower fragrances, which complicates comparisons of data, even from the same plant species, that have been sampled or analyzed using different methods. If you are planning comparative work, be sure to use consistent methods.

Although the human nose can detect many floral odors, it is probably much less sensitive than the receptive organs of most flower visitors, and it doesn't lend itself well to quantitative studies. Thus qualitative and quantitative studies of floral odor are much more recent than studies of other morphological traits that are more easily measured. To describe qualitative and quantitative aspects of floral odors we must turn to sophisticated instrumentation, usually gas chromatography in combination with mass spectrometry (MS). A gas chromatograph (GC) equipped with flame ionization or nitrogen/phosphorus detectors, and combined GC-infrared spectrometers may be used to identify floral odor compounds. We will review briefly various techniques for collecting fragrance samples, and even more briefly some of the gas chromatography techniques that have been used. The references we provide should be consulted for additional information for this laboratory technique. Hills and Schutzman (1990), Dobson (1991), and Kaiser (1991) give overviews of methods and potential pitfalls of sampling floral fragrances.

If you don't have access to equipment for analysis by GC or GC-MS, there is an inexpensive (but less quantitative) alternative, involving a "nose bioassay." To use this technique it may be necessary to concentrate the odor. Enclose freshly picked flowers in a jar, for a minimum of 15–30 minutes on a warm day, and then smell the inside of the jar when you remove the lid (technically, this is a method known as "headspace concentration"). Faint floral odors may become detectable that way (Buchmann 1983). If you have collected the flowers during the time of day when the flower is actively emitting odor, the smell in the jar may be quite strong. The smell can also give you a good idea of what sorts of classes of chemical compounds are in the floral odor. Although these odors are usually a complex mixture of compounds, it is often possible to classify to categories such as terpenoid, benzoid, or aminoid (J. Patt, *personal communication*). This technique can also be used to identify which floral parts emit the odor (e.g., D'Arcy et al. 1990).

If you need to collect floral odors to do behavioral experiments, such as determining the significance of odor alone as an attractant, you may also find that putting freshly picked flowers in a container is a convenient way to present an odor. Kirk (1985) used a similar technique to trap thrips that were attracted to various test scents. He put the test chemicals in glass tubes (50 mm long, 10 mm in diameter) with a 20-cm wick (a piece of a dental roll; the kind your dentist puts between your teeth and cheek while drilling) projecting above the top of the glass. A strip of filter paper running the length of the tube acted as a wick between the dental roll and the chemical. See Chapter 7, Section K.4, for additional methods of testing insects' responses to floral odors.

The first step in more sophisticated analyses is the collection of an odor sample, which can be accomplished in a variety of ways. Williams (1983) reviews several techniques (1–5 below; see his figures 3–12 for diagrams of the trapping and desorption devices he uses), and others are suggested by Bergström et al. (1980; 6–7 below), Armbruster et al. (1989; 8 below), and Nilsson (1978, 1981; 9 below). Dobson (1991) and Kaiser (1991) also show figures of different devices for trapping floral odors.

1. Make a Plexiglas chamber to contain an inflorescence while it is still on the plant (use an adhesive that will not emit volatiles for a long time after it is dry). Make a slot in the bottom or side panel through which the stem can be placed. Close the slot with foam weatherstripping tape. Insert Swaglok fittings with gas chromatograph septa in one side through which air samples can be withdrawn with a gas-tight syringe. Leave the inflorescence in the chamber for at least 30 minutes at a time when odor is being released by the flowers, and then take the gas sample, which can be injected directly into a GC. An advantage of this technique is that it requires little in the way of special sampling equipment. A disadvantage is that it provides a very small sample of the fragrance compounds, and therefore works

only with strong fragrances (H. Dobson, *personal communication*).

2. Saturate a piece of glass filter paper with a thermally stable oil (e.g., Convalex 10 pump oil) dissolved in ether. After the ether is driven off, purge the paper with a flow of helium at 200°C for 18 hours to get rid of any substances that would volatilize under the gas chromatography conditions to be used later. Store the purified paper in a jar or sealed in aluminum foil. Collect floral fragrances by putting fresh flowers into a jar with a strip for a period of a day or longer. Remove the sample strip, wrap it in aluminum foil, and return it to the laboratory for analysis. The injection port of the GC can be modified to accept paper strips (Holman and Heimermann 1973 and 1976, cited in Williams 1983). By heating them to 170°C, the volatile fragrances are driven off and fed into the GC. An advantage of this technique is that the sampling strips can be mailed or carried anywhere and that sampling itself requires little preparation. Disadvantages are the elaborate process of preparing the strips and the necessity of a specialized injection port. However, Williams (1983) has obtained good results by placing fragrance-loaded sample strips in a cartridge and desorbing them. (See also Thien et al. 1975, who used a similar method.) It is difficult to standardize this method (J. Patt, *personal communication*), and other methods should be used if possible.

3. Use a glass collection tube packed with one of the following: (1) Chromosorb G 60/80 mesh, (2) Porapak Q 60/80 mesh, (3) Porapak Q 50/80 mesh adsorbent, (4) Tenax, or (5) activated charcoal. Attach the collection tube to a larger container of flowers and circulate air from the flower chamber through the collection tube for a period of several hours to several days. After sufficient material has been collected, place the collection tube in the GC inlet and heat it for 20 minutes at 150°C to drive the sample into the GC. This method has the advantage of not harming the flowers. It is also easy to use and concentrates good sample sizes of fragrance but has the disadvantage of requiring a special

inlet system on the GC. However, heating might alter fragrance components and desorption with a solvent is probably preferable (H. Dobson, *personal communication*). This method, with solvent extraction, is probably the most commonly used method now (H. Dobson, *personal communication*).

4. Suspend flowers in a sealed chamber over dishes filled with cyclohexane. The odor will be dissolved in the solvent, which can then be injected into the GC (Armbruster and Webster 1979). Advantages of this technique are that it is inexpensive and can be carried out in the field. Disadvantages are that the sample is not very concentrated and that the fumes will kill the flowers quickly. Another disadvantage is that this method is difficult to standardize (J. Patt, *personal communication*).

5. Place flowers in a collection box as in method 1. Circulate air from the chamber (using a vacuum pump or an aspirator connected to a water faucet) through a collection tube filled with Porapak Q, or Tenax GC (with or without activated charcoal), separated by a plug of glass wool, with the ends also plugged with glass wool. Pass the air through the Tenax first. Continue collection for an hour to a day. To collect the sample, place the collection tube in a copper or stainless steel tube wrapped with an electrical heating tape, and heat it to 200°C while purging (from the charcoal end first) with a stream of nitrogen gas. (Solvent extraction may be preferable.) Pass the gas through a glass capillary tube that is cooled by liquid nitrogen to condense the floral fragrance. Elute the fragrance with solvent and inject it into the GC. Advantages of this system are that it does not require a special inlet on the GC and that the combination of activated charcoal and another adsorbent may trap more types of fragrance compounds than a single adsorbent. It also provides a large sample in a liquid form that can be stored. It is not a good field method as it requires access to liquid nitrogen and other laboratory equipment.

6. A combination of cold trapping and sorption is an accumulating method that provides a relatively accurate picture of natural proportions of fragrance components. A cold trap

can be kept at 0°C (ice), -70°C (dry ice/ethanol), or -170°C (liquid nitrogen). Bergström et al. (1980) used a glass spiral as a cold trap, attached to a vessel containing the flowers, and circulated air through the system at a rate of 5–10 mL/minute. A disadvantage of this technique is that it also results in trapping of water vapor; the inside diameter of the glass spiral should be at least 8 mm to prevent blocking with ice (Bergström et al. 1980). Other methods are probably more convenient and more accurate.

7. Extraction is also possible by using organic solvents, such as pentane or hexane, carbon disulfide, diethyl ether, methanol, chloroform/methanol, or kaotron (a chlorofluorocarbon). Put whole fresh flowers in the solvent for 15 minutes. This has the advantage of being a good field method, as the sample can be returned to the laboratory for analysis (Bergström et al. 1980). It also facilitates use of the extracted compounds for experiments because the solvent can be placed on artificial flowers. A potential major problem with this method is that the solvent may extract extraneous compounds that are not fragrance components, especially if the flowers are left in it for long periods. A sample collected with this method could contain literally thousands of extraneous compounds along with floral odor compounds, and the method should only be used if you have already conducted a thorough GC-MS analysis of floral odors so that you can use a GC-MS to decipher the myriad of peaks that will result (J. Patt, *personal communication*).

8. Place fresh flowers in a sealed chamber with powdered Tenax or Porapak Q held in folded filter paper. After four hours' exposure, place the Tenax in a glass ampule and seal it. Return it to the laboratory for analysis with GC (Armbruster et al. 1989). Pellmyr et al. (1991) put 150 mg of Porapak Q 80/100 mesh on an aluminum tray inside a polyacetate bag with cycad cones for 15–46 hours and then put the adsorbent in a vial for transport to a lab for analysis. Patt (*personal communication*) found that this method did not collect enough sample and suggested that the adsorbent method works best when an airstream containing the volatiles is passed through a column containing the adsorbent.

However Dobson (*personal communication*) had success in enclosing flowers with adsorbent inside a container formed by loosely wrapped aluminum foil; the Porapak Q was then eluted with solvent. This method may favor the uptake of compounds with low volatility (Dobson 1991).

9. Place fresh flowers inside a glass vessel or tube, which is coupled to a glass pre-column packed with an adsorption material (Chromosorb G 60/80 mesh treated with 10% silicone grease, or Porapak Q 50/80 or 80/100 mesh). Circulate air from the flowers through the collection column for a few hours to 2 days (e.g., at 150 mL/minute, with a battery-operated membrane pump; Knudsen and Tollsten 1991). Then attach the collection pre-column to the GC inlet, and heat it to 150°C for 20 minutes to extract the fragrance sample (but use of heat may affect the fragrance compounds, so solvent extraction is preferable). If Tenax is substituted as an adsorption material, the fragrance sample is extracted with pentane or diethyl ether, and concentrated by evaporation to a few µl. The fragrance can also be desorbed from Porapak by extraction with 2 mL of redistilled diethyl ether (Thien et al. 1985) or pentane (Pellmyr et al. 1987, Dobson et al. 1990, Knudsen and Tollsten 1991), or 4 mL of hexane (Nilsson et al. 1985). Groth et al. (1987) used a polyacetate bag (Bosco) to hold inflorescences for this method. Gerlach and Schill (1989) used a similar method, but with activated charcoal in the adsorbing cartridge and elution with acetone (Nilsson 1978, 1981).

10. The following "headspace" or "purge and trap" method was developed by chemists at the Rutgers Center for Advanced Food Technology for trapping trace quantities of volatile contaminants, but it was modified by Patt (*personal communication*) for collecting floral volatiles both in the field and in the greenhouse. It can be used to collect odors from flowers still attached to a plant, or from cut flowers. The cost (excluding pump) for the collection setup (and 25 trap tubes) is about $250.

Place a three-necked flask (available in a variety of sizes from Kontes Scientific Glassware; see Appendix 2) over the inflorescence

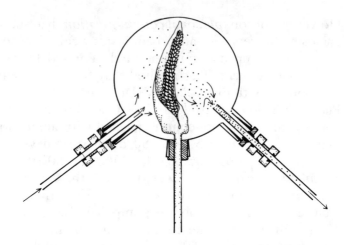

Figure 3–1. Apparatus for collecting scent from an inflorescence. From J. Patt, *personal communication.*

or flower, support it with a ring stand, and seal the neck around the stem with a cork collar or piece of Parafilm (Figure 3–1); if floral odor is weak, beware of contamination from the cork or Parafilm. If inserting a fully opened flower would damage it, try inserting a bud and letting it open in the flask. Use a flask large enough so that floral parts will never contact the wall, which might cause damage and the release of extraneous volatile compounds. To collect odor compounds, pass clean dry air from a pressurized gas cylinder into one of the other necks; flow rate should be controlled with a bubble-type flowmeter. To collect odor compounds, a trap cartridge is inserted into one end of a Swaglok union fitting containing either graphite or Teflon ferrules, the other end of the union is attached to a purge line of 0.25-inch (outside diameter) Teflon tubing, and then the open end of the trap tube is inserted into a glass adapter that fits into the third neck on the flask. The adapter has an O-ring that can be tightened to make a seal around the trap tube. Attach the purge line to a vacuum pump or faucet aspirator; monitor the airflow with a bubble-type flowmeter (Patt recommends a rechargeable portable vacuum pump from SKC that provides a very accurate flow rate and can be used in the field; see Appendix 2). The flow rate of both airlines should be equivalent to avoid backpressure buildup.

To make the trap cartridge, cut a 7–10 cm length of 0.25-inch (outside diameter) borosilicate glass tubing, flame-polish the ends of the glass tube, and then silanize it to remove any contaminants. (Silanization is a process that removes reactive sites from glass surfaces. Typically one treats the glass with a 5% solution (v/v) of dimethyldichlorosilane in toluene; J. Patt, *personal communication*). Fill the tube with adsorbent: Tenax TA (60/80 mesh, 30 mg) and Carbotrap (Supelco, 100 mg). Patt et al. (1988) demonstrated that a single adsorbent may not trap all fragrance components. They used both charcoal and Tenax, and neither alone effectively trapped the full range of fragrance components. Seal the ends of the tube with glass wool plugs and condition it in a gas chromatograph oven (heating it to 250°C while purging it with nitrogen or helium at a low flow rate). Trap tubes can be used many times as long as they are conditioned after each use. Although manufactured tubes can be purchased (e.g., from SKC), they are much more expensive and are not always preconditioned sufficiently for collecting trace amounts of floral odor compounds (J. Patt, *personal communication*).

Sampling time should be about 1–4 hours (longer times may result in the adsorbents' becoming saturated and loss of sample via breakthrough). If you need to sample continuously, for example, to determine the temporal pattern of odor production, the trap cartridges can be replaced easily. Remove the cartridge and seal it with Swaglok column caps fitted with graphite or Teflon ferrules. Store samples at -20°C or colder if possible, and analyze them as soon as possible to avoid decomposition on the adsorbent (known to be a problem with charcoal). The floral odors can be eluted from the trap tubes either by using solvents (ether) or a thermal desorption device. Two rinses of 2.5 mL of ether should suffice. Condense the eluate to 0.5 mL or less by evaporation under a stream of nitrogen, and then inject it into a GC injection inlet, or store it in a glass vial with Teflon-lined cap at -20°C. A homemade desorption unit can be made from an old GC injection port and heating tape. If you use thermal desorption, it is a good idea to chill the GC column with chipped dry ice to retain the odor compounds in the column during the desorption procedure.

If you are sampling in the field, you may be able to use ambient air instead of a pressurized gas cylinder for the air inlet, but you should still analyze a control sample of background air to identify possible background contaminants. If you are sampling in bright sunlight, place a nylon screen over the flask to prevent overheating in the flask.

Once the floral fragrance samples have been collected, they can be analyzed by GC or GC-MS. The combined techniques permit identification and confirmation of many fragrance constituents. Alternatively, high-performance liquid chromatography (HPLC) can be used to separate liquid fragrance samples. This method may prove particularly useful for obtaining material for a structural confirmation with nuclear magnetic resonance spectroscopy (Williams 1983). However, all of these techniques are beyond the scope of this book. For additional information see references cited in the descriptions of the ten techniques just listed.

Fragrance samples can be used for behavioral experiments with flower visitors. Brantjes (1978) extracted odors with heptane and applied the extract to strips of absorbent paper. The responses of moths in cages to these strips were then investigated, using different concentrations of odor. Pellmyr et al. (1990) identified the four components of the floral odor of *Zygogynum viellardii* (Winteraceae) and then examined the responses of insects to artificial mixtures of the components after adding them to artificial flowers. Ten mL of the test blend was put into a 50-mL vial with a wick, and paper petals were added. The 10 mL evaporated in 60–120 minutes, so this technique obviously produced a stronger odor than do natural flowers.

These same techniques should also work for analyses of pollen volatiles, which can be a significant component of floral odors. Dobson et al. (1987, 1990) found that only one-third of the compounds they identified from whole flowers and from pollen of *Rosa rugosa* (Rosaceae) were detected in both samples. Most of the major pollen volatiles were contained in the pollenkitt. Dobson (1988) surveyed pollen lipids and found that pollenkitt may provide pollen with species-specific odors. A few studies (e.g., Dobson et al. 1990, Knudsen and Tollsten 1991) have described differences in floral odors of flower parts. For example, petals and stamens of *Moneses uniflora* (Ericaceae) have quite different floral odors (Knudsen and Tollsten 1991).

It may also be possible to use a stain, such as neutral red, to determine the site of odor production in a flower. Buchmann and

Buchmann (1981), and Slater and Calder (1988) used a technique originally described by Vogel (1963) to try to localize the parts of flowers that were releasing floral fragrance. They put flowers in an aqueous solution of neutral red:water (1:10,000, with Aerosol OT added as a surfactant) for 15 minutes. This stains lipids red and may therefore indicate the presence and distribution of volatile compounds responsible for the floral odor. Stern et al. (1986) placed fresh flowers in a stain bath of 1:1,000 neutral red:tap water for 2–10 hours, rinsed them in tap water, and then examined for staining (Knudsen and Olesen 1992 left the flowers for 1–5 hours). Tissues that stained deep red were assumed to be areas of osmophoric function. This identification by staining may be necessary because there is "little or no special modification of floral parts to indicate the site(s) of fragrance emission" (Stern et al. 1986). Stern et al. found that the same areas were also stained with Sudan black B in flowers embedded in epoxy resin (put floral tissue in 2% [w/v] paraformaldehyde and 2.5% [v/v] glutaraldehyde in 0.1 M cacodylate buffer [pH 7.2], postfix in buffered 1% [w/v] osmium tetroxide, dehydrate in an ethanol/acetone series, and embed the sample in epoxy resin ERL 4206). Although this study looked only at orchid flowers the terpenoids the stains appeared to be indicating are components of many floral fragrances so the method may work for many species. Knudsen and Tollsten (1991) found that neutral red also correctly indicated the presence of osmophores of a species whose scent consisted almost exclusively of isoprenoids.

A bioassay might also be possible by putting different floral parts into different containers and looking at the responses of pollinators to them. For example, Pellmyr and Patt (1986) tested the responses of beetles to different inflorescence parts by putting them in petri dishes with holes in them to let the odor escape. (See the Chapter 7, Section K.4, for additional methods of testing insects' responses to floral odors.)

E. Flower Color

For almost two centuries flower color has been interpreted as an adaptation by which flowers attract or guide pollinators (see references cited by Waser and Price 1981), signaling not only their presence but also more subtle cues such as the location of nectar rewards (via nectar guides, first described by Christian Konrad Sprengel in 1793, as Manning 1956a mentions) or even their absence (via age-related color changes). Gori (1983, 1989) and Weiss (1991) discuss age- or pollination-related color changes in flowers and their significance for flower visitors. There also appear to be associations between particular flower colors and particular groups of pollinators (Kevan 1983). Flower color can be used as a genetic marker for experimental studies (e.g., Snow and Mazer 1988) or to experiment with the behavior of flower visitors (see Chapter 7, Section K.12). A variety of methods have been used to quantify floral colors.

1. Visual Appearance to Humans
Although the visual pigments of insect, avian, or mammalian pollinators do not match those of humans, it is still useful to be able to describe floral colors in terms relevant to other human observers. Because of the subjective nature of color description, an objective means of standardizing descriptions is important. One possibility is to give careful descriptions of factors such as hue, saturation, and luminance, but a simpler method is to use a color chart with a sufficient number of color samples that one can closely match any floral color. Both Scogin (1980) and Lamont and Collins (1988), for example, used the Royal Horticultural Society's Colour Charts (see Appendix 2). An alternative color guide (Naturalist's Color Guide) is published by the American Museum of Natural History. Tucker et al. (1991) surveyed available color charts.

The visible floral pigments are chemical compounds, including flavonoids, carotinoids, and betalains (Scogin 1983). Scogin (1980, 1983) reviews procedures (mostly chromatography) for extracting and characterizing them.

2. Measuring Visible Light Reflectance Levels
A study by Reese and Barrows (1980) found evidence that light reflectance alone might affect visitation by pollinators. They found greater activity by bees on east- and west-facing slopes, although there were no differences in ambient temperature. To measure light reflectance they used a Gossen Luna Pro photographic light meter with a Variable Angle Spot Meter Attachment.

Reflectance can also be quantified by comparison of flowers (or parts of flowers) photographed with a calibrated reflectance scale (see below for suggestion on how to construct one). Typically this is done while using a series of photographic filters that match the visual pigments of the pollinators of interest; for example, mono-chromatic filters matching the primary colors for honeybees and bumblebees, ultraviolet (about 350 nm), blue (about 450 nm), and yellow (about 580 nm). These studies were pioneered by Daumer (1958), and one method is described in detail by Kevan (1983); he recommends the following series of broad-band monochromatic filters and gives details about their use:

Filter	Color	Wavelength
Kodak 18A	ultraviolet	300–400 nm
Kodak 35	violet	320–470 nm
Kodak 98	blue	390–500 nm
Kodak 65	blue-green	440–570 nm
Kodak 61	deep green	480–610 nm
Kodak 90	dark greenish amber	540–650 nm
Kodak 25	red	580 nm and higher
Kodak 70	dark red	650 nm and higher

3. Measuring Spectral Reflectance

A more quantitative method for color description involves use of a spectrophotometer with an integrating sphere or diffuse reflec-tance unit (e.g., Unicam SP 8,000 recording spectrophotometer with an SP 890 diffuse reflectance unit; J. Olesen, *personal communica-tion*) to generate a spectral reflectance curve, which shows the percentage of the available light reflected (or luminance factor) for each wavelength. The same techniques that have been used for studies of insects can be applied to flowers. Roland (1978) used a recording spectrophotometer with a diffuse reflectance accessory to record absolute reflectance curves for butterfly wings, and Horovitz and Harding (1972) and Macior (1983) examined fresh corollas in the same way. Magnesium carbonate or barium sulfate are usually used as a white standard for calibration and assumed to be 100% reflective at all wavelengths measured.

A combination of reflectance spectrophotometry and photographic analysis may be appropriate in some cases. Macior (1978) measured the spectral reflectance of *Dicentra* (Fumariaceae) corollas for ten wavelengths in the visible spectrum by using reflectance spectrophotometry (Bausch and Lomb Spectronic 20 spectrophotometer equipped with a reflectance attachment). He complemented this analysis with an examination of ultraviolet reflectance patterns by photography (illumination under artificial ultraviolet light, a fused quartz lens, and a filter that only passed ultraviolet). The flowers were photographed against a Kodak gray scale with the following percent reflectance equivalents for the corresponding scale values:

Scale Value	% Reflectance
0.00	40.5
0.10	31.5
0.20	19.2
0.30	12.0
0.50	7.0
0.70	5.0
1.00	3.0
1.30	2.5
1.60	2.0
1.90	1.5

Nilsson (1983, 1984) dispensed with photographic analysis in the ultraviolet by using a spectrophotometer (Perkin-Elmer 554 UV/vis with external integrating sphere attachment) to cover the range of 360–800 nm. He placed individual petals or other floral parts so that they covered the 10 x 10–mm opening of the sphere. Haslett (1989) transported flowers from his field site to the laboratory, where he scanned whole inflorescences through the sample window and recorded wavelengths from 325 to 700 nm (Unicam SP800 spectrophotometer fitted with an SP890 reflectance attachment). Patt et al. (1989) recorded the spectral composition of reflected light from an orchid inflorescence between 350 and 750 nm with a Beckman DK-2A ratio-recording spectroreflectometer, illuminated with an angle of incidence of 3°.

If you do not have access to a spectrophotometer for your field study, an alternative is still photography, using a series of gray scales and photographic filters (e.g., Penny 1983; Lindsey and Bell 1985, using methods of Kevan et al. 1973). If one knows the characteristics of the visual pigments of the pollinator of interest, it is possible to use a set of photographic filters that matches the pigments. An ordinary glass camera lens will work for the portion of the spectrum visible to humans and partway into the ultraviolet, but in order to photograph ultraviolet as insects can see it, one must use a lens made with a material other than glass (glass lenses don't pass ultraviolet below about 350–380 nm; McCrea and Levy 1983). Some plastics will work, as will fused quartz or fluorite, or you can use a pinhole camera that doesn't require a lens at all (e.g., Lindsey and Bell 1985). The object is to photograph the flowers through a series of mono-chromatic filters (see previous section), each passing only a re-stricted range of wavelengths, together with a scale that has been calibrated for reflectance at those wavelengths (e.g., Penny 1983). A suitable gray scale can be made by combining carbon black and magnesium oxide in different proportions; Kevan et al. (1973) describes how to construct and calibrate such a scale. Alternatively, Penny (1983) used plaster and soot, with the darkest element being filter paper soaked in Nigrosin black, and Guldberg and Atsatt (1975) constructed an ultraviolet-gradient standard from a paint sample sheet (Sears no. 0826) that had sections coated with silver metallic paint or flat black paint.

The reflectance of flower parts can be determined by visual comparison with the gray scale on negatives or prints, or more precisely by measuring density of the negatives. Most black-and-white films, as well as videotapes, appear to be receptive over the range of wavelengths (including ultraviolet) that are of interest for pollination studies (but not infrared; Kevan 1983).

Kevan (1983) describes in detail the methods necessary for photographic analysis. In brief, once you have a calibrated gray scale and an appropriate camera and lens, you photograph the flower through a sequence of broad-band monochromatic filters to deter-mine the reflectance at each wavelength of interest. Although natu-ral light can be used, artificial light will also work. Penny (1983)

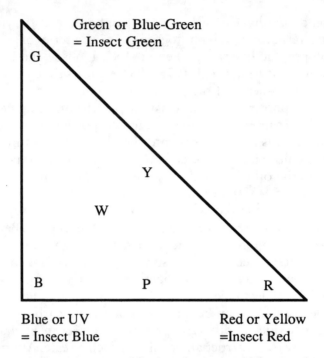

Figure 3–2. Representation of flower color in a trichromatic color space reflecting the visual pigments of honeybees. Modified from Kevan 1979.

illuminated flowers with tungsten bulbs supplemented with a mercury vapor lamp, which is rich in ultraviolet. Electronic flash will also provide a close match to the spectrum of daylight.

The object of these photographic methods is to try to quantify some aspects of the appearance of flowers to the flower visitors. Thus the filters you use should match the spectral sensitivity of the visual pigments of the flower visitor of interest. A convenient way to summarize information about the visual appearance of flowers to pollinators with trichromatic color vision (e.g., honeybees, bumblebees) was proposed by Kevan (1978). In these color triangles, each corner represents one of the three color receptor types (e.g., for the honeybee, 340, 440, and 540 nm; see Figure 3–2). The color of a flower, or flower part, can be represented in this color space by indicating the proportion of reflectance in each of the three colors (e.g., Nilsson 1983). Bear in mind, however, that the interaction between reflectance curves (e.g., in areas of overlap), which may be an important component of how the flowers are perceived, is not really addressed by these photographic or analytical methods (O. Pellmyr, *personal communication*). Menzel (1990) also points out that the color space is non-Euclidian in nature, which complicates its use.

McCrea and Levy (1983) described an inexpensive photographic technique for visualizing how honeybees see flowers. It involves reproducing the ultraviolet component in a color that is visible to humans. It produces an approximation (dichromatic) of floral colors in the insect visual system that is qualitatively interpretable by humans, as the ultraviolet component undergoes a wavelength shift to a color we can see. One disadvantage of the method is that it does not readily distinguish blue and ultraviolet (P. Kevan, *personal communication*). The method employs a combination of filtered daylight and blacklight fluorescent light sources to achieve an equiproportionate near-ultraviolet:blue:green balance that is recorded on color slide film. Because the method is rather complex, we will not reproduce the details here but refer the reader to the original paper.

4. Ultraviolet Reflectance Patterns

Although in principle ultraviolet reflectance patterns are no different from those of other wavelengths, they have often been accorded special treatment because they are not visible to humans. Thus the methods for observing and quantifying them may be a bit different. Certainly our appreciation of them and their significance is more recent than that of other color patterns. Daumer's work with honeybees in the 1950s (cited in Jones and Buchmann 1974) was the first to demonstrate the biological significance of ultraviolet floral patterns as orientation guides.

For still photography, the same techniques are used as described above for visible colors, but the filter used will appear black because it does not transmit visible light. Filters that have been used for this include the Schott-Filter U.V. 1 (Giesen and Van der Velde 1983), Hoya 360 (Nilsson et al. 1986), and Wratten 18A (transmits 300–400 nm; Mulligan and Kevan 1973, Primack 1982, Macior 1986). Hill (1977) and Primack (1982) describe ultraviolet photography; Primack cautions that not all electronic flash units put out sufficient ultraviolet light for this type of work.

Eisner et al. (1969) described how to use a videocamera in order to visualize ultraviolet reflectance patterns. For this technique, as for still photography, it is important to use a lens that will transmit ultraviolet. This technique was used by Jones and Buchmann (1974) to design and monitor experiments on the significance of ultraviolet floral patterns. It has the advantage that you don't have to wait for film to be developed before you can look at the patterns of interest. A combination of video and still photography works well for observation and documentation (e.g., Jones and Buchmann 1974, Guldberg and Atsatt 1975).

Eisner et al. (1973) found that floral ultraviolet reflectance patterns could also be observed in herbarium specimens, or in freshly dried flowers, as visible fluorescent images induced by ultraviolet illumination. Thus photographic methods may not be necessary if your goal is simply to confirm the existence or absence of ultraviolet patterns. Casper and La Pine (1984) compared two types of photographic analysis of ultraviolet characteristics with measurements using an integrating sphere/spectroradiometer. Photographs taken with no filter in fluorescent light did not reveal patterns that appeared when a combination of fluorescent and ultraviolet light (or ultraviolet light alone) was used for illumination. Although the integrating sphere/spectroradiometer (which measured percent of diffuse and specular ultraviolet light reflected at 5-nm intervals from 290 to 400 nm) indicated an average reflectance from 1.21% to 6.95%, photographs taken with a Wratten 18A filter in sunlight showed the flowers as totally black (highly absorbing). Thus the photographic methods were not as sensitive.

Kevan (1979) cautions against reporting ultraviolet reflectance or absorption patterns without quantification, or without reference to the background coloration and ambient lighting. Without this context, it is possible to make serious errors in interpreting the functional significance of these floral colors and color patterns. By including a gray scale in a photograph, you can minimize this risk of this misinterpretation.

Ultraviolet reflectance patterns can also be manipulated. Jones and Buchmann (1974) removed some nonreflective portions of flowers, glued reflective parts of one flower on top of nonreflective parts of another, and covered reflective parts with yellow felt-tip-marker ink to make them absorptive. Waser and Price (1985a) painted sepals and petals of *Delphinium nelsonii* (Ranunculaceae) with artists' acrylic paints that absorbed ultraviolet light. This combination of techniques, with observations of flower visitors, may allow you to discover what are the most important visual orientation cues.

Additional references describing ultraviolet reflectance patterns include Horovitz and Cohen (1972), Kevan (1972b), Mulligan and Kevan (1973), Utech and Kawano (1975), Hill (1977), Clark (1979), Silberglied (1979), Willmer and Corbet (1981), Joel et al. (1985), Macior (1986), Inouye and Pyke (1988), Menzel (1990), and Menzel and Backhaus (1991).

F. Gynoecium

The gynoecium, or stigma, style, and ovary, is important as the female floral component. The stigma's importance in pollination lies in its role as the site of deposition of pollen grains, which germinate there. Thus the stigma's secretions and its receptivity to pollen germination are often the object of study. The style's significance may lie in providing a screening ground for the viability and vigor of pollen grains (Mulcahy 1979, Mulcahy et al. 1983). Ovaries are sometimes the subject of study and manipulation, primarily because of the ovules or developing seeds they contain.

1. Stigma Secretions

Mature stigmas can be categorized as either wet (with a free-flowing secretion) or dry (with a hydrated, proteinaceous extracuticular layer or pellicle, but no free-flowing secretion) (Heslop-Harrison and Shivanna 1977). This condition is correlated with the type of self-incompatibility system: most species with gametophytic self-incompatibility systems have wet stigmas, whereas sporophytic self-incompatibility is associated with dry stigmas (Heslop-Harrison and Shivanna 1977; this reference reports characteristics of almost 1,000 species of about 900 genera from about 250 families). Some of the wet stigmas produce such copious secretions that flower visitors may visit flowers to collect them (Baker et al. 1973). These secretions can play an important role in pollen germination and have a complex chemistry that may change with the age of the flower. In general the same methods used for chemical analyses of nectar can be used for stigmatic secretions (see Chapter 5).

To examine stigmatic secretions for the presence of lipids, a solution of Sudan black or auramine O can be used (Ciampolini et al. 1990). This can be applied to freshly squashed stigmas, to thin sections of the stigma (Lord and Kohorn 1986, Ciampolini et al. 1990), or to a piece of filter paper onto which stigmatic secretions have been collected. Baker et al. (1973) used an aqueous solution of osmium tetroxide to test for unsaturated lipids; when this solution is applied to stigmatic solutions on chromatography paper, a black color develops in the presence of lipids. Coomassie blue (Lord and Kohorn 1986) or 1-ANS (1-anilinonaphthyl-sulphonic acid, a fluorescent protein "probe"; Heslop-Harrison et al. 1974) are appropriate stains for proteins; they can be used in the same way as the other stains. Heslop-Harrison et al. used 1-ANS at 0.001% in 0.01 M phosphate buffer at pH 6.8 containing 15% methanol. Stigmas were transferred into this medium for 3–5 minutes, then into water or 15% methanol before viewing with fluorescence microscopy. These are not quantitative analyses but will suffice to determine the presence or absence of lipids or proteins.

Baker et al. (1973) also examined stigmatic exudates for the presence of antioxidants, which in every case accompanied lipids. These substances give "a pink color, rapidly bleaching white, when 2-6-dichlorophenol-indophenol (a 0.1% ethanolic solution of the sodium salt) is added to an exudate spot on chromatography paper" (Baker et al. 1973). Ascorbic acid (a reducing agent) was identified by chromatography.

Lord and Kohorn (1986) and Ciampolini et al. (1990) describe techniques for optical or electron microscope studies of pistils. Lord and Kohorn fixed gynoecia for histological studies in 2.5% glutaraldehyde in 0.025 M phosphate buffer at pH 7.0. The fixed material was dehydrated rapidly in DMP (2,2-dimethoxypropane), stored in 100% acetone, and then embedded in glycol methacrylate and sectioned at 2–3 μm. The sections were stained with auramine O to detect lipids and the cuticle, and examined with fluorescence microscopy. Coomassie blue was used to stain for proteins, and Periodic acid–Schiff's (PAS) reagent was used to detect insoluble polysaccharides. For scanning electron microscopy, stigmas were dehydrated in acetone, critical-point dried with carbon dioxide, and coated with gold/palladium.

Ciampolini et al. (1990) fixed pistils for optical microscopy in 3% glutaraldehyde in 0.006 M cacodylate buffer at pH 7.2 for 30 minutes at room temperature, dehydrated them through an ethanol series, and embedded them in historesin. They cut 2–3 μm sections and localized cuticle with 0.02% aqueous auramine O, pectinaceous material with alcian blue and ruthenium red, insoluble polysaccharides with PAS reagent, lipoidal material with Sudan black, and callose with aniline blue fluorescence. Fresh stigmas, uncoated, were mounted and observed with a scanning electron microscope. For transmission electron microscopy, pistils were fixed in glutaraldehyde as above, postfixed in 1% osmium tetroxide, dehydrated through an ethanol series, and embedded in Spurr's resin. Sections were stained with uranyl acetate and lead citrate for 10 minutes and 3 minutes, respectively.

Other techniques that screen for the presence of protein (e.g., brom-phenol blue) or alkaloids (e.g., iodoplatinate test) in nectar can also be used for stigmatic exudates. See Chapter 5 for details.

2. Stigma Receptivity

For pollination to occur, not only must pollen be transferred to the stigma, but it must also be deposited during the period of stigma receptivity (although in some cases pollen deposited before the period of receptivity may remain viable long enough to germinate when the stigma becomes receptive; O. Pellmyr, *personal communication*). If the stigma is not receptive, the pollen may not adhere or may not germinate. Thus for some studies it is important to determine the time of receptivity. There are several ways to do so. In some species the stigmas have lobes that remain clasped or grooves that remain closed until the onset of receptivity, but in most cases it is not this simple to determine. The oldest method, although also the most tedious and the slowest to provide results, is to hand-pollinate flowers at different times of day, or over a period of days, and then wait to see whether seeds are produced. For example, Thomson and Barrett (1981) and Bertin (1982) hand-pollinated stigmas at different developmental stages and then observed fruit set and maturation, and Morse (1987) used a 3 x 3–factorial design with pollinium age and receptor flower age as variables to check the role of pollen age in follicle production in *Asclepias syriaca*. This process might be shortcut a bit by examining the stigma for germinated pollen grains or the style for growth of pollen tubes. Vaughton and Ramsey (1991) classified pollinated stigmas with at least one germinated pollen grain as receptive. Preston (1991) measured the growth rate of pollen tubes at different phenological stages and found that it declined as flowers senesced and stamens abscised.

Chemical tests have been developed to detect receptivity more quickly. Two of these tests involve enzymatic reactions, under the assumption that enzyme presence reflects stigma receptivity. One test stains for the presence of peroxidase enzymes and the other for esterases. Note that most of these enzymatic tests will give false positives if there is pollen present on the stigma or if the style or stigma is damaged. To test for peroxidase enzymes, use a freshly made benzidine solution (1% benzidine in 60% ethanol:hydrogen peroxide:water, 4:11:22 v/v/v; Galen et al. 1985, adapted from King 1960). The stain works by oxidation (and accompanying color change) of the benzidine when hydrogen peroxide is broken down by the presence of peroxidase. In the presence of peroxidase enzymes there may be a period of bubbling at the stigma surface, followed by a color change in the stigma, for example, from white to blue (Galen et al. 1985) or pale green to dark brown (Galen and Plowright 1987). To monitor the reaction to benzidine solution place freshly picked pistils in a depression slide and add enough of the test solution to submerge the stigma. Color change can be scored with the help of a dissecting microscope or hand lens. A relative indication of peroxidase activity can be obtained by the amount of time (to the nearest second) required for the onset of color change, or the percent of stigma surface area that exhibits color change within 3 minutes (Galen and Plowright 1987). Arnold (1982), Macior (1986), Herrera (1987), and Osborn et al. (1988) applied 3% hydrogen peroxide to stigmas and used the presence of a bubbling action on the stigma as an indication of peroxidase activity, or inferred receptivity (a technique described by Zeisler 1938). For tiny stigmas, put hydrogen peroxide solution in a capillary tube, put the style in the tube, and watch under a microscope for about 5 minutes for bubbling (J. Thomson, *personal communication*).

Another way to assay for peroxidase is with a test paper (e.g., Macherey-Nagel Peroxtesmo KO peroxidase test paper; Motten 1982) or other indicator (e.g., peroxidase indicator solution from Sigma Chemical Co.). Motten (*personal communication*) cautions that test papers won't work on dry stigmas (e.g., Cruciferae). Galen and Plowright (1987) found that significant increases in pollen adhesion and germination on stigmas of *Pedicularis canadensis* (Scrophulariaceae) corresponded with the initiation of peroxidase activity. For *Clintonia borealis* (Liliaceae), which shows peroxidase activity throughout the period of receptivity, changes in peroxidase level also tracked the net amount of pollen germinated on stigmas in flowers of different age classes.

Stigmatic esterase activity is another commonly used enzymatic indicator of receptivity, which can also be used to identify the location of the receptive stigmatic surface (Bernhardt 1983). However, this method may not work with some species of plants (Ramsey and Vaughton 1991). Mattsson et al. (1974) found that esterase activity was best demonstrated using α-naphthyl acetate as a substrate, with either fast blue B salt or hexazotised pararosanilin (also used by Ducker and Knox 1976, and Slater and Calder 1988) in a coupling reaction; as the naphthyl alkyl group is liberated by the esterase it couples with the dye. Lord and Kohorn (1986) used the following method for fast blue B (Mattsson et al. 1974):

1. Dissolve 2.5 mg of the substrate (α-naphthyl acetate) in 3 drops of acetone, and mix this with 5 mL of phosphate buffer (0.1 M, pH 7.0).
2. To this solution add 12.5 mg of o-dianisidine blue (tetrazotized; also known as fast blue) with vigorous shaking.
3. Place fresh stigmas in a drop of this solution, which is active for 15 minutes, and look for a strong red coloration on receptive areas of the stigma. A control procedure can be carried out with the same technique but eliminating the substrate.

The following recipe (Bancroft 1975) can be used for the pararosanilin method, which has the advantage that pararosanilin is a fluorochrome (excitation max 570, emission max 625; Pearse 1980, p. 373):

1. Mix 50 mg of α-naphthyl acetate in 5 mL of acetone (solution A).
2. Mix 400 mg of sodium nitrite in 10 mL of distilled water (solution B).
3. Mix 2 g of pararosanilin hydrochloride with 50 mL of 2 N-hydrochloric acid, heat the solution gently, cool to room temperature, and filter (solution C).
4. Add 0.4 mL of solution B to 0.4 mL of solution C and mix.
5. Add the mixture to 0.25 mL of solution A, 7.25 mL of 0.2 M phosphate buffer, and 2.5 mL of distilled water.
6. Adjust the pH to 7.4 if necessary with more of the phosphate buffer.
7. Incubate stigmas at 37°C for 2–20 minutes, and then look for reddish brown color indicating the presence of esterase.

A final method for studying the period of stigma receptivity involves decolorized aniline blue fluorescence and is based on the appearance of callose in pistil tissue, which signals the end of the period of receptivity. Thus for this method it is important to fix pistils rapidly to minimize any deposition of callose in a wound response (Dumas and Knox 1983). Appropriate fixation techniques include ethanol-acetic acid (3:1, v/v for 1 hour at room temperature followed by storage in 70% ethanol); or by FAA (formalin:acetic acid:70% ethanol in the proportions 5:5:90 v/v, respectively; Dumas and Knox 1983). Fix the pistils for 1 hour or more at room temperature, after which they can be stored in fixative (which should be prepared fresh before each use). After fixation it is best to clear the pistils to facilitate preparation of squashes and to remove components that may quench the aniline blue fluorochrome reaction in the pollen tubes. Clearing can be accomplished by treating the pistils with 8 N or a saturated solution of NaOH for 8 hours or longer at room temperature, or by autoclaving them in 50 g/L sodium sulfite for 10 minutes to 1 hour at 121°C.

The aniline blue fluorescence technique relies on a fluorochrome that is actually an impurity in the aniline dye that complexes with callose and other cell wall polysaccharides (Dumas and Knox 1983). For additional information on this technique, see Chapter 4, Section K.1.a.

Once stigma receptivity has been examined with one of these techniques, it may be possible to associate it with some developmental change. For example, Madden and Malstrom (1975) found that stigma receptivity was reliably indicated when the stigmatic surfaces became glossy, and when a light coat of pollen applied with a blower to the stigma would remain adhered when gently blowing across it. Gori (1989) found that the change in color of the banner spot of *Lupinus argenteus* (Fabaceae) flowers corresponded with loss of receptivity and pollen viability. Neff and Simpson (1990) found that styles of sunflower florets retracted after successful pollination, providing a quick visual indicator that is faster than waiting for seed production.

3. Techniques for Observing Ovules
Lisa Rigney (in preparation) has devised a technique for observing ovules in developing fruits of *Erythronium grandiflorum*. Approximately 3 weeks after pollination, she uses a razor blade to remove a portion of the ovary wall of a developing fruit. In most cases fruit and seed maturation progress as in other, unaltered fruits. These ovary windows permit her to monitor ovary development and to remove (for genotyping via electrophoresis) ovules that are not growing as large as their neighbors and are presumably in the process of being aborted. If the ovules were not sampled at this stage, they might die and then be unavailable for electrophoresis.

There are also a variety of lab techniques for observing harvested ovules, by clearing them. Here are some representative methods:

- Herr (1971) developed a technique for studying ovule development that involves a combination of clearing and squashing. Remove pistils from flowers and fix them for 24 hours in FPA_{50} (formalin, propionic acid, 50% ethanol; 5:5:90); they can then be stored in 70% ethanol. Transfer the pistils to a clearing fluid of lactic acid (85%), chloral hydrate, phenol, clove oil, and xylene (2:2:2:2:1, by weight) for 24 hours at room temperature. Palser et al. (1989) also tried adding benzyl benzoate (4.5, by weight). The pistils will then appear nearly transparent, and ovules can be dissected and transferred with some of the clearing fluid to a microscope slide for examination (e.g., with phase contrast optics, or differential interference contrast optics; Palser et al. 1989). By putting pressure on the ovule with a coverslip, you can facilitate observations of different cytological features. Guth and Weller (1986) used this technique to observe ovules in concavity slides using a microscope with Nomarski optics.

- Harper and Wallace (1987) harvested seed pods before dehiscence, cleared them in lactic acid at 64°C for 24 hours, washed them in running water, and then stored them in sealed tubes in 70% alcohol with glycerol. The pods maintained their flexibility and clarity and could be examined microscopically to determine the state of the ovules within the pods.
- Leduc, Douglas, et al. (1990) studied in vitro pollination in *Trifolium* (Fabaceae), including observation of cleared ovules:
    1. Fix ovules in 3% gluteraldehyde and preserve in 0.05 M phosphate buffer.
    2. Postfix with 2% osmium tetroxide.
    3. Dehydrate in an alcohol series and embed in resin.
    4. Cut section 10 μm thick and stain with Paragon (270 mg of basic fuchsin + 730 mg of toluidine blue powders in 100 mL of 30% ethanol solution; gentle heating of the sections was necessary for staining).
- Stelly et al. (1984) described a stain-clearing technique for observations of whole ovules:
    1. Fix inflorescences, floral buds, or ovaries for 24+ hours in FAA or CRAF V (see recipes below); slicing ovaries or buds once will improve penetration of fixatives.
    2. (a) Hydrate FAA-fixed specimens in a series consisting of 50%, 25% ethanol, H_2O, H_2O, for 15+ minutes each, and then H_2O for 2–24 hours. (b) Rinse CRAF V–fixed specimens in tap water for 2–24 hours.
    3. Stain the samples in Mayer's hemalum (using Sass's modification of 20 g instead of 50 g of alum per liter), a regressive hematoxylin stain, for 1–2 days.
    4. Destain with 0.5% to 2.0% acetic acid for 1–2 days, until proper stain intensity is achieved.
    5. Rinse with tap water or 0.1% sodium bicarbonate for 2–24 hours.

6. Dehydrate in an ethanol series of 25, 50, 70, 95, 100, 100% ethanol for 15+ minutes each, then 100% for 2–8 hours.
7. Infiltrate with xylene, in the series 2:1, 1:2 (ethanol:xylene), xylene, xylene, xylene for 15+ minutes each.
8. Clear with methyl salicylate in the series 2:1, 1:2 (xylene:methyl salicylate), methyl salicylate, methyl salicylate, methyl salicylate for 15+ minutes each.

Alternatively, you can omit step 7, and replace the xylene:methyl salicylate series in step 8 with an ethanol:methyl salicylate series.

This method produces clarity, resolution, and contrast within ovules that is comparable to those of sectioned, paraffin-embedded ovaries (Stelly et al. 1984) yet requires much less time and labor.

Recipe for FAA fixative (from Berlyn and Miksche 1976):

1. 10 mL formaldehyde (37–40%)
2. 50 mL 95% ethyl alcohol
3. 5 mL glacial acetic acid
4. 35 mL water

Recipe for CRAF V fixative (from Berlyn and Miksche 1976; also known as Nawaschin fixative):

1. 50 parts 1% chromic acid
2. 35 parts 10% acetic acid
3. 15 parts formalin

Palser et al. (1989) found that for satisfactory clearing they had to pretreat ovules to remove starch and tannin. To do this they used amyloglucosidase (4% in acetate buffer, pH 4.5, at 37°C for 4–5 days) to remove starch, and Stockwell's solution (90 mL water, 1 g potassium dichromate, 1 g chromic acid, and 10 mL glacial acetic acid) for 20 hours at room temperature to remove tanniferous material. Their method combined these treatments in the following sequence:

1. Fix ovaries in FAA for 24–36 hours, then transfer to 70% ethanol for storage.
2. Hydrate samples gradually.
3. Treat with amyloglucosidase.
4. Rinse thoroughly in water.
5. Treat with Stockwell's solution.
6. Rinse thoroughly with 0.1% sodium bicarbonate.
7. Rinse in water.
8. Stain in Mayer's hematoxylin for two days.
9. Destain in 2% acetic acid to desired color.
10. Wash repeatedly in 0.1% sodium bicarbonate over several hours.
11. Dehydrate gradually to 100% ethanol.
12. Clear using Herr's method or Stelly et al.'s method (see above).

They found that the preparations lasted up to 2–3 weeks, with less deterioration using Stelly et al.'s method (slides had to be sealed or have clearing solution replenished every few days).

If you want to clear whole flowers, try the method of Peterson and Fletcher (1973); Sage et al. (1990) used this method to clear whole *Asclepias* flowers. Fix flowers in a 1:3 mixture of glacial acetic acid:ethanol (95%) for 24 hours; this mixture is very effective at decolorizing tissues, so that tissues containing chlorophyll become white and opaque. Rinse the flowers in water and clear them in 85% lactic acid in a boiling water bath for 1 hour. Rinse with water. The flowers could now be stained if necessary; for example, staining with aniline blue will indicate sieve tubes and pollen tubes because of the callose in them.

If the ovaries or fruits you are working with are thin enough, you may also be able to count, measure, or monitor ovules by holding a bright light behind them; D. Schemske (*personal communication*) uses this technique for work with rapid-cycling *Brassica* (Brassicaceae). Rocha and Stephenson (1990) used an otoscope to shine light through the ovary in order to see the location and outline of each ovule.

4. Ovule Viability

If determination of stigma receptivity or pollen tube growth do not provide sufficient detail for a study of the gynoecium, it may be possible to study ovule viability. For example, loss of viability (receptivity for pollen tubes) is indicated by the appearance of callose and its spread across the cells of unpollinated ovules of avocado and some species of *Prunus* (Rosaceae) (Dumas and Knox 1983). This phenomenon could be monitored with the method for decolorized aniline blue fluorescence described above in Section E.2. Alternatively, it may be possible to use some of the techniques described in the previous section to monitor morphological changes in the ovules.

G. Suggestions for Planning Studies

1. Effects of flower morphology on visitation. What might seem like insignificant differences in a floral trait, such as the degree of exsertion of stigmas or anthers, can in fact have important effects on processes such as pollen transfer (e.g., Campbell 1991a).

2. Effects of bagging flowers. Bagging flowers to prevent pollination or to conduct controlled pollination is likely to affect the microclimate of the enclosed flowers. For example, Gross and Werner (1983) found that hand-pollination reduced seed production late in the season only when the hand-pollinated flowers were bagged. Abraham and Gopinathan Nair (1990) found that bagging with paper bags reduced fruit set of hand-pollinated female flowers by 50%. Mesh or netting bags are less likely to have an effect on microclimate than enclosures of paper or dialysis tubing, but a caging control treatment is always necessary.

3. Self-pollination in bagged flowers. Dole (1990, 1992) reported that when pollinators were excluded from pollinating *Mimulus guttatus* (Scrophulariaceae) flowers, 82% of seed produced was from pollen transfer effected as the corolla was shed. Direct anther-stigma contact following curling of the lower stigmatic lobe into the anthers was another cause of autofertility. Beware of these as mechanisms for self-pollination in bagged flowers of other species.

4. Position effects within ovaries. If you are concerned with pollination at the level of individual seeds, keep in mind the potential for position effects. Rocha and Stephenson (1990) showed, for example, that the probability that an ovule of *Phaseolus coccineus* (Fabaceae) will produce a mature seed varies significantly with position in the linear arrangement of ovules in the flattened ovary.

5. Temporal effects. In some species of plants, carpel number can vary. Pellmyr (1987) reported that carpel number per flower decreased significantly with flowering date; he found this trend at individual, population, and among-population levels. Ovule number can vary, too; Thomson (1985) found that ovule number per flower increased throughout the blooming period. The number of ovules per carpel and ripe seed weight also declined with later flowering dates. Kang and Primack (1991) also found significant temporal variation in seed number per fruit during the growing season. Bertin (1985) found that early in a season plants were more selective about donors they would accept pollen from (as measured by the parentage of mature fruits); this selectivity was mediated by number of prior pollinations and developing fruits in an inflorescence. A similar effect may result from flower age as well. Beware that floral odors may also change with floral development. For example, Patt et al. (1992) found that the ratio of major floral odor components changed with the sexual stage in *Peltandra virginica* (Araceae).

6. Fungus infections. An infection by an anther smut fungus may result in the production of fungal spores instead of pollen grains by anthers (e.g., Jennersten 1983, Jennersten and Kwak 1991).

7. Tag or flag color. If you are using colored tags, flags, or flagging to mark different treatments in an experimental study, be careful that their color doesn't affect your results. For example, hummingbirds are attracted to red markers more than other colors and might be more likely to pollinate flowers marked with this color.

4. Pollen

Pollen is the major attractant for many pollinators, an important part of the diet of many flower visitors, and an essential component of sexual reproduction and gene flow. Thus it is not surprising that so much effort has been devoted to developing techniques to identify, count, analyze, mark, and follow pollen.

The pollen grain is a very reduced male gametophyte. The angiosperm pollen grain is made up of the sculptured exine (predominantly of lipoprotein; Richards 1986), the intine or cell wall, and the internal cellular material. The first mitosis in the haploid microspore produces a generative and a vegetative nucleus; the generative nucleus undergoes mitosis to produce two male gametes. Depending on the taxon, this may occur before (trinucleate pollen) or after (binucleate) the pollen is shed. One of the gametes can fuse with an egg cell and the second then fuses with the primary endosperm nucleus to form the usually triploid endosperm in the process of double fertilization.

Pollen grains of conifers also have an intine and a sculptured exine, although the exine pattern is generally not characteristic of genera or species as it is in angiosperms (Scagel et al. 1969). Pollen grains from most members of the Pinaceae and Podocarpaceae are saccate, bearing small bladderlike structures for buoyancy. All conifers are wind-pollinated, but the number of cells in the pollen grain at the time when it is shed differs among genera (Scagel et al. 1969). See Sterling (1963) for further information on the structure of gymnosperm pollen.

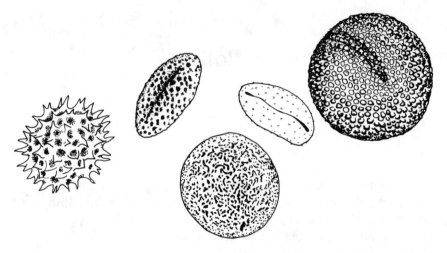

Figure 4–1. Some examples of the variety of pollen morphologies.

A. Pollen Identification

There are many reasons why one might need to identify pollen in a pollination study. Collections of pollen from flower visitors can provide some evidence about the variety of species visited. The identification of pollen on stigmas can indicate whether most of the pollen deposited is conspecific or whether the potential exists for stigma clogging by pollen from other species, pollen allelopathy, or other such effects. Pollen identification is also important in studies measuring temporal patterns of male and female pollination success. Researchers studying anemophily may need to identify the species represented in pollen samples in order to see the percentages of different pollen types and to determine the allergenic species in the air. Identification of fossil pollen provides a record of the flora and the climate of the past.

Pollen from different angiosperm species can be very distinctive (Figure 4–1). Identification is based on characteristics such as size, sculpturing, and number and position of pores. Pollen can be identified from fresh or prepared material. When working with fresh material, one should keep in mind that the degree of hydration of the pollen will affect its appearance and should be controlled. Stead et al. (1979) measured dry pollen grains that were ellipsoidal and then exposed the grains to relative humidities in excess of 75%. The pollen absorbed water and became more spherical. Volumes were calculated from the equation

$$V = \frac{4}{3} \pi a^2 b ,$$

where a and b are the major and minor axes of the ellipse. Hydration produced only slight changes in volume, but addition of nutrient medium caused changes in both shape and volume.

Pollen in reference collections has often been prepared by acetolysis (see next section). Acetolysis removes the protoplasm and thus the problems associated with hydration, leaving only the exine. This technique was developed for use in palynology, the study of structure, formation, dispersion, and preservation of pollen grains and spores (Moore et al. 1991). Palynology is primarily concerned with the readily preserved exine (Erdtman 1969) as pollen microfossils can indicate past floras and climates. Palynologists employ specialized terminology describing the structure, sculpture, and morphology of pollen grains; some of these terms are included in the Glossary (Appendix 5).

The palynological literature is extensive. General references, major journals of interest to those studying pollen, and taxonomic references for specific families or geographical regions appear in Appendix 1. Shivanna and Rangaswamy (1992) have recently completed a laboratory manual of procedures for studying live pollen.

1. Preparing Pollen for Viewing
In addition to the descriptions we give here, many of the references listed in Appendix 1 provide details for pollen preparation.

a. Light microscopy
Light microscopy has been the standard means of viewing fresh pollen, acetolyzed pollen, and pollen mounted in various stains and mounting media.

i. Acetolysis (Erdtman 1954, 1960, 1969). Acetolysis was introduced by Erdtman in 1934 as a way of preparing pollen grains from organic deposits. It relies on the fact that the exine of most pollen grains is highly resistant both to strong acids and, to a lesser degree, bases, so that treating a sample that includes other organic debris will dissolve the extraneous material yet not affect the pollen grains. The same technique can be used with herbarium specimens or fresh samples as a way of preparing reference collections. This method leaves the exine very clean, which makes the pollen grains well suited for studies of sculpturing by light microscopy or scanning electron microscopy (SEM), although grains sometimes collapse in the vacuum of the SEM (Lynch and Webster 1975). The method has also been used to prepare samples from insect guts or bee cocoons, and samples from anthers (Wolfe and Barrett 1989) or stigmas (Ganders 1974, Ornduff 1975a), as the treatment will dissolve other tissue and extraneous matter and leave the pollen grains unharmed (e.g., Benedict et al. 1991).

Erdtman (1969) reported that pollen grains of some taxa are often destroyed or badly damaged by acetolysis, including the genus *Populus* (Salicaceae) and families Lauraceae, Ceratophyllaceae, Juncaceae, Thruniaceae, Rapateaceae, Cannaceae, Musaceae, Zingiberaceae, and Zannichelliaceae. For these taxa, methods such as chlorination or warming in dilute (2–5%) potassium carbonate or sodium carbonate are alternatives.

Materials needed:
    Pollen sample
    Acetic anhydride
    Fume hood
    Concentrated sulfuric acid
    Water bath
    Squeeze bottles
    Centrifuge tubes (15 mL) (for single flowers plastic microcentrifuge tubes work well [Dudash, *personal communication*])
    Centrifuge (for 2,400 rpm)
    Glacial acetic acid
    Distilled water
    Stainless steel wire mesh (0.14-mm squares)
    70% ethanol
    5% KOH

Acetolysis mixture:
Prepare this mixture fresh each day, using a fume hood. Add 9 parts of acetic anhydride to 1 part of concentrated sulfuric acid, adding the acid very slowly (e.g., a drop at a time) to the anhydride. As a precaution, because this causes an exothermic reaction, set the beaker of anhydride in a cool water bath while you are adding the acid. The mixture can be stored in a plastic squeeze bottle.

For herbarium specimens (Erdtman 1954):
1. Grind dried flowers or stamens through a small mesh screen held over a funnel.
2. Wash the plant powder into a centrifuge tube by pipetting acetolysis solution one drop at a time.
3. Fill the centrifuge tube to 5 mL with acetolysis mixture and continue as with fresh material (step 4 below — start with stirring and heating in a water bath in a fume hood).

For fresh material:
1. Add anthers directly to a glass tube containing glacial acetic acid. They should be left there for a minimum of a few hours but can also be stored for years that way.
2. Pour off the acid and add a few mL of acetolysis mixture. Use a glass rod to crush the anthers against the wall to release the pollen grains.
3. Transfer the acetolysis fluid and sample material to a centrifuge tube.
4. Add 5–10 mL of acetolysis mixture to the centrifuge tube, stir it with a glass rod, and heat the tube in a water bath (inside a fume hood). If you start with a water bath between room temperature and 70°C, heat it slowly to boiling, and remove the tubes after 1–2 minutes at 100°. If the water bath is already boiling at the time you put the tubes in it, leave them for about 15 minutes. Stir the contents of the tube with a glass stirring rod a few times during the heating process. If you break the centrifuge tube while it is in the water bath the resulting reaction may splash the water around.
5. Cool the tubes for a few minutes and then centrifuge them again (at about 2,400 rpm). Erdtman (1969) recommends putting some cotton padding at the bottom of the centrifuge

tube holders to protect the glass tubes and to soak up the liquid if a tube breaks.

6. Pour off the acetolysis mixture.

7. Add 5 mL of distilled water and wash the sediment in the tube by shaking it in a mixer; if a mixer is not available stir with a glass rod. (Erdtman [1969] found that rinsing and centrifuging with glacial acetic acid prior to the distilled water rinse was not necessary.)

8. After mixing or stirring, add 5 more mL of distilled water, centrifuge, and pour out the water.

9. Add distilled water again, stir, and then pour the water through a stainless steel (or bronze) mesh into a clean centrifuge tube to filter out pieces of anther, etc. If you have access to a variety of mesh sizes, you can match the mesh size to the size of the pollen grains in the sample; otherwise a mesh with 0.14-mm squares works well.

10. Centrifuge the sample and pour off the water. The next step you take will depend on what you intend to do with the pollen grains. If you intend to make permanent glycerin-jelly microscope slides, use 11a and 12a; if you are preparing pollen for wet storage or for SEM analysis, use steps 11b and 12b.

   (a) If you find upon examination that there is still extraneous material on or mixed in with the pollen grains, you can add 5% KOH to the centrifuge tube, boil for 5 minutes in a water bath, centrifuge, and then pour off the solution (D. Livingstone, *personal communication*).

   (b) If the pollen grains have turned dark from the acetolysis, they can be bleached to facilitate examination of details of the exine. Add (in a fume hood) 2 mL of glacial acetic acid, 2–3 drops of saturated sodium chlorate solution, and 1–3 drops of concentrated hydrochloric acid. Stir the mixture with a glass rod for a minute or less (as chlorine is released into the mixture), centrifuge, and pour off the liquid. Then wash the sample twice with distilled water.

11. (a) Add about 12 drops of a glycerin:water mixture (1:1). Let the mixture stand for at least 15 minutes, centrifuge, pour off the liquid, and then stand the tubes upside down on filter paper or a paper towel to drain for 2–24 hours.

(b) Add about 12 drops of a water:ethanol mixture (3:1). Let the mixture stand for at least 15 minutes, centrifuge, pour off the liquid, and then stand the tubes upside-down on filter paper or a paper towel to drain for 2–24 hours.

12. (a) See the description below on how to prepare glycerin-jelly slides.
(b) For wet storage or to prepare specimens for SEM, add 3 mL of 70% ethanol, stir well, and transfer to a storage vial. Samples for SEM can be removed with a pipette, deposited on a stub, and left for the alcohol to evaporate.

ii. Glycerin (Glycerol) jelly slides (Erdtman 1960, 1969). This technique was developed for use with samples prepared by ace-tolysis, but it also works with untreated samples (Dudash, *personal communication*). You can also use this procedure for squashes of stigmas for pollen counts. Stain can be added to the glycerin jelly if stained pollen grains are desired.

Materials needed:
Pollen samples
Slides
Coverslips
Hot plate set at 100°F, or Bunsen burner or alcohol lamp
Dissecting microscope
Glycerin jelly (e.g., see recipe for basic fuchsin jelly)
Forceps and needle
Liquid glycerin

1. Clean two slides and one coverslip per sample (e.g., wash with detergent, rinse in 70% ethanol).
2. If the pollen sample is stored in 70% ethanol, put a drop of liquid glycerin on the slide and mix in a drop of pollen suspension. Let the mixture sit and evaporate for about 10 minutes.
3. Clean the point of a dissecting needle by washing it in alcohol or heating it in a flame.
4. Spear a small piece (about 2 mm^3) of solid glycerin jelly (which may contain stain) on the point of the needle.

5. Touch the piece of jelly to the glycerin-pollen mixture on the slide to pick up pollen on the jelly. If the pollen sample is still in the bottom of a centrifuge tube (following acetolysis), touch the jelly onto the pollen in the tube.

6. Place the piece of jelly with the pollen on it onto the second clean slide, about 2.5 cm from one end of the slide (leaving room for a label at the other end).

7. Warm the slide gently on a hotplate or over a flame (Bunsen burner or alcohol lamp) until the jelly melts. Don't allow the glycerin jelly to boil, which would introduce air bubbles that are difficult to remove.

8. Place the slide under a microscope and spread the pollen grains out while stirring the jelly carefully with a clean needle.

9. Rewarm the slide gently, and lower the coverslip carefully over the molten jelly. The jelly should make a circle 2–3 mm in diameter.

10. For permanent slides you can seal the sample under the coverslip with paraffin wax. This will prevent the glycerin jelly from drying out and will keep mold or fungi from growing in it. However, even without sealing, slides will last for a long time. Place a small piece of paraffin on the slide next to the coverslip; heat the slide to melt the wax so that it is drawn under the coverslip and surrounds the glycerin jelly. If there are air bubbles trapped under the coverslip try gently warming the slide again until they disappear.

11. While the slide is still warm, place it face down on a stand that supports it at both ends, without letting anything touching the coverslip. This allows the pollen to settle near the coverslip while the jelly is still molten.

12. Scrape off excess wax on the slide with a razor blade. Clean the slide with xylene if necessary (in a fume hood).

13. Label the slide.

iii. Hoyer's medium. An alternative medium for making reference slides is Hoyer's medium (Radford et al. 1974). This mounting medium produces unstained, fully expanded pollen grains.

Materials needed:
    50 g distilled water
    30 g Arabic gum
    200 g chloral hydrate
    20 g glycerin

1. Soak Arabic gum in water for 24 hours.
2. Add chloral hydrate. Let the solution stand until the chloral hydrate is completely dissolved (this may take several days).
3. Add glycerin, and the medium is ready for mounting pollen.

Augspurger (1980) used Hoyer's medium in a study of floral constancy of *Melipona* bees visiting *Hybanthus prunifolius* (Violaceae). She mounted pollen removed from 10 bees in Hoyer's medium and examined 500 grains per bee. Since *Hybanthus* pollen has a very distinctive shape, she was able to determine that 99.7% of the pollen grains on the bees belonged to that species.

iv. Polyvinyl lactophenol mounting medium. Pollen can be permanently mounted in polyvinyl lactophenol. This is a water-based medium, and grains do not need to be dehydrated. It can be purchased or made with this recipe:

1. Prepare stock solution: in a water bath, dissolve 15 g polyvinyl alcohol (viscosity 24–32 centipoise works best; see Appendix 3) in 100 mL distilled water.
2. Mix 56 mL of stock solution, 22 mL of lactic acid, and 22 g of phenol.
3. Keep this clear solution in a dark container. Aniline blue may be added directly to this mounting medium.

b. Scanning electron microscopy

Scanning electron microscopy is being used increasingly to record and study surface details of pollen. SEM studies of fossilized pollen have used clean, acetolyzed material. Many studies of living pollen have used the same process, which permits comparison of modern and ancient pollen types. However, because acetolyzed pollen no longer has the intine and cellular contents, it often collapses in the vacuum of the SEM. In addition, acetolysis destroys nonresistant colpar structures that can aid in identification. Lynch and Webster (1975) developed a technique for use with live pollen that prevents pollen collapse, retains colpar structures, and provides good surface details of the exine. Buchmann and Buchmann (1981) describe this method of pollen preparation of fresh material for electron microscopy (modified from Lynch and Webster 1975):

1. Centrifuge pollen in 1:1 acetone:distilled water.
2. After 1 hour in an acetone solution, place the centrifuge tube in an ultrasonic bath for 1 minute. Replace the fluid with fresh acetone:water and allow it to stand for another hour to remove pollenkitt.
3. Rinse the pollen in distilled water for 10 minutes and run it through an ethanolic dehydration series. Allow the pollen to remain in the final 100% ethanol for a minimum of 4 hours.
4. Replace the ethanol with amyl acetate and allow it to remain for at least 1 hour.
5. Critical-point dry the pollen with carbon dioxide (detailed in Lynch and Webster 1975).
6. Mount pollen on glass-capped specimen stubs using Microstik (see Appendix 3).
7. Coat the pollen with 300 Å of gold:palladium (60:40) using a sputter-coater.
8. Examine with a SEM. Buchmann and Buchmann (1981) used an ETEC Autoscan SEM with accelerating voltage 10 kv and 7–8 mm working distance.

For SEM of pollen from herbarium material, use the following procedure (Lynch and Webster 1975):

1. Remove flowers or anthers from herbarium sheets and place them in a 3% aqueous solution of Aerosol-OT (a biological detergent) until the tissue is pliable (2–3 hours).
2. Under a microscope, remove pollen from the anthers, taking care to keep it free from tissue debris. Use a clean pipette to suck up pollen from a dissection in a depression slide.

3. Place the pollen in 15-mL centrifuge tubes and keep it in Aerosol-OT solution for 5 more days to rehydrate it completely.
4. Pipette the Aerosol-OT solution from the tubes and add distilled water for 10 minutes.
5. Remove the water and proceed from step 1 for fresh material.

Smith and Tiedt (1991) developed another technique of preparing pollen for SEM that involves fixing anthers with osmium tetroxide vapor. With this procedure, pollenkitt remains intact, and pollen grains retain their aperture structure and natural state of hydration.

1. Fix fresh anthers in a petri dish containing a few drops of 2% OsO_4 (anthers should not contact the liquid). Close the dish and place it in a fume hood. NOTE: **Osmium tetroxide is toxic and vapors can cause blindness. Use with extreme caution.**
2. Mount anthers on SEM stubs with conductive carbon cement.
3. Carbon-flash evaporate stubs and sputter-coat with gold-palladium (3-nm thickness).

Leffingwell and Hodgkin (1971) illustrate a micromanipulator that can be built and mounted to a microscope objective in order to place pollen exines precisely in position for SEM. They coated a glass coverslip with adhesive and positioned the pollen on the coverslip with the hair tip of the micromanipulator. Many pollen grains were attached to the coverslip, which was then attached to an SEM stub and coated with 100Å of gold-palladium for SEM. The adhesive used on the coverslip must be sticky enough to pull the pollen from the hair, firm enough so the pollen does not become embedded, transparent, and resistant to heat in the vacuum evaporator used to coat the pollen for SEM. After testing 23 adhesives, they found the best was International Coatings Company's polyvinyl chloride adhesive 5V229CL.

Langford et al. (1990) are working on a method to automate pollen identification using texture analysis of SEM photographs. Texture analysis is often superior to human recognition of patterns.

The procedure uses a video camera attached to a computer that digitizes SEM photographs. A matrix is produced for each digitized pattern and a program developed to identify each matrix. A VAX computer accurately identifies pollen 94.3% of the time, in 10 seconds. This technique, if it becomes more widely available, has the potential to make pollen analysis much faster and easier.

Feuer (1987, 1990) examined morphological detail of sectioned pollen with SEM. Pollen was osmicated in 1% aqueous osmium tetroxide and washed in distilled water. Pollen was then embedded in 1.5% purified agar to form a pellet. Half of this pellet was used for SEM and half for transmission electron microscopy (TEM, described below). She mounted the pollen/agar pellet portions for SEM on cryosectioning stubs, embedded the pellets in Tissue-Tek, and froze them with liquid nitrogen. Tissues were sectioned with a cryocut microtome and picked up on cover glasses. The cover glasses were placed in small beakers and Tissue-Tek boiled away in water. Cryocut pollen was mounted on SEM stubs and coated with gold for SEM. Pollen was viewed with an ISI DS-130 microscope at 5 kv or 10 kv to minimize charging and contrast problems often seen with cryocut pollen.

SEM is also useful for studying pollen adhering to insects. Migratory patterns and feeding behavior of nectar-feeding moths can be studied from their pollen loads (Turnock et al. 1978). Removing pollen from moths for examination with light microscopy resulted in removal of scales and debris that contaminated samples (Bryant et al. 1991). SEM identification of pollen on moths was more accurate than studies in which pollen was removed for light microscopy. Because pollen grains were confined to the anterior regions of moths, the head and prothorax of individuals were mounted on aluminum stubs and dried in a warming oven before coating. Several mounting media were tried, and the best was a nail polish (Nailslick's Golden Russet made by Cover Girl–Noxell). Moths were sputter-coated with gold-palladium in a Hummer vacuum coater. On average, preparation, examination, and reporting on each specimen took 3.5 hours. Turnock et al. (1978) used a similar method, directly mounting the moth and sputter-coating it for SEM observation. They split the proboscis so pollen in the tube could be seen as well as surface pollen.

88

c. Transmission electron microscopy

If it is necessary to look at sections of pollen grains, TEM is a good technique to use. For TEM, Miller and Nowicke (1990) fixed acetolyzed pollen with osmium tetroxide, stained it with uranyl acetate, and embedded it in Spurr's medium for sectioning. After sectioning, lead citrate stain was applied and then pollen was viewed with a JEOL 1200EX TEM.

Feuer (1990) prepared pollen/agar pellets (described in the SEM section above). The portions of these pellets for TEM were cut into cubes less than 1 mm^3, subjected to an ethanolic dehydration series, embedded in Araldite 6005 resin, and sectioned using a diamond knife. The sections were picked up on 75 x 300–uncoated grids and stained with uranyl acetate and Reynolds lead citrate.

B. Pollen Size

Size is an important character for identifying pollen. Most species have a limited size range (Radford et al. 1974). An exception to the limited size range occurs in heteromorphic species. Pin and thrum flowers of distylous species (Ornduff 1975b, 1980) and the three morphs of tristylous flowers (Weller 1981) have significantly different sizes of pollen. This variation may present opportunities for studies of pollen dispersal. The pronounced pollen trimorphism in *Pontederia* (Pontederiaceae) has enabled Barrett and Glover (1985) to quantify legitimate between-morph pollen transfer. However, heteromorphic species in the Lythraceae and Oxalidaceae produce different sizes of pollen from different anther levels, resulting in size overlap among morphs (Barrett and Glover 1985). Therefore these species are not as useful for studies of dispersal.

Erdtman (1945) lists six pollen size classes:

Very small	<10 µm
Small	10–25 µm
Medium	25–50 µm
Large	50–100 µm
Very large	100–200 µm
Gigantic	>200 µm

Pollen diameters range from 5 μm to more than 210 μm, with most falling in the 15–60-μm range (Simpson and Neff 1983). Pollen collected from temperate bees generally falls in the range of 10–100 μm with an average of 34 μm (Roubik 1989). Hilsenbeck (1990) also presents data on the range of sizes of pollen grains for several species.

Pollen grains can be measured with an ocular micrometer on a microscope. Usually, many (e.g., several hundred from different plants) fully expanded grains are measured along both the equatorial and axial plane (McKone and Webb 1988). Fresh pollen can be suspended in a 3:1 mixture of lactic acid and glycerin and mounted on a slide for measuring (McKone and Webb 1988). Size of pollen prepared this way is not affected by the amount of time the pollen is in suspension, so it can be measured days later. If pollen grains have been dried, they can be rehydrated and mounted in Hoyer's medium (see recipe, Section A.1.a.iii) before measuring (Buchmann and Buchmann 1981). Pollen mounted in glycerin jelly will continue to expand for several days, so measurements should be made after stabilization (Radford et al. 1974).

Mean pollen grain size can also be calculated as a weighted mean based on size classes measured with the Coulter Counter (Harder et al. 1985; Thomson, McKenna, and Cruzan 1989; see Section D.2 for more information on Coulter Counters). This method is potentially much faster than optical measurements. Pollen grains can also be measured on the electronically produced scale line made by an electron microscope (Buchmann and Buchmann 1981).

If you can measure pollen diameter, or the length of the axes for nonspheroidal shapes, you can use simple trigonometric formulas to calculate the volume. Pollen grain surface areas of spheres, and prolate and oblate spheroids can be determined from volumes by using tables developed for the fruit industry (Turrell 1946).

C. Collecting Pollen

Careful collection of pollen samples is an important aspect of many pollination studies. Anemophilous pollen collections show which species are shedding pollen, providing an indication of the phenology of flowering, the abundance of pollen in the air, and a record of the allergenic species present. Pollen from anthers is routinely collected in preparing reference collections, testing viability, and for performing hand-pollinations. Pollen can also be removed from pollinators to determine how many species they visit, the magnitude of the loads they carry, and the location of body surfaces that contact anthers (see Chapter 7). When determining pollen-ovule ratios (Cruden 1977), or allocation to male function, counts of the total amount of pollen per flower or per anther may be required. To determine the amount of pollen removed, pollen remaining in anthers after an insect visit can be compared with counts from intact flowers (Young and Stanton 1990b, Dudash 1991). All these operations require careful pollen collection.

1. Anemophilous Pollen — Sticky Traps

Airborne pollen can be collected with pollen traps or air samplers. Gregory (1973) compares a number of different methods and samplers for collecting airborne particles. Pollen traps consist of a flat, sticky surface to which pollen adheres. They can be made by coating microscope slides with petroleum jelly, glycerin, or clear tape (Stanley and Linskens 1974). Captured pollen grains are identified by comparison with prepared slides of known species. Densities are determined by counting pollen grains collected in a limited time period. One problem with these traps is that precipitation or heavy dew can wash pollen off slides or embed it in the substrate, making identification difficult. Stanley and Linskens (1974) describe a pollen trap designed for use under these conditions.

Several types of air samplers are available. Samplers may be placed in open meadows free from obstruction that would bias pollen samples, or at canopy level for sampling tree pollen. Smith et al. (1988) sampled tree pollen using a Kramer-Collins 7-Day Drum Spore Sampler (see Appendix 2) placed at canopy level. This apparatus collects pollen on double-coated cellophane tape wrapped around a 15.24-cm drum (Kramer et al. 1976). The drum rotates past an opening in the body of the apparatus at a uniform rate, permitting determination of pollen collected per unit time. The sampler body rotates so the opening is always directed into the wind. A 115-volt or 230-volt AC vacuum pump maintains suction and air flow rates can be adjusted from 0.5–28.3 L per minute (Kramer et al. 1976). Smith et al. (1988) used the sampler to collect pollen at

20-minute intervals for 17–21 days in several stands. Relative pollen densities were calculated by averaging counts for the peak 7 days in each stand.

Dafni and Dukas (1986) made pollen traps from 20 x 20–cm perspex (Plexiglas) boards coated with gelatin-glycerin 5%. They exposed traps from midnight to 6 A.M. to avoid insect pollination activity. During this time period, they collected pollen on their traps as well as on stigmas. Sado (1990) sampled airborne pollen with a suction trap called a Cascade Impactor. Air passing through a folded tube is accelerated as the tube narrows until particles in the air are trapped on sticky panels.

Rotorod samplers (see Appendix 2) have retracting collector heads with collecting rods that are coated with silicon grease. The rods can be removed at the end of the desired sampling interval, stained with Calberla stain, and examined under the microscope for pollen and fungal spores (Buck and Levetin 1985). They can be set to cycle at 2,400 rpm for a limited number of seconds in a given time period (e.g., with a 5% cycling period, the sampler might collect pollen for 45 seconds each 15 minutes, sampling over 3,000 L of air per day; Buck and Levetin 1985). Rotorod size and pollen size affect sampling efficiency. One can calculate "P," a dimensionless particle parameter determined from a formula involving Rotorod arm velocity, diameter and density of spore, viscosity of air, width of collector, and a dynamic shape factor of the particle. Sampling efficiency is maximized when P is greater than or equal to 10 (Edmonds 1972).

To determine whether pollen from a plant is airborne, you can set out glass microscope slides coated with glycerol in the area near the plant (e.g., Kaplan and Mulcahy 1971, Green and Bohart 1975). Kaplan and Mulcahy (1971) put glycerol-coated slides at 2-foot intervals from test plants in the field and later examined the slides for pollen of that species. Additional plants were tested in a closed room with a periodically cycling electric fan. See Chapter 2, Section D.2, for more about techniques for studying anemophily.

2. Collecting from Anthers
Several techniques allow removal of pollen from anthers. Some of these are appropriate only for collection for counting or microscopic

examination; others produce live pollen that can be used for hand-pollinations.

1. Remove orchid or milkweed pollinia with toothpicks.
2. Pollen can be collected directly from anemophilous plants for pollen viability tests, hand pollinations, or pollen quantification. Bag catkins before pollen dehiscence, or cut catkin-bearing branches and hold them in the laboratory until pollen is shed. Cumming and Righter (1948) illustrate a pollen extractor for holding picked catkins until they shed and a pollen-collecting funnel.
3. The vibrations produced by a tuning fork can remove pollen from buzz-pollinated flowers (flowers with poricidal anthers). Use a 512-Hz tuning fork to sonicate pollen from flowers. (Note: A drop of Elmer's Glue-All, applied to the apical pores of anthers with a toothpick, prevents pollen removal by bee sonication or tuning fork vibrations; Buchmann and Cane 1989).
4. A high-tech method of collecting pollen from buzz-pollinated flowers uses an optical tachometer (Corbet et al. 1988). A small loudspeaker attached to a wire probe directs the variable reference frequency output of the tachometer. The probe is used to vibrate anthers, which then release pollen. Corbet et al. varied the frequency from 156–1076 Hz to remove pollen from air-dried anthers of *Actinidia* (Actinidiaceae) flowers.
5. Collect anthers by picking them with forceps shortly before anthesis and place them in small polypropylene centrifuge tubes until the pollen is released. Leave the centrifuge tubes open to inhibit mold formation. Addition of 70% ethanol preserves the grains until they can be counted (Thomson, McKenna, and Cruzan 1989). Although this works well for *Erythronium* (Liliaceae), anthers of other species may not release pollen completely without some type of manipulation.
6. Macerate undehisced anthers in filtered sea water and dilute to a total volume of 10 mL. Agitate the suspension for 30 seconds and then filter it through a 50-μm screen to remove tissue debris (Tomlinson et al. 1979).

7. Sonicate the anthers in vials filled with a liquid medium to aid in complete pollen removal (Young and Stanton 1990b).
8. Collect anthers by rubbing flowers over a 2-mm mesh before the anthers dehisce. Allow anthers to dehisce. Rub dehisced anthers over a 0.2-mm wire mesh to separate pollen (Raff and Knox 1982).
9. Acetolyze (see Section A.1.a.i) undehisced anthers, wash them to remove acid and supernatant, and resuspend them in 95% ethanol until they can be counted (Buchmann and Shipman 1990).

D. Pollen Quantification

Some of the many reasons why researchers count pollen grains include estimation of male reproductive output, comparison of functional gender of different plant morphs, quantification of pollen removed by a pollinator visit, and estimation of levels of allergenic airborne pollen.

Cruden (1977) suggested that plants with efficient pollen transfer mechanisms produce less pollen than those in which much pollen is wasted. His data, collected from approximately 100 species, indicated a correlation between pollen-ovule ratios (pollen grains per flower divided by ovules per flower) and plant breeding systems (see discussion in Chapter 6, Section J). He found that cleistogamous flowers had relatively low pollen-ovule ratios, selfing plants somewhat higher ratios, and outcrossing plants even higher ratios. Cruden found little intraspecific variation in pollen production among populations of 14 plant species. However, interpopulation differences were very large in four other species. In one species, pollen-ovule ratios were related to flower size. In another species, there were notable variations among individual flowers on a plant as well as between plants. Other authors have sometimes found major differences in pollen production among and within plants, sometimes correlated with anther length (Section D.3).

In heterostylous species, there may be large differences in pollen production among morphs. Barrett and Wolfe (1986) found significant differences in pollen production among morphs and among anther levels within morphs in *Pontederia cordata*. Short- and mid-level anthers of the S morph produced twice as much pollen as the same levels of the L morph.

Counting large numbers of pollen grains under the microscope can be very tedious. Several tricks can facilitate manual counting:

1. Collect pollen, anthers, or pollinia.
2. For loose pollen grains, disperse the sample in a known quantity of ethanol. (Pollen that is to be counted can be stored indefinitely in ethanol.) Place a small subsample on a microscope slide. Use of a digital pipette with a depressing plunger (such as Pipetman; see Appendix 2) increases precision when dealing with pollen in small volumes of liquid.
3. Evaporate the ethanol and count the pollen with a microscope. Replicate readings may vary if pollen rapidly settles out of suspension while subsampling.
4. If necessary, add glycerin to help keep particles in suspension, or detergent to break down clumps of pollen.
5. Whole anthers can be crushed with a glass rod in a stain solution and vortexed for total pollen counts (Weller 1981, Ritland and Ritland 1989).
6. Mount pollinia in polyvinyl lactophenol aniline blue (see recipe in Section A.1.a.iv) and crush them under a coverslip to separate pollen grains (Koptur 1984).
7. Gori (1989) used chloroform to wash pollen from anthers. The chloroform was evaporated and the pollen suspended in a known volume of ethanol.

Once the pollen grains are in suspension or on a slide they can be counted with a hemacytometer or by scanning fields of the microscope slide.

Some species of plants may produce variable numbers of anthers per flower (e.g., *Lindera benzoin* [Lauraceae]; Niesenbaum 1992). Beware of this variation if, for example, you are sampling to determine the number of pollen grains per anther and then multiplying by the number of anthers per flower to estimate total pollen production per flower.

1. Hemacytometers

Developed for use in counting blood cells, hemacytometers can be used for counting pollen (Shore and Barrett 1984, Barrett 1985, Pyke et al. 1988, Dudash 1991). Pollen is dispersed in a liquid medium, stained, and a known dilution is placed on the grid of the hemacytometer for counting (the hemacytometer chamber holds a standard quantity of liquid under the coverslip). Barrett (1985) quantified pollen by crushing anthers in lactophenol-glycerin with aniline blue and making replicate counts using a hemacytometer.

Wolfe and Barrett (1989) used a hemacytometer to quantify the amount of pollen left in anthers after a bee visit. Inflorescences of *Pontederia cordata* were bagged to prevent insect visits. When an observer was present, a bag was opened and a single bee allowed to visit all five flowers of the inflorescence. Four inflorescences (20 flowers) from each of the three floral morphs received bee visits. Then anthers were removed and acetolyzed to destroy everything except pollen exines. Anthers from unvisited flowers of each morph were also acetolyzed. Pollen was counted with a hemacytometer and the number of grains removed determined from the difference between the counts of visited and unvisited flowers of each morph.

2. Particle Counters

The modern alternative to manual counting involves using an electronic particle counter (e.g., Coulter Counter). It is still advisable to count a few samples manually to confirm the accuracy of the count.

M. Cruzan explains the basics of Coulter Counter operation in his unpublished paper "The Coulter Counter Idiot Manual: An Introduction to the Use of the Coulter Counter" (unpublished manuscript; available from Kearns). The Coulter Counter passes an electric current through a small orifice through which a saline solution is drawn by a vacuum pump. When a particle suspended in the saline passes through the orifice a change in resistance is recorded. The total number of particles in a standard dilution of saline can be counted in seconds. Model B Coulter Counters count particles within a specified size class. Model TA counters can produce frequency distributions of different size classes of particles. Samples should be free of debris and clumped pollen for accurate

counts. The tube bearing the aperture can be changed to accommo-
date different size particles. Most pollen can be counted with a
280-μm aperture or possibly a 400-μm aperture for large grains (M.
Cruzan, *personal communication*). Harder et al. (1985) used the
Coulter Counter TAII to estimate the amount of pollen produced
by individual plants. They suspended pollen in 200 mL of saltwater
(0.1% NaCl) and counted it in 2-mL aliquots using a 400-μm
aperture.

Other types of electronic particle counters are also used to
expedite counting pollen. Buchmann and Shipman (1990) used the
HIAC-ROYCO PS-320 particle size analyzer (Pacific Scientific,
Inc.) with a 0–300-μm (model No. CMH-300) visible light sensor.
A glass syringe positioned about the light sensor served as a gravity
feeder. Pollen collected from individual bees was suspended in
ethanol and counted from 50-mL samples. Additional references
using the Coulter Counter include Tomlinson et al. (1979) and
Thomson, McKenna, and Cruzan (1989). Additional references
using the Elzone 180XY particle counter (Particle Data, Elmhurst,
Illinois) include Young and Stanton (1990b) and Harder (1990).

3. Other Methods

If exact pollen counts are not necessary, relative anther weights
may be compared as estimates of allocation to male reproduction
(Osborn et al. 1988). Weight times anther number has been used as
an index of male function (Osborn et al. 1988). Snow and Roubik
(1987) were able to use anther weights to estimate the amount of
pollen removed by vectors.

Anther length can sometimes be used to estimate the number of
pollen grains (Harder et al. 1985, Thomson and Thomson 1989,
Harder and Thomson 1989). Thomson and Thomson (1989) estimated the number of pollen grains in an anther from a regression
equation based on the previously determined relationship between
the length of an undehisced anther and the pollen it contained.
McKone (1990) also used anther length to estimate pollen production in grasses. Although there was a great deal of variance in pollen
production among plants (the largest 10% of the plants produced
more than one-third of all the pollen), simple measurements of
tussock area and number of flowering shoots were good predictors
of pollen production.

E. Pollen Viability

Pollen viability is one measure of male fertility. Viability tests are often conducted in breeding experiments in agriculture and to monitor the condition of stored pollen (Heslop-Harrison et al. 1984). They may also be used to identify hybrids in the many instances where hybrid pollen is shrunken and nonviable (Hauser and Morrison 1964). Pollination biologists who are conducting hand-pollinations can conduct viability tests to be sure they use live pollen so as not to confound compatibility tests.

There are direct and indirect measures of pollen viability. Direct tests consist of depositing the pollen on receptive stigmas and determining whether seeds are produced. Such testing has the advantage of providing an unequivocal measure for the population of pollen grains deposited on the stigma, but it has several disadvantages (see below), such as its time-consuming nature. Pollen germination can also be scored in vitro. Indirect methods rely on the correlation between ability to fertilize an ovule and some physiological or physical characteristic that can be determined more rapidly. Indirect methods that correlate with pollen germination include (1) the fluorochromatic procedure (FCR), (2) testing pollen for enzyme activity, and (3) testing stainability of vegetative cells. The correlation is greatest for FCR and lowest for stainability (Heslop-Harrison et al. 1984).

When testing pollen viability, one must keep in mind that time and method of collection and storage age of pollen affect viability (Shivanna and Johri 1985, Stanley and Linskens 1974). Pregermination relative humidity probably affects the internal solute potential and wall properties of the pollen grain and has a subsequent effect on pollen germination (Shivanna and Johri 1985, Corbet and Plumridge 1985, Digonnet-Kerhoas and Gay 1990). Any work requiring maximally viable pollen should be conducted shortly after anther dehiscence. Germination is most successful immediately after anthesis, and viability deteriorates rapidly in most species. J. Thomson (*personal communication*) found that *Erythronium grandiflorum* pollen viability decreases significantly within an hour of exposure to the air after dehiscence. This is also true for some grass species. In addition, there may be differences in pollen viability among individuals. Oni (1990) looked at flowers from three trees of the African *Triplochiton scleroxylon* (Sterculiaceae) and found that the percentage of flowers with viable pollen ranged from 25% to 65%. There were also significant differences in the amount of viable pollen among flowers. Thomson (*personal communication*) points out that pollen viability is often treated as a dichotomous condition, when it is probably a continuous variable. Thomson (*personal communication*) and L. Delph (*personal communication*) are both conducting studies on the reproductive success of stressed pollen in competitive situations.

Pollen can be shed in the binucleate (tube nucleus and generative nucleus) or trinucleate (tube nucleus and two sperm nuclei) stage. Binucleate pollen germinates fairly easily (Percival 1965). Trinucleate pollen (for example, pollen of the Asteraceae and Gramineae) has a very short life and is difficult to germinate in vitro (Percival 1965; Leduc, Monnier, and Douglas 1990). Not all pollen is short-lived; some Rosaceous and Liliaceous pollen can remain viable for 100 days (Leduc, Monnier, and Douglas 1990). In general, pollen longevity is affected by the temperature and humidity (Shivanna and Johri 1985, Shivanna et al. 1991a). If pollen is not tested immediately, it should be dried in darkness to retain maximum viability. Grass pollen is an exception that does not tolerate drying and loses its viability within 1 day when dried (Stanley and Linskens 1974). Smith-Huerta and Vasek (1984) found that *Clarkia* (Onagraceae) pollen viability decreased with time at room temperature both on the plant or in the lab, but that pollen stored at 5°C retained its ability to fertilize ovules for longer periods.

Viability of pollen collected at different times of the day (Stanley and Linskens 1974) or at different stages of floral development (Leduc, Monnier, and Douglas 1990) may differ. Tree pollen from several species showed the same pattern. Exposure to ultraviolet light and ozone decreased viability of some species (Feder and Shrier 1990; see below). High or low temperatures can inhibit pollen tube growth, which is usually optimum between 25° and 30°C (Richards 1986, Shivanna et al. 1991a). Even naturally occurring volatile compounds produced by flowers, leaves, and fruits can inhibit pollen germination (Hamilton-Kemp et al. 1991). Some types of external stimuli can be used to overcome self-incompatibility mechanisms so that pollen can germinate (see Chapter 6, Section G.5).

Shivanna et al. (1991a, 1991b) demonstrated that *Nicotiana* (Solanaceae) pollen as well as pollen from other species subjected to high temperatures (38°–45°C) and high relative humidities may be viable according to FCR tests. Although this pollen is viable, it takes longer to germinate and pollen tube growth is slow. Pollen kept at 45°C germinated on stigmas but would not germinate in vitro. Shivanna et al. suggest that researchers may wish to evaluate both vigor and viability when using stored pollen. They developed the semi-vivo technique (Shivanna et al. 1991a) for assessing pollen vigor (see Section E.1.c). Pollen vigor can also be assessed from in vivo pollinations with stressed and control pollen and subsequent determination of pollen tube growth rates.

Although germination and pollen tube growth generally indicate healthy pollen, it is startling to note that germination and pollen tube penetration of ovules can occur even when the male nuclei have been lethally irradiated (Pandey et al. 1990). Pandey et al. irradiated pollen with 0.5–0.9 kGy to induce parthenogenic production of homozygous kiwifruit (*Actinidia deliciosa*, Actinidiaceae) seeds. Diploid homozygous plants are useful for breeding new strains of commercial crops. Because dead pollen can produce pollen tubes, you should be cautious about interpreting results of germination tests. In addition, the different procedures for testing viability often give different results, so it is probably wise to try several methods and interpret results carefully (N. Waser, *personal communication*). The stylar tract that provides a medium for pollen tube growth is also capable of transmitting nonliving particles. Sanders and Lord (1989) placed small latex beads on stigmas or in the extracellular matrix of styles of three species and found that the beads traveled unidirectionally down the style at a rate similar to the growth of pollen tubes. Some beads even entered micropyles. The exact mechanism of bead movement is unknown.

1. Germination Tests

a. In vivo

The ultimate test of pollen viability lies in its ability to fertilize ovules (Heslop-Harrison et al. 1984). However, measuring viability from seed set delays test results for prolonged time periods. In addition, in vivo assays may be biased if incompatibility reactions suppress pollen tube growth (Stanley and Linskens 1974), if pollen donors have differential success with different recipients (Bertin and Peters 1991, Snow and Spira 1991a) or if other postpollination phenomena compromise seed set (Heslop-Harrison et al. 1984).

Thomson (1989a) is fortunate in having a system in which natural germination rates can be determined on the plant by a pollen color change. Some *Erythronium grandiflorum* plants produce red pollen that contrasts with the white stigmas of the flowers. When pollen begins to germinate, the hydrated grain becomes colorless. Closer examination reveals a swollen grain with a short pollen tube. Grains that remain red have neither hydrated nor produced pollen tubes. Note that sometimes anther color can vary even though pollen grain color doesn't. For example, most *Arenaria conjesta* (Caryophyllaceae) around the Rocky Mountain Biological Laboratory have white anthers, but some have pink anthers. Pollen appears to be white in both morphs (DWI, *unpublished*).

b. In vitro

In vitro pollen germination can take place (1) in water (some gymnosperms), (2) in a sucrose solution, or (3) in a sucrose solution on agar or gelatin as a medium (Stanley and Linskens 1974). Appropriate sucrose solutions range from 2% to 40% depending on the optimum for the species, which must be established empirically. A 20% solution has been used successfully with *Brassica* (Brassicaceae) pollen, whereas a 2.5% solution was better for *Microtis* (Orchidaceae) pollen (Peakall and Beattie 1989). The optimal sucrose solution can be determined by germination trials at different sucrose concentrations (e.g., 1, 2.5, 5, 10, 20, 30%; R. Peakall, *personal communication*). If the concentration is too low, germination is poor and pollen tubes are short. At extremely high or low concentration, pollen will burst or shrink (R. Peakall, *personal communication*).

In nature, water, sugar, and amino acids are supplied by the style to nourish the growing pollen tube. For many species, boron and calcium are also required for pollen tube growth. Boron, which is provided by stigmas and styles of these species, facilitates sugar uptake and has a role in pectin production in the pollen tube (Richards 1986). Calcium, found on the surface of some pollen grains, is often required for germination and has been implicated in the successful germination of large numbers of pollen grains in instances where only a few pollen grains cannot germinate. The total amount of calcium from a small number of pollen grains may be insufficient for germination (Richards 1986). Density effects have been demonstrated in 39 families, with optimal germination occurring in the presence of 75–1,250 pollen grains per 10-μL drop of medium (Brewbaker and Kwack 1963). Calcium is also involved in pectin synthesis and control of osmotic conditions, and calcium gradients can regulate pollen tube growth down the style (Richards 1986). At least 79 genera require a medium containing sucrose, boric acid, and calcium for in vitro germination (Brewbaker and Kwack 1963; Vasil 1964 cited in Leduc, Monnier, and Douglas 1990). See Shivanna and Johri (1985) for more details on the physiological requirements for in vitro germination.

Unfortunately, appropriate media that give consistent results have not been established for several plant families (Heslop-Harrison et al. 1984). In vitro pollen tube growth is generally slower than in vivo and tubes do not get as long. In vitro germination can be affected by time of pollen collection and storage conditions as well as pollen density on the culture medium (La Porta and Roselli 1991). A serious problem with viability tests using a sugar or sugar-agar medium is that larger numbers of pollen grains increase germination percentages and pollen tube lengths (Brewbaker and Kwack 1963). This problem can be rectified by adding calcium and boron to stimulate germination (see discussion above). The medium developed by Brewbaker and Kwack (1963) corrects this deficiency.

Brewbaker-Kwack Medium:
    10% sucrose (adjust according to species)
    100 mg/L H_3BO_3 (boric acid)
    300 mg/L $Ca(NO_3)2 \cdot 4H_2O$ (calcium nitrate)
    200 mg/L $MgSO_4 \cdot 7H_2O$ (magnesium sulfate heptahydrate)
    100 mg/L KNO_3 (potassium nitrate)
    dissolved in distilled H_2O

Moisten a piece of filter paper with water and place it in a petri dish. Set a microscope slide with a drop of nutrient medium on top of the filter paper. Sprinkle pollen onto the drop, or mix it in with a toothpick. Because pollen germination may continue over a period of time, incubate slides in the humid chamber for 24–48 hours before germination assessment (Peakall and Beattie 1989). Too much pollen can obscure results as pollen tubes become tangled. Yeast and fungal growth can be inhibited by refrigerating the petri dish (this slows tube growth so the pollen should be incubated longer; A. Montalvo, *personal communication*).

Brewbaker-Kwack medium can be adapted for use on microscope slides (Williams et al. 1982, Beattie et al. 1984). Dip slides two times into a medium composed of 0.3 g agarose, 5 mL Brewbaker-Kwack 10% medium, 20 mL of 30% sucrose, and 25 mL of water. Withdraw the slide and position it horizontally until an even film develops. Wipe the back of the slide with tissue. Place pollen on slides about 20 minutes after the medium solidifies, and then incubate at 22°C for 18 hours in a moist environment. Fix pollen tubes in FAA to arrest growth.

Brewbaker-Kwack medium has been used successfully for many species with binucleate pollen, but other species, particularly those in the Cruciferae and Compositae, with trinucleate pollen require a different medium (Leduc, Monnier, and Douglas 1990). Leduc, Monnier, and Douglas (1990) got only 16% germination of *Capsella bursa-pastoris* (Brassicaceae) pollen with Brewbaker-Kwack medium but were able to develop a medium that increased germination to 47%. Their medium, presented below, adds substances (marked with asterisks) absent from the Brewbaker-Kwack medium.

Macronutrients(mg/L)

$MgSO_4 \cdot 7H_2O$	370.00	(magnesium sulfate, heptahydrate)
KNO_3	950.00	(potassium nitrate)
$H_2PO_4K^*$	85.00	(potassium phosphate)
$CaCL_2 \cdot 2H_2O^*$	880.00	(calcium chloride, dihydrate)
$NH_4NO_3^*$	412.50	(ammonium nitrate)
KCl^*	175.00	(potassium chloride)
Na_2EDTA^*	7.45	(ethylene diamine tetraacetic acid)
$FeSO_4 \cdot 7H_2O^*$	5.55	(ferrous sulfate, heptahydrate)

Micronutrients

H_3BO_3	50.00	(boric acid)
$MnSO_4 \cdot H_2O$*	16.80	(manganese sulfate, monohydrate)
$ZnSO_4 \cdot 7H_2O$*	10.50	(zinc sulfate, heptahydrate)
KI*	0.83	(potassium iodide)
$Na_2MoO_4 \cdot 2H_2O$*	0.25	(sodium molybdate, dihydrate)
$CuSO_4 \cdot 5H_2O$*	0.025	(cupric sulfate, pentahydrate)
$CoCl_2 \cdot 6H_2O$*	0.025	(cobalt chloride)

Vitamins

B1*	1.00
B2*	1.00

Osmoticum

PEG 4000	120,000.00

PEG (polyethylene glycol) replaces the usual sucrose in germination media. Leduc, Monnier, and Douglas found that germination was somewhat more successful at pH 8.0 than 4.0. The pH was adjusted with 1 N NaOH or HCl.

Leduc, Monnier, and Douglas (1990) placed about 100 pollen grains on a microscope slide and covered them with 10 μL of medium. A small viewing hole was cut in a piece of filter paper and the paper was then soaked with medium and placed in the bottom of the petri dish. The slide was set on top of a U-shaped piece of glass tubing in a petri dish. Germination was scored after 1 hour. A grain was considered to have germinated when the length of the pollen tube was at least as great as the radius of the pollen grain.

Raff and Knox (1982) scored pollen germination in 20% sucrose in hanging-drop cultures. Hanging-drop cultures can be made by suspending a drop of medium on a coverslip suspended above a microscope slide by an O-ring (Williams et al. 1982). Advantages to this technique include the small amounts of pollen and medium necessary and the fact that pollen tube growth can be observed continuously by leaving the slide under the microscope. It can, however, be difficult to use this technique when many cultures must be run simultaneously, as movement may disturb the drop. Also, pollen tubes do not grow straight with this method.

Hewitt et al. (1985) developed a modified multiple hanging-drop technique that permits pollen tubes to grow straight. The procedure was used to test the effect of plant growth promoters and inhibiters on pollen tube growth. They used PVC microtitre trays with 96 wells to hold culture medium. Brewbaker-Kwack medium containing the appropriate sucrose concentration was added to a pollen suspension plus test substances to produce a final volume of 200 μL. The 200-μL solution was placed in a well and replicated in a second microtitre tray. After all test wells were ready, the microtitre trays were inverted and the test solutions hung from the wells by surface tension. The trays were set on supports within containers lined with filter paper and then these containers were closed for incubation. Incubation temperature, light, and relative humidity in the containers could be controlled. After incubation, trays were turned over and sampled with a pipette. For sequential sampling of the same cultures, 15–20 μL were drawn off. For single samples, 180 μL were drawn off, and the remaining volume was placed on a microscope slide for viewing. A drop of 1% w/v Alcian blue 8 GX in 3% v/v acetic acid could be added to stain the walls of the pollen tubes. Pollen tube growth was arrested by a drop of formaldehyde added to the slide. To fix all the pollen tubes in a microtitre tray simultaneously, formaldehyde was added to the filter paper in the incubation container and the container resealed for another 15–20 minutes.

Multiple "spot tests" can be made by spotting petroleum-jelly rings on the bottom of a petri dish using the end of a glass tube (Stanley and Linskens 1974). Fill each ring with nutrient medium and dust pollen over the medium. Place a piece of moist filter paper in the lid of the petri dish to create a humid chamber. If a plastic petri dish is used, the petroleum jelly may be unnecessary because surface tension will maintain the droplets of medium.

Some types of pollen will germinate on dialysis tubing suspended in a moist chamber (Stanley and Linskens 1974).

Bamberg and Hanneman (1991) discovered that aerating *Solanum* (Solanaceae) pollen tube cultures enhanced pollen tube germination and growth in vitro. This technique will probably be useful for other types of pollen as well. Pollen was cultured in 1 mL of a lactose medium in a 10 x 50–mm test tube. Cultures were aerated by (1) forcing tiny bubbles of air into the medium through drawn-out pipettes; (2) sealing test tubes and rocking them horizontally on a shaker at 90 cycles per minute; (3) sealing test tubes and rotating them end over end at 6 rpm. All three techniques worked equally well.

c. Semi-vivo growth of pollen tubes

Shivanna et al. (1991a, 1991b) use the semi-vivo technique to study pollen vigor. They point out that stressed pollen (subjected to environmental stress such as high humidity and high temperature) may germinate and will thus be considered viable by FCR tests. However, although this pollen may germinate, the time required for germination and for pollen tube growth may have greatly increased. This loss of vigor is not detected by standard germination tests but can be examined with the semi-vivo method.

Flower buds are picked shortly before anthesis and the cut ends are placed in water. After anthesis, pistils are hand-pollinated. After 3 hours (or an appropriate time depending on the species) the stigma and a portion of each style are cut with a sharp razor and implanted in an agar germination medium (0.8% agar plus germination medium) in a petri dish. The dish is placed in a dark, humid environment. After pollen tubes emerge from the cut end of the style into the medium, pistils are fixed in glacial acetic acid:ethanol (1:3) and the pollen tubes are counted. Tubes may then be stained with DAPI (4',6-Diamidino-2-phenylindole) so the generative cells may be observed with fluorescence. Alternatively, pollen tubes can be fixed with acetic alcohol instead of acetic acid:ethanol and cleared with NaOH and observed with the aniline blue fluorescence technique (Section K.1.a).

2. Fluorochromatic Procedure

The fluorochromatic procedure was developed by Heslop-Harrison and Heslop-Harrison (1970; see also Heslop-Harrison et al. 1984) for assessing pollen viability by enzymatically induced fluorescence. Fluorescence microscopy is a technique in which specimens are treated so that they (or parts of them) will fluoresce, or absorb light at one wavelength and reemit it as a longer wavelength. For example, callose, a constituent of pollen tubes, can be treated with a fluorochrome (fluorescent dye) such as aniline blue to facilitate identification and counting of pollen tubes in a style. In order to produce the fluorescence, a fluorescence microscope relies on an excitation filter, placed somewhere between the light source and the specimen, and a barrier filter, placed between the object and the eye. The excitation filter ensures that light of only a particular wavelength (or narrow band of wavelengths) reaches the fluorescent dye, whereas the barrier filter prevents light of the excitation wavelength from reaching the observer's eye. Pearse (1980, Chapter 10) presents a good discussion of the mechanics of this technique.

Fluorescein is commonly used as a fluorochrome. Fatty-acid esters of fluorescein enter living cells (e.g., pollen grains) and are hydrolyzed to fluorescein. When the cell membrane is intact, the rate of entry of esters exceeds the rate of escape of fluorescein, so that fluorescein accumulates in the cell and the cell fluoresces. A positive test, indicated by fluorescence, demonstrates an intact cell membrane and the presence of active esterase in the cell (Shivanna and Heslop-Harrison 1981). This procedure works for both bi- and trinucleate pollen (Shivanna et al. 1991b).

Shivanna and Heslop-Harrison (1981) found a high correlation (r = 0.86–1.00) between FCR and in vitro germination of some species, supporting the idea that the ability to germinate is dependent on the condition of the plasmalemma. La Porta and Roselli (1991) also found that FCR and in vitro germination were highly correlated (r = 0.99) and determined that FCR was an easier and faster technique. However, FCR tests demonstrate the potential for pollen germination and so may overestimate the actual number of grains that germinate in vitro (Shivanna and Heslop-Harrison 1981).

Procedure:
1. Prepare a stock solution of fluorescein diacetate (FDA): 2 mg FDA/mL acetone. R. Mitchell (*personal communication*) suggests using fresh FDA solution because it decays with time and solution decay results in artificially low estimates of pollen viability.
2. Determine the appropriate sucrose concentration for your pollen (see Section E.1.b). Sucrose concentration (w/w) should be the lowest concentration that prevents pollen from immediately bursting.

3. Drop FDA solution into a few mL of sucrose solution until a persistent milkiness appears.

4. Place pollen on a microscope slide and add a few drops of the FDA-sucrose solution, mix, and add a coverslip.

5. Wait 10 minutes before examining the slide for fluorescence. Fluorescence will last about 15 minutes. Viable grains fluoresce yellow-green.

6. Examine slides with a transmitted-light ultraviolet system (e.g., Osram HPO 200 mercury-arc light source with Bausch and Lomb 7-37 excitation filter and T2 barrier filter). Fluorescence can be measured with a photomultiplier tube (EMI IP28; Heslop-Harrison et al. 1984). A microscope with a regular tungsten filament bulb can also be used with an appropriate filter (e.g., Bausch and Lomb 7-60).

Cautions:

1. Immature pollen gives false high estimates of percent germinability.

2. This procedure can be used with aged pollen. However, proper rehydration is crucial for accurate estimates of germination (Heslop-Harrison et al. 1984). Heslop-Harrison et al. (1984) present the results of several pollen pretreatments on different species, including rehydration at 22°–24°C at 80% to 95% relative humidity. Partially desiccated pollen from *Primula vulgaris* (Primulaceae) flowers behaved quite differently depending on the morph; partially desiccated thrum pollen germinated poorly, whereas desiccated pin pollen was only slightly affected.

3. Tests for Enzyme Activity

Enzyme tests depend on the presence of oxidation and reduction reactions that are correlated with cell respiration. Dyes used in these tests are colorless until reduction or oxidation changes them to the colored form. Reduction or oxidation occurs in the presence of active cellular enzymes.

a. Nitro blue tetrazolium (Hauser and Morrison 1964)
Medium:
17 mL 0.06 M Sorensen's phosphate buffer (pH 7.4)

17 mL 0.2 M sodium succinate (0.066 M in total medium)
17 mL nitro blue tetrazolium (NBT; 1 mg/mL)
12.65 mg sodium amytal (1mM in total medium)

1. Place pollen on a 0.01-mL hanging-drop slide. Add medium and incubate at 37°C for 30–45 minutes.
2. Fix material in FAA and pipette it to a microscope slide. Grains containing active succinate dehydrogenase enzyme will stain. Stained grains are viable and unstained grains are not.

b. Triphenyl tetrazolium chloride (Cook and Stanley 1960)
1. Prepare 0.5% solution of 2,3,5-TTC in 12% sucrose.
2. Place a drop of solution on a microscope slide and add pollen. Add a coverslip immediately to exclude oxygen.
3. Incubate the slide at 60°C for up to 3 hours before examining. Pollen grains stain red in the presence of reductases, indicating the presence of active enzymes. Red grains are considered viable.

Stanley and Linskens (1974) use a 1% TTC solution with 0.15 M tris-HCl buffer at pH 7.8. This solution will keep for 3 months if stored in a dark bottle at 5°C.

c. Peroxidase reaction (King 1960)
Although the peroxidase reaction has been used extensively for crop plants in the past, superior tests are available today (Hauser and Morrison 1964, Heslop-Harrison et al. 1984). The reaction tests for the oxidation of benzidine by peroxidase in the presence of hydrogen peroxide, but the specificity of the reaction is low, and pollen of different taxa react differently, making this a poor test for viability (Hauser and Morrison 1964). In addition, the substrate is a health hazard.

4. Stains
Stains specific to pollen components can also be used to test for viability. Aniline blue in lactophenol stains callose; acetocarmine stains chromosomes; phloxin-methyl green stains cytoplasm and

cellulose (Stanley and Linskens 1974). However, immature or non-viable pollen sometimes contains enough of these elements to cause staining, and viable pollen of some species does not stain well (Stanley and Linskens 1974). Therefore, stains provide (at best) only rough estimates of viability (Stanley and Linskens 1974, Heslop-Harrison et al. 1984).

Dyes such as acetocarmine in glycerol jelly and aniline blue in lactophenol stain nonabortive pollen but not abortive pollen. However, Alexander's stain differentially colors abortive and nonabortive pollen; malachite green stains cellulose in pollen walls, and acid fuchsin stains protoplasm. Thus abortive (or germinated) grains appear green, and grains with protoplasm appear pink. Alexander's stain can also be used to distinguish self versus outcross pollen on stigmas in species where the incompatibility mechanism acts at the stage of pollen germination and self-pollen retains its cytoplasm (M. Zimmerman, *personal communication*).

a. **Alexander's stain** (Alexander 1969, 1980)
1. Mix together:
    20 mL ethyl alcohol (95%)
    2 mL of 1% solution of malachite green in ethanol
    50 mL distilled water
    40 mL glycerol
    10 mL of 1% water solution of acid fuchsin to which 1 g of phenol has been added
    5 g phenol
2. Add the mixture to lactic acid; quantity of lactic acid depends on the thickness of the pollen walls: for thin walls use 0.5–1 mL lactic acid; for medium walls use 1–2 mL; for thick walls use 2–4 mL.
3. Store the stain in a dark bottle for 8–10 days before using.
4. Place pollen on a slide, add 1–2 drops of solution, and stir. Wave the slide over a flame four or five times, cover it with coverslip, and let it stand for 5 minutes. Examine it under a microscope.

There are some disadvantages to this technique. It assumes that the presence of protoplasm means a grain is viable. Oily grains or grains that clump do not stain well.

b. Lactophenol–aniline blue (Maneval 1936)
1. Prepare lactophenol by mixing:
    20 mL of phenol
    40 mL glycerin
    20 mL water
    20 mL lactic acid
2. To 100 mL lactophenol, add 1–5 mL of 1% aqueous solution of aniline blue.
3. Place pollen on a microscope slide in a drop of stain. Fertile pollen grains stain dark blue, and sterile grains stain faintly or not at all. Be aware that in an aqueous solution, empty pollen grains will tend to move to the edge of the coverslip, so counts in the center of the slide will be biased.

Mayer (1991) used this procedure to evaluate male fertility of *Wikstroemia* (Thymelaeaceae) hybrids by scoring a minimum of 300 grains per sample. Darkly stained grains were considered viable. The accuracy of this method was confirmed with a nitro blue tetrazolium stain for dehydrogenase activity.

c. Acetocarmine jelly (Radford et al. 1974)
1. Prepare Brandt's glycerol jelly:
    (a) Soak 40-g sheet of granulated gelatin for 2–3 hours in cold water.
    (b) Drain off water.
    (c) Melt hydrated gelatin in water bath.
    (d) Add 60 mL glycerol and 1–2 g phenol crystals; mix.
    (e) Filter through glass wool.
    (f) Bottle.
2. Add 0.25 g of powdered carmine to 50 mL melted Brandt's glycerol jelly. Mix thoroughly to form an even suspension. This concentrated stock jelly can be further prepared as needed.
3. Add 40 mL of 45% acetic acid to 20 mL of melted stock jelly. Boil gently until the carmine dissolves and the solution is clear.
4. Add 4 mL of saturated solution of ferric acetate in 45% acetic acid. This produces a solution with a deep port wine color. The medium is liquid at room temperature.

5.   Macerate an undehisced (but almost ripe) anther in a drop of medium on a microscope slide. Add a coverslip. Do not allow the medium to overflow beyond the coverslip. The medium sets to a permanent jelly and viable (stained) grains can be scored at any time.

5. Nuclear Magnetic Resonance

Dumas et al. (1982 cited in Ladyman and Taylor 1988) used proton nuclear magnetic resonance (NMR) as a nondestructive means of estimating pollen viability from its correlation with pollen water content. After testing, pollen can still be stored and used for pollination treatments (Dumas et al. 1985). Ladyman and Taylor (1988) found this did not give reliable estimates because dead pollen can rehydrate and inflate results. However, phosphorus (^{31}P) NMR can be used to determine the amount of organic phosphorus present as an estimate of the amount of ATP (adenosine triphosphate) present. Ladyman and Taylor used an NMR spectrometer with a 1-mm broad-band VSP probe at 4,500 scans/15.38 minutes. They found a good correlation between pollen germination and phosphorus NMR results. However, a large amount of pollen (100–400 mg) was required. ATP has also been estimated using a luciferase-luciferin assay (Shivanna and Johri 1985).

F. Storing Pollen to Maintain Viability

Many types of pollen retain viability well when they are freeze-dried or stored in liquid nitrogen (Shivanna and Johri 1985, Vithanage 1984). To freeze-dry fresh pollen, Vithanage placed it in glass ampules and froze it at -15°C for 8–12 hours prior to freeze-drying. Then the pollen was freeze-dried at -60°C and 0 mm Hg in a freeze-drying unit. The pollen ampules were sealed with an oxygen flame and stored at -15°C until they were needed. Vithanage notes that different species of pollen store best under somewhat different conditions of freeze-drying.

To store pollen in liquid nitrogen, Vithanage (1984) wrapped it in aluminum foil and then placed it inside drinking straws that had steel ball bearings at one end. The ends of the straws were dipped in polyvinyl alcohol and then moistened with water to seal them. The straws were placed in the liquid nitrogen (Vithanage 1984).

Some *Rhododendron* (Ericaceae) breeders maintain pollen banks to facilitate hybridization of species that bloom several months apart (Rouse 1985). Interspecific crosses may require storage of pollen for nearly a year. Rouse collected anthers from unopened or bagged flowers to avoid pollen contamination. Anthers were wrapped in tissue paper and placed in paper envelopes and then dried in the envelope over calcium chloride at 4°C for 2 days. After 2 days, the envelopes were stored until use at -20°C over calcium chloride. Although fresh pollen should be used when it is available, this technique works for many *Rhododendron* species as well as *Kalmia* and *Ledum* (both Ericaceae). The American Rhododendron Society does not attempt to store pollen for more than 1 year (Rouse 1985).

Pollen banks may eventually be important for allergists as well as plant breeders. Efforts are under way to develop cereal grass pollen banks that could provide both genetic material for crop breeders as well as a live source of pollen extracts to use as antigens in desensitization procedures for pollen allergies (Dumas et al. 1985).

Shivanna and Johri (1985) present a table listing taxa for which successful storage methods have been worked out. In addition to freeze-drying and storing over liquid nitrogen, pollen has been successfully stored in organic solvents (benzene, petroleum ether, acetone, etc.).

G. Testing the Effects of External Factors on Pollen Viability

Feder and Shrier (1990) demonstrated that a combination of ultraviolet light and ozone have negative effects on pollen tube growth. They placed pollen in petri dishes on 0.75% agar with a modified Brewbaker-Kwack medium and 12% sucrose. Pollen was then treated with ozone and ultraviolet-B. One mL of Brewbaker-Kwack medium was added after exposure, and then plates were kept at 27°C for 18 hours. Pollen tube growth was arrested by dropping I/KI (Lugol solution) onto the petri dish. Pollen was blotted onto a filter-paper disk, the petri dish drained of liquid, and then pollen transferred back onto the medium for measurement. Pollen tube length was measured by attaching a video camera to a microscope and sending the image to a monitor. Tubes were traced onto clear plastic and then the traced images measured with a digitizing tablet. The ultraviolet and ozone treatments resulted in pollen tubes that were 70% to 79% shorter than controls.

Williams et al. (1982) tested the effects of style components on pollen tube germination. Slides were coated with a nutrient/agar medium (see Section E). Style components to be tested were soaked onto Whatman 3MM filter-paper strips. A line of pollen was applied down the middle of the slide. Two filter-paper strips soaked in the same solution were placed 2 mm away from the pollen line on either side. Slides were incubated in a humid chamber formed by lining a petri dish with moistened filter paper and covering the dish with waterproof film.

Beattie et al. (1984) and Peakall et al. (1990) were interested in the effects of ant secretions on pollen viability and pollen tube growth. Ants produce exocrine secretions of antibiotics that inhibit fungi and bacteria (Harriss and Beattie 1991). These antibiotics appear to inhibit pollen germination as well. Beattie et al. used culture slides (microscope slides dipped in pollen tube growth medium; Williams et al. 1982; see Section E.1.b) to compare the lengths of pollen tubes of control pollen and pollen that had been exposed to ants. They used the fluorochromatic procedure to evaluate pollen viability: pollen-carrying ants were allowed to walk through a drop of fluorescein diacetate in 10% sucrose on a microscope slide. Pollen that washed off the ants was evaluated with fluorescence microscopy.

Peakall et al. (1990) introduced *Myrmecia* ants into tubes and applied pollen to their thoraces for 30 minutes. The test pollen and control pollen collected from the same anther and exposed to air for 30 minutes were then tested for germination. The germination medium consisted of Brewbaker-Kwack solution prepared with Hepes buffer (10 mL of concentrated Brewbaker-Kwack stock solution [10x concentration] in a solution of 20 mL Hepes, 70 mL distilled water, and 20 g sucrose for a total volume of 100 mL). A second species of pollen that germinated poorly in vitro was evaluated with the fluorochromatic procedure.

Harriss and Beattie (1991) compared the effects of exposing pollen to ant integuments and to the integuments of two bee species and a wasp. Honeybees and *Vespa germanica* wasps had little effect on pollen germination. Some *Trigona carbonaria* bees inhibited pollen germination, but not as severely as ants. This inhibition may be attributable to plant resins with microbial activity on the bees' integument. Brood cells of social bees often contain antibiotic substances. Ants do not produce brood cells, and the surface secretions may serve the same function as the substances in bee brood cells. Harriss and Beattie suggest that disease control substances may be important in social insects. That those substances are isolated in a brood cell for bees and on the surface integument of ants probably had significance in the evolution of pollination systems involving the Hymenoptera. Few ant pollination systems have been documented.

H. Counting Pollen Grains on Stigmas

Determination of the number and types of pollen grains on stigmas is important for some studies. For example, this information may be necessary for a study of pollen carryover, or to determine the frequency of deposition of pollen from different morphs in a heterostylous species. There are two components to pollination quality: intensity and purity (Beattie 1971a). Intensity refers to the number of pollen grains delivered. Large amounts of (compatible) pollen help ensure that there is a pollen grain to fertilize each ovule. Pollen tube germination is often enhanced when many pollen grains are present on a stigma. Purity of the pollen load is a reflection of the foraging behavior of pollinators and the availability of flower species. The presence of heterospecific grains can sometimes inhibit ovule fertilization through allelopathic effects. One inherent difficulty in simply counting and identifying pollen grains on stigmas is that it may be hard to distinguish self and outcross grains, leading to overestimates of outcrossing.

A widely used technique for studying pollen deposition is to expose virgin stigmas to a single visit by a flower visitor, and then collect it to count the number of pollen grains deposited (e.g., Bertin 1982). This can be done in the field, using previously bagged plants with unvisited flowers, or under more controlled conditions in a flight cage (Thomson and Plowright 1980) or laboratory (Thomson et al. 1986). If pollen grains are large enough and not too numerous, and the stigmatic surface is smooth enough, it may be possible to estimate the number of pollen grains with a hand lens. For example, Kearns (1990) could detect pollen grains (about 55 μm in diameter) on stigmas of *Linum lewisii* (Linaceae) and was able to determine which flowers had not yet received pollen and therefore were still suitable for experiments on pollen deposition by different species of flower visitors. Mulcahy et al. (1983) used a 10X hand lens to check for pollen grains of *Geranium maculatum* (Geraniaceae), and Snow (1986) counted pollen tetrads on stigmas of *Epilobium canum* (Onagraceae) with a similar hand lens. In other cases, it may not be necessary or possible to examine stigmas on flowers in the field.

If visitation rates are low, and you find yourself spending a lot of time waiting for pollinators to visit the flowers with clean stigmas, you can pick flowers and present them to foragers you find on other flowers. This works well for flies (Kearns 1990), honeybees (Silander and Primack 1978), and bumblebees (Thomson 1981b). You may be able to hold cut flowers in your hand, place them next to a flower where a forager is occupied collecting nectar or pollen, and have them move right to the flower you are holding. Thomson put cut flowers in florists' flower holders that were attached to the end of wooden rods, and then presented them to foraging bees. This method works better for insects (e.g., bumblebees) that may be disturbed if you get too close to them. Picked flowers can also be used in laboratory studies (Thomson et al. 1986).

Schmid-Hempel and Speiser (1988) modified this method. They picked focal flowers with low pollen loads on their stigmas (freshly opened stigmas), and at the beginning of a trial cut two of the four stigmatic lobes off the style. They placed the lobes in a glass tube, together with a wet piece of paper to keep the humidity high. At the end of the experiment they removed the whole style, placed it with the other two lobes, and froze the sample for later analysis (they stained and counted pollen grains on all four stigmatic lobes). The difference between the numbers of pollen grains found on the pairs of stigmatic lobes, multiplied by 2, served as an estimate of pollen deposition during the experimental period.

If pollinators are in short supply or otherwise unavailable, it may be possible to substitute a dead one in order to approximate the effect of a live one. Paton and Ford (1977) used a stuffed New Holland Honeyeater to study deposition of pollen in the laboratory. They inserted the bird's head into freshly picked flowers and then counted pollen on the stigmas. If flowers are in short supply, you can also try removing pollen from stigmas of flowers you have already used once, in order to reuse them; Paton and Ford cleaned pollen from stigmas with a cloth or brush.

Species of plants with different colors of pollen (e.g., *Erythronium grandiflorum* [Liliaceae]; Thomson 1986) or different sizes of pollen (e.g., heterostylous species) facilitate studies of pollen deposition and carryover. Harder and Thomson (1989) introduced flowers with red pollen into a large stand of flowers with yellow pollen. A bumblebee queen was allowed to visit the red-pollen donor and was then followed to identify the next 20–40 flowers she visited. They collected stigmas from these flowers and then examined them microscopically to count the number of red pollen grains. If you can't work with such a system, you may be able to use a pollen analogue or marked pollen to investigate pollen flow (see Section L.2).

Stigmas can be stained with a few drops of basic fuchsin or mounted in basic fuchsin gel (Beattie 1971a; see recipe in Chapter 7, Section E.1.a) and examined under a microscope for pollen counts (Snow 1982, Thomson 1986, Bertin 1990a, Waser and Price 1991a). To prepare basic fuchsin gel or glycerin-jelly slides in the field, use a cigarette lighter to melt a small amount of jelly on a clean slide (M. Dudash, *personal communication*). Place the stigma on the jelly and add a coverslip. Store the slide horizontally until the jelly has cooled. If the stigmatic surface is large and obscures pollen grains in whole mounts of stigmas, the stigma can be squashed gently or dipped into a chunk of basic fuchsin gel to remove pollen (Thomson and Plowright 1980). J. Thomson (*personal communication*) found that when making stigma mounts with glycerin jelly, it helps to let the stigma soften in the warmed jelly before squashing it firmly to spread out the grains and make the stigmatic tissue so thin that all the grains can be seen. To remove pollen, melt three small pieces of gel on a microscope slide and press the stigma into each in order, so that the absence of pollen on the third piece serves as a check that all the pollen has been removed. This technique must be used immediately after pollen deposition, before pollen tubes anchor the grains to the surface (Thomson and Plowright 1980; see also Thomson et al. 1981). In the field, stigmas can also be "mounted" on microscope slides with cellophane tape and then later pollen grains can be counted under a dissecting scope (Brown and Kodric-Brown 1979, Feinsinger et al. 1986, Murcia 1990).

Bertin (1982) determined how many pollen grains were necessary to initiate fruit set by applying different amounts of pollen to stigmas and then harvesting stigmas 24 hours later and staining with acid fuchsin. The numbers of grains were ranked in range categories. Fruits were allowed to develop for later examination.

Ornduff (1975b) compared two methods for determining the numbers of two sizes of pollen grains deposited on stigmas. He collected flowers and placed them in plastic jars with cotton wadding saturated with FAA. This fixed the tissue and minimized the disturbance to pollen loads on the stigmas. Later he excised the stigmas, mounted them on slides, and examined them microscopically. To determine the accuracy of these counts he then acetolyzed the stigmas for 10 minutes in 9:1 acetic anhydride:concentrated sulfuric acid, centrifuged them, washed them, and then counted the number and sizes of pollen grains. He found that this method produced counts of pollen grains about three times higher than those obtained by counting grains on intact stigmas, although the proportional representation of the two sizes of pollen grains was approximately the same with both methods. Because he lumped stigmas for the acetolysis analysis, it had the disadvantage of not permitting analysis of the frequency of grains on individual stigmas. However, it should be possible to conduct the acetolysis count on individual stigmas.

Cruden et al. (1990) prepared stigmas for viewing with epifluorescence using the staining procedure described (Section K.1.a) for visualizing pollen tubes with decolorized aniline blue. Pollen grains on stigmas were then counted. Bernhardt et al. (1980) used a similar procedure to observe adherence and germination of pollen grains on stigmas.

Some pollen grains on untreated stigmas mounted on microscope slides fluoresce naturally under ultraviolet and B2 filters in the Nikon epifluorescence system. To increase the brightness of fluorescence, place the stigma on a slide without a coverslip, apply a drop of 9 M NaOH on the stigma, and place the slide overnight in a humidified chamber or petri plate. Squash the treated stigma under the coverslip for viewing under epifluorescence. Grains fluoresce much more brightly and often fluoresce different colors. (Technique developed by W. H. Busby and K. G. Murray; P. Feinsinger, *personal communication*.)

Microscopes with Hoffman Modulation Contrast optics can also be used for counting and identifying pollen on stigmas (Feinsinger et al. 1986). Hoffman optics are "the poor person's Nomarski," providing three-dimensionality and surface texture of objects at about one-tenth the cost of Nomarski optics (see Appendix 2). Hoffman optics can be used on lab-quality microscopes (P. Feinsinger, *personal communication*).

If the pollen grains are too small to count or identify with the previous methods, it is possible to make these determinations with a scanning electron microscope. Jennersten et al. (1988) used SEM to examine pollen on stigmas when two or more species with small pollen could not be distinguished with light microscopy. Pellmyr (1984) also used SEM to count pollen of three species on stigmas of *Actaea spicata* (Ranunculaceae). Although this technique is more time-consuming because of the preparation of specimens that is involved, the pictures that result can be spectacular.

J. Hand-pollination

Hand-pollinating flowers allows the researcher to control the type and quantity of pollen a flower receives. Hand-pollinations are used for many different reasons, such as developing new hybrids, providing pollen in the absence of pollinators, and supplementing natural pollination to ensure fruit development. Pollination biologists often conduct tests to determine the relative success of self- versus outcross pollen, and pollen from different genotypes, different populations, or different distances within populations. Success is sometimes measured by seed set or inferred from pollen germination or pollen tube growth. Remember that in most systems, more than one pollen grain is required to initiate seed production. For example, Spira et al. (1992) found they needed 2.6 pollen grains/ovule to get 100% seed set. Larger amounts are needed in other systems. So be sure to use enough pollen in hand-pollinations if you're trying to get full seed set.

1. Application

Many methods have been used to collect pollen and apply it to receptive stigmas for hand-pollination experiments:

- Collect pollen from dehiscing anthers with toothpicks (Price and Waser 1979, Preston 1991), needles, small paint-brushes (Stanley and Linskens 1974, Waser and Price 1991a), or forceps. Use separate toothpicks for each flower when you must be sure of the pollen donor (A. Snow, *personal communication*).
- Rub an entire anther, capitulum, or a male-phase flower over the recipient stigmas (Marshall and Ellstrand 1985, Sobrevila 1989). Locking forceps facilitate this procedure (Thomson, *personal communication*).
- Collect pollen from several flowers on a small piece of tissue wrapped around forceps (Marshall and Ellstrand 1985).
- Gregg (1991) constructed hand-pollination tools from wooden sticks (0.6 x 6 cm) with fabric glued to the end. The fabric chosen had a 3-mm thick pile that simulated bee hairs. The tool was manipulated as if it were a pollinator, so that it first became covered with sticky stigmatic fluid that picked up pollen tetrads when the tool was removed.
- For transporting pollen collected on toothpicks, poke the clean end of a toothpick through the lid of a plastic vial, and then close the vial around the end with the pollen (Price and Waser 1979).
- Collect pollen by tapping dehiscing anthers over a petri dish. Apply this pollen to stigmas using a small piece of felt attached to a toothpick (Fenster 1991a).
- Mimic pollen application by birds or insects by using dead animals to transfer pollen. Introduce the beak of a bird or head of a bee into a pollen donor and a receptive flower sequentially (Waddington 1981, Pyke 1982a, Zimmerman and Pyke 1988). Construct "bee sticks" from dead bees and cocktail sticks by mounting the bee thorax on the stick with glue (Tomkins and Williams 1990).

- Use hypodermic syringes, equipped with rubber bulbs and 16-gauge needles to pollinate anemophilous flowers. Fill the syringe with pollen, insert the needle into a bag covering the female flowers, and then eject the pollen (Cumming and Righter 1948).
- Scoop anthers out of the flower with the sharp tip of a pencil. Use the pencil tip to insert the excised anther onto a recipient stigma (Abraham and Gopinathan Nair 1990).
- M. Cruzan (*personal communication*) uses high-test nylon fishing line to apply pollen. He collects pollen in a glass vial and then uses the fishing line to get the amounts needed to pollinate flowers. The electrostatic charge on the nylon causes pollen to adhere without excessive clumping (see Section R).
- Shore and Barrett (1984) applied small numbers of pollen grains in approximately equal quantities to all stigmas on a flower. They removed pollen from donor anthers with a fine dissecting needle and used a second needle to transfer pollen. Part of each corolla was removed and the pollen was applied to stigmas with a fine needle while viewing the stigma under a dissecting microscope.
- When larger amounts of pollen (2–100 gm) are needed, a milliliter hypodermic syringe can be modified by replacing the piston with a rubber atomizer (Stanley and Linskens 1974). This device directs pollen application to stigmas. Pollination of large areas such as fields and orchards can be accomplished by spraying with an air-pressure hose, pole duster, or back pump. Airplanes or helicopters can apply pollen over extensive areas. Such mass pollination programs usually use a carrier (talc, dead pollen, or *Lycopodium* spores) to dilute the pollen (Stanley and Linskens 1974).
- Alspach et al. (1991) used pollen collected by bees to test supplemental pollen treatments for the production of kiwifruit. A large kiwifruit of commercial value is only produced when many seeds are developing inside, and supplemental pollen increases the number of seeds. Pollen pellets were collected from bee traps at hives. Pollen was subsequently suspended in an aqueous solution or mixed with powder diluents (*Lycopodium* spores, and two commercial diluents; see Appendix 2). The aqueous solution was applied with an

atomiser and the powders were applied with a calibrated Max pollen application gun.

- Some flowers have multiple stigmas, and some stigmas have multiple lobes. These morphologies may simplify some types of experimental studies, for example, keeping track of different types of pollen used in experimental pollinations. Snow (1982) found that pollen placed on any of the three stigmas of *Passiflora vitifolia* (Passifloraceae) was capable of fertilizing ovules in any part of the ovary. This is also true for many umbellifers (Borthwick 1931, Endress 1982, Schlessman *personal communication*). Gorchov (1988) found that the five styles of *Amelanchier arborea* (Rosaceae) appear fused but are actually distinct and diverge at the stigmas, so that each of the five carpels can be pollinated separately. Cruzan (1990) showed that the tubes from pollen placed on different stigmatic lobes of *Erythronium grandiflorum* were mixed by the time the tubes grew down to the ovary, although they remained in the separate sections of the style for much of its length. Snow and Spira (1991a) used the three stigma lobes of *Hibiscus* (Malvaceae) for adding pollen from three different pollen donors. Montalvo (1992) found that the multiple locules of *Aquilegia caerulea* (Ranunculaceae) flowers each had a series of stigmas and styles attached to them. Thus she was able to use each locule for a separate treatment, such as self- or outcross pollination, bypassing potential problems associated with using different flowers on a plant for different treatments; in this species, there are significant interflower differences in resource allocation that would have confounded experimental effects. Waser and Price (1991b) applied self-pollen to one stigma lobe of *Ipomopsis aggregata* and outcross pollen to two stigma lobes of the same flowers.
- F. Imbert (*personal communication*) washed flowers on which self-pollen was deposited automatically before stigma receptivity. By removing self-pollen she assured that hand-pollination treatments contained only pollen from the correct experimental donor.

2. Estimating Numbers of Pollen Grains Applied

Galen et al. (1987) developed a technique to apply specific quantities of pollen to stigmas. Small amounts of pollen were brushed into inked rings drawn on microscope slides. The total amount of pollen was counted under a dissecting microscope at 50 power. The slide was carried in a covered petri dish to the recipient flower and the stigma was wiped over the pollen within the inked ring. Slides were rechecked under the scope so any remaining pollen grains could be subtracted from the total to determine the number applied.

Richardson and Stephenson (1991) used two pollen loads (high and low) in experiments. To check the amount of pollen applied in the high and low treatments, 42 nonexperimental flowers (21 high load, 21 low load) were hand-pollinated using the same procedure as on the experimental plants. After pollination, each pistil was collected and placed in a vial with FAA to prevent pollen germination. Upon returning to the lab, the FAA was evaporated, 5 mL of 1% NaCl were added, and each vial was sonicated for 5 minutes to remove any pollen remaining on the stigma. Subsamples of the solution were counted under the microscope to determine the average number of grains in the high and low treatments.

Waser and Price (1991b) found it impossible to apply exactly equal pollen loads when hand-pollinating with toothpicks or dissecting needles. Twenty-six hours after pollination, styles were excised at the top of the ovary. Seeds continued development and were counted later. Styles were stained for examination of pollen tubes with epifluorescence. Pollen grains on the stigma and pollen tubes in the style were counted. There was considerable variation in pollen loads on stigmas. Because pollen quantity and quality can confound conclusions about pollination treatments, they suggest applying a range of biologically realistic loads and comparing the differences among treatments across the entire range of loads applied. Comparisons should involve examination of residuals and comparison of goodness of fit (rather than linear regression).

K. Pollen Tubes

Microscopic study of pollen tubes permits assessment of pollen germination, rates of pollen tube growth, and pollen tube attrition. Pollen tube studies can be used to assess the effectiveness of hybrid fertilizations in breeding programs and to evaluate some forms of incompatibility. Relative growth of pollen tubes from different pollination treatments can be used to compare the success of self- and outcross pollen and of pollen from different donors or from donors at different distances from the recipient plant. Differences in pollen tube growth create a potential environment for pollen tube competition (Mulcahy 1979) and differential pollen tube attrition.

The rate of pollen tube growth varies widely among species. If you are looking at how many pollen tubes reach the ovules, or comparing growth rates or attrition of two genotypes, you may need to run some trials, collecting styles at several different times after pollination to determine how long to allow pollen tube growth for your experiment. For example, if all pollen tubes have reached the ovules after 10 hours, you will not be able to discern differences among various treatments. Mulcahy et al. (1983) collected and fixed stigmas at 30-minute intervals after hand-pollinations to observe pollen tube growth rate in *Geranium maculatum* (Geraniaceae). Aizen et al. (1990) used 2-hour intervals (2–10 hours) for *Dianthus chinensis* (Caryophyllaceae).

Pollen tube attrition can be compared by counting the percentage of pollen tubes that reach the ovary or by scoring the number of tubes that reach each zone of the style (zones are arbitrarily determined by the distance from the stigma, or as some fraction of style length (M. Cruzan *personal communication*, Williams and Knox 1982). A. Snow (*personal communication*) points out that in monocots, whole, unsquashed tubes can be seen in the stylar canal, but in squashed styles of dicots it is more difficult to see the tips of the tubes. For species with many ovules, Cruzan (1986) made serial sections of styles at 24 hours (while pollen tubes were still growing) and examined the distribution of pollen tube lengths for different crosses. Alternatively, the number of pollen tubes crossing the base of the style can be counted at a point in time when some tubes have "crossed the finish line" to effect fertilization (M. Cruzan, *personal communication*).

For species with one ovule it may be sufficient to measure the length of the longest pollen tube or group of longest tubes. Ideally, individual tubes should be measured, but the tips are usually difficult to see, and the presence of multiple tubes in the style makes measurements difficult (M. Cruzan, *personal communication*). Raff and Knox (1982) averaged the length of the longest pollen tube in each of 5–20 styles after 30 minutes and used this value to compare among pollination treatments. Anderson and Barrett (1986), working with *Pontederia cordata*, measured the longest pollen tube in each style. Since *Pontederia cordata* has only one ovule per flower, they assumed that the length of the longest pollen tube would be related to fertilization success. Aizen et al. (1990) measured the third longest pollen tube in each flower to produce a time curve of pollen tube growth and then counted the number of pollen tubes present at 1-mm intervals down the length of the style to compare the success of self- and outcross pollen. Differences between self- and outcross pollen tubes were compared with analysis of variance of the average tube length, the coefficient of variation, and the square-root transformed number of pollen tubes.

Waser et al. (1987) used several measures of pollen performance: (1) the number of pollen grains adhering to stigmas, (2) number of germinating grains, (3) mean tube length determined from the number of tubes present at four levels down the style, (4) number of tubes at the base of the ovary and (5) number of tubes entering ovaries. Snow and Spira (1991a) prepared stylar tissue for fluorescence microscopy and then took cross sections of the styles. They were able to count the number of tubes and the number of callose plugs in each section.

If you have access to a digitizing tablet, you can use it to measure pollen tubes (Kahn and DeMason 1988, Feder and Shrier 1990). Kahn and DeMason (1988) used a graphics tablet and Graphic Tablet System Tools (area and distance program) to trace pollen tube lengths at standard time intervals to determine the rate of pollen tube growth.

When cultured pollen tubes are long and intertwined, direct measurement is difficult. An alternate procedure is to isolate the pollen tube mass with filtration and to weigh the mass (Tupý et al. 1977, Shivanna and Johri 1985). Differences in weight can be compared over time intervals to determine the rate of growth.

Germination and pollen tube growth are sensitive to atmospheric pollution (Paoletti and Bellani 1990). Acid rain in industrial areas may have a pH of 4.5 and fog or mist may further absorb S and N acids, producing pH values lower than 2.0. Paoletti and Bellani (1990) demonstrated that pollen tube germination and growth are inhibited by exposure to sulfuric and nitric acids applied in vitro to mimic acid fog.

1. Light and Fluorescence Microscopic Examination of Pollen Tubes

The most widely used procedure for looking at pollen tubes involves the use of aniline blue stain and examination of pollen tubes under fluorescence microscopy (Martin 1959). This method has been used for over 25 years and is an easy, convenient procedure for routine counts, measurements, and identifications of pollen tubes. As they grow down the style, pollen tubes from many species periodically deposit callose plugs that may serve to separate the protoplast in the tip from the empty tube above (Williams et al. 1982). Aniline blue stain with a pH of 6 or 7 fluoresces under ultraviolet light and fluorescence microscopy illuminates the callose. There are several variations on the fluorescence procedure that may produce different results with different species. The basic steps are fixing, softening and clearing tissues, staining, and viewing under epifluorescence.

a. Aniline blue epifluorescence

Fixing. Stigmas, styles, and locules can be collected and treated. Long styles can be cut into pieces to facilitate examination, and thick ovaries and styles may be slit longitudinally. After collection, the tissues are fixed to prevent further pollen tube growth. Tissues may be fixed in one of the following solutions:

- formalin–acetic acid–alcohol (FAA = 1 formalin:1 acetic acid:18 50% ethanol) (Anderson and Barrett 1986, Feinsinger and Busby 1987, Fenster and Sork 1988, Murcia 1990). Tissues fixed in FAA should be transferred after 24–36 hours to 70% ethanol for storage if examination will be delayed (Copland and Whelan 1989, Palser et al. 1989).

- 1:3 acetic acid:ethanol (v/v) for 1–24 hours (Bernhardt et al. 1980; Williams et al. 1982; Copland and Whelan 1989; Galen et al. 1989; Thomson, McKenna, and Cruzan 1989) followed by 30 minutes in 50% ethanol (Feinsinger and Busby 1987). After fixing, tissues may be stored in 70% ethanol until further processing (Bernhardt et al. 1980).
- 70% ethanol (Mulcahy and Mulcahy 1982, Aizen et al. 1990).
- Carnoy's solution (Smith 1991, Waser and Price 1991a): glacial acetic acid:absolute ethanol:chloroform (1:6:3).

Softening tissues and clearing tissues.
- Some tissues require softening before staining. This can be accomplished by autoclaving in 5% to 10% (w/v) sodium sulphite (Na_2SO_3) for 20 minutes at 15 psi (Palser et al. 1989, Jeffries and Belcher 1974, Williams et al. 1982) and then rapidly rinsing in distilled water. Bernhardt et al. (1980) used 0.4 M sodium sulfite (Na_2SO_3) and autoclaved for 10 minutes at 121°C.
- Kho and Baër (1968) found that treatment with NaOH (sodium hydroxide) may be adequate to soften tissues. One hour in 1 N NaOH at room temperature is adequate for fine styles (e.g., tomato), whereas stronger concentrations and longer time periods may be necessary for tougher styles. Anderson and Barrett (1986) softened *Pontederia cordata* (Pontederiaceae) styles for 24 hours in 8 M NaOH. Styles can be left in 8 or 10 N NaOH overnight (Thomson, McKenna, and Cruzan 1989; Aizen et al. 1990), for as long as 48 hours (Fenster and Sork 1988), or for 20 minutes to 1 hour at 60°C (Mulcahy and Mulcahy 1982; A. Snow, *personal communication*). The time required varies among species (A. Snow, *personal communication*).

Staining. Dissolving aniline blue dye in K_2HPO_4 or K_3PO_4 decolorizes it; after 1–2 hours at room temperature the solution becomes essentially colorless (Currier 1957). There is apparently some variation in the composition of aniline blue dye from different suppliers, and some brands may decolorize more readily than others. Use one that decolorizes well, or the dye itself may stain the tissue and interfere with the fluorescence. Slight variations in the composition of the stain medium may work better for different species.

- Rinse tissues in water and stain them in decolorized aniline blue (Martin 1959, Kho and Baër 1968, Weller and Ornduff 1989).
- Rinse tissues with water and then mount in decolorized aniline blue 2 hours to overnight (Palser et al. 1989) with a drop of 50% glycerin (Thomson, McKenna, and Cruzan 1989) or clear corn syrup (Cruzan, *personal communication*) or decolorized aniline blue containing 10% glycerin (Bernhardt et al. 1980, Copland and Whelan 1989).
- Anderson and Barrett (1986) prestained tissues with 0.05% toluidine blue in benzoate buffer for 1–2 minutes, rinsed tissues in tap water, and then stained in 0.01% aniline blue in 0.01 M K_3PO_4 for at least 4 hours. Tissue was then mounted in a drop of stain on a slide and squashed with a coverslip.

Viewing. Place tissue on a microscope slide within a ring of petroleum jelly before adding a coverslip to prevent evaporation (Palser et al. 1989) or squash with a coverslip and view with an epifluorescence microscope. Many epi-illumination fluorescence microscope systems are available, and microscope companies will provide information to help you set up a system that best suits your needs.

Counterstaining with DNA probes ethidium bromide (EB; NOTE: **Use with caution, as it is carcinogenic**) or Hoechst 33258 (bis-benzimidazole derivative) in conjunction with aniline blue fluorochrome permits visualization of nuclear events within the pollen tubes (Hough et al. 1985). Waser et al. (1987) used ethidium bromide counterstain simply to make *Delphinium nelsonii* (Ranunculaceae) pollen tubes visible. In addition, ethidium bromide (0.01%) in phosphate buffer will enhance fluorescence of pollen grains attached to the stigma (Dumas and Knox 1983). Counterstaining the decolorized aniline blue with acridine orange (0.01%) in 0.01 M phosphate buffer, pH 7.5 or 8.5 for 10 minutes aids visualization of cytological features (Alves et al. 1968 cited in Hough et al. 1985).

The aniline–blue/fluorescence procedure works well for apple pollen tubes, because tubes from compatible pollen contain compact callose plugs and those from incompatible pollen have extensive callose deposits over long regions of the pollen tube (Jefferies and Belcher 1974). Not all species produce pollen tubes with large amounts of callose, however. Jefferies and Belcher used the fluorescent brightener Calcofluor White M2R New (Cyanamid of Great Britain Ltd.). The brightener stains cellulose, and the entire pollen tube fluoresces. They recommend using a mixture of calcofluor and aniline blue prepared in the following manner:

1. Dissolve each of the following components separately in small amounts of distilled water:
   1 g aniline blue
   1 g calcofluor
   3.5 g tribasic potassium phosphate
2. Mix the three components together and bring the volume up to 1 L.
3. Autoclave pistils in a solution of sodium sulphite (50 g/L) until they soften.
4. Rinse tissues with water and put them onto a microscope slide.
5. Add a drop of the combined stain and then crush the style with a glass rod.
6. After 10 minutes, add a coverslip; the slide is ready to view under ultraviolet light. A BG12 exciter filter in combination with a yellow or orange barrier filter reduces the brightness of the calcofluor, which can tend to obscure the effect of the aniline blue.
7. If calcofluor still fluoresces too brightly, rinse the slide with water or aniline blue.
8. For permanent slides, euparal may be used as a mounting material to maintain fluorescence for long periods of time (Ramanna 1973).

Williams et al. (1982) used the aniline blue staining procedure to trace pollen tubes through transverse sections of style. Thick sections cut with a razor worked best for preserving the structure of the pollen tubes. Sections were placed in depression slides with a drop of decolorized aniline blue for 30 minutes and viewed with epifluorescence (Zeiss microscope with filter combination KP490, KP500, Rfl, 510, LP 528).

Bernhardt and Calder (1981) used the fluorescence method to study hybridization in mistletoes. After pollination, flowers were implanted on a 10% sucrose, 2% agar medium in a humid chamber. Twenty-four hours after implantation on the growth medium, styles were collected, fixed, and stained for examination with epifluorescence. Adherence of pollen grains to the stigma, pollen germination, and pollen tube growth were evaluated.

Five other methods for pollen tube examination do not require epifluorescence:

b. Basic fuchsin/fast green (Levin 1990)
1. Fix pollen tubes in 3:1 v/v alcohol:acetic acid for at least 24 hours.
2. Stain them in 1% basic fuchsin:1% fast green (4:1) for a minimum of 24 hours.
3. Destain and soften tissue in lactic acid for 12 hours and then squash under a coverslip. Pollen tubes stain maroon against a white background under white light.

c. Acidified aniline blue/acetocarmine (M. Cruzan, *personal communication*)
1. Cut styles at the base for pollen tube examination. If ovaries are left intact and sufficient time has elapsed since pollination, seed and fruit formation may continue.
2. Slit styles longitudinally and stain them for 10 minutes in a drop of acidified 0.1% aniline blue and a drop of acetocarmine. (Aniline blue solution in either water or phosphate buffer is acidified with HCl until it turns dark blue.)
3. Acetocarmine is prepared by saturating 40% acetic acid with carmine. Boil the acid solution in a fume hood. The acetocarmine darkens pollen tubes.
4. Wet mounts of the stained pollen tubes viewed under a stereoscope can be manipulated with a dissecting pin to facilitate counting. Cruzan uses this procedure for

Erythronium (Liliaceae) pollen tubes, which do not contain much callose.

d. Acetocarmine/basic fuchsin (Prakash 1986)
1. Split the style longitudinally with fine pins and spread it flat on a slide.
2. Add a drop of acetocarmine, followed by a drop of 3% aqueous basic fuchsin.
3. Blot excess stain.
4. Destain tissues with a drop of absolute alcohol and absorb excess with filter paper.
5. Add a few drops of glycerin before placing a coverslip. Pollen tube cytoplasm stains red against a lighter background.

e. Aniline blue in lactophenol (D'Sousa 1972)
1. Fix styles in 95% ethanol:lactic acid for a minimum of 15 minutes (style may remain in this solution for a few days).
2. Stain is composed of 1 g aniline blue in 100 mL lactophenol.
3. Place a drop of stain on a microscope slide and place rinsed style in stain for 5–20 minutes.
4. Add a drop of 40% acetic acid to destain stylar tissue (10–20+ minutes). Aniline blue stains style nuclei and stigma lobes obscuring pollen tubes; destaining remedies these problems.
5. Wash styles in a petri dish full of water.
6. Mount styles in 100% lactic acid.
7. Crush tissue lightly with a coverslip. Pollen tubes stain blue. Slides keep for several months.
8. Kambal et al. (1976) modified the stain by adding 1% trypan blue in lactophenol to the cotton blue stain for viewing *Vicia faba* (Fabaceae) pollen tubes. Forty-five percent acetic acid was used for destaining styles.

f. Alexander's combination of four stains (Alexander 1987)
1. Prepare 1% stock solutions of malachite green, acid fuchsin, and aniline blue in distilled water, and orange G in 50% alcohol.

2. Combine ingredients in the following order:
    78 mL lactic acid
    4 mL 1% malachite green
    6 mL 1% acid fuchsin
    4 mL 1% aniline blue
    2 mL 1% orange G
    5 g chloral hydrate
    The solution should be stored in a dark bottle.
3. Prepare clearing and softening solution by combining:
    78 mL lactic acid
    10 g phenol
    10 g chloral hydrate
    2 mL 1% orange G
4. Prepare mounting medium by combining:
    50 mL lactic acid
    50 mL glycerol
5. Fix pollen tubes in Carnoy's fluid (absolute alcohol:chloroform:glacial acetic acid 6:4:1) for 12 hours.
6. Rehydrate styles through a descending alcohol series ending with water.
7. Place pistils in stain and incubate at 45°C for 12 hours. Large pistils can be sectioned longitudinally.
8. Transfer tissue to clearing and softening medium for 24 hours at 45°C.
9. Transfer tissue to clean clearing and softening medium and hydrolyze in oven for 30 minutes at 58°C.
10. Rinse twice with lactic acid.
11. Store the tissue in lactic acid until you are ready to mount and view it.
12. Mount tissue on a microscope slide in mounting medium and examine with a light microscope.

2. Electron Microscopy of Pollen Tubes Following Semi-Vivo Culture

Rao and Kristen (1990) allowed pollen tubes to grow in semi-vivo culture, long enough to produce short compact tubes in the culture medium, and then processed them to view ultrastructure with the electron microscope. Flowers were hand-pollinated and pollen tubes began growth down the style. Some hours later, the pistils were cut below the level of the growing pollen tubes and the stylar ends incubated in culture medium composed of 10% sucrose, 0.1% boric acid, 3 mM $Ca(NO_3)_2$ (calcium nitrate) in distilled water with a pH of 6. Pollen tubes grew to the end of the cut style and then out into the culture medium. Rao and Kristen added test substances to the culture medium to determine their effects on pollen tube growth. Subsequently, pollen tubes were viewed with both fluorescence microscopy and SEM. They fixed tissues for fluorescence microscopy with the following procedure:

1. Fix tube bundles and a small portion of stylar tissue in phosphate-buffered 2.5% glutaraldehyde at pH 7.0 and 4°C for 3 hours.
2. Rinse several times with buffer.
3. Postfix tissues in phosphate-buffered 1% osmium tetroxide at 20°C for 4 hours and then rinse in 0.1 M sodium acetate.
4. Stain tissue with 0.5% uranyl acetate for 12 hours and rerinse in 0.1 M sodium acetate.
5. Dehydrate tissues in an acetone dehydration series and embed in Spurr's resin.
6. Section tissues.
7. Stain with lead citrate to observe pollen tube tips, or treat with periodic acid-thiocarbohydrazide silver proteinate to show polysaccharides.

For SEM, fix and postfix using the same method given above. Then dehydrate tissues in an acetone series over ice. Next, critical-point dry them with liquid CO_2 and gold coat for SEM viewing.

L. Pollen Dispersal

An understanding of patterns of pollen dispersal is important for studies of (1) gene flow and plant population structure, (2) effectiveness of different pollinators, (3) the evolution of plant breeding systems, and (4) evolution of floral traits (Waser and Price 1982). However, direct measurements of pollen dispersal are often difficult. Direct measurements have depended on a pollen heteromorphism (Richards and Ibrahim 1978) or pollen color variation.

Thomson and colleagues (Thomson and Plowright 1980, Thomson and Stratton 1985, Thomson 1986, Thomson et al. 1986, Thomson 1989a, Thomson and Thomson 1989) used two forms of *Erythronium grandiflorum* (Liliaceae), one producing yellow and one producing red pollen, to examine pollen carryover and to map gene flow and pollen shadows. Even in this system, some red grains in contact with stigmatic papillae may lose their color upon hydration and germination, making them difficult to distinguish (Thomson 1986).

1. Anemophilous Systems

Windborne pollen dispersal is affected by source strength, aerodynamics, and biophysics of deposition (Di-Giovanni and Kevan 1991). Timing of pollen release, volume of pollen, and the position of the pollen source on the plant can affect the reproductive success of a plant (Roberds et al. 1991, Di-Giovanni and Kevan 1991). After pollen production is initiated, individual conifers generally alternate between high-production and low-production years (Di-Giovanni and Kevan 1991). Once pollen is released, wind speed, direction, and turbulence affect the deposition site. Turbulence tends to be greater among tree crowns than above open ground and is generally greatest in the afternoon. Pollen size, pollen density, and the presence of bladders or flotation devices will affect terminal velocity. Final deposition is affected by impaction (on vegetation), sedimentation, precipitation, and electrostatic and thermal processes (Di-Giovanni and Kevan 1991). Di-Giovanni and Kevan review factors that affect pollen dispersal and state that no studies have yet combined the theoretical, biological (gene flow), and physical (pollen monitoring) aspects of anemophilous pollen transport. Studies incorporating all these aspects are particularly important for seed orchards, where pollen contamination from external sources can result in loss of genetic improvements initiated by orchard managers.

Many of the techniques described below under entomophilous systems can be adapted for monitoring the movement of anemophilous pollen. For more information on anemophily see Chapter 2, Section D.2.

2. Entomophilous Systems

One method of evaluating pollen movement is to measure foraging distances traveled by pollinators. But because all pollen from a donor flower is not transferred to the next flower visited, distances based on pollinator movement will underestimate actual pollen flow. In fact, pollen carryover to sequential flowers can be extensive (Waser and Price 1982).

Thomson and Plowright (1980) were able to estimate carryover by allowing bees to visit a series of hand-held flowers. The initial dehiscent flowers presented served as pollen donors. Sequential emasculated flowers served as pollen recipients, and stigma loads were counted to determine pollen decay. Feinsinger and Busby (1987) used trained hummingbirds to visit flowers for carryover studies. Waddington (1981) stained pollen on dehisced anthers and inserted a dead bumblebee with protruding mouthparts into a stained flower followed by six receptive flowers. Stained grains on stigmas were counted under a microscope.

Indirect measurements of pollen carryover include the use of pollen stains, pollen-mimicking dye powders (e.g., Stephens and Finkner 1953, Linhart 1973, Linhart and Feinsinger 1980, Price and Waser 1982, Waser and Price 1982, 1983, Webb and Bawa 1983, Campbell 1985, Murawski and Gilbert 1986, Campbell 1991b), radioactive labeling of pollen, and genetic markers.

a. Stains

Peakall (1989) used aqueous histochemical stains to label pollinia: brilliant green (1% w/v), Bismarck brown (1%), methylene aniline blue (1%), orange G (10%), rhodamine (0.2%), and trypan red (2%). One–2 μL of each stain were injected into anther flaps supporting the pollinia lobes with a 10-μL syringe. Individual pollen grains took up the stain. Labeled pollinia were traced by examining all flowers in the population. No difference was observed in pollinium removal or in pollinia cohesiveness between labeled and unlabeled flowers. Dilution of stains by rain was minor. Orange G was diluted the most, and the distinction between methylene blue and brilliant green became less noticeable. According to Peakall, the accuracy of measuring pollen flow with this procedure compares favorably with radioactive-labeling methods. In addition, the process costs little and multiple labels can be made by using different colors. Tremblay (1991) used Peakall's 1989 technique to label orchid pollen and found (*personal communication*) that crystal violet killed the flowers, but other stains (e.g., toluidine blue O, acid fuchsin, Sudan black) worked fine. He injected the stain (dissolved in water or, if necessary, with a little alcohol) into pollinia with a hypodermic needle.

Powdered stains (methylene blue, Evans blue, neutral red, and Bismarck brown) adhere to sticky pollen (Stephens and Finkner 1953, Linhart 1973). They absorb small amounts of moisture on the pollen, creating bright spots of color that can be used as markers (Linhart 1973). These marks are usually visible to the naked eye, but sometimes use of a 14X hand lens is required.

Waddington (1981) stained pollen with basic fuchsin dissolved in 95% ethanol. Stelleman and Meeuse (1976) tried spraying pollen with aqueous solutions of methylene blue and neutral red. Only pollen near the site of dehiscence acquired stain. They had more success marking pollen by collecting pollen in the field and bringing it to the lab:

1. Spread pollen evenly on filter paper in the bottom of a petri dish and spray with dye using a flower sprayer.
2. Allow 2–3 hours for pollen to dry in the uncovered dish.
3. When grains are dry, they should be loose enough to move with a dissecting needle.
4. Store dyed pollen in a glass tube.
5. To mark flowers in the field, insert dry flowering spikes into the tube; marked pollen will adhere to anthers (and also some other parts of the inflorescence; damp plants will cause pollen to stick to the tube). The normal fly visitors readily visited marked plants, consumed pollen, and carried pollen on their bodies.

b. Fluorescent powdered dyes

Fluorescent powdered dyes have been used as pollen mimics. These powders form the base for fluorescent paints and are available commercially from the companies in the following list (see Appendix 2 for addresses). Numbers and color names indicate specific powders that have been used in pollination studies.

Radiant Color Corporation — Series R-103-G (Campbell and Waser 1989). Dye particles are irregular in shape and size (longest linear dimension: x = 6.9 μm, s = 5.3, N = 20; Thomson et al. 1986).

U.S. Radium Corporation — brand name Helecone; fluorescent pigment numbers 1953, 2205, 2225, and 2267, green, blue, red, and yellow (Stockhouse 1976).

Day-Glo Color Corporation — Saturn Yellow A-17N; Arc Yellow (orange) A-16; blue A-19 (Ordway et al. 1987).

Dyes can be applied with an atomizer or insufflator (De Vilbiss Company; Stockhouse 1976) or to individual dehiscing anthers with toothpicks (Thomson et al. 1986, Campbell and Waser 1989). Paper shields used in conjunction with an atomizer can direct dust to specific flower parts (Stockhouse 1976).

The dye particle movement is tracked under the assumption that dye mimics pollen. An insect that visits a marked flower is followed to successive flowers that are then flagged. Flagged flowers are examined for dye particles. For carryover studies, stigmas or flowers are harvested and the presence or absence of dye is noted. The fluorescent powders glow brilliantly under ultraviolet light, so floral parts are best viewed with a black light (ultraviolet) source (for example, Blak-Ray portable lamp ML-49 or Model #C-70 Chromato-Vue cabinet from Ultraviolet Products Inc.); or dye particles can be counted under a dissecting scope at 50X (Waser 1988). If you prefer not to harvest recipient flowers for examination in the lab, try using a portable ultraviolet lamp at night to look for the presence or absence of dye particles (you will not be able to count particles; also see caution below about reliability of presence/absence scores).

The fluorescent powders can be used to examine dispersal. A male flower near the center of the population is marked, or flowers on several different central plants are marked with different colors at the time of anthesis. Subsequently, all possible recipient flowers in the population are numbered and their distance from the marked flower(s) is recorded. When flowers close they are harvested and checked for powder (Ordway et al. 1987, Campbell and Waser 1989, Campbell 1989), or flowers can be examined in the field at night by using a portable ultraviolet light. Thomson (1982) used an aqueous suspension of the powder to prevent the wind from carrying dye particles. He collected test flowers in individual glassine envelopes to prevent any contamination with dye from other flowers and then examined the flowers using a microscope with an ultraviolet light source.

Pollinators that visit marked flowers carry fluorescent powder on their bodies that is visible upon capture (Stockhouse 1976). Insects can also be marked directly with the powders (see Chapter 7, Section J), and the flowers they visit are often marked (Kearns, *personal observation*).

Palmer et al. (1988) studied seasonal effects on neighborhood size using fluorescent powders. A circle with a radius of 20m was used to study short-distance dispersal. Four wedges radiating 60 m from the radius were marked as a comparable area for studying long-distance dispersal. Plants in the circle and wedges were mapped. Four plants at the center of the circle were dusted repeatedly with different colored fluorescent powders. The distance of powder movement from the source plants was measured several times a week by examining all mapped plants under ultraviolet light at night.

Craig (1989), attempting to measure carryover in the field, found that birds (tuis) avoided flowers with stamens that were marked with yellow and orange fluorescent dyes. However, captive zoo birds did not object to feeding at the marked flowers. Dudash (1991) found that pollinators moved green more often than pink or white fluorescent dye. It may be important to test several colors or to alternate colors to take into account pollinator preferences.

The value of fluorescent powders in pollination studies depends on the similarity of powder and pollen transport. This similarity has been tested (reviewed by Thomson et al. 1986) for *Ipomopsis aggregata* (Polemoniaceae; Waser and Price 1982), *Stellaria pubera* (Caryophyllaceae; Campbell 1985), *Brassica campestris* (Brassicaceae; Handel 1983a), *Erythronium grandiflorum* (Liliaceae; Thomson et al. 1986), *Delphinium nelsonii* (Ranunculaceae; Waser 1988), and *Sabatia angularis* (Gentianaceae; Dudash 1991). Waser and Price counted both powder particles and pollen on stigmas, and determined that dispersal distance and carryover were similar when captive hummingbirds were presented with *Ipomopsis* flowers (pollen grain diameter about 50 µm, papillose stigmas). Dudash (1991) was unable to obtain quantitative data from dye transport because of clumping of dye particles, but she found that presence/absence scores were informative. Campbell found that powder served as a good analogue when examining pollen and powder movement by solitary bees and bee flies (Bombyliidae) visiting *Stellaria* (Campbell 1985). She used potted *Stellaria* plants to compare pollen and dye movement by pollinators. Plants were set out in areas where the species did not occur naturally. Fluorescent dye was applied to dehiscing anthers on several intact plants. All other plants were emasculated and numbered. Insects were prevented from visiting emasculated plants until they had visited a treated anther. Campbell then kept track of the sequence of insect visits after they visited an anther with dye powder. Stigmas from sequential flowers were collected and examined for the presence of pollen or dye. Carryover (C_x) was characterized for each flower in the sequence as the number of pollen grains deposited divided by the number of grains deposited on that flower. She compared pollen and dye carryover using regression coefficients for sequential deposition and by comparing Spearman rank correlations of pollen deposition.

However, not all studies have found a close correspondence between pollen and powders. Handel (1983a) scored presence or absence of dye and deduced pollen transfer from progeny marker genes and concluded that plants were more likely to receive powder particles than pollen; Thomson et al. (1986) concluded that only some aspects of pollen and powder transport were similar on *Erythronium*. The total number of dye particles was greater than the number of *Erythronium* pollen grains, and dye was transferred to stigmas that did not receive pollen, whereas the opposite was true for *Delphinium nelsonii* flowers (Waser 1988), making presence/absence scores unreliable (Waser and Price 1982, Thomson et al. 1986). In carryover studies, powder decayed at a greater rate than pollen. Relative powder and pollen movement followed similar patterns, so that powders could be used in experiments examining qualitative effects of different experimental treatments.

When the reliability of dye as a pollen analogue can be demonstrated, the number of dye particles on stigmas can be counted to estimate the number of pollen grains (Svensson 1986, Campbell and

Waser 1989). Several studies have suggested that fluorescent powders (Series R-103-G, Radiant Color Corp.) can be used as pollen analogues for *Ipomopsis aggregata* (Campbell and Waser 1989, Waser and Price 1982, Waser 1988). (Series R-103-G dyes are no longer made but have been replaced by Series R-105, which is supposed to have similar properties. The company lists the dye particle size as 6–7 μm, but there is considerable variation in size; Waser, *personal communication*.) However, since carryover sequences are highly variable and often contain lots of zeroes, it is difficult to determine if there are real differences between pollen and dye particle deposition with small to medium sample sizes (Thomson, *personal communication*). Thomson (*personal communication*) cautions that qualitative comparisons of dye and powder movement may be useful but that absolute quantitative comparisons may be misleading.

Rather than comparing just dye and pollen movement, Campbell (1991b) compared dye movement with actual gene flow in natural populations of *Ipomopsis*. Different colors of dye were applied to four central plants in three populations. The morning after application, stigmas of all flowers in a 10 x 10–m square around these plants were examined and the number of dye particles counted. Subsequently, genotypes of seeds from each plant in the square were assayed with electrophoresis at eight polymorphic loci. Next, genotypes of the maternal plants and all plants within a 10 x 10–m area around each were determined and paternity analysis followed. Based on pollen movement, the 10-m^2 area around each seed was expected to contain 99.9% of the pollen parents. However, based on allozyme data and paternity analysis, 16% of seeds had no possible paternal parent within that area. Because dye and pollen dispersal in previous studies were so similar for *Ipomopsis,* Campbell suggests that postpollination events such as biparental inbreeding depression, stylar discrimination of pollen tubes, or seed abortion may have been responsible. Thus actual gene flow and neighborhood size were greater than predicted by dye dispersal.

c. Radioactive labels
By incorporating radioactive labels into pollen, pollen transport can be measured directly with scintillation counters. However, use of radioactive labels requires special precautions and permission to

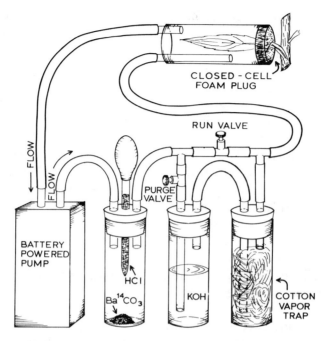

Figure 4–2. Apparatus for radioactive labeling. $^{14}CO_2$ is generated in the first vial by adding 6 mol/L HCL to Ba$^{14}CO_2$ with the dropper. Gases are circulated through the system with a modified aquarium pump. All connections are made with rubber tubing and stoppers. The labeling chamber has a closed-cell foam plug with a slit in it for inserting a leaf, but this could be modified to hold a flower. The cells are supported by stakes inserted into the ground to hold them in place. After labeling, the purge valve is opened and the run valve closed so that gas passes through the KOH solution and residual label is removed. Reprinted from McCrea et al. 1985.

release radioactive substances into the environment. A radiation safety officer should be consulted before using this method of labeling, as it poses both health and environmental hazards.

Pleasants et al. (1990) and Pleasants (1991) labeled pollen with a technique originally developed for physiological studies labeling plant ramets (Figure 4–2). The label was incorporated into a plant under a hood in the laboratory. Every 30 minutes, 6 N HCl was added dropwise to 37 or 74 GBq Ba$^{14}CO_3$ (barium carbonate; specific activity 2153 GBq mmol^{-1}) until all carbonate had reacted. The labeled $^{14}CO_2$ produced by the reaction was pumped through a closed circulating system containing a potted plant. The plant was kept in the chamber for 6 hours to incorporate the label. Flowers opened 1–19 days later, and radioactive label could be detected in pollinia, individual pollen grains, and nectar. The labeled plant was placed in the study plot where all the plants had been mapped. All umbels with open flowers were collected at the end of each day and the flowers placed in scintillation vials, scintillation cocktail added, and radioactivity assessed.

Other radioactive labeling procedures have used ^{131}I (Turpin and Schlising 1971, Schlising and Turpin 1971), and ^{32}P (Colwell 1951). Although the methodology for ^{32}P is relatively simple, the dangers of working with an isotope as radioactive as ^{32}P in the field probably outweigh any advantages the technique might have. ^{131}I has the advantage of a relatively shorter half-life (8.05 days), and it can be immediately attached to pollen grains. Check with your radiation safety officer, if you have one (most universities do), before considering using either of these techniques.

d. Neutron activation analysis

One of the major disadvantages of radioactive labeling methods is the potential health hazard associated with working with radioactive materials. The safest of these methods is probably neutron activation analysis (NAA), as it involves use of a nonradioactive marker element that can be exposed later to neutron radiation in order to make it radioactive. This method has been used in only a few studies and seems promising, but it does require access to sophisticated analytical equipment.

NAA relies on the fact that an element not normally found in the target plant, or only found in very low concentrations, can be used to label the plant or some part of it, such as the pollen. The element can be incorporated as a solution applied to the roots or poured into a hole in the trunk of a tree near the crown (Fendrick and Glubrecht 1967 cited in Gaudreau and Hardin 1974), injected into the cambium (McElwee 1970 cited in Gaudreau and Hardin 1974), or applied externally to staminate spikelets (Handel 1976). A variety of marker elements have been used, including samarium, gold, europium, manganese, and rhodium.

Gaudreau and Hardin (1974) used NAA to trace pollen dispersal and examine the flower constancy of pollinators in a plant community. They used samarium as a marker, which they prepared by dissolving small quantities of samarium sesquioxide in warm HNO_3 and adding distilled water to produce a final concentration of 100 mg per mL. They buffered this solution with a 0.1 M solution of EDTA and samarium, and then raised the pH to 5.6 with 1 M NH_4OH. (Handel 1976 used 1 M NaOH instead, and a pH of 5.7). The final solution was used to label composite flower heads by applying microliter volumes to entire capitulae. Handel (1976) also used samarium (15 or 25 μL) of the same type of mixture used by Gaudreau and Hardin (1974) to label staminate spikelets of two *Carex* (Cyperaceae) species, by applying it directly to the spikelets prior to anthesis; he used a 0.15% solution of Tween 20 (polyoxyethylene [20] sorbitan monooleate) as a wetting agent. Marker solutions could potentially be applied solely to anthers if they are large enough. However, it is unclear from the papers we have read whether the solution would penetrate the anther wall and label grains, or whether it would have to be applied to grains in dehisced anthers.

Samples (e.g., individual stigmas, flowers, or capitulae) suspected to have received pollen containing the marker element are collected and can be stored indefinitely before the next step of the analysis, irradiation. Gaudreau and Hardin (1974) subjected their samples to a flux of 5×10^{11} neutrons per cm^2 per second. The marker element is thereby made radioactive, and as these radioactive atoms disintegrate, high-energy γ and other rays are emitted; these rays are counted and analyzed to provide evidence of the marker element. For example, unstable isotopes of samarium such as 47-h ^{153}Sm β decay to ^{153}Eu, which in turn emits γ-rays. The dominant emissions, 70 keV and 104 keV γ-rays, can be counted on a NaI scintillation detector (Handel 1976). After allowing the radiation produced by their samples to decay for 24 hours, Gaudreau and Hardin counted them for 480 seconds in a Harshaw 3 x 3–inch NaI (well-type) scintillation counter, which was connected to an RIDL 400 Multichannel Analyzer. If a sample exhibits the appropriate γ levels, it can be assumed that the marked pollen has reached it.

One of Gaudreau and Hardin's experiments resulted in the transferal of very low concentrations of samarium. In order to enhance the probability of detection, these samples were subjected to more intense neutron radiation: 1 hour at a flux of 3 x 10^{13} neutrons per cm^2 per second. These samples were then counted for 400 seconds on a 16-mm Ortec Ge(li) Low Energy Photon Detector (LEPD), which was connected to a Nuclear Data 2200 Multichannel Analyzer.

The advantages of NAA are that some of the marker elements are inexpensive (samarium oxide is \$17.50 for 10 g from Sigma Chemical Co.), very small amounts of labeled pollen can be detected, samples need not be analyzed immediately, and radioactive materials need not be introduced into the environment (Handel 1983a). Problems with the technique include the assumption that pollen flow is not altered by the labeling process (probably a reasonable assumption), and the "need for close collaboration with a reactor and scintillation counter facility, which is often busy and requires financial compensation that may be beyond the budget of many pollination biology investigations" (Handel 1983a).

The requirement for a scintillation counter might be avoided by using the technique proposed by Reinke and Bloom (1979) for using liquid emulsion autoradiography for detecting labeled pollen grains. They suggest using "track autoradiography" in which a thick photographic emulsion records the β emissions from labeled pollen, in the form of tracks of silver grains. Although they tested this method with ^{14}C labeling, it might work with NAA as well.

e. Genetic markers

Handel (1982, 1983a, 1983b) studied gene flow from dispersal of marker genes in experimental gardens. He planted two strains of *Cucumis melo* (Cucurbitaceae), one with a homozygous dominant allele producing green cotyledons, and the second with recessive alleles producing yellow cotyledons (Handel 1982). Eight plants carrying the homozygous dominant allele were grown in the center of a test garden, surrounded by plants carrying the recessive allele. Plants were naturally pollinated and fruits developed. The location of all melons was mapped, and the seeds from each were planted. Upon germination, the F1 progeny were scored for cotyledon color to determine the spatial movement of the dominant allele through pollen transport. Similar experiments were formed with two strains of cucumber (*Cucumis sativus*), one bearing a dominant allele conferring a bitter taste to fruits (Handel 1983b). The disadvantage to this procedure is the amount of time involved from flowering to germination of the F1 seedlings.

Numerous researchers have used electrophoresis and quantitative measures of allelic frequencies to study gene flow in plant populations (see Chapter 6).

f. Backscatter SEM and X-ray microanalysis

Wolfe et al. (1991) have developed a new technique for monitoring inter- and even intrafloral pollen movement. As a test, pollen of *Astrophytum* (Cactaceae) was labeled by spraying a 1-g/mL slurry of micronized tin powder (Fisher Scientific Co.) over the anthers with an atomizer while shielding the rest of the flower with paper. Several hours later, pollen was collected on double-sided tape, and gold-palladium coated for SEM viewing. Then an experiment was conducted to determine if *Solanum rostratum* (Solanaceae) had fodder anthers as earlier literature indicated. Putative fodder anthers were labeled with micronized zinc (Mallinckrodt Chemical Works) and putative pollinating anthers with micronized tin. Later in the day, flowers on nearby plants were collected to see if fodder pollen was transferred to stigmas. Styles were cut and air-dried for 24 hours and then stigmas were mounted on double-sided tape to anchor them to copper boats. Stigmas were coated with gold-palladium for SEM viewing.

The presence of micronized powders was detected by viewing the samples with a JSM 880 SEM equipped with a LaB_6 emitter, and an accelerating voltage of 20kv. A solid-state, divided annular-type backscatter detector was used for imaging, and a Kevex Delta class energy dispersive spectrometer system with a beryllium window aided in X-ray microanalysis. Individual pollen grains were clearly visible. The use of two metals with different atomic numbers enabled Wolfe et al. to distinguish pollen from two types of anthers within the same flower and to determine that pollen from putative fodder anthers is actually transferred between flowers in similar quantities to the pollen from putative pollinating anthers. This new technique appears to have much potential for future studies of pollen movement.

M. Pollen Interference: Mechanical and Biochemical

Inconstant pollinators may transfer heterospecific pollen to stigmas (Waser 1978), or the high concentrations of anemophilous pollen (e.g., 1,000 pollen grains/cm^3; Raynor et al. 1972 cited in Murphy and Aarssen 1989) may result in similar contamination by heterospecific pollen. In these cases there is the potential for either mechanical or biochemical interference with pollination by homologous pollen.

Galen and Gregory (1989) found that although application of *Mertensia* (Boraginaceae) or *Castilleja* (Scrophulariaceae) pollen to *Polemonium viscosum* (Polemoniaceae) stigmas had no effect on conspecific pollen adherence, pollen tube germination and ovule fertilization were reduced. This suggests that the presence of foreign pollen can decrease female reproductive success. Pollen deposition to *Ipomopsis aggregata* results in the closure of stigma lobes, reducing further receptivity (Waser and Fugate 1986). This is a mechanical effect that occurs whether foreign pollen or *Ipomopsis* pollen is applied. Deposition of self-pollen along with outcross pollen in the self-sterile *Ipomopsis aggregata* results in a 42% reduction in seed set (Waser and Price 1991b). Self-pollen deposition occurs commonly through autodeposition and geitonogamy. Self-pollen tubes grow, and often enter ovules, but presumably because of a late-acting self-incompatibility mechanism these ovules do not develop into seeds (Waser and Price 1991b).

Relatively few studies have considered the potential for pollen allelopathy. Sukhada and Jayachandra (1980) published the first documentation of this phenomenon, which has now been tested for in other species, including *Agropyron repens, Phleum pratense, Poa pratensis, Zea mays* var. *chalquinoconico* (all Poaceae), *Ambrosia artemisiifolia, Erigeron annuus, Hieracium aurantiacum, H. floribundum, H. pratense* (all Asteraceae), *Melilotus alba, Vicia cracca* (Fabaceae), and *Brassica oleraceae* (Brassicaceae) (Murphy 1992). Allelopathic chemicals in pollen appear to be phenolics, terpenoids, alkaloids, and possibly nonprotein amino acids (Murphy 1992). Murphy (1992) reviews the methods for extracting, isolating, and characterizing pollen allelopaths.

Thomson et al. (1981) demonstrated a significant depression in fecundity of *Diervilla lonicera* (Caprifoliaceae) when pollen from

Hieracium floribundum (Asteraceae) was mixed in the *Diervilla* pollen; only 5% to 10% of the foreign pollen was necessary to observe this effect, and 50% *Hieracium* pollen led to almost complete reproductive failure. They pointed out that the allelopathic pollen was produced by species of Asteraceae with allelopathic vegetative parts. Murphy and Aarssen (1989) found that pollen of 1 of the 5 species they tested inhibited growth of 37 of 40 species on which it was tested.

N. Pollen as a Resource

Relatively few studies have quantified pollen as a resource for insects or analyzed its nutritional content. This contrasts strongly with the number of studies quantifying and analyzing nectar. More research is needed in this area, especially as pollen provides significant nutrition for many pollinators.

Petanidou and Vokou (1990) determined the energy available from pollen by using bomb calorimetry. Flowers were covered to prevent insect visits that would contaminate pollen samples. Dehisced anthers were collected and air-dried in small petri dishes until pollen was completely released. Pollen was separated from anther tissue by using a series of plastic sieves of decreasing mesh (236 µm, 132 µm, 95 µm, and 50 µm) and then removing any additional foreign matter with fine forceps under the stereoscope. Pollen was then stored, for no more than 40 days, in corked vials at about 15°C and 40% relative humidity. Preceding calorimetry, pollen was dried (at least 2 days) in a warm (35°C) oven to avoid the loss of volatile oils that would occur at higher temperatures. Pollen was then vacuum-dried for 12 hours, compressed into tablets, and the tablets weighed.

Tablet caloric content was determined with a Phillipson oxygen microbomb calorimeter that was calibrated by burning standard benzoic acid tablets. Correction for platinum fuse wire residue (No. 45C10 fuse wire for Parr oxygen bombs) was 1,400 $cal \cdot g^{-1}$ or 2.3 $cal \cdot cm^{-1}$. Ash-free caloric content of pollen was estimated in $kcal \cdot g^{-1}$ and ranged from 4.4068±0.0539 (S.E., n = 3 or 4)–6.2954±0.0441, mean 5.6105±0.0588.

In addition to providing energy, pollen may contain from 5% to 40% protein (Grogan and Hunt 1979). Honeybee preferences for different species of pollen are strong, and the choice of pollens affects the life span of a bee under laboratory conditions (Schmidt et al. 1987). Both the quantity of protein in the pollen and the amount of pollen of a plant species that are consumed are correlated with increased life span. However, mixtures of pollens are preferred by honeybees and promote the longest life spans.

Pollens also contain several proteolytic enzymes. Based on the quantities of these enzymes in pollen, and in bee guts, Grogan and Hunt (1979) conclude that these enzymes may be physiologically active in digestion of pollen protein by bees. Pollen also contains phagostimulants (Schmidt 1985, Schmidt et al. 1987) that may play a role in bees' ability to discriminate between pollen and nonpollen (Schmidt et al. 1987). When pollen extracts are added to sugar candy, consumption by honeybees increases (Schmidt 1985).

O. Pollen as an Attractant

Pollenkitt, the external lipid coating derived from the tapetum of the anther, provides an adhesive layer that aids in adhesion to insects and stigmas. Dobson (1988) suggests that this pollenkitt may also serve as an attractant or olfactory cue for pollinators. Many pollens are odoriferous, and bees are attracted to isolated pollenkitt. Volatile chemical attractants are generally lipid-soluble, and it is likely that they are localized in the pollenkitt. Pollenkitt may be involved in the pollen specificity of oligolectic bees. Pollenkitt can be isolated and analyzed with TLC or GC, and insect behavior can be examined in the presence of pollenkitt extracts (Dobson 1988).

P. Pollen Presentation Schedules

One of the few studies examining the quantity and time of pollen presentation was published by Percival in 1955. Percival examined 87 plants visited by honeybees and found that anther dehiscence was variable among species; dehiscence could occur in the bud stage or later, and dehiscence of all anthers could be simultaneous or sequential, spread over as many as several days. A single anther might release all pollen at once, over the course of hours, or over days. The time of day that pollen was presented also varied, and she categorized species as:

22 early morning crops
33 chiefly morning crops
 3 midday crops
19 all-day crops
 6 chiefly afternoon crops
 2 afternoon crops
 1 night crop
 1 day and night crop

Bees might arrive at the time of pollen presentation, or there might be a time lag before arrival. Percival determined the amount of pollen (mg) available from each flower per day and throughout the period the flower was open.

The schedule of pollen presentation should have effects on male reproductive success (Thomson and Thomson 1992). Harder and Thomson's (1989) model predicts that if pollinator visitation rates are low, then pollen presentation should be synchronous, whereas if pollinator visitation rates are high, pollen presentation should be staggered. Staggered presentation should result in pollen application to more pollinators and subsequently to several different stigmas. In addition, staggered presentation will prevent weather-related deterioration of the entire investment in pollen. Percival's (1955) data indicate that staggered pollen presentation is common. Some of the forms it may take include pollen packaging (pollinia), gradual "unzipping" of anthers, staggered opening of anthers within a flower or flowers within an inflorescence, and gradual squeezing of pollen from an anther pore (Harder and Thomson 1989). Even nectar production may have an effect on pollen removal rates in those species where large nectar volumes result in longer pollinator visits and greater pollen removal (Harder and Thomson 1989). Contrary to predictions from the Harder and Thomson model, *Erythronium grandiflorum* (Liliaceae) has low visitation rates yet staggers pollen presentation. Thomson and Thomson (1992) used empirical data from *E. grandiflorum* to develop a new model that takes into account pollen viability, as well as visitation rates and pollen depletion rates by different pollinators. Low pollen viability after exposure may account for staggered presentation.

Q. Staining Pollen Components

Potassium iodide (I/KI) can be used to stain starch in pollen grains (Olesen and Warncke 1989a). A stock solution of 80 g KI and 10 g I in 100 mL distilled water can be prepared in advance. When testing pollen, add 2 mL of stock solution to distilled water to a total volume of 100 mL (J. Olesen, *personal communication*). Place pollen in a small quantity of this solution. Starch will stain blue-black.

Auromine O can be used to stain the sporopollenin of pollen grains. A 0.01% solution of stain in tris-HCl buffer at pH 7.2 stains the sporopollenin orange-yellow when viewed under fluorescence (Course manual. 1990. Botany 606-305 lab manual, School of Botany, University of Melbourne).

R. Electrostatic Effects on Pollen Transfer

Electrostatic forces may be involved in the transfer of pollen from anthers to insects and from insects to stigmas (Corbet et al. 1982). Foraging honeybees can carry electrostatic potentials of 450 volts, and localized charges on bee surfaces may exceed this figure (Corbet et al. 1982). Corbet et al. (1982) observed the behavior of pollen grains in an electric field produced by charged electrodes. They tested the effects of moving a charged pin toward a grounded anther impaled on a pin and moving pollen on a charged pin toward an impaled stigma. Finally, they tested the movement of pollen from an anther to a freshly killed bee impaled on a charged electrode, as well as the movement of pollen from the impaled bee to a stigma. Their experiments demonstrated that electrostatic forces can cause pollen to "jump" through air across a distance of about 0.5 mm. Electrostatic forces could increase the chance of pollen transfer when an insect approaches but does not directly contact anthers or stigma.

S. Suggestions for Planning Studies

1. **Effects of pollen load size.** Size (and genetic diversity) of pollen loads on stigmas may or may not (reviewed in Snow 1990) have an effect on the quality of offspring produced by a pollination treatment. Although seed production is usually assumed to be a positive monotonic function of pollen deposition or pollinator visitation, Young and Young (1992) found that not infrequently seed production declines with increased pollen loads.

2. **Potential for significant variation in pollen viability.** Individuals may vary significantly in pollen viability and vigor (e.g., Oni 1990). Some of this variation may be due to differences in the quality of the environment in which plants are growing (Young and Stanton 1990a). Pollen age can also affect viability; an hour or even minutes can make a notable difference (Shivanna and Johri 1985, Corbet 1990, Thomson and Thomson 1992).

3. **Potential for changes in pollen size.** The size and shape of pollen grains can change with the degree of hydration (Stead et al. 1979).

4. **Different results with different tests of pollen viability.** It is probably wise to try several methods and interpret results carefully (N. Waser, *personal communication*). Germination and pollen tube penetration of ovules can occur even when the male nuclei have been lethally irradiated (Pandey et al. 1990). The stylar tract that provides a medium for pollen tube growth is also capable of transmitting nonliving particles (Sanders and Lord 1989).

5. **Effects of age, storage conditions, and degree of hydration on pollen viability.** Pregermination relative humidity probably affects solute potential and wall properties of the pollen grain and has a subsequent effect on pollen germination (Digonnet-Kerhoas and Gay 1990, Shivanna and Johri 1985, Corbet and Plumridge 1985). Any work requiring maximally viable pollen should be conducted shortly after anther dehiscence. Viability differences can also be found among flowers of different age or from different individuals (Stanley and Linskens 1974; Leduc, Douglas, et al. 1990).

5. Nectar

observe the pollination vector
pursuing nectar by the hectare.
equipped with tongue, tarsi, or beak
the vector has such good technique.

it needs no p of .05
to tell it just how not to thrive.
and it would rather be unseen
than be an error of the mean.

it does not pause, it does not tary
it does not use a capillary.
it wants no refractive index
to show it what to visit next.

and though chromatographs unfurled
may tell us "it's amino world"
the vector in innate resolve
is pleased to simply coevolve.

— Candace Galen

Nectar is the primary floral reward for many pollinators. It is produced from phloem sap by active secretion that results in changes in composition from the original fluid. Nectar is primarily a sugar solution with sucrose, fructose, and glucose in varying proportions as major constituents (Baker and Baker 1982). Other sugars are found in some nectars (see Baker and Baker 1982), with maltose and melezitose the next most common and all others occurring in smaller quantities or rare. Many other substances are found in nectars, though sometimes only in trace quantities (Table 5–1). The components of a nectar solution impart a specific taste and odor that may be important in attracting specific groups of

Table 5–1. Nectar Constituents

Monosaccharides	Alkaloids
fructose	Pyrrolizidine alkaloids
glucose	Thiamin
Disaccharides	Riboflavin
sucrose	Nicotinic acid
Oligosaccharides	Pantothenic acid
maltose	Pyridoxine
melezitose	Biotin
raffinose	Folic acid
melibiose	Mesoinositol
Amino acids	Organic acids
Proteins	ascorbic, fumaric, succinic,
Enzymes	malic, oxalic, citric,
Transfructosidases	tartaric, α-ketoglutaric, gluconic,
Transglucosidases	glucuronic
Lipids	Allantoin
Phenolics	Allantoic acid
Glycosides	Dextrin
Inorganic materials	

Sources: Baker and Baker 1982; Percival 1961.

insects (Southwick 1990). The presence or absence of the nonsugar components is also responsible for the differences that humans notice in the flavor and color of honey derived from different plant species (Southwick 1990).

Because nectar functions solely as a reward for pollinators, it is likely to be subject to selection pressures imposed by pollinators (Baker and Baker 1982). Nectar characteristics tend to be similar for plants pollinated by the same animal taxa, resulting in nectar differences among closely related plants with different pollinators (Pyke and Waser 1981, Baker and Baker 1982).

Measurements of nectar volume, composition, concentration, and spatial distribution are common in pollination research. Knowledge of these variables is important in understanding pollinator energetics and nutrient requirements, pollinator behavior and movement in floral patches, and in conducting studies of optimal foraging. Because nectar distribution affects pollinator movements in plant populations, it also affects pollen dispersal and mating patterns in plant populations.

A. Locating Nectaries

Active floral nectaries can be located by staining inflorescences with neutral red (Meeuse 1982). Nectary cells selectively accumulate this stain. However, this will not work for extrafloral nectaries (Southwick, *personal communication*). Dissolve neutral red stain (1:10,000) in distilled water. Submerge the entire inflorescence in the solution for 2–3 hours and then examine it under a microscope. Red spots appear on the surface of nectaries.

Patt et al. (1989) located nectar production sites by dabbing flowers with glucose indicator paper (Lilly Tes-Tape, Glucose Enzymatic Test Strip, USP). Subsequently, tissues suspected of secreting glucose were placed in NaOAc phenylhydrazine-HCl solution for 48 hours, which resulted in formation of osazone crystals. Tissues were examined under the microscope for the presence of crystals. Prepare phenylhydrazine reagent by mixing 2.2 gm of phenylhydrazine hydrochloride and 3.3 gm sodium acetate in 22 mL of distilled water and heating to 60°C (Morrow and Sandstrom 1935). Cool the solution before use. In the presence of this reagent, glucose is converted into glucosazone in the form of yellow crystals. Other hexoses and pentoses also form yellow or orange crystals.

B. Nectar Volumes and Standing Crop

The standing crop of nectar is the quantity and distribution of nectar determined by randomly sampling patches of flowers (that have not been protected from pollinators by bagging). It is a measure of resource availability at a single point in time (Possingham 1989). Standing crop is usually presented as mean nectar per flower (sometimes including the standard deviation), but only rarely is the frequency distribution or spatial distribution of the nectar described (Pleasants and Zimmerman 1983). Possingham warns that standing crop may not equal encountered crop (the quantity and distribution encountered by a forager) in the presence of systematic foragers.

Measures of standing crop increasingly underestimate encountered crop as foragers become more systematic. Pleasants and Zimmerman (1983) state that the amount of nectar present in a flower at any time (standing crop) is equal to the product of the production rate multiplied by the time elapsed since the last visit.

$$\text{Mean standing crop} = t \cdot \frac{(npr)}{p},$$

where npr = average nectar production rate; t = handling time + travel time between flowers; and p = insect to flower ratio.

The variance of the mean reduces to the term

$$(1-p) \cdot \left(\frac{t\,(npr)}{p}\right)^2 \quad \text{or} \quad (1-p)(\text{mean})^2.$$

Because p is usually small, 1-p approximates 1 and thus the mean standing crop approximates the standard deviation. Empirical results show this is roughly true, but the standard deviation is slightly larger than the mean in most cases. This is explained by variability in nectar production rate among flowers, by nonrandom foraging by pollinators (Pleasants and Zimmerman 1983), and also by effects of evaporation (Southwick et al. 1981).

Nectar distribution can be patchy within plants or within a population, but relatively little is known about the level at which patchiness actually occurs (Shmida and Kadmon 1991). Patchiness can have a large effect on pollinator foraging movements. Southwick (1982a) measured the available nectar in three species of flowers and determined that late in the day when most flowers were depleted, some flowers contained volumes twice the average for nectar samples throughout the day. These flowers, designated "lucky hits," could keep pollinators active, as foraging remains energetically rewarding. To support this premise Southwick (1982a) presents data on the caloric expenditure of foraging bees and the energetic rewards provided by these "lucky hits." Southwick (1983) even detected patchiness within single *Asclepias* flowers. Some cuculli consistently contained more nectar even when pollinators were excluded, indicating that some nectaries produced at a higher rate (Southwick 1983).

At the population level Pleasants and Zimmerman (1979) discovered patchiness in *Delphinium nelsonii* (Ranunculaceae) nectar dispersion. Inflorescences with nectar were called "hot," those without "cold," and a nearest-neighbor analysis indicated the presence of hot and cold patches of plants. Data were analyzed with a chi-square test for independence using a 2 x 2 contingency table:

Subject Inflorescence

        hot   cold

2 Nearest Neighbors
at least 1 hot

both cold

Waser and Mitchell (1990) analyzed populations of the same species, *Delphinium nelsonii,* and were unable to demonstrate the presence of hot and cold spots. They used a spatial autocorrelation statistic (Moran's *I*) to analyze standing crop data from transects through populations. They also used a contingency table analysis similar to that of Pleasants and Zimmerman (1979) for comparison. Spatial autocorrelation is somewhat more realistic in that nectar volume is treated as a continuous rather than discrete variable. In addition it permits pattern characterization on several spatial scales (Waser and Mitchell 1990). They suggest avoiding contingency analysis in favor of autocorrelation.

Autocorrelation statistics are used to examine the dependency of a character at one location on the character state at another location, to analyze patterns in space (Sokal and Oden 1978). Character data can be nominal, interval, or ranked data. The null hypothesis states that the autocorrelation between the character at locations A and B = 0. Biologists may be more familiar with the use of autocorrelation statistics for examining genotype or gene frequency patterns within a population.

Sampling nectar volumes or production rates is not as straightforward as one may think. Secretion may change diurnally and with flower age (Pyke 1978a, Corbet 1978a, Southwick and Southwick 1983, Possingham 1989, Búrquez and Corbet 1991). Brink and deWet (1980) studied nectar production in relation to flower age by dividing flowers into five "age classes" based on the percentage of anther dehiscence. Nectar volumes from the two floral nectaries were averaged for flowers in each age class. They regressed these volumes on mean population nectary depths for each age class and the results indicated that production is also correlated with nectary depth. Pyke (1978a) showed that nectar production increased with age (relative position on the stalk) by a factor of 1.6 in *Aconitum*, 2.3 in *Delphinium barbeyi*, and 7 in *D. nelsonii* (all Ranunculaceae) (see also Cruden and Hermann 1983). Although nectar production is not affected by flower age in some other species (e.g., *Ipomopsis aggregata* [Polemoniaceae]; Pleasants 1983 and references therein) production rate may vary over the course of the season. In addition to these temporal effects, spatial differences also complicate sampling. Flowers within a patch and even on a single plant may vary in nectar production. Southwick and Southwick (1983) determined that nectar production in *Asclepias syriaca* (Asclepiadaceae) was age dependent. Because flower anthesis within an umbel was asynchronous, pollinators would usually be rewarded by high volumes of nectar in several flowers. Feinsinger (1983), working in Trinidad, determined that the average *Heliconia psittacorum* (Heliconiaceae) flower produced 66.4 μL of nectar, but of 215 flowers, 29 secreted less than 10 μL and 24 secreted more than 120 μL.

Another factor that may confound studies of nectar production is reabsorption of nectar. Reservoirs of nectar sitting in flowers may be reabsorbed into nectaries and sometimes reincorporated into a more concentrated nectar (Corbet 1978b). Although there is no reabsorption in some species (Pleasants 1983, Búrquez and Corbet 1991), it is significant in others (Búrquez and Corbet 1991). Búrquez and Corbet suggest that nectar may not be reabsorbed if it is stored in an area remote from the nectary or if nectaries are abscised along with the corolla shortly after fertilization. They used removal of nectar solutes as a measure of reabsorption so they would not have to account for changes in nectar volume due to condensation or evaporation. "Apparent reabsorption" can be determined from the difference between the apparent secretion rate (rate of change of solute content of nectar in undisturbed flowers) and gross secretion rate (rate of change of solute content of nectar of repeatedly sampled flowers). This is an approximation of the true reabsorption rate (the rate of net solute loss from unvisited flowers). Southwick (*personal communication*) cautions that this may not hold if measures of gross secretion are affected by the sampling rate.

In some flowers, surfactants on the surface of the nectar pool retard nectar evaporation (Corbet and Willmer 1981). The presence of a surfactant can be demonstrated with a tiny grain of camphor. When the camphor is placed on a water surface without surfactant, the grain "dances." If surfactant is present, the grain is motionless (Corbet and Willmer 1981).

Nectar secretion often shows a daily peak (Pleasants 1983, Búrquez and Corbet 1991). Apparent secretion is probably affected by irradiance, temperature, and water balance, and therefore a noon or midafternoon peak for secretion is fairly common (Pleasants 1983, Búrquez and Corbet 1991). However, this pattern is not universal. Southwick and Southwick (1983) found peak nectar production in *Asclepias syriaca* flowers occured 50 hours after anthesis. And in plants that rely on carbohydrate reserves, or immediate photosynthesis, severe midday water deficits may reduce translocation of photosynthate or reduce total midday photosynthesis and shut down nectar secretion (Búrquez and Corbet 1991). Nocturnal nectar secretion, often found in CAM (crassulacean acid metabolism) plants (including Crassulaceae, Cactaceae, and Agavaceae), may be linked with stomatal opening and translocation occurring at night (Búrquez and Corbet 1991). Nocturnal secretion is also found in several bird-pollinated plants and may be a result of competition for early-morning pollinators (Pleasants 1983).

Gill (1988) found that periodic nectar harvesting from some tropical hummingbird flowers appears to increase total nectar yield. Early-morning sampling increased this effect. Other studies have also shown that total nectar production from flowers sampled repeatedly often exceeds the volume from flowers sampled once at the end of the same period, indicating that removal of nectar stimulates production (Corbet 1978b). However, Pleasants (1983) found no significant differences in nectar production between *Ipomopsis* flowers that were repeatedly sampled and flowers sampled once at the end of a 27-hour period.

Although there appears to be a strong genetic component to nectar production characteristics (Pedersen 1953, Hawkins 1971), microenvironment also exerts effects. Low temperatures and low light intensity may decrease nectar production, so sampling under different weather conditions may bias estimates of nectar production rates (Pleasants 1983). Wyatt et al. (1992) found a twofold variation in nectar volume before and after watering *Asclepias* plants with the equivalent of 10 cm of rain. Nectar concentration did not change substantially, so overall sugar production also increased. They suggest that researchers consider watering plants before collecting nectar if large volumes of nectar are needed for experimental analyses.

When measuring nectar production rates, researchers often enclose flowers in mosquito netting, gauze, paper bags, or cages, or plug flowers with cotton balls that prevent insect visitation as well as evaporation (Marden 1984a). Cruden and Hermann (1983) found that paper pollination bags raised the temperature around the flower 6°–14°C, which is enough to alter nectar production (see Chapter 3, Section B). Mosquito netting had the least effect on floral temperature. The value of exclosures should therefore be weighed against their tendency to alter nectar production by changing temperature and humidity in the flower (Corbet 1978a, Cruden and Hermann 1983). To control for any cage effects, Armstrong (1991) covered plants in one treatment group with an open-ended cage to show that the effects on nectar production were from the exclusion of pollinators and not from artifacts of caging.

1. Collection Techniques

Collecting nectar for quantitative and/or qualitative analyses is a bit of an art. If you are fortunate enough to be working with large flowers that have open corollas and lots of nectar, you're in luck. But if you are working with minute flowers, or composite flower heads such as Asteraceae, you will need patience and dexterity. Small volumes (measured in microliters) and restricted working space are the primary problems you will encounter. Some of the techniques described below may help. Most of these involve working with flowers in the field, but you may also get satisfactory results by bringing flowers back to the lab and placing them in water to prevent wilting (see Chapter 7, Section K.3.a.iii, for tips on preventing wilting). If you are doing physiological studies on nectar secretion, you will probably need to work with plants in controlled conditions in an environmental chamber. We present a variety of sampling techniques here, as different floral morphologies or sampling regimes will call for different techniques.

a. Capillary tubes

Nectar is commonly extracted from flowers with microcapillary tubes that collect the nectar by capillary action. Tubes of 1–20 µL volume work well for most flowers. You can use tubes up to 50-µL volume if you are careful to prevent nectar from running out. To be assured of collecting all the nectar, you should use a bulb on this size tube to create suction (Southwick, *personal communication*). Compute nectar volume with this equation (Cruden and Hermann 1983):

$$\frac{\text{mm of nectar in the pipette}}{\text{mm total length of tube}} \cdot \text{calibrated volume of pipette} = \text{volume of nectar.}$$

This calculation assumes a constant bore size in the pipette, which micropipettes of this small size seem to have. You can purchase calibrated micropipets (Clay Adams Accufill 90 Micropipet — see Appendix 2; Southwick et al. 1981) that attach to an aspirator to ensure complete nectar collection.

There are some problems associated with the capillary extraction technique (Roberts 1979, McKenna and Thomson 1988). The pipette may not collect all the nectar and may damage the flower, resulting in an inaccurate estimation of nectar production or incorrect qualitative analyses due to impurities from the damaged flower. Capillary action alone is often insufficient to remove viscous nectar of high sugar concentration (Southwick, *personal communication*). (Zimmerman and Pyke [1988c] found that by adding 40 µL of water to very viscous nectar and leaving it for 10 minutes they were able to extract the diluted nectar with capillary tubes and then correct for the added water.) Nectar removal from small flowers is difficult, and use of this method on multiple flowers is tedious. Listed below are alternative techniques.

b. Syringes

Willson and Bertin (1979) measured nectar volumes to the nearest 0.1 µL with calibrated syringes. Pleasants and Chaplin (1983) collected milkweed nectar by inserting a 10-µL Hamilton microsyringe (blunt needle) into each of the five floral hoods (milkweed flowers have five nectaries that secrete into a central chamber that funnels into the staminal hoods). The volume of nectar collected is read from the syringe, or by transferring nectar to a calibrated capillary tube if air bubbles interfere with direct readings. A Hamilton microsyringe can be fitted with a narrow polyethylene tubing tip attached with superglue. The tubing is narrow enough to enter the flower without damage to the nectary and removes the detectable nectar (Carpenter 1983). However, Southwick (*personal communication*) has discovered bees collecting additional nectar from flowers sampled with this method.

c. Centrifugation

Swanson and Shuel (1950) collected nectar by centrifuging flowers. They trimmed flowers to facilitate nectar extraction, then placed them in calibrated centrifuge tubes. A stiff wire support and an insect pin inserted into the cork stopper held the flower in place. These authors centrifuged red clover flowers at 2,540 rpm. The radial distance to the tubes was 7 cm, so flower heads were subjected to 465 G. Nectar volume was read from the height of liquid in the calibrated tube after 6 minutes of centrifugation. Centrifugate increased after 6 minutes, but the additional liquid was probably from the flower rather than nectar. Because inclusion of dew and sepal guttates will increase the volume of centrifugate, flowers to be centrifuged should be collected at a time of day that will minimize these sources of error. Small insects present on flowers may be excluded from the liquid with a mesh screen.

In a variation on this centrifugation technique, Furgala et al. (1976) extracted nectar from sunflower florets by cutting "floral plugs" from sunflower heads with a cork borer. Corollas of open florets were trimmed and unopened florets removed. The floral plug was fastened to a 64-mm cork backing by pushing the shank of a three-pronged fishhook through the cork and then the plug until the hooks gripped the cork. Masking tape wrapped around the plug added support. Plugs were centrifuged at 1,600–1,800 rpm for 10 minutes in Bauer-Schenk calibrated sedimentation tubes with 80-mesh copper screen to exclude particulates from the centrifugate. Nectar volumes were recorded and percent concentration determined with a refractometer. Using this procedure, Furgala et al. found volumes of 35–86 µL per 100 florets in commercial sunflowers.

These methods may provide reasonable quantitative estimates of nectar volumes, but the samples they provide are probably not suitable for qualitative analyses, as they will probably be contaminated with cell contents of cells damaged during sample preparation.

Armstrong and Paton (1990) used centrifugation as one means of sampling nectar from *Banksia* (Proteaceae) inflorescences, which have numerous tiny flowers that are closely packed together. An entire inflorescence was placed in a sturdy plastic bag, and a 1-m rope was tied around the opening. The bag was swung for two 15–20-second bouts and nectar accumulated in a corner of the bag. The nectar could be removed from the bag with a pipette or capillary tube. A film of nectar remained on the inflorescence, and further sampling with an aspirator removed additional nectar. However, the technique was very useful for qualitative comparisons of the energy available in different inflorescences.

d. Wicks

Filter-paper wicks can blot small quantities of dilute nectar, or viscous nectar when capillary tube extraction is difficult (McKenna and Thomson 1988; Thomson, McKenna, and Cruzan 1989; Harder and Cruzan 1990). Cut uniform-sized wicks from Whatman number 1 filter paper with an insect-pinning point punch (the shape is triangular, or ovoid tapering to a point, 2 x 8 mm; Thomson, McKenna, and Cruzan 1989). Forceps, or an insect pin pushed through the point, aid in manipulating the point into the flower. A sampling design sheet can be taped to a styrofoam block so pinned wicks can be pushed into the appropriate position for later analysis. Alternatively, wicks can be air-dried and stored in cell culture trays for further analysis (Ashman and Stanton 1991). Several wicks can be used in flowers with abundant nectar. In some cases, a wick can be left in place to sample a nectary completely. Filter-paper wicks can also be used to remove nectar in order to facilitate experimental manipulations with nectar. Thomson (1986) removed nectar with filter-paper wicks and then used a Hamilton dispensing syringe to introduce sucrose solutions of known volumes and concentrations.

If nectar samples collected with wicks will be used for amino acid analyses, be careful not to contaminate the paper by touching it with your fingers. Use only clean forceps or pins to handle the wicks.

Filter paper can also be used for a semiquantitative analysis of nectar volume. DesGranges (1979) squeezed floral nectar onto filter paper and compared the size of the nectar spot to a series of spots produced from known volumes of 20% sucrose. Núñez (1977) determined nectar quantity by weighing a flower and then placing filter-paper strips into the flower and squeezing the nectar onto the filter paper by rolling the flower over a piece of 0.5-mm polyethylene cannula. Nectar quantity was determined by the difference in flower weight after nectar removal. Roberts (1979) claims that all the sugar may not be removed by this method and that excess squeezing can release sap, causing overestimation of sugars.

To get an approximate idea of the amount of nectar in a small flower such as a composite floret, you may want to try the "dipstick" method (J. Thomson, *personal communication*). Mix library paste with cobalt chloride. Rub this into cotton sewing thread. Bake or dry the thread in a desiccator and then cut it into 1-cm lengths. Store the pieces in a vial with a package of silica gel desiccant. To use a dipstick, insert the stiff piece of thread into the floret. The tip will turn color to the depth of the nectar. Cobalt chloride changes from pink to blue with changes in relative humidity.

e. Rinsing flowers

You can rinse nectar from flowers with distilled water (Núñez 1977). Weigh a micropipette, fill it with about 20 µL of distilled water, and reweigh it (Núñez draws his own micropipettes from thin-walled capillary tubes). Attach a plastic cannula fitted to a mouth adapter to the micropipette and expel water over floral nectaries. Draw up the rinsing fluid and force it over the nectaries again two more times, without removing the micropipette from the flower. This can be repeated on several flowers with the same rinse water. After the final flower is rinsed, weigh the micropipette and determine the solution concentration with a refractometer. Divide the sugar weight by the number of flowers rinsed. The mean error created by losing small amounts of rinsing fluid can be estimated by rerinsing the same flowers (nectar removed) with distilled water following the same procedure. Additional fluid gained from "dry" runs showed that Núñez lost an average of 0.4 µL per flower.

Cresswell and Galen (1991) found that nectar extraction with capillary tubes damaged *Polemonium viscosum* (Polemoniaceae) flowers, preventing multiple readings necessary for estimating the nectar production rate. They developed another technique for rinsing nectar from flowers: place a fine-gauge hypodermic needle (25 gauge, 16 mm) deep into the corolla without puncturing tissues. Flush 4 mL of water through the corolla and collect it through a funnel under the corolla lip. (To increase the water pressure attach the syringe to a handpumped garden sprayer. Several meters of plastic tubing between the portable sprayer and syringe allow maneuverability in the field.) Cresswell and Galen tested this method by rinsing a sample of flowers twice to see if additional sugar was removed. A single rinse removed all sugar. They also tested whether the procedure removed nonnectary sugars by forcing a small (0.3 mL) quantity of water directly over the nectaries of cut flowers. This "mini-rinse" was repeated 10 times, and then 4 mL of water were forced into the corolla. Little or no sugar from the stigma or from pollen grains was found in the 4 mL of water.

f. Homogenization

Núñez (1977) extracted nectar by placing the nectar-bearing region of flowers together with 30 µL of distilled water into a homogenizer. He measured volume and concentration of the solution and calculated the original nectar volume and concentration. The quantity of sugar extracted was determined with this equation:

$$M_{az} = \frac{30 \cdot C_d}{1 - \dfrac{C_d}{C_c}},$$

where
 M_{az} = quantity of sugar in mg;
 C_d = concentration of extracted solution;
 C_c = concentration of original nectar;
 30 = total volume of distilled water and flowers (µL).

Roberts (1979) points out that homogenization may result in contamination of the nectar sample from other floral tissues.

g. Micrometer measurements

Cartar (1991) used a micrometer to measure nectar levels in minute seablush flowers. The height of the nectar was measured to the nearest 0.1 mm through the semitransparent spur. The volume was subsequently calculated from an empirically derived regression equation. Kato et al. (1991) also determined the relationship between nectar level and volume of *Impatiens* (Balsaminaceae) flowers and simply backlit flowers to measure nectar level. This is a noninvasive sampling technique that allows nectar levels to be measured repeatedly.

2. Rate of Secretion

Some authors measure the rate of nectar secretion by emptying all nectar from the same set of flowers at regular intervals. These flowers may be bagged to exclude insects (but see cautions listed above). Schemske (1980a) bagged flowers in the bud stage with dialysis tubing. As flowers opened, they were sampled at dawn and at 4-hour intervals until dusk. Southwick and Southwick (1983) determined that most nectar production of *Asclepias* flowers occurred between midnight and 6 A.M. and used that information in designing their sampling scheme. Flower buds were bagged with mesh to exclude pollinators and then observed to determine the time of anthesis. Thus only flowers of a known age were marked for sampling. Flowers were sampled once a day at either 7 A.M. or 8 A.M. (at two different sites). Nectar samples were collected daily for 7 days and production patterns were clearly revealed.

Instead of repeatedly sampling the same flower, Corbet (1978a) sampled each flower once and marked flowers to avoid subsequent sampling. Flowers were not bagged. Estimates of nectar secretion thus reflected the amount of nectar available to insects, and error due to rapid replenishment of drained flowers was avoided. The absolute quantity of nectar secreted by unprotected flowers cannot be determined by this method because visitors may be removing nectar. However, nectar in protected flowers can be reabsorbed, creating the same problem. It is important to weigh these factors when designing a sampling technique. If you are interested in the volume of nectar available to insects, your sampling scheme will differ from a scheme designed to estimate maximum nectar output.

Because a single insect visit may not deplete nectar in a flower, it is wise to remove floral nectar by hand when determining secretion rates, rather than assuming that a recently visited flower is empty. Zimmerman (1983a) found that the residual nectar left after bumblebee visits to *Aconitum columbianum* was quite variable and often represented a large fraction of the total daily nectar production. Hodges and Wolf (1981) determined that queen bumblebees left nectar in *Delphinium nelsonii* flowers with large volumes but depleted flowers with lower nectar volumes.

3. Cost of Nectar Production

There is still some uncertainty about how significant the cost of nectar production is in a flowering plant's energy budget. The few studies that have examined this have found a wide range of values, but there is some evidence (Harder and Barrett 1992) to suggest that for most bee-pollinated plants, nectar production involves little energetic cost. In some cases this cost may be met in large part by green flowers, which may produce most of the energy and carbon required for flower structure and seed production (Bazzaz et al. 1979, Pyke 1991).

Southwick (1984) sampled nectar from 1,500 *Asclepias syriaca* flowers and then calculated the energy value of nectar secreted per day by all flowers on a single stem. He measured daily photosynthesis and then expressed the energy in nectar as a percentage of photosynthesis per ramet. Energy in nectar ranged from 4.3% to 36.6% of daily photosynthetic assimilate. Minimum estimates for the total energy in nectar divided by the total photosynthate of the entire growing season ranged from 0.7% to 2.8%. Southwick (1984) also calculated that the energy contained in the total nectar production of alfalfa blossoms was almost twice the amount of energy contained in the seeds and about 10% of the energy contained in the alfalfa forage. Pleasants and Chaplin (1983) state that nectar production exhausts about 30% of the energy devoted to flowering in *Asclepias quadrifolia*. Thus for some species nectar production can be a significant energy sink. Harder and Barrett (1992) calculated that nectar constitutes only 3% of the energy content of *Pontederia* (Pontederiaceae) flowers.

167

Pyke (1991) demonstrated that the cost of nectar production in *Blandfordia nobilis* (Liliaceae) can be measured in terms of reduced seed production. Pyke hand-pollinated flowers from two treatment groups: (1) flowers that were sampled repeatedly for nectar and (2) flowers that were probed with plugged (dummy) capillary tubes so that no nectar was removed. Flowers were bagged to prevent insect visits. *Blandfordia* flowers that are sampled repeatedly produce more total nectar than flowers sampled once at the end of the same total time period. Seed production of repeatedly sampled plants was significantly lower than that of plants that did not have nectar removed. Nectar attracts pollinators that effect seed set and pollen dispersal. However, there must be some point at which further investment in nectar production is limited by reduced energy available for seed set (Pyke 1991).

4. Determining Organic Carbon Content
Carbon content of nectar samples can be analyzed with a total carbon analyzer. Elmqvist et al. (1988) used a Beckman Model 915-B Total Carbon Analyzer and compared sample results with organic (potassium biphthalate) and inorganic (sodium bicarbonate) standards.

C. Sugar Concentration

Sugar concentration varies among species and many consider it adapted to pollinator type. However, ratios of sucrose to hexose (Baker and Baker 1979, Southwick 1982b) and nonsugar components of nectar may be more important in determining which pollinators are attracted to a particular type of nectar (Southwick 1982b, 1990). For example, *Crataegus* (Rosaceae) flowers have a very high sugar concentration (48%) and are visited by bee flies and wasps (Southwick 1982b, 1990). Honeybees are not attracted to these flowers despite the high sugar concentration. Plowright (1987) finds that corolla tube characteristics are important in determining nectar concentration. Nectar evaporates faster from flowers with short corolla tubes, increasing concentration. In experiments with *Aquilegia* (Ranunculaceae) he cut 4 x 1 mm–windows in four of the five floral spurs, at distances of 1–4 mm from the distal end and measured the effect on nectar concentration. His results led him to suggest that any adaptation of nectar concentration to pollinator type is secondary to effects produced by corolla tube length. According to this hypothesis, the watery nectar of hummingbird flowers is a reflection of corolla size, as hummingbird flowers have deep corollas (Plowright 1987).

Because of evaporation and changes in production rates, nectar concentration within a single flower can vary over the course of the day (Corbet 1978b; Corbet, Unwin and Prŷs-Jones 1979; Corbet et al. 1979). If you plan to collect nectar samples and read concentrations at a later time in the lab, Reader (1977) suggests collecting nectar in microcapillary tubes to minimize evaporation. The tubes can be sealed with Crit-O-Seal (see Appendix 2).

Sugar concentration of nectar is most commonly measured with a refractometer in the field. Laboratory methods for measuring concentration include the spot-staining technique (Baker 1979) and the anthrone method (McKenna and Thomson 1988).

1. Refractometers

The easiest way to determine sugar concentration is to use a light refractometer, which measures the refractive index of a liquid sample. Park (1933) was the first to propose this method as an easier and more accurate technique than the microchemical methods that had been used previously. The refractive index of nectar is used as a measure of sucrose equivalents, as this method does not take into account the variety of sugars present. Glucose and fructose have similar effects on the refractive index (Baker and Baker 1982), but their presence in nectar introduces a small error (Corbet 1978a; Southwick, *personal communication*). Hainsworth and Wolf (1972a) point out that the refractive indices of equimolar concentrations of glucose and fructose are approximately half that of equimolar sucrose, so the refractive index can be used to determine the caloric value of nectar. A 1-molar sucrose solution contains 1,349.6 cal/mL = 56.41 joules/mL (1 mg sucrose provides 16.48 joules), and using this value introduces an error of only 3% to 4% for glucose and fructose solutions.

Refractometer readings are usually not modified for the presence of hexose sugars in nectar, but Southwick adds 1.41 to the temperature-corrected refractometer reading for solutions high in hexose (e.g., honey; Southwick, *personal communication*). The presence of amino acids, electrolytes, and other materials in nectar also contributes to the refractive index of the solution (Inouye et al. 1980, Hiebert and Calder 1983) and introduces further error. Some refractometers are temperature compensated, and others need to be corrected for temperature. Those that are not temperature compensated sometimes have a thermometer attached, which is calibrated to provide the necessary correction factor.

Several brands and models of refractometers are available (Bausch and Lomb, Zeiss, Leica [formerly American Optical], Bellingham and Stanley — see Appendix 2) differing in range, weight, and price. Cruden and Hermann (1983) caution that all refractometers should be checked with a known sugar concentration to confirm accuracy. You may have to use a pair of refractometers to cover the range of sugar concentrations you encounter in nectars, although some models have multiple ranges built in. Most models require relatively large samples to fill the gap between the prism and sample-bearing surface. Some (e.g., Leica) can be modified by adjusting the hinge to work with smaller volumes. Filing down the acrylic cover plate reduces the angle between the plate and the glass, leaving a smaller space for the sample. The angle of light refraction is independent of the distance through the plastic plate, so this should not affect your reading (Southwick, *personal communication*). Bellingham and Stanley make and modify refractometers for volumes less than 0.5 μL.

Cruden and Hermann (1983) found that refractometers measuring sugar concentrations in the range of 0% to 50% are useful for most nectars, but those measuring from 35% to 75% or 80% will be more useful if you are dealing with bee-pollinated flowers that often have concentrations in the higher range. Nectar can be diluted with distilled water (measured carefully with a micropipette) should concentration exceed the range of the instrument.

Most refractometers measure the percent of sucrose or sucrose equivalents on the BRIX scale (BRIX scale = weight of sugar per volume of solution at a given temperature; after A.F.W. Brix, nineteenth-century German inventor). Many authors then calculate the amount of sugar present in the nectar by multiplying the volume extracted from the flower (in μL) by the percentage (mg sugar/mg solution) as read on the refractometer (Pleasants and Chaplin 1983).

This method of calculation introduces a small error at low percentages, but an increasingly larger error at high percentages (Bolten et al. 1979). The error should be corrected by converting mg sugar/mg solution to a more useful measure, mg sugar/100 mL nectar (Bolten et al. 1979; Cruden and Hermann 1983; Southwick, *personal communication*). Total sugar per sample is then calculated by multiplying mg sugar/mL by total nectar volume. A conversion table is duplicated in Table 5–2.

When nectar quantities were very small or viscous, Corbet et al. (1979) determined the amount of sugar per flower from the change in concentration when a μL of distilled water was pipetted into the flower, mixed with the nectar, and removed for a reading. With this technique, however, it is important not to leave the water in the flower for too long, as the amount of sugar released can increase two to six times if left for as long as 1 hour.

2. Spot-Staining Technique

Irene Baker (1979) developed a spot-staining technique for estimating volume and sugar concentration of small volumes of nectar. Nectar was collected in drawn-out micropipettes. Small known volumes of 10% and 50% sugar (w/w) were spotted onto Whatman number 1 chromatography paper, and the diameter of the spots was measured under a magnifier. Both concentrations produced spots of similar diameter when volumes were small (diameter of spots ≤ 12 mm. Diameters and their corresponding volumes were determined from a regression of the empirical data and the results tabulated (Table 5–3).

To determine sugar concentration of a sample:

1. Determine the volume of nectar from the diameter of the spot.
2. With a paper punch, punch out the piece of chromatography paper bearing the nectar spot and a second piece bearing no nectar as a control.
3. Put each paper spot in a 75 x 10–mm test tube filled with 50 μL of distilled water for 30–60 minutes.
4. Add 150 μL of PABA (p-aminobenzoic acid) stain, mix, and allow to dry (a vacuum desiccator will speed this step). Prepare PABA stain immediately before use by mixing

Table 5–2. Conversion table for changing % sugar
(g sugar/g solution) to g sugar/L nectar

% Sucrose	g/L	Molarity	% sucrose	g/L	Molarity
0.50	5.0	0.015	24.00	263.8	0.771
1.00	10.0	0.029	26.00	288.1	0.842
1.50	15.1	0.044	28.00	312.9	0.914
2.00	20.1	0.059	30.00	338.1	0.988
2.50	25.2	0.074	32.00	363.7	1.063
3.00	30.3	0.089	34.00	389.8	1.139
3.50	35.4	0.103	36.00	416.2	1.216
4.00	40.6	0.118	38.00	443.2	1.295
4.50	45.7	0.134	40.00	470.6	1.375
5.00	50.9	0.149	42.00	498.4	1.456
5.50	56.1	0.164	44.00	526.8	1.539
6.00	61.3	0.179	46.00	555.6	1.623
6.50	66.5	0.194	48.00	584.9	1.709
7.00	71.8	0.210	50.00	614.8	1.796
7.50	77.1	0.225	52.00	645.1	1.885
8.00	82.4	0.241	54.00	838.6	1.975
8.50	87.7	0.256	56.00	707.4	2.067
9.00	93.1	0.272	58.00	739.3	2.160
9.50	98.4	0.288	60.00	771.9	2.255
10.00	103.8	0.303	62.00	804.9	2.352
11.00	114.7	0.335	64.00	838.6	2.450
12.00	125.6	0.367	66.00	872.8	2.550
13.00	136.6	0.399	68.00	907.6	2.652
14.00	147.7	0.431	70.00	943.1	2.755
15.00	158.9	0.464	72.00	979.1	2.860
16.00	170.2	0.497	74.00	1015.7	2.967
17.00	181.5	0.530	76.00	1053.0	3.076
18.00	193.0	0.564	78.00	1090.9	3.187
19.00	204.5	0.598	80.00	1129.4	3.299
20.00	216.2	0.632	82.00	1168.5	3.414
22.00	239.8	0.701	84.00	1208.2	3.530

Source: Adapted from the CRC Handbook of Chemistry and Physics 1978–1979, Weast 1978.

Table 5–3. Volumes (μL) corresponding to nectar sample spot diameters (mm)

mm	μL	mm	μL	mm	μL
0.1	.01	3.7	.525	7.3	2.605
0.2	.02	3.8	.550	7.4	2.690
0.3	.03	3.9	.575	7.5	2.775
0.4	.04	4.0	.600	7.6	2.860
0.5	.05	4.1	.650	7.7	2.945
0.6	.06	4.2	.700	7.8	3.030
0.7	.07	4.3	.750	7.9	3.114
0.8	.08	4.4	.800	8.0	3.200
0.9	.09	4.5	.850	8.1	3.300
1.0	.10	4.6	.900	8.2	3.400
1.1	.11	4.7	.950	8.3	3.500
1.2	.12	4.8	1.000	8.4	3.600
1.3	.13	4.9	1.050	8.5	3.700
1.4	.14	5.0	1.100	8.6	3.800
1.5	.15	5.1	1.160	8.7	3.900
1.6	.16	5.2	1.220	8.8	4.000
1.7	.17	5.3	1.280	8.9	4.100
1.8	.18	5.4	1.340	9.0	4.200
1.9	.19	5.5	1.400	9.1	4.310
2.0	.20	5.6	1.460	9.2	4.420
2.1	.275	5.7	1.520	9.3	4.530
2.2	.230	5.8	1.580	9.4	4.640
2.3	.245	5.9	1.640	9.5	4.750
2.4	.260	6.0	1.700	9.6	4.860
2.5	.275	6.1	1.765	9.7	4.970
2.6	.290	6.2	1.830	9.8	5.080
2.7	.305	6.3	1.895	9.9	5.190
2.8	.320	6.4	1.960	10.1	5.450
2.9	.335	6.5	2.025	10.2	5.600
3.0	.350	6.6	2.090	10.3	5.750
3.1	.375	6.7	2.155	10.4	5.900
3.2	.400	6.8	2.220	10.6	6.200
3.3	.425	6.9	2.285	10.7	6.350
3.4	.450	7.0	2.350	10.8	6.500
3.5	.475	7.1	2.435	10.9	6.650
3.6	.500	7.2	2.520	11.0	6.800

mm	μL	mm	μL	mm	μL
11.1	6.990	11.5	7.750	11.8	8.320
11.2	7.180	11.6	7.940	11.9	8.510
11.3	7.370	11.7	8.130	12.0	8.700
11.4	7.560				

Source: Baker 1979.

solution A (75 mg oxalic acid in 15 mL ethanol) with solution B (150 mg p-aminobenzoic acid in 25 mL chloroform and 2 mL glacial acetic acid).

5. When paper is dry, heat it for 10 minutes at 100°C and a brown color develops.
6. Redissolve the spot in 1 mL methanol:water (1:1 v/v) and collect solutions in semi-micro cuvettes.
7. Use a spectrophotometer to read absorption at 470 nm, using the control paper as a blank. Compare with a standard calibration curve to determine the concentration (w/v).

3. Photometric Procedures

Thomson et al. (1989) point out that the capillary/refractometer method doesn't work well for nectar volumes less than 1 μL. Only refractometers with closely set prisms will yield clear concentration readings from such small volumes. (Southwick [*personal communication*] is able to get clear readings with volumes as small as 0.1 μl using American Optical [now Leica] refractometers). The anthrone technique is useful for small volumes. The method is more time-consuming than using a refractometer and requires more elaborate equipment, but it is suitable for flowers with small amounts of nectar and can be performed in the lab at your convenience.

The anthrone method is based on the reaction of anthrone reagent with fructose to produce a solution whose optical density can be measured with a spectrophotometer. Anthrone and fructose react at room temperature, and optical density of the resulting solution is proportional to the amount of fructose present. To measure amounts of sucrose, a disaccharide of glucose and fructose, one can hydrolyze the disaccharide and then measure the resulting

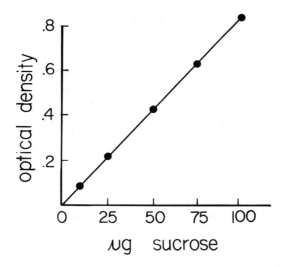

Figure 5–1. Calibration curve for calculating μg sucrose. Redrawn from Van Handel 1968.

fructose concentration. But the presence of free fructose in addition to that resulting from hydrolysis would confound the determination of sucrose, so it must first be removed; this is accomplished by treatment with hot KOH. The hydrolysis is then effected by the addition of acidic anthrone reagent, and the amount of fructose is measured.

Preparation of sample (from Van Handel 1967, 1968)

1. Collect nectar samples (two are needed).
2. Prepare anthrone reagent (400 mg anthrone in 200 mL concentrated H_2SO_4).
3. Evaporate the nectar sample down to 0.1 mL in a test tube at about 120°C.

Determination of sucrose component

4. Add 0.1 mL 30% aqueous KOH, maintain at 100°C for 10 minutes.
5. Cool to room temperature and add 3 mL anthrone reagent. Maintain at 40°C for 10–15 minutes.
6. Measure optical density at 620 nm.

7. Use a calibration curve (Figure 5–1) to calculate μg of sucrose. To determine the free fructose component from the second nectar sample, follow steps 8–11. Because this sequence does not include addition of KOH, it will give you a measure of the combined free fructose and fructose derived from hydrolysis (from addition of acid anthrone reagent). By subtracting the value for the sucrose component alone (from step 6) from this combined value, you can calculate the free fructose.

Determination of total fructose (free and from hydrolysis)

8. Add 3 mL anthrone reagent to the second nectar sample.
9. Measure optical density at 620 nm.
10. Use a calibration curve (Figure 5–1) to calculate μg of sucrose.
11. Subtract the value from step 7 from the value from step 10. This is the amount of free fructose in the nectar sample.

McKenna and Thomson (1988) and Thomson et al. (1989) extracted nectar from flowers using filter-paper wicks and then used the anthrone method to determine sugar concentration. The procedure follows:

1. Collect nectar with filter-paper wicks (see Section A.1.d) and dry them for subsequent analysis in the lab.
2. Vortex (swirl gently) wicks in 5 mL boiling distilled water for 1 minute to dissolve dried sugars.
3. Prepare sugar standards for comparison of absorption. Thomson et al. used equal amounts of fructose and glucose in standards ranging from 10 to 300 μg of total sugar per 2 mL of standard. Analysis proved relatively insensitive to the proportions of fructose and glucose. Standards keep up to 14 days if refrigerated.
4. Put 2-mL samples of the test solution, reagent blanks, and sugar standards into screw-cap test tubes in an ice bath.
5. Add 4 mL anthrone reagent (400 mg anthrone in 200 mL concentrated H_2SO_4).
6. Cap and vortex tubes.

7. Place test tubes in boiling water for 10 minutes.
8. Allow tubes to cool to room temperature.
9. Read absorbances on spectrophotometer.

This procedure gives an estimate of total carbohydrate but not of volume. Identification of the nectar sugars could follow. If your sample's concentrations fall outside the range of the standards, you can either dilute the samples or create a new range of standards.

Following paper chromatography (see Section D.1.), Watt et al. (1974) used the anthrone method of photometric analysis modified from Mokrasch (1954) to quantify nectar sugars.

This procedure is suitable for concentrations ranging from 5 nanomoles to 5 micromoles:

1. Chill 116 mL demineralized water (or ice) in a water bath and carefully add 400 mL concentrated reagent grade sulfuric acid while stirring to dissipate heat.
2. Cool the solution to room temperature.
3. Add 400 mg anthrone.
4. Place 6 mL of prepared anthrone reagent into screw-cap Pyrex vials and store in a dark, cool place.
5. To analyze sugar samples, place 1 mL of each sample (i.e., sugars dissolved from chromatograms) in a prepared reagent vial. Do not stir until all vials are prepared.
6. Mix contents of each vial and place them on a dry heating block for 45 minutes at 80°C. Color will develop.
7. Rapidly cool vials to room temperature to stop reaction. Color is stable for at least 1 hour.
8. Pipette samples into cuvettes and read optical density at 620 nm (Watt et al. use a Gilford 240 spectrophotometer).
9. Compare optical densities with curves developed from standards of known concentrations of each sugar.

Roberts (1979) used a different photometric method for analysis of sugars (adapted from Dubois et al. 1956). It involved rinsing a flower and then analyzing the rinsate (but the same method works for analysis of the sugars in nectar removed from flower visitors, or contained in flower visitors that are crushed in a vial with water). Advantages of this method include its accuracy and sensitivity to small concentrations of sugar, the speed with which samples can be collected in the field, the fact that analysis can be deferred, the speed of analysis (ca. 500 samples/person/day), and the low cost of reagents (Roberts 1979). A disadvantage is the possibility of contamination of the sample with cellulose from lint or plant tissue, which would be hydrolyzed and affect the results. Roberts used the following procedure:

1. Place a flower in a vial with a measured amount (e.g., 4 mL) of distilled water. Shake it to make sure the nectaries are contacted by the water.
2. Let the flower soak for 45–60 minutes. Remove the flower and store the rinsate in a freezer until you are ready to analyze it.
3. Thaw the rinsate and place a 1-mL subsample in a cuvette.
4. Add 1 mL of a 5% solution of phenol.
5. Add 5 mL of concentrated sulphuric acid (use a fast-delivery pipette to facilitate mixing and avoid spattering).
6. Mix the hot solution thoroughly and let it sit for 45 minutes for the color to develop.
7. Measure absorbance in a spectrophotometer at 490 nm and calculate the sugar content from a standard curve.

This method is most accurate if the absorbance is kept between 0.10 and 0.90. If the sample is too concentrated to fall within this range, try this dilution process (Roberts 1979):

1. Place 1 mL of the rinsate solution in each of three cuvettes.
2. Add the phenol and sulphuric acid (steps 4–6 as described above) to two of the cuvettes.
3. If the absorbance of the solution in the two cuvettes is between 0.10 and 0.90, proceed to the calculations shown below. If the absorbance is above 0.90, add 6 mL of distilled water (i.e., dilution factor of 7) to the third cuvette, and repeat steps 1–3 as many times as necessary to achieve the desired range of absorbance.

To calculate the amount of sugar in the flower (or insect, if you are obtaining samples from crushed insects; see also Section H), use the following calculations:

$$\text{weight of sugar } (\mu g) = \frac{(A_1 + A_2)}{2} \cdot RDS$$

where:

A$_1$ and A$_2$ = absorbances of the two cuvettes (original and diluted);

R = the volume (mL) of rinse water in the original vial;

D = the number of serial dilutions you carried out times the dilution factor (7); if you did not have to dilute the solution, omit this term from the calculation;

S = the slope of the regression line from a series of dilutions (10% to 90% absorbance) of a standard sugar solution.

Roberts used a 1:1:1 mixture of glucose, fructose, and sucrose as his standard, but if you know the relative concentrations of sugars in your sample, you can mix the appropriate standard solution.

Käpylä (1978) assayed the total amount of sugar in nectar samples photometrically. Flowers were rinsed with a given volume of water (5–10 mL) and the solution filtered through Whatman number 1 filter paper. One mL of the filtrate and 1.5 mL of 3,5-dinitrosalicylic acid reagent were mixed in a test tube and placed in a boiling water bath for 5 minutes. After cooling, 2 mL of water were added and absorbance read at 540 nm against a reagent blank. Absorbance was compared with a standard of equimolar solution of fructose and glucose. A 3,5-dinitrosalicylic acid reagent is prepared as follows (Clark 1964):

1. Dissolve by warming 5 g 3,5-dinitrosalicylic acid in 100 mL 2 N NaOH (8 g NaOH/100mL).
2. Add 150 g sodium potassium tartrate to 250 mL water and dissolve by warming.
3. Mix the two solutions together and add water to a total volume of 500 mL.

A colorimetric method developed for microanalysis of blood sugar (Nelson 1944, Somogyi 1945) can also be used for quantitative nectar sugar analysis (Patt et al. 1989) following separation of sugars by paper chromatography. Alternate known sugars and the nectar sample on a paper chromatogram. Run the chromatogram with a solution such as butanol/acetic acid/water. Cut out the vertical strips with the knowns and unknowns. Spray the chromatograms with one of the visualizing reagents described in Section D.1.b. Identify the positions of the known sugars. Cut the corresponding spots for each sugar from the lanes with the nectar sample and redissolve the sugars by placing the filter-paper spot into 2 mL of distilled water in a test tube. A colorimetric technique is used for the quantitative determination of nectar concentrations (Nelson 1944, Somogyi 1945):

1. Place equal amounts (approximately 2 mL) of Somogyi's copper-phosphate reagent (see below) and nectar sample in a test tube; cover the test tube. Use distilled water as a blank, and a known sample as a standard.
2. Immerse the test tube in a boiling water bath for 10 minutes.
3. Cool the test tube in a pan of cold water. Add approximately 2 mL of Nelson's reagent (see below).
4. Dilute the test solution to 10–25 mL with distilled water. The degree of dilution required will depend on the sugar concentration. Mix the solution.
5. Read absorbance at 400 or 520 nm in a spectrophotometer.

Somogyi's copper-phosphate reagent

1. Dissolve 28 gm of anhydrous disodium phosphate and 40 gm of sodium potassium tartrate in about 700 mL of distilled water.
2. Add 100 mL of 1 N sodium hydroxide.
3. While stirring, add 80 mL of a 10% copper sulfate solution.
4. Add 180 gm of anhydrous sodium sulfate.
5. After all the ingredients have dissolved, dilute the solution to 1.0 L with distilled water.

6. Allow the solution to stand for a day or two. Decant the top part of the solution and filter the rest through filter paper. The solution will keep indefinitely.

Nelson's arsenomolybdate reagent

1. Dissolve 25 gm of ammonium molybdate in 450 mL of distilled water.
2. Add 21 mL of concentrated H_2SO_4 and mix.
3. Add 3 gm of $Na_2HAsO_4 \cdot 7H_2O$ dissolved in 25 mL of distilled water. Mix.
4. Place the solution in an incubator at 37°C for 24–48 hours. Store it in a glass-stoppered brown bottle.

4. Freezing-Point Depression

Stiles (1976) attempted to use the freezing-point depression of nectar samples as a way to determine sugar concentrations. This method is adversely affected by the presence of other nectar constituents. In practice Stiles found the data difficult to interpret and not comparable to refractometer readings.

However, Corbet and Willmer (1981) were able to use freezing-point depression to measure water gradients in the nectar of tubular flowers. Nectar properties can be studied by plotting the refractive index (r) of nectar within the flower against ambient relative humidity (P) and comparing the plot with the r/P curve of a pure sugar solution (Corbet and Willmer 1981). The plot of the r/P curve for a nectar solution may diverge from that of a pure sugar solution (prepared from the same constituent sugars). There are two possible reasons for the divergence: the presence of additional solutes in the nectar and the nectar's equilibrium with a relative humidity other than that measured for the ambient air. The first reason occurs because relative humidity is a colligative property and when a solution is in equilibrium with the air, P is lowered by the presence of solutes in the solution. The second reason occurs when the microclimate within the flower differs from ambient conditions (microclimate protection). If the plot for nectar diverges from that of the sugar solution because of the presence of additional solutes, then the plot of r against freezing-point depression, another colligative property, will also diverge. If the plot of r against freezing-point depression is the same for the nectar and the sugar solution, then divergence in the r/P plots must be due to microclimatic effects. Comparison of the r/P and r/freezing-point curves will indicate the relative importance of chemical versus microclimatic effects in determining nectar concentration (Corbet and Willmer 1981).

Unwin and Willmer (1978) give instructions for making a field cryoscope-osmometer to determine the freezing point of small volumes of liquid such as nectar. A refrigerant in an aerosol can is used to cool the fluid to below its freezing point.

D. Sugar Identifications

Identification of the chemical constituents of nectar, such as sugars and amino acids, is in principle easy to accomplish, as a variety of analytical techniques are available for these chemicals. In practice, however, the analysis is constrained by the very small amounts of nectar that are usually available, especially at the level of individual flowers.

To verify the presence of nectar sugar, you can use a dropper to place a few drops of concentrated H_2SO_4 and then a few drops of 5% phenol in the flower (Schemske et al. 1978, Casper and La Pine 1984). The solution becomes orange in the presence of sugar. Test papers that are sensitive to glucose are routinely used for urinalysis (see Appendix 2). Several researchers have used these strips to test nectar. Nilsson (1979) checked for nectar secretion in small orchid flowers by using glucose test papers made by Clinistix, Ames Company. Sowig (1989) sampled nectar from flowers with Glucose-Testpapier, a paper that changes from yellow to green in the presence of glucose. Nectar volume was estimated from the size of the green spot, and the calibration equation:

$$y = (1001.3) \, x^{0.634},$$

$r^2 = 0.95$; where $x = \mu L$ nectar; $y =$ the area size of the spot in relative units measured under a dissecting microscope.

Patt et al. (1989) used glucose indicator paper to locate glucose production sites on floral and raceme surfaces. Búrquez (*personal communication*) uses Uristix to determine glucose concentration in

Table 5–4. Uristix calibration scale

Color Patch	mM/L	Color	Concentration (w/w)
negative	0		0
I	5.5	blue	0.1
Ia	10.0		0.2
II	14.0		0.3
IIa	21.0	green	0.4
III	28.0		0.5
IIIa	42.0		0.8
IV	56.0		1.0
IVa	84.0	brown	1.5
V	112.0		2.0

Source: Búrquez, *personal communication.*

nectar. These strips change from blue to brown with increasing glucose concentration. He has calibrated the Uristix so they can be used to determine the percentage of glucose in nectar (w/w) as well as mM/L. The strips come with a color guide with six color patches ranging from neutral (no glucose) to brown (111 mM/L). By adding intermediate levels to the color scale, Búrquez produced a calibration table (Table 5–4).

Sugars (and amino acids) can be separated, identified, and quantified with chromatography. Chromatography is the process of separating compounds based on their different solubilities in solvents, on a porous, inert framework. Once the sugars have been separated and identified, they can also be quantified. Paper chromatography is the simplest, followed by thin-layer chromatography. Gas and high-performance chromatography require more complex apparatus, usually available in a university biology or chemistry department. Some of the chemicals used for chromatography are toxic or flammable. Use caution and appropriate techniques when working with them, and dispose of them properly.

1. Paper Chromatography

a. Methods

Paper chromatography uses porous chromatography paper or filter paper as a framework (Pritham 1968). A glass column with a thick glass lid is saturated with solvent. The solutions to be analyzed are applied as spots on the porous paper a few cm from one end. Known standards may be spotted on the same paper. Spotting is done with a microcapillary tube. The porous paper is introduced into the chromatography tank with one edge in the solvent and the solvent migrates through the paper at a specific rate under controlled temperature and saturation conditions. Components of the solution travel up the paper with the solvent at different rates. Once the migration is complete, the paper is generally treated (often sprayed with a chemical solution and then heated) so that the components of interest become colored spots. Under controlled conditions, substances can be identified by the R_f value,

$$R_f = \frac{\text{(the distance of component movement up the paper)}}{\text{(the distance of solvent movement up the paper)}} .$$

Turning the paper 90° and running the sample again with a second solvent can resolve two components that do not separate well in the first solvent (two-dimensional chromatography; Figure 5–2).

Hough et al. (1950) separated mixtures of sugars using n-butanol saturated with water as a solvent. Rapid separation of sugars occurs on Whatman number 1 and number 54 filter paper at 37°C. They present R_f values for controlled conditions. Using the basic procedure of Hough et al., Baskin and Bliss (1969) detected mono-, di-, and trisaccharides using paper chromatography for a rough analysis, which they followed with gas chromatography. They ran 15 µg spots of sugar mixtures and sugar standards on Whatman number 1 paper with an ethyl acetate–pyridine–water (8:2:1) solvent. Sugars were detected with p-anisidine HCl. They compared R_g values with the standards to determine qualitative composition and determined relative concentration by color intensity and size of spots when examined under ultraviolet light at 360–400 nm.

Percival (1961) used Whatman number 1 filter paper with a solvent of ethyl acetate, acetic acid, and water (9:2:2). Sugars ran

Nectar

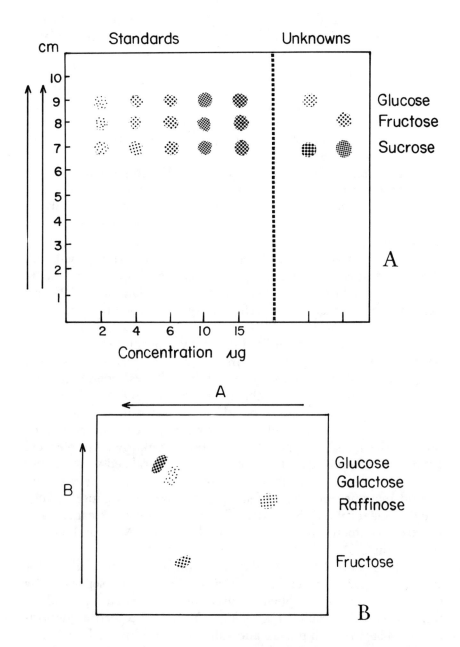

Figure 5–2. One-dimensional chromatograms of standards and unknown sugars (A). Two-dimensional chromatograms of sugars (B). Redrawn from Jeffrey et al. 1969.

for 24–36 hours and were visualized by spraying with 2% anisidine hydrochloride in butanol and heating in front of an electric heater.

Grant and Beggs (1989) separated sugars using paper chromatography on number 4 chromatography paper, or on Whatman number 1 filter paper. They used three descending solvent systems: A, ethyl acetate–pyridine–water (8:2:1, v/v); B, n-butanol–glacial acetic acid–water (3:1:1, v/v); C, 1-propanol–ethyl acetate–water (7:1:2, v/v). Dried chromatograms were treated with p-aminobenzoic acid (Saini 1966; see below) to detect sugars.

Watt et al. (1974) used Partridge's solvent (organic phase of a 4:1:5 v/v/v mixture of n-butanol–glacial acetic acid–water). Chromatograms ran for 48–72 hours for best separation, and the solvent was allowed to run off the paper. Chromatograms were dried with hot air and sprayed with bromcresol purple indicator spray (below) to reveal monosaccharides. These spots were circled in pencil, and chromatograms redried and sprayed with anthrone–acetic acid–ethanol (below) or an acidified aniline-diphenylamine (ADPA) reagent to color other sugars. After spraying, chromatograms were heated for 10 minutes at 100°C.

Watt et al. (1974) quantified sugars by spotting sugar standards on the same paper as the nectar sample, running in the same solvent, and drying chromatograms. The standard "lanes" were cut from the paper and treated with anthrone solution or ADPA to determine by comparison the relative position of sugars in the untreated sample and to divide the unknown strip into zones representing the position of each sugar type. The zones on the unknown sample were cut apart and the sugar eluted in 2 mL demineralized water. Paper fibers were removed from the eluate first by centrifugation and second by squeezing a 1-mL sample through a 25-gauge syringe needle. Sugars were then quantified using anthrone reagent and spectrophotometry in a procedure modified from Mokrasch (1954; see Section C.3).

Dubois et al. (1956) found that the anthrone method is fine for standard sugar solutions, but when used for sugars separated by chromatography using phenol-water solvents, residual phenol interferes with the results. They suggest using photometric quantification of sugars with phenol and sulfuric acid. This method can be used for sugars, methyl derivatives of sugars, oligosaccharides, and polysaccharides. It is suitable for sugars separated by paper chromatography with phenol solvents or volatile solvents such as butanol–ethanol–water, ethyl acetate–acetic acid–water, or methyl ethyl ketone–water. This method was employed by Roberts (1979) and Immers (1964). Details are presented by Roberts (1979) under photometric procedures (Section C.3).

b. Additional reagents for visualizing chromatograms

Aniline-oxalate. This spray turns aldohexoses and oligosaccharides brown, aldopentoses bright red, and ketoses a pale brown (Meeuse 1982, adapted from Partridge 1951).

1. Prepare a 0.2-M aqueous solution of oxalic acid.
2. Mix 0.93 g aniline in 50 mL ethanol.
3. Mix the 0.2 M oxalic acid and the aniline solution and spray chromatograms.
4. Dry chromatograms for 10 minutes at 110°C.

Aniline-phthalate. This spray colors aldohexoses olive-brown and ketoses and oligosaccharides a weaker shade (both with yellow fluorescence). Aldopentoses turn red-brown with red fluorescence (Meeuse 1982, adapted from Partridge 1949).

1. Dissolve 0.93 g aniline in 100 mL n-butanol previously saturated with water.
2. Add 1.66 g phthalic acid to the solution and spray chromatograms.
3. Dry chromatograms for 5 minutes at 105°C.

Anthrone-acetic acid-ethanol. This is a general reagent for detecting carbohydrates (Block et al. 1958).

1. Dissolve 300 mg anthrone by warming in 10 mL glacial acetic acid.
2. Add 20 mL ethanol, 3 mL phosphoric acid (H_3PO_4), and 1 mL water. This solution is fairly stable when stored in the refrigerator.

3. Apply to chromatograms and then heat them for 5 minutes at 108°C.

Bromcresol purple. Bromcresol purple spray reveals monosaccharides as yellow spots on a purple background under ultraviolet light (Block et al. 1958).

1. Mix 40 mg bromocresol purple and 100 mg boric acid in 100 mL methanol.
2. Add 7.5 mL 1% aqueous solution of borax and apply to chromatograms.
3. Spots are examined immediately under ultraviolet light as fading is rapid.

Naphthoresorcinol-trichloracetic acid. This spray colors ketoses and ketose-containing oligosaccharides bright red, whereas pentoses and uronic acids turn blue (Meeuse 1982, adapted from Partridge and Westhall 1948).

1. Prepare a 2% solution of naphthoresorcinol in ethanol.
2. Prepare a 2% aqueous solution of trichloracetic acid.
3. Immediately before use, mix equal parts of the two solutions for spraying chromatograms.
4. Dry chromatograms at 100°C to reveal the red color.
5. Expose chromatograms to water vapor at 70°C to reveal the blue color.

P-aminobenzoic acid. Saini (1966) used p-aminobenzoic acid as a dip reagent for producing chromatogram color. It is prepared as follows:

1. Slowly stir 3 g p-aminobenzoic acid into 5.0 mL hot phosphoric acid until it is completely dissolved.
2. Add 300 mL of n-butanol–acetone–water (10:5:2 v/v) to the solution. Discard any residue. The reagent keeps for long periods of time at temperatures below 25°C. If it turns deep orange, discard it, but yellowing of the solution will not affect its action.

3. Dip chromatograms in the reagent and then heat them at 105°–110°C for a few minutes.
   Pentoses stain dark red and other sugars stain dark brown.

Saini used this dip reagent following paper chromatography with an isopropanol–tert-butanol–n-butyl alcohol–water (4:2:2:2) solvent.

Silver nitrate. Trevelyan et al. (1950) used a silver nitrate solution to visualize reducing sugars (sugars with a free aldehyde group such as glucose) on developed chromatograms:

1. Dilute 0.1 mL of saturated aqueous silver nitrate solution to 20 mL with acetone.
2. Add drops of water and shake the solution until the silver nitrate redissolves (it comes out of solution in acetone). Pass dried, developed chromatograms through this reagent. Sugars are not very soluble in acetone (0.014% at 23°C), so spreading of spots is restricted.
3. Spray dried chromatograms with 0.5 N NaOH in aqueous ethanol (dilute saturated aqueous sodium hydroxide solution with ethanol). Spraying turns chromatograms brown upon the formation of silver oxide. Reducing sugars produce black spots on the brown background.
4. After complete reduction, dissolve excess silver oxide by dipping the paper in 6 N ammonium hydroxide.
5. Wash paper in running water for 1 hour.
6. Dry paper in oven. Sugar spots appear dark on a white background and turn deep black on brief exposure to hydrogen sulfide.

Triphenyltetrazolium chloride (TTC). Trevelyan et al. (1950) also presented a second procedure for visualizing reducing sugars using a TTC reagent.

1. Prepare reagent consisting of 0.5% solution of triphenyltetrazolium chloride in chloroform. Pass dried developed chromatograms through this reagent.
2. Evaporate solvent at room temperature.

3. Spray paper with an alcoholic solution of NaOH.
4. Allow chromatograms to stand overnight (or heat 5 minutes at 100°C in a moist oven). Reducing sugars produce red spots on a pink background.
5. Flush excess reagent with water.

Sucrose and other nonreducing sugars do not react.

2. Thin-Layer Chromatography (TLC)

Thin-layer chromatography is similar in principle to paper chromatography (Pritham 1968). A thin layer of an adsorbent substance (silica gel, diatomaceous earth, alumina, powdered cellulose, etc.) combined with a binding medium is spread on a glass plate or plastic sheet, dried, and heated. Sample solutions are spotted on the plate and dried. The glass plate is then placed on edge in a tank saturated with solvent. The solvent and sample components migrate up the plate at different speeds. Plates are sprayed with an indicator solution to color the spots. Two-dimensional chromatography is also possible, as in paper chromatography.

Thin-layer chromatography offers several advantages over paper chromatography. Because the silica-gel matrix is more uniform, separation of compounds is sharper, and because the pore size is usually quite small, the more restricted diffusion of compounds produces smaller spots. TLC is also faster than paper chromatography. High-performance TLC (HPTLC) relies on very thin layers with even more uniform pore size in the silica gel. A sample can be separated into its components on a 10 x 10–cm (or even smaller) plate in an hour, and precoated plates are available commercially.

Jeffrey et al. (1969) detail a procedure for qualitative and quantitative determination of di-, tri-, and tetrasaccharides using TLC. Prepare plates by mixing 30 g Silica Gel G (Merck) with 60 mL of 0.02 M boric acid and spread the mixture on 20 x 20–cm glass plates with a Desaga applicator to a thickness of 0.25 mm. Next, place plates in a forced draft chromatography oven for 1 hour at 100°C, then store in a desiccator until needed.

For qualitative analysis, spot samples 2 cm from the bottom of plates with 1 - 5 µL pipettes. Develop plates for 45 minutes–1 hour in a standard 8 x 15 x 15–cm TLC jar containing 150 mL of solvent. Most sugars develop in solvent A, a methyl ethyl ketone–acetic acid–methanol (6:2:2) solvent. Solvent B, acetone-water (9:1), is used to separate fructose, glucose, and galactose (run first in solvent B, followed by solvent A in the second dimension). Double developing sharpens separation. Plates are air-dried after the first development and then developed again in the same direction with the same solvent. Double development in two dimensions with solvent A or A and B further resolves problem sugars (those that overlap on the plate).

For quantitative analysis, Jeffrey et al. (1969) used solvent A and double-developed plates. Sugars were visualized with ADPA (aniline-diphenylamine), modified by the addition of 0.66 g benzidine dissolved in 110 mL acetic acid to increase stability and distinction of colors. Colors were developed by heating for 10 minutes at 100°C. After heating, spots were immediately scraped from the plate and redissolved in HAM (1 mL of 1% HCl and 0.1% ascorbic acid in methanol) in test tubes. Test tubes were centrifuged at 1,800 G for 15 minutes to remove silica gel. To control for background color, a blank region of the plate was also eluted and the absorption of the supernatant was determined. Background color becomes stronger with time, so samples should be eluted soon after development. The supernatant was placed in 1.5-mL cuvettes and absorption determined with a Beckman DB-G Spectrophotometer. Absorption maxima were compared with curves prepared from reference standards on the same plates.

3. Gas Chromatography
Gas chromatography is generally used for compounds that volatilize at low temperatures. For analyzing nectar for sugars or amino acids, this requires creation of volatile derivatives prior to analysis. Gas chromatography requires a glass or metal column, packed with a finely ground inert solid such as diatomaceous earth. A nonvolatile liquid (liquid phase; lubricating oil, silicones, or high-molecular-weight esters) coats the inert particles. An inert gas is forced through the column at temperatures from 170° to 300°C. The volatile compounds to be analyzed are flash-evaporated at one end and forced through the column by the moving gas. Different components partition differently between the gas and liquid phase and thus move at different rates through the column. The unknown

compounds are thus separated and can be detected and quantified as they exit the column.

Sweeley et al. (1963) developed a trimethylsilyation procedure for creating volatile derivatives of sugars. Unlike previous methods for creating volatile sugar derivatives, this procedure occurs rapidly at room temperature:

1. Dissolve 10 mg of carbohydrate in 1 mL of anhydrous pyridine kept over KOH pellets.
2. Add 0.2 mL hexamethyldisilazane and 0.1 mL trimethyl-chlorosilane in a 1-dram plastic-stoppered vial.
3. Shake the mixture for 30 seconds and then allow to stand for 5 minutes before chromatography (precipitate that forms will not interfere with the procedure). If the carbohydrate seems insoluble, warm the vial at 75°–85°C.
4. Use 0.1–0.5 µL for injection into the gas chromatograph.

Baskin and Bliss (1969) modify Sweeley's procedure so that low-molecular-weight sugars are not obscured by the pyridine solvent peak:

1. Dissolve 10 mg of standard or nectar in 1 mL of pyridine.
2. Add 0.3 mL hexamethyldisilazane and 0.1 mL of trimethyl-chlorosilane.
3. Shake vigorously for 20 seconds and heat for 3 minutes at 60°–70°C.
4. Add 3 mL $CHCl_3$ (chloroform), shake, and add 3 mL distilled water. Shake again and then discard the aqueous-pyridine phase. Repeat this extraction three times.
5. Evaporate the $CHCl_3$ solution to dryness.
6. Residue is taken up in $CHCl_3$ and 2-mL volumes are prepared for sugar standards and 50–200 µL for unknowns.

Baskin and Bliss used an F & M 810 Research Gas Chromatograph with hydrogen flame ionization detector and dual coiled copper-alloy columns. Columns contained 3% SE-30 on Chromasorb W 80–100 mesh. Column temperature was 250°C, flash heater at 260°C, and detector at 270°C. Flow rate was 75 mL per minute.

Loper et al. (1976) also used silyated sugars for assays. Materials were frozen until analysis. Samples of 1 mg were weighed and dried in a vacuum oven and allowed to react with silylating agents (TRI-SIL or STOX) overnight to produce trimethylsilylated derivatives of sucrose, glucose, and fructose. Samples were chromatographed on an SE-30 column or OV-17 column.

Today, analyses of TMS sugars with gas chromatography will probably use fused-silica columns with bonded stationary phases rather than packed columns (C. Elliger, *personal communication*).

4. High-Performance Liquid Chromatography

High-performance liquid chromatography also relies on a column, through which a solvent (e.g., 60% methanol in water) moves the sample. HPLC can also be used to identify phenolics, alkaloids, lipids, or ascorbic acid, all of which are found in some nectars (Linskens and Jackson 1987). Linskens and Jackson's book (1987) is devoted to HPLC techniques. Another good source of technical information for HPLC is manufacturers of equipment for this technique. They usually have technical assistants who can advise you about the best columns or solvents for a particular application.

HPLC is an easy, accurate method for quantifying specific sugars (Southwick, *personal communication*). Southwick et al. (1981) used HPLC on a Tracor 980A to detect fructose, glucose, and sucrose. Nectar samples of 1–2 μL were dissolved in a solvent consisting of 85% acetonitrile, 15% water and adjusted to pH 4.5 with dilute phosphoric acid. An ultraviolet xenon variable wavelength detector at 195 nm was used with a 25 cm x 4.6–mm Whatman Partisil P x S 10/25 PAC column. Flow rate was 1.5 mL per minute at 360–400 psi.

Neff and Simpson (1990) analyzed sugar composition of nectar and bee stomach samples with a Beckman 330 isocratic HPLC system. They used a BioRad HPX 87C carbohydrate column and water solvent. Freeman and Wilken (1987) collected nectar on filter paper and eluted it in 20 μL of water for analysis with HPLC (modified method of Freeman et al. 1984).

E. Identifying and Determining Amino Acid Concentrations

For about 25 years the laboratory method of choice for amino acid analysis has been classical ion-exchange column chromatography followed by post-column derivatization of the amino acids by either ninhydrin or a fluorogenic reagent so they can be detected. This method, although it resolves most amino acids with good detection limits, isn't usually suitable for nectar analyses because of the small sample volumes that are typically available. Most studies of nectar have instead used paper or thin-layer chromatography, or HPLC, all of which work well with small samples. There are a number of handbooks devoted exclusively to techniques for characterization of proteins and amino acids (e.g., Shively 1986, Franks 1988) or particular techniques such as HPLC (e.g., Linskens and Jackson 1987).

The oldest method for visualizing amino acids is probably the use of ninhydrin, which reacts with the free NH_2 group of amino acids to produce a bluish violet color after exposure to a temperature of 100°C. A commonly used technique for visualizing amino acids through fluorescence has been the production of fluorescent derivatives with dansyl chloride. This reaction also relies on an interaction with the NH_2 group. Neither of these techniques produces a permanent record, as the ninhydrin color and fluorescence both fade. Baker and Baker (1976a, 1976b) recommended the dansyl chloride technique, but there are also other fluorescent techniques available that do not appear to have been tried in nectar analyses, such as treatment with o-phthaldialdehyde (Jones 1986).

Gottsberger et al. (1990) found that the nectar of *Hibiscus rosa-sinensis* L. (Malvaceae) shows an increase in amino acid concentration with the aging of flowers, correlated with an increase in the number of detectable amino acids. Puncturing floral tissue or contaminating nectar with pollen dramatically increased amino acid concentrations, so careful technique is required when collecting nectar for amino acid analysis. Corbet and Willmer (1981) found that nectar-robbing by insects increased amino acid concentrations in nectar. Baker and Baker (1973) point out that nectar produced by carrion- and dung-fly flowers is rich in amino acids.

Table 5–5. The histidine scale, or concentration of aqueous
histidine solution required to produce depth of color scoring
0–10 in dried spots when ninhydrin solution is added

Score	Concentration of aqueous solution of histidine (μM)
0	<49
1	49 (7.58 μg/ml)
2	98
3	195
4	391
5	781
6	1.56 mM
7	3.13 mM
8	6.25 mM
9	12.50 mM
10	23.00 mM (3.90 mg/ml)

Source: Baker and Baker 1975.

1. Ninhydrin Test for Total Amino Acid Content

Baker and Baker (1973) used this method to analyze samples of
nectar spotted on filter paper with microcapillary tubes. Spots were
treated with a drop of 0.2% ninhydrin (triketohydrindene hydrate)
in acetone. Color developed in 24 hours at room temperature. Most
alpha-amino acids produced violet spots. Exceptions were proline
and hydroxyproline (yellow) and asparagine (orange-brown). Although many amino-bearing compounds produce colors with this
treatment, at room temperature only alpha-amino acids are likely
to react. To estimate concentration, a histidine scale (Table 5–5) was
produced by spotting different concentrations of aqueous histidine
solution on filter paper and subjecting them to the ninhydrin
treatment. This resulted in a series of dried spots with different
depths of color. Samples were compared to the histidine scale and
matched by intensity of the spots to determine approximate amino
acid concentration. This technique provides information on total
amino acid content but not about individual amino acids.

Baker and Baker (1973) recommended checking sugar concentration before using the ninhydrin test. If sugar concentration is
extremely high, sugar will not spread far on the filter paper and

ninhydrin penetration is incomplete. Highly concentrated sugars can be diluted before testing.

2. Paper Chromatography

Baker and Baker (1973) used two-dimensional paper chromatography to produce preliminary identifications of the amino acids present in nectar of over 40 species. The solvents used were butanol–acetic acid–water and phenol-water. A series of chromatograms run with known standards were required to identify the unknowns.

3. Thin-Layer Chromatography

Baker and Baker (1977) spotted drops of nectar on Whatman number 1 filter paper and tested for amino acids with ninhydrin staining. If nectar proved weak in amino acids (based on the histidine scale), several drops were loaded on one spot by drying the paper between drops. Amino acids were dansylated and then separated with TLC on polyamide plates with three solvent systems (Baker and Baker 1976a, 1976b, 1977).

Dansylation produces highly fluorescent amino acids that can be identified and quantified under ultraviolet light (Baker and Baker 1976a, 1976b). To dansylate the amino acids, the paper strips bearing the concentrated nectar spots are cut into small pieces and each piece is placed in a 6 x 50–mm culture tube with 10–30 µL of distilled water and refrigerated for a few hours. The eluted extract is transferred to a new tube and adjusted to a pH of 8.5 by adding 0.1 M sodium bicarbonate. Fresh dansyl chloride (1-dimethylaminonaphthalene-5-sulfonlyl chloride in acetone) is added to the tube, in concentration determined by the strength of color produced in the ninhydrin test. The solution is left overnight at room temperature, or at 37°C for 1 hour, during which time the extract turns from a yellow to a colorless liquid. The extracts are concentrated in a vacuum desiccator and refrigerated if further analysis will not follow immediately.

Baker and Baker carried out chromatography of dansylated amino acids on polyamide plastic sheets in three solvent systems: (1) formic acid–water (1.5:98.5 v/v); (2) benzene–acetic acid (9:1 v/v); (3) ethyl acetate–methanol–acetic acid (20:1:1 v/v/v; Baker and Baker 1976a, 1976b). Amino acids were identified under ultraviolet light by their color and position on the chromatogram in comparison to standards. Fluorescence of dansylated amino acids is proportional to concentration and may be used for rough quantitative estimates. For more precise quantification of concentrations, use a filter fluorometer (e.g., Turner, model 111) with an automatic TLC scanner. This procedure is more sensitive than paper chromatography and ninhydrin staining and can be used for quantities of nectar as small as 1 µL and concentrations as low as 10 pmol. Be aware that with this procedure phenylalanine can be obscured by the presence of leucine.

Willmer (1980) looked at amino acids in the field, using two-dimensional chromatography on Gelman miniaturized ITLC silica plates. The first solvent was butanol:acetic acid:water; either phenol:water or butanol:acetone:diethylamine:water was used for the second. Amino acids were visualized with ninhydrin. Willmer confirmed his results in the lab on larger plates. He claims this procedure is about as sensitive as the dansylation method, although separation of phenylalanine and leucine and of aspartate and threonine is poor.

Hanny and Elmore (1974) used a combination of gas chromatography and TLC for amino acid analysis. One aliquot of amino acid solution elute (derived by chromatography on an ion-exchange column) was dansylated and separated with two-dimensional TLC on Cheng-Chin polyamide sheets and ninhydrin assay followed.

4. Gas Chromatography

Preceding gas chromatography, amino acids in nectar (Hanny and Elmore 1974, Bosi and Battaglini 1978) or honey (Gilbert et al. 1981) can be isolated by chromatography on an ion-exchange column of cationic resin. Gilbert et al. (1981) used this method of isolation to study the amino acids in honey:

1. Dilute 1 g of honey with 10 mL of distilled water.
2. Add an internal standard of 50 µg of norleucine.
3. Pass the solution through a column (20 cm x 11 mm internal diameter, filled with 3 g of wet Dowex 50 W x 8 H+ form resin, 200–400 mesh) at a rate of 1 mL per minute.
4. Wash the resin with three 10-mL portions of distilled water and discard the effluent (which contains interfering components).

5. Elute the amino acids with 25 mL of 7 M ammonium hydroxide, followed by 10 mL of distilled water.
6. Collect the eluent in a round-bottomed flask.
7. Dry the eluent at 40°C on a rotary evaporator and transfer it to a 3-mL Pierce Reactivial together with approximately 2 mL of 0.1 M hydrochloric acid.
8. Dry the contents on a rotary evaporator at 45°C using a manifold and adaptors, and then dry them for 1.5 minutes at 100°C under a flow of nitrogen.

Amino acids must be transformed into volatile derivatives for gas chromatography. Derivatives are formed through esterification (producing n-butyl esters) and acetylation (N(O)-trifluoroacetyl) of the polar groups (Bosi and Battaglini 1978) or formation of n-acetyl n-propyl esters (Hanny and Elmore 1974). Production of n-acetyl and n-propyl esters of arginine and histidine requires prior ozonolysis conversion (Hanny and Elmore 1974). Acetylation hydrolyzes glutamine and asparagine, so these procedures detect the free acids (Hanny and Elmore 1974).

Derivatized amino acids are analyzed with gas chromatography. Elmqvist et al. (1988) used a fused-silica capillary column (25 m x 0.2 mm internal diameter) coated with SE-54 (Altech Associates) and a Varian 3700 instrument with a flame ionization detector. The injector temperature was 250°C and detector temperature was 300°C. Oven temperature was programmed from 80°–230°C at a rate of 5°C per minute. Hydrogen gas flow was 70 cm per minute and make-up gas flow (nitrogen) 30 mL per minute.

Gilbert et al. (1981) used the following procedure, which assumes that the dried sample of amino acids is in a Reactivial:

1. Dry and redistill all reagents; store them in a desiccator (containing phosphorus pentoxide) at -10°C.
2. Dispense reagents with glass micropipettes to avoid contact with metal surfaces.
3. Add 400 μL of 6.0 M HCl in propanol and heat the mixture for 4 minutes at 150°C on a heating block.
4. Cool the mixture and dry it on a rotary evaporator at 45°C.

5. Add 100 µL of a solution of 2, [6]-Di-tert-butyl-p-cresol (BHT; 0.03% wt/wt in ethyl acetate) and 100 µL of heptafluorobutyric anhydride.
6. Heat the mixture for 12 minutes at 150°C.
7. Cool the solution and blow it with nitrogen until dry.
8. Redissolve it in a 1:1 mixture of ethyl acetate and acetic anhydride prior to analysis with gas chromatography.

For analysis, Gilbert et al. (1981) used a Perkin-Elmer Sigma 2 gas chromatograph with a 360 cm x 2–mm packed column of 3% SE 30 on Chromosorb Q (80–100 mesh). Nitrogen gas flow rate was 20 mL per minute and oven temperature was programmed from 70°C at 4°/minute to 250°C, then isothermal. Injector and detector temperatures were 270°C. Peaks were quantified with a Perkin-Elmer Sigma 10 data station. Gilbert et al. (1981) prepared calibration standards for each amino acid from samples of a standard mixture containing 2.5 to 4.5 µM total amino acids (or 0.25 to 1.25 µM of each).

Hanny and Elmore (1974) used a Perkin-Elmer Model 900 gas chromatograph with 2 ft. x 1/8 in. dual stainless steel columns packed with Chromosorb G (80–100 mesh) coated with 0.7% PEG (6 M)–0.05% TCEPE, and equipped with flame ionization detectors and Perkin-Elmer Model 56 recorder. Helium gas flow was 60 mL per minute. Injector temperature was 230°C and detector temperature was 260°C. Column temperature was programmed from 100°–250°C at 12° per minute. Hanny and Elmore found overlap in retention times of some amino acids and similar R_f values for other amino acids using TLC. Both methods were used to resolve this problem.

5. High-performance Liquid Chromatography
In the past several years, ion-exchange chromatography has been supplanted by a faster, more sensitive technique, precolumn derivatization with phenylisothiocyanate (PITC) followed by reversed-phase high-performance liquid chromatography. This newer method correlates well with older ninhydrin-based techniques (Davey and Ersser 1990) and offers analysis times as short as 10 minutes and detection limits under 1 pmol (Cohen and Strydom 1988).

Gottsberger et al. (1990) found that nectar samples can be preserved by mixing them with ten times the volume of 70% alcohol and refrigerating until HPLC analysis.

6. Amino Acid Analyzers

Nectar from several flowers may be pooled for analysis with an amino acid analyzer (Gottsberger et al. 1984, Rust 1977). Gottsberger et al. (1984) stored nectar in a known quantity of 70% alcohol and then determined amino acids directly from the solution using a Jeol JDH-6AH modified for this purpose (for details see Linskens and Schrauwen 1969, Schrauwen and Linskens 1974, Welte et al. 1971). Rust (1977) froze nectar samples at -30°C until analysis on a Durram D-500 Amino Acid Analyzer with five lithium-citrate buffers.

F. Other Nectar Constituents

Baker (1977) reminds us that nectar is not merely sugar water, providing energy for pollinators. Nectar constituents vary among species, and he suggests there is an adaptive reason for the presence of these components that bears further investigation. Table 5–1 lists the variety of nectar constituents, which vary greatly among species. We present some methods for detecting and identifying these nectar constituents. For further information on alkaloid, flavonoid, and glycoside analysis with TLC, see Wagner et al. (1984).

1. Alkaloids

Masters (1990) reports that ithomiine butterflies derive protection from predators from the pyrrolizidine alkaloids found in nectars of some flowers they frequent. Earlier tests by Baker and Baker (1975) indicated that alkaloids were present only in bee nectars and they submitted that perhaps adult Lepidoptera were intolerant of alkaloids. If this were the case, nectars containing alkaloids might discourage inconstant butterfly pollinators without discouraging bees.

Deinzer et al. (1977) examined alkaloids in honey by two different methods. In the first, they extracted an ammoniacal solution of honey with chloroform, and then followed Mattocks' spectrophotometric procedure (see below) with solvent modifications. In a second experiment, honey was diluted with water and acidified by adding hydrochloric acid for extraction with chloroform. The resulting emulsion was centrifuged, and ammonium hydroxide was added to the aqueous component, which was again extracted with chloroform in an iterative process until all waxes were removed. The final clean extraction was concentrated and assayed with gas chromatography and mass spectrometry.

Mattocks (1967) presented two methods, standard and quick, for estimating alkaloids or alkaloids + N-oxides. Both of these methods are intended for dry samples, containing 5–30 µg of alkaloid or alkaloids + N-oxides in the basic form. He suggested using analytical-grade chemicals to make the following reagents for the standard method:

Oxidation reagent. Stabilize 30% w/v aqueous hydrogen peroxide by dissolving it in sodium pyrophosphate (5 mg/mL). Dilute 0.1 mL of this solution to 25 mL with absolute methanol, and use it the same day it is prepared.

Diglyme (bis-2-methoxyethyl ether). To test the reagent, use acidified potassium iodide. If peroxides are present, iodine will be released from the solution. To purify the reagent shortly before it is used, pass it through a short column of activated alumina. If you are using freshly purchased reagent, it is probably not necessary to test it.

Acetic anhydride. Redistill analytical-grade reagent and collect the fraction that boils at 136°–139°C.

Modified Ehrlich reagent. Dilute 10 mL of methanolic boron trifluoride (containing about 14% BF_3) to 100 mL with methanol. Dissolve in it 2 g of 4-dimethylaminobenzaldehyde. Keep this reagent in the dark; light causes it to darken and to give high readings in blanks. In the dark it is stable for several weeks.

1. For the analysis, put the sample in a test tube (e.g., 15 x 120–mm), and add 0.5 mL (± 5%) of the oxidation reagent.

2. Immerse the lower half of the unstoppered test tube in a boiling water bath for 20–30 minutes; after 10 minutes all

the methanol will have evaporated. Don't let water condense on the inside of the tube.

3. Add 1 mL (± 10%) of Diglyme, then 0.1 mL (±5%) of acetic anhydride.

4. Reheat the test tube in the boiling water bath for 1 minute ± 10 seconds. Avoid contamination of the solution.

5. Cool the solution to room temperature and then add 1 mL modified Ehrlich reagent.

6. Heat the solution in a water bath at 55°–60°C for 4–5 minutes to develop the color. The color will not reach its maximum below 50°C but will fade quickly at high temperatures, so carefully monitor the temperature.

7. Cool the solution, dilute it with acetone to 4.0 mL, and measure the absorbance with a spectrophotometer.

8. Create a blank by duplicating this procedure but omitting the sample. If the nature of the alkaloid is not known, or if the spectrophotometer is not calibrated well, the wavelength of maximum absorption may have to be determined experimentally.

If the samples are too concentrated they can be further diluted with acetone. If they cannot be examined immediately, they should be stoppered and stored in the dark; a decrease in absorbance of about 1.4% per hour occurs at room temperature.

To estimate only the N-oxide component, omit steps 1 and 2. If you suspect that the sample contains pyrrole or indole derivatives, or other compounds that might produce a color with Ehrlich reagent, treat a duplicate sample with steps 5–7 only (start by adding Ehrlich reagent) and measure absorbance at 565 nm (which may not be the wavelength of maximum absorbance). Subtract this from the absorbance obtained by the full procedure.

Mattocks also suggests a "quick" method, which is much faster than the standard method, but the strength of color and reproducibility of results are "somewhat decreased." Only a single reagent is required. This method is suitable for total quantities of alkaloid + N-oxide of 5–30 µg in a dry form; alkaloids must be in the basic form.

1. Mix 0.01 mL of 30% (w/v) aqueous hydrogen peroxide and stabilize it with sodium pyrophosphate (20 mg/mL).

2. Make the solution up to 20 mL with Diglyme (purified through alumina); the reagent contains about 150 μg H_2O_2 per mL. This reagent will keep for several hours.
3. Add 1 mL of the reagent to the dry sample in a test tube.
4. Add 0.25 mL of acetic anhydride.
5. Develop the color by heating at 55°–60°C for 4–5 minutes and then measure it with a spectrophotometer as described for the standard procedure.
6. To estimate N-oxides and/or the background color in the presence of alkaloids, use the same procedure but omit the hydrogen peroxide.

Baker and Baker (1975) tested for alkaloids with the iodoplatinate test (Smith 1969). This test involves running a chromatogram and then using the iodoplatinate as a dip reagent. A blue color indicates a positive test for the presence of a tertiary amine group, and a white color indicates a primary amine group that is visible on the white background only if it is present in large quantities. To prepare iodoplatinate reagent, mix 5 mL of 5% aqueous solution of platinic chloride, 45 mL of a 10% aqueous solution of potassium iodide, and 100 mL of water (Smith 1969). This reagent will keep for several months. Baker and Baker (1975) retested nectars that were positive for alkaloids with a second reagent called Dragendorff's reagent (bismuth nitrate in acetic acid with aqueous potassium iodide). Positive results were indicated by a change from a yellow to orange color.

2. Vitamins
Vitamin C has been identified in the nectar of several species; its presence can be detected with ascorbic acid test paper (see Appendix 2). Baker and Baker (1975) suggest that reducing agents identified in nectar may not always be ascorbic acid, and they offer a generalized test for antioxidants, including vitamin C. Nectar spots on chromatography paper are tested with 2,6-dichlorophenolindophenol (1% ethanolic solution of the sodium salt) reagent. The initial red color is bleached to white or pink in the presence of a reducing substance. The test is substantiated when bleaching occurs from applying 0.05% aqueous potassium permanganate reagent to nectar spots. Sixty-three of 210 species tested positive. Antioxidants

may prevent oxidation of lipids, as the two often occur together in nectar. Most nectivorous insects do not require vitamin C, so the presence of ascorbic acid is probably not related to pollinator nutrition. Chromatography can be used to ascertain that the antioxidant is vitamin C.

For historical reasons, it is probably worth describing briefly how early studies on the vitamin content of nectar were conducted. Ziegler and Ziegler (1962) and Ziegler et al. (1964) used a microbiological technique, employing strains of bacteria or molds that require a particular vitamin in order to grow. When the vitamin is present in a nectar sample added to the growth medium (which does not contain the vitamin in question), the test organism will show a growth response that can be measured acidimetrically (e.g., lactic acid titration), turbidimetrically, or linearly (colony size, mycelium dry weight). Difco Laboratories (1984; see Appendix 2) maintains a wide range of bacteria, yeasts, and molds that can be used for assays for vitamins or amino acids.

Ziegler et al. (1964) used these microbiological techniques to survey 23 species of floral nectar and demonstrated the presence of vitamins thiamin (B_1; in all 14 species examined), riboflavin (B_2; in all 14 species), pyridoxine (B_6; in 16 of 17 species), niacin (in all 16 species), pantothenic acid (in all 16 species), folic acid (in 16 of 17 species), mesoinositol (in 11 of 15 species), and biotin (in all six species) in a variety of floral nectars. They also used a chemical assay (colorimetric assay with 2,4-dinitrophenylhydrazine; Lüttge 1962) for vitamin C, which they found in seven of 23 species. The concentrations of these vitamins were of the same order of magnitude for samples of honeybee honey (0.01–810 µg/g sugar). They cautioned that the source of these vitamins could have been contamination from pollen, or cell contents from damaged cells. Significant differences in vitamin and sugar concentrations between nectar collected by centrifugation and by microcapillary tubes led them to suggest that centrifugation is not a good collection technique for either quantitative or qualitative analyses. Samples from centrifugation had more dilute sugar concentration and higher vitamin concentrations (which they speculate came from contamination with pollen).

A more modern alternative to microbiological analysis for vitamin analysis is HPLC. One alternative for water-soluble vitamins (such as vitamin C) is a Rainin Microsorb C18 column (see Appendix 2; Rainin Instrument Co., *personal communication;* see also Vandemark and Schmidt 1981). There do not appear to have been any studies of vitamins in floral nectar yet using this technique.

3. Proteins

Enzymes (transglucosidases and transfructosidases, esterases, malate dehydrogenase) have been detected in nectar. Other proteins may be present as well, in addition to protein contaminants from pollen, microorganisms, or glandular secretions from bees in the nectar (Baker and Baker 1973). Some honeys also have high protein contents.

Protein presence is detected with electrophoresis, photometry, spot-staining techniques, or commercial test strips.

a. Electrophoresis

Baker and Baker (1973) used starch gel electrophoresis to discern proteins (see description of electrophoresis in Chapter 6, Section B). The gel buffer was tris-citrate, and the electrode buffer was lithium hydroxide-boric acid. Gels were stained with bromphenol blue and amido black.

Chrambach et al. (1967) stained polyacrylamide gels with Coomassie blue, which is very sensitive to low concentrations of many different proteins. It stains proteins of high concentrations without producing thick, dark bands that would obscure other, less-concentrated proteins. The staining procedure works with acid, neutral, or basic buffer systems:

1. After electrophoresis, immerse polyacrylamide slab or cylinder in 10–40-fold volume of 12.5% TCA (trichloroacetic acid) and agitate. Leave for 30 minutes.
2. Place gel in fresh stain made by a 1:20 dilution in 12.5% TCA of 1% aqueous stock solution of Coomassie blue. Leave 30 minutes to 1 hour.
3. Transfer gel to 10% TCA. Band intensity increases for about 48 hours. Gels may be photographed using a red filter.

Scogin (1979) used 7.5% polyacrylamide gel electrophoresis according to the methods of Maurer (1971) and Shaw and Prasad

(1970). Protein was stained with Coomassie blue. Spectrophotometry and a commercial assay prep (Bio-Rad Laboratories) were used to determine protein content (Scogin 1980).

Shaw and Prasad (1970) compiled recipes for buffers and stains used for detecting specific enzymes with electrophoresis.

b. Photometric procedures

Lowry's colorimetric analysis of proteins (Lowry et al. 1951) is sensitive to low protein concentrations. For proteins in solution, such as nectar, this procedure is adopted:

1. Place sample of 5–100 µg of protein in 0.2 mL of solution into a 3–10–mL test tube.
2. Add 1 mL alkaline copper solution.
3. Mix and allow to stand at least 10 minutes at room temperature.
4. Rapidly add 0.10 mL Reagent E (diluted Folin reagent prepared by titrating Folin-Ciocalteu phenol with NaOH to a phenolpthalein endpoint and then diluting approximately twofold to make 1 N acid). Mix.
5. After at least 30 minutes, read sample in colorimeter or spectrophotometer. Weak solutions (5–25 µg in total solution) are read at 750 nm and stronger solutions read at 500 nm. Calculate protein weight from standard curves. If protein is less than 25 µg per mL, 0.5 mL of solution may be mixed with 0.5 mL of double strength alkaline copper solution.

Bradford (1976) quantified protein with spectrophotometry based on the fact that Coomassie Brilliant Blue G-250 changes from red to blue on binding with protein. Binding occurs rapidly in about 2 minutes and color persists for about 1 hour. The method is sensitive to 5 µg protein/mL:

1. Dissolve 100 mg Coomassie Brilliant Blue G-250 in 50 mL 95% ethanol.
2. Add 100 mL 85% (w/v) phosphoric acid and dilute resulting solution to 1 L (final concentration: 0.01% [w/v] Coomassie

Brilliant Blue G-250, 4.7% [w/v] ethanol, and 8.5% [w/v] phosphoric acid).

3. Pipette into test tubes up to 0.1 mL protein solution (containing 1–10 µg protein). Adjust total volume to 0.1 mL with an appropriate buffer.

4. Add 1 mL of Coomassie Brilliant Blue solution and mix. Wait 2 minutes to 1 hour.

5. Measure absorbance at 595 nm against a blank prepared from 0.1 mL of buffer and 1 mL of Coomassie Brilliant Blue solution.

6. Compare absorbance with protein standards using a curve produced by plotting standard protein weights versus absorbance.

Kevan et al. (1983) used Flores' (1978) photometric procedure for protein analysis. A solution of 0.0075% bromphenol blue in 15 mL of 95% ethanol and 2.5 mL glacial acetic acid is adjusted to a volume of 100 mL by adding distilled water. Bromphenol blue exists in a yellow form below pH 3 and a blue form above pH 4.6. When the acidic form is combined with protein, the absorbance of the protein-dye complex at 610 nm is proportional to the amount of protein present if protein remains in the range of 10–80 µg (the response curve is parabolic but roughly linear within this range). Varying the proportions of the initial dye solution can result in protein precipitation or interference in absorption with some proteins. The pH of the dye is also critical and must be maintained just below the transition level for the color change for the test to be sensitive. The presence of free amino acids does not interfere with this assay.

c. Spot staining

Baker and Baker (1975) spotted nectar on chromatography paper, applied 0.1% bromphenol blue in methanol for 1 hour, and then rinsed in 5% acetic acid for 15–30 minutes. Papers were then dried, and a blue color indicated the presence of protein.

d. Protein test strips

Willmer (1980) used protein test strips that detect levels as low as 0.1 µg/µL. The commercial test strips (Albustix: Ames Co.) use tetrabromphenol blue to detect protein. In his study, only flowers visited by nectar robbers contained protein (0.1–0.3 µg/µL in *Phaseolus* (Fabaceae) flowers and up to 0.5 µg/µL in *Lamium* (Lamiaceae) flowers). This probably reflects the addition of protein from damaged floral tissue.

4. Phenolics
Buchmann and Buchmann (1981) spot-tested nectar with p-nitroaniline, which yields a deep orange-red in the presence of phenolic compounds. TLC confirmed positive results.

5. Lipids
Baker and Baker (1975) spotted nectar onto chromatography paper and then applied a 1% aqueous solution of osmium tetroxide. Lipids originating from unsaturated fatty acids stained black. Osmium tetroxide is believed to oxidize double bonds in unsaturated fatty acids to form a black product (Jensen 1962). Large numbers of double bonds, as in phospholipids and sterols, produce a strong reaction. Be aware that prolonged exposure to osmium tetroxide will result in staining of many other tissue substances in addition to lipids. Jensen suggests sectioned tissue can be stained for lipids in a 1% solution of osmium tetroxide in water for an hour at room temperature. NOTE: **Extreme caution must be used in using osmium tetroxide as its vapors can cause blindness.**

Baker and Baker (1975) also placed drops of nectar on glass slides and applied Sudan III or Nile Blue to stain for the presence of lipids. Sudan dyes (especially Sudan IV and Sudan black B), Nile Blue, and fluorescent dyes are soluble in lipids, and their effectiveness is based on the principle of greater solubility in lipids than in the solvent containing the lipid (O'Brien and McCully 1981, Jensen 1962). Sudan IV stains fats, oils, waxes, and free fatty acids but not phospholipids. Sudan black B stains all lipids. The stains are often used on lipid-containing tissues that have been preserved in ethanol.

Lipoidal material occurred in 75 of 220 species tested, with no apparent taxonomic or growth habit pattern characterizing their presence (Baker and Baker 1975). The Bakers point out that both Hymenoptera and Diptera have digestive lipases and could use lipids as a minor energy source.

6. Organics

Erickson et al. (1979) used HPLC with an ultraviolet absorbance detector to locate nectar constituents containing double carbon bonds. One μL nectar samples were analyzed with a Waters Associates Model 202 liquid chromatograph furnished with a Schoeffel SF-770 variable-wavelength ultraviolet absorbance detector. They used a 30 cm x 4–mm Bondapak C_{18} (reverse phase) column and detector wavelength of 240 nm at 0.04 or 0.1 absorbance units full-scale sensitivity. The solvent was 25:72 acetonitrile:water at 0.5 mL/minute and 56 kg/cm^2 (800 psi) and the 10 mV recorder was operated at 0.5 cm/minute.

7. Microorganisms

Yeasts and bacteria can grow in nectar, particularly in nectars of low concentration that remain exposed for long periods (Baker and Baker 1975). Microorganisms might also be introduced through contamination by floral visitors (Gilliam et al. 1983). *Anthomyces reukaufii* is a common floral yeast (Meeuse 1982). In some cases, yeasts can hydrolyze the sucrose in nectar to yield glucose and fructose, reducing the likelihood of nectar crystallization (glucose plus fructose solutions tend not to crystallize; Meeuse 1982). The presence of microorganisms such as yeasts in nectar can also increase the amino acid levels and add alcohols and other fermentation products (Kevan et al. 1988). Kevan et al. (1988) were unable to detect any differences in foraging behavior of bees visiting milkweed flowers with and without yeast in their nectar.

Gilliam et al. (1983) cultured nectar samples to test for the presence of microbes. Duplicate nectar samples were tested in petri dishes containing Czapek solution agar (Difco), YM-1 agar (Wickerham 1951), and nutrient agar (Difco), and on YM-20 to test for osmophilic species. Samples were placed in tubes of thioglycollate medium without indicator (Difco) to test for anaerobes. One container of each medium was incubated at 25°C and the other at 37°C for 2 weeks. Any colonies that appeared were stained with gram stain and observed under the microscope. The concealed nectar of cotton flowers, which last only 1 day, harbored no microorganisms. Three of 23 citrus nectar samples contained gram-negative and rod-shaped bacteria. Saguaro nectar also contained these bacteria as well as gram-positive rods and cocci. Prickly-pear nectar contained no microorganisms.

Honey often harbors microorganisms. Spencer et al. (1970) identified yeasts in honey using proton magnetic resonance spectroscopy on the characteristic mannose-containing polysaccharides produced by different species.

8. Flavonoids

Mabry et al. (1970) and Wilkins and Bohm (1976) present methods for identifying flavonoids using chromatography, spectroscopy, and proton nuclear magnetic resonance spectroscopy. These techniques are so specialized that we will not attempt to cover them here.

9. Iridoids

Through a bioassay Stephenson (1982) was able to determine that *Catalpa* (Bignoniaceae) nectar contained iridoids. Iridoids act as a phagostimulant for larvae feeding on *Catalpa* leaves; in their absence, feeding does not occur. Stephenson found that larvae would take *Catalpa* nectar but not a sucrose solution of the same concentration. TLC confirmed iridoid presence.

10. Acids

Buchmann and Buchmann (1981) spot-tested nectar drops dried on chromatography paper with dichlorophenol-indophenol reagent. Spots turned red in the presence of acid.

11. Ions

Some floral nectars can have significant concentrations of ions, such as potassium. Waller et al. (1972) used an atomic absorption spectrophotometer to assay nectars for potassium, sodium, magnesium, and calcium. They found that onion (*Allium cepa;* Liliaceae) nectar contained 3,600–13,000 parts per million of potassium, whereas all of the other cations were present at much lower levels (e.g., Na 130–136 ppm, Mg 15–24 ppm, and Ca 34–94 ppm). The biological significance of these cations for pollinators is not clear (see Chapter 7, Section I.3), but the high levels of potassium do deter honeybees from collecting onion nectar.

G. Nectar Viscosity

In 1975 Baker pointed out that viscosity of sugar solutions rises exponentially with increased concentration and increases with decreasing temperature. Changes in viscosity change the rate of capillary flow of nectar, which can affect nectar uptake by capillary feeders such as hummingbirds. Highly concentrated nectar can become difficult for insects to handle. Honeybees dramatically decrease the rate of sucking when nectar concentration rises about 50% to 60%. The position of the mouthparts changes while sucking and watery saliva is released into the flower.

Kingsolver and Daniel (1983) modeled nectar uptake by capillary feeders and by suction feeders such as butterflies. They derived the following regression equations (which explain over 90% of the variation) for fluid density (ρ), surface tension (γ) and viscosity (μ) as a function of sugar concentration (S, measured as a percentage):

$$\rho = 1000 + 5.37\,S \qquad\qquad [\text{kgm}^{-3}],$$

$$\gamma = 7.18 \times 10^{-2} + 7.11 \times 10^{-5}\,S \qquad [\text{Nm}^{-1}],$$

$$\mu = \exp\,[0.00076\,S^2 + 0.0125 - 6.892] \qquad [\text{kgm}^{-1}\text{s}^{-1}],$$

They determined that the optimum nectar concentration for capillary feeders licking nectar from flowers is 35% to 40% sucrose for large nectar volumes and 20% to 25% for small volumes. For suction feeders the optimum is 20% to 25%.

Heyneman (1983) estimated nectar viscosity by a simple method: timing how long it took 2 µL of nectar to descend 50 mm along the central section of a vertical 10 µL capillary tube. These measurements ranged from about 1 second for pure water to over 20 seconds for nectars with more than 50% sugars. Ten or more measurements were made for each sample; measurements were reproducible within a few hundredths of a second (standard deviation usually less than 5% of the mean). Measurements can also be corrected for temperature differences by normalizing them to 15°C (using data on temperature dependence of viscosity for aqueous sucrose solutions as given in Fasman 1975). Heyneman converted descent times to viscosity using the equation:

$$\mu = \frac{gr^2}{8h} \cdot \rho t_d,$$

where r is the internal radius of the capillary tube; g is the acceleration due to gravity; h the test distance traversed by the nectar in time t_d, and ρ is the fluid density (assumed to be equal to the density of a pure sucrose solution of the same concentration).

Heyneman discusses some of the flow complications of this method and suggests that it may overestimate true viscosities by about 7%, but it is probably sufficient for most purposes.

H. Nectar in Insects

Various tests have been used to determine whether insects have been imbibing nectar, what volumes they consume, and the sugar concentration in the gut (see Chapter 7, Section F, for techniques for collecting nectar from insects). The contents of the honeysac in bees are of interest in nutrition and foraging studies.

Mosquitoes or other insects that feed at flowers can be identified by the presence of fructose in the gut. Unfed and blood-filled mosquitoes will not contain fructose. Van Handel's method (1972) indicates the presence of free fructose or the fructose component of sucrose (see also Magnarelli 1979 and Magnarelli et al. 1979 for detecting nectar in insects).

1. Pour 380 mL concentrated sulfuric acid into 150 of mL distilled water (cool the mixture while pouring cautiously).
2. Add 150 mg of anthrone to 100 mL of the dilute sulfuric acid. The reagent is good for a week.
3. Place the mosquito in a small vial or test tube and crush the abdomen to allow the reagent to penetrate the waxy outer layer. (Adding a drop of 1:1 chloroform-methanol removes the wax and crushing is not necessary.)
4. Add 0.5 mL of fructose reagent to the vial. Within a few hours at room temperature, the reagent turns blue-green in the presence of fructose. The color is stronger with large quantities of nectar. With prolonged exposure to fructose reagent (12–24 hours at room temperature), starch, paper,

212

and glycogen will produce color and the test will no longer be a valid indicator of the presence of fructose. If mosquitoes cannot be tested immediately, gut enzymes that break down fructose must be inactivated by freezing at -60°C or destroyed by heating at 80°–90°C for 15 minutes.

5. Color standards can be prepared from known sugar quantities, or samples can be matched to standards prepared from 10%–20% glycerol and food coloring matched to sugar standards.

Magnarelli and Anderson (1977) adapt Van Handel's procedure by adding 0.75 mL reagent to an insect crushed in a depression slide and score the reaction after 1 hour. They also use TLC to test nectar sugars. Insects were crushed in 95% ethanol on a ceramic spot plate and the liquid evaporated. Ten μL of 95% ethanol was added and then collected into 10-μL microcapillary tubes. Samples were spotted onto silica gel plates (EM Reagents No. 60F-254) along with individual 1-μL standards of fructose, glucose, and sucrose and a 2-μL standard containing 0.33% of each sugar. Sugars were double developed in one direction in 9 butanol:6 glacial acetic acid:3 ethyl ether:1 distilled water v/v, sprayed with aniline diphenylamine and heated to 110°C for 5 minutes. Sugars were evaluated by R_f values and colors.

I. Other Floral Rewards: Oils and Resins

Floral oils and resins can serve the same function as nectar in attracting pollinators (e.g., Vogel and Machado 1991, Steiner and Whitehead 1991a). Elaiophores are special oil-secreting floral glands that produce lipids attractive to female anthophorine bees (Anthophorinae, Anthophoridae). These bees have modified tarsal or abdominal structures for collecting oil, which is used, at least in part, to feed their larvae. Oil contains 9 cal/gram, which exceeds the caloric value of pure sugar (4 cal/gram) and is thus a concentrated energy source (Simpson and Neff 1983). Oils have been identified as saturated, β-acetoxy free fatty acids, or diacylglycerols (Simpson and Neff 1983).

Buchmann and Buchmann (1981) collected floral oil from *Mouriri myrtilloides* (Melastomaceae) with 5-μL capillary tubes and determined the weight on a semi-micro balance. About 0.02 mg of the bright yellow oil were collected from each gland. The color is believed to be due to carotenoids. Samples were frozen until analysis could be performed. Oils were spotted on Whatman number one filter paper and eluted by soaking in chloroform:methanol (2:1). Constituents were separated with two-dimensional TLC and then identified with gas chromatography after methyl esters of the fatty acids were prepared. The methyl esters were prepared using the following procedure:

1. Put the eluted samples into glass centrifuge tubes with glass stoppers. Add 1.5 mL of boron trifluoride (BF_3), 14% in methanol.
2. Heat the solution for 3 minutes in a water bath at 100°C.
3. Cool the mixture. Add 1 mL each of distilled water and chloroform.
4. Mix the contents thoroughly; remove the lower (water) layer to a centrifuge tube.
5. Dry the solvent with calcium chloride spheres.
6. Transfer the solvent to a Reactivial, dry it under dry nitrogen, and then redissolve the contents in n-pentane for injection into the gas chromatograph (1.0–1.5 μL samples).

Buchmann and Buchmann used gas liquid chromatography (Varian aerograph model 274010-20; Supelco glass and stainless steel packed columns, GP 3% SSP-2310/2% SP-2300 on 100/120 chromosorb) and mass spectroscopy to identify the methyl esters of the fatty acids. The spectra were analyzed by comparison with a computerized data bank of authenticated mass spectra. Analysis indicated saturated fatty acids with carbon chain lengths from C_{12} to C_{24}.

Simpson et al. (1977) characterized lipids in floral oils with TLC, gas chromatography, infrared spectroscopy, nuclear magnetic resonance, and mass spectroscopy. Extracts from all plant species were similar in nature, consisting predominately of free, saturated, fatty acids with a β moiety and methyl ester chain lengths ranging from C_{16} to C_{20}. Simpson et al. (1979) determined that oils of seven different species of *Krameria* (Krameriaceae) were made up of identical compounds. An initial extraction of materials preserved in the field indicated the presence of methyl esters of acetoxy fatty acids, which were not found in subsequent analyses. They suggest lipids be preserved in chloroform only, as preservation in methanol-chloroform may produce artifacts. (See also Seigler et al. 1978 on analysis of floral oils.)

Several unrelated species of tropical flowers offer resin as a floral or floral-bract reward for bees. Resins are used in nest construction by some types of bees in the Apidae (euglossine, meloponine, and honeybees) and Megachilidae (*Hypanthidium, Heriades*) (Armbruster and Webster 1979, Armbruster 1984, Armbruster and Mziray 1987). The resins provide a waterproof construction material as well as having bactericidal and fungicidal properties. In the process of collecting resin, the bees are dusted with pollen. Unlike pollen-collecting bees, the bees (Apids) do not groom the pollen into their corbiculae, which are used for carrying resin, so pollen may be readily transferred to the next flower. At least one genus of resin flowers, *Clusia* (Guttiferae), secretes nectar instead of producing resin in high-elevation areas where resin-collecting bees are absent (Armbruster 1984). Resins have been analyzed with TLC, mass spectrometry, and NMR. Floral resins appear to be composed of triterpenes that do not harden immediately after secretion but remain liquid.

J. Suggestions for Planning Studies

1. **Effect of nectar removal on nectar accumulation.** If you are trying to determine nectar production throughout the lifetime of a flower, periodic removal may have the effect of increasing nectar production (Gill 1988, Búrquez and Corbet 1991, Pyke 1991).
2. **Effect of variation in nectar production.** Individual plants may vary significantly in rates of nectar production, and this variation may have important effects on reproductive success (e.g., fruit set; Real and Rathcke 1991). Individual plants may also vary from year to year in their levels of

nectar production (Real and Rathcke 1991). Relative humidity and soil moisture may also significantly influence characteristics of nectar.

6. Mating Systems

An understanding of pollination biology is crucial for interpretation of plant mating systems. Mating systems determine the pattern of genetic transmission and affect the organization of genetic variation in a population. Inbreeding restricts heterozygosity and gene migration through pollen flow, reducing the variation within populations and increasing the variation among populations. In contrast, outcrossing, which promotes gene flow, reduces the likelihood of microgeographic differentiation and population substructuring. There are five basic types of plant mating systems (Brown 1990), although Jain (1984) stresses that these form a continuum. The basic types are: (1) predominantly selfing, (2) predominantly outcrossing, (3) mixed mating, (4) partial apomixis, and (5) partial selfing of gametophytes (in ferns or artificial dihaploid breeding; Brown 1990).

Self-fertilization has evolved repeatedly in the plant kingdom. About 20% of higher plants are predominantly selfing (Brown 1990), and in these species occasional outcrossing has important consequences. Effective gene flow in selfing species is very small, and low levels of outcrossing dramatically increase gene flow. In contrast, selfing in predominately outcrossing species may result in high inbreeding depression (Lande and Schemske 1985). Yet even in outcrossing populations, some self-fertilization combined with near-neighbor mating and limited seed dispersal can lead to reproductive isolation and subpopulation formation (Levin and Kerster 1974, Brown 1990). Studying patterns of gene transmission in natural populations is important to understanding the evolution and maintenance of self-fertilization, dioecy, and other breeding systems that restrict the number of available mates. In fact, understanding the evolution of selfing, under the opposing force of inbreeding depression has been one of the major reasons for studying plant mating systems (Holsinger 1991).

Knowledge of mating systems also has important practical applications in crop, forest, and orchard management (Brown et al. 1985) and potential application in management of rare and endangered plant species to conserve their genetic variability. For example, management plans for these species will depend on quantitative assessment of neighborhood size, effective population size, and gene migration (Brown et al. 1985).

Mating systems are affected by environmental factors (Barrett and Eckert 1990, Mitton 1992), and because they are regulated by genes they are also subject to selection (Brown 1990). Consequently, the mating system can differ among populations (Hamrick 1982, Schoen 1982, Holtsford and Ellstrand 1990, Lyons and Antonovics 1991). Even within populations, individuals can exhibit differences in the amount of selfing or outcrossing (Clegg and Epperson 1985, Brown et al. 1985, Lyons and Antonovics 1991, Mitton 1992). For instance, the plant flowering at one end of the phenology curve may be predominantly selfed whereas those during peak flowering are predominantly outcrossed; the degree of stylar exertion can affect the percentage of self-pollen reaching the stigma (Thomson and Stratton 1985); rare color morphs receiving fewer pollinator visits may self more (Horovitz and Harding 1972, Brown and Clegg 1984); plants in the center of a dense population may be mostly outcrossed whereas those on the outskirts are mostly selfing; plant density may result in different mating systems in different populations (Levin and Kerster 1969, Ellstrand et al. 1978, Farris and Mitton 1984); populations where pollinators are common may show different patterns from those where pollinators are rare. Within individual trees, lower portions of the crown may show more selfing than higher portions (Shaw and Allard 1982). Levels of inbreeding can be manipulated by altering the timing and duration of the pollination period (El-Kassaby and Davidson 1991).

A. Field Tests for the Ability to Self-Pollinate

Until the 1970s, mating systems were determined by studying plant morphology, performing breeding experiments to test for

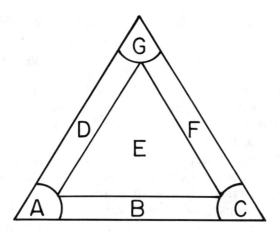

Figure 6–1. The reproductive triangle. A= cross-fertilized plants (including self-incompatible, dichogamous, and dioecious plants); B = partially selfing species; C = predominantly selfing; D, E, F = facultative apomicts that fall closer to one of the preceding categories when sexual; G = obligate apomicts; H = apomicts, but inadequate information is available to classify them more specifically as D, E, F, or G. Redrawn from Fryxell 1957. Permission granted from New York Botanical Garden (Botanical Reviews) and author.

self-compatibility and apomixis, and studying the movement of marker genes between generations. In 1940 East reviewed the occurrence of self-incompatibility in different orders of plants, and in 1957 Fryxell summarized the literature and presented mating system information on more than 1,200 species, giving each species a designation somewhere on a schematic triangle (Figure 6–1).

For many pollination studies, the researcher simply needs to know whether a plant can set seed in the absence of pollinators. Features like dioecy, protandry, and protogyny eliminate or reduce self-pollination, but in many systems with perfect flowers and little or no temporal separation of sexual phases, breeding behavior must be determined experimentally. Experiments of this type are usually easy to conduct, but there is some room for misinterpretation if the appropriate experimental treatments are not included.

The standard approach has been to cage flowers prior to anthesis in order to exclude pollinators (for techniques, see Chapter 2, Section C; Chapter 3, Section B). This protocol, when combined with transfer of self-pollen from anther to stigma of perfect flowers, will provide answers to the following questions: (1) Are flowers able to self-pollinate and self-fertilize? (However, when performing these self-pollination treatments, beware of the possibility of cryptic self-fertility; see Section G.4.) (2) Is a pollinator required for seed set? It will not, however, permit distinction between facultatively xenogamous (outcrossing) plants and facultatively autogamous (selfing) plants. R. Cruden (*personal communication*) suggests that an open-pollinated emasculation treatment should be included with the above protocol in order to make this distinction. Because autogamous flowers are visited infrequently by pollinators, emasculated, autogamous flowers will rarely set fruit. Facultatively xenogamous flowers will be visited by insects more frequently even when emasculated and are more likely to set fruit (R. Cruden, *personal communication*). For this test, only a few flowers on multiflowered plants should be emasculated, so that pollinators do not alter their foraging behavior. If the normal pollinators of the flower are foraging for pollen, the absence of anthers on emasculated flowers may alter their behavior, resulting in fewer or shorter visits. Kearns (1990) and Cruden and Lyon (1989), both working with species of *Linum* (Linaceae), found that some types of potential pollinators (muscoid flies, small bees) remained on emasculated flowers for shorter periods of time than they did on unaltered flowers. In addition, emasculation may alter the posture of the pollinator so that it misses the stigma. The emasculation of all the flowers on multiflowered plants will demonstrate whether cross-pollination can occur, as seed produced must be from cross-pollination unless apomixis is occurring. Cruden and Lyon (1989) offer a cautionary note about the interpretation of emasculation treatments, pointing out that they may not distinguish between xenogamy and geitonogamy. Table 6–1 presents the results that could be expected from a field protocol involving five treatments (R. Cruden, *personal communication*).

Whether or not seed will result from the caged autogamous flowers depends on whether there is some mechanism for transfer of pollen to the stigma in the absence of flower visitors. Some plants, such as some orchid species, will deposit the pollinium on the stigma as a flower wilts, thereby ensuring seed production by selfing in the absence of cross-pollination. The difference in seed production

**Table 6–1. Field tests for the ability to self-pollinate.
Pluses indicate seed set, minuses indicate no seed set**

Treatment	Breeding System		
	Autogamous	Facultatively Xenogamous	Xenogamous
Caged	+/-	+	-
Caged and emasculated	-	-	-
Caged and self-pollinated	+	+	-
Emasculated and open-pollinated	-	+	+
Open-pollinated	+	+	+

Source: R. Cruden, *personal communication.*

between the caged and caged, emasculated autogamous categories may give an indication of the frequency of this sort of self-pollination. If all treatments result in seed set or caged, emasculated flowers produce seeds, apomixis may be responsible.

It may also be important to control for the activity of aphids and thrips during studies of breeding systems. Although these insects are not normally thought of as pollinators, they can affect the results of some studies (R. Cruden, *personal communication*), and it may be necessary (or prudent) to include a cage treatment with insecticide in the experimental protocol. Observation of the target species should reveal quickly whether these groups of insects are flower visitors.

Although the experiments described above will provide information on the potential methods by which seed set can be effected in a particular species, for species with a mixed selfing and outcrossing mating system, they do not reveal how much of the seed set normally results from these different possibilities. Additional work may be required to determine quantitative aspects of plant mating system parameters. If available, appropriate genetic markers can be used to make this determination for researchers working with parents of known genotype and looking at progeny genotype arrays that result from pollinator visits (e.g., Ennos and Clegg 1982).

Charlesworth (1988) published an arithmetical procedure for estimating outcrossing rates in populations of self-compatible plants based on differences in viability of zygotes produced by open pollination, hand-outcrossing, and hand-selfing. The procedure is based on the ability to measure a difference between selfed and outcrossed seeds shortly after fertilization (e.g., seed size corrected for any correlation between seed number per fruit and seed size; or the fraction of nonaborted seeds of all fertilized seeds per flower when this differs between selfed and outcrossed flowers). The selfing rate S is determined from p_x (the measure for hand-outcrossed flowers), p_s (the measure for hand-selfed flowers), and p_o (the measure for open-pollinated flowers in a natural population).

$$S = \frac{p_x - p_o}{p_x - p_s}.$$

A method for calculating the variance of the estimate S is presented in Charlesworth (1988).

B. Quantitative Studies of Mating System Parameters

The study of patterns of gene inheritance in plant populations is in a stage of rapid growth. Because this book is intended to present experimental techniques useful to pollination biologists, it cannot cover all the theoretical and statistical models that are important to the study of plant population genetics. Several excellent books (Brown et al. 1990, Soltis and Soltis 1989, Lovett Doust and Lovett Doust 1988) review the literature of the field. This section is intended as an overview of the subject of patterns of gene inheritance.

Researchers may be interested in comparing estimates of mating systems of individuals or estimating the mating system on a population level. On the individual level, the following questions are asked (Brown et al. 1989):

Is autogamy complete or partial?

Is apomixis complete, or does some normal sexual reproduction
    occur? Is there pseudogamy?
Are the seeds in a single fruit all from the same pollen parent?
Are matings related to the spatial distribution of parents?
What are the proportions of chasmogamy and cleistogamy?
Is there cryptic self-incompatibility or gametophytic competition?

At the population level, the researcher may want to trace genotype
frequencies in space or time and perform quantitative tests to
compare populations to mating system models (Brown et al. 1989).

1. Electrophoresis and Molecular Techniques

The electrophoretic revolution made it possible to extract more
detail on patterns of gene inheritance through the study of allozyme
variation. Most plant species have several polymorphic allozymes,
whereas species with morphological markers are harder to find. In
addition, many morphological markers exhibit dominance and
obscure heterozygosity in progeny. Allozymes of heterozygotes
express both alleles on the electrophoretic gel.

Electrophoresis was the first molecular tool to gain widespread
use for examining genetic variation. It is based on the principle that
proteins carrying different charges will migrate at different rates
through a medium in an electric field. A gel (starch, polyacrylamide,
or agarose) serves as the medium; starch gels are usually used for
mating system analysis, as they are easy to use, inexpensive, and
nontoxic. Polyacrylamide gels offer high resolution, but the acry-
lamide used to make them is a neurotoxin until polymerized. In
addition a single polyacrylamide gel is used to test for only one
enzyme, whereas starch gels may be sliced for multiple stains so that
three or four enzymes can be tested per gel. Most labs pour their
own gels.

Wells containing buffer solutions are placed at both ends of the
gel (Figure 6–2), and thin sponge cloths are set in each well with
one end in buffer and the other resting on the gel. Electrodes
connected to the sponges create a charge that passes through the
gel.

Tissue homogenates to be loaded onto gels are prepared by
grinding or sonicating samples in buffer solutions. Homogenates
are frozen at very low temperatures if further analysis will be

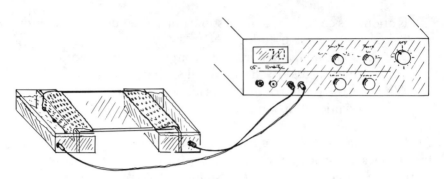

Figure 6–2. Apparatus for gel electrophoresis.

delayed. For starch gel electrophoresis, the liquified tissue samples are wicked onto small pieces of filter paper cut to uniform size. A groove is cut in the gel and the paper wicks are inserted, spaced evenly across its length. The gel is connected to the buffer/electrode system and current applied for several hours. Then the gel is stained for a specific enzyme. Enzymes of the same shape, size, and electrical charge migrate at the same rate (the gel acts with a "molecular sieving effect"; Maurer 1971). The proteins coded for by each allele are both expressed on the gel, so in a two-allele system, there will be either a "fast," "slow," or "fast and slow" mark on the gel. The words "fast" and "slow" indicate the relative speed of enzyme migration across the gel, so the "fast" enzyme will be the one farther from the origin. If there is only one "fast" mark, the sample is homozygous with both alleles coding for the faster migrating protein. Including a standard of a known heterozygote on each gel will help verify interpretations (Fenster, *personal communication*).

Shaw and Prasad (1970), Soltis et al. (1983), Rieseberg and Soltis (1987), and Kephart (1990) provide recipes and buffer systems that have been successful for specific plant enzymes. Weir (1990) presents methods for statistical analysis of genetic data. See also Soltis and Soltis (1989).

Allozyme techniques can only assess differences in alleles that code for enzymes (Clegg 1990). Molecular techniques that detect DNA sequences can examine any mutations and provide substantially more information about genetic diversity (Clegg 1990). Sequencing techniques are more time-consuming than electrophoresis and require more sophisticated, expensive equipment but promise

224

to revolutionize our current understanding of genetic diversity. See Clegg (1990) for a discussion of restriction fragment length polymorphisms (RFLPs), DNA sequencing, organelle DNA, and genetic fingerprinting and their future for plant molecular genetics.

2. Models

Early quantitative studies of the genetics of mating systems were based on Sewall Wright's F statistics and calculated inbreeding coefficients as deviations from random mating (Hardy-Weinberg equilibrium). With the advent of electrophoresis, more elaborate statistical models have been developed, and there has been a boom of research in the area of plant mating systems (Clegg 1980). Other models that employ estimates of outcrossing rates include the mixed mating and effective selfing models, analysis of correlated matings, estimates of male fertility, and paternity analysis of progeny (Brown 1990).

a. Inbreeding coefficients

The inbreeding coefficient, F, can be derived from data collected from one generation sampled at one point in time (Mitton 1992). Calculations of F are based on the assumption that the population is in inbreeding equilibrium and that self-fertilization is the only form of inbreeding. Although this model is acceptable for preliminary estimates, it is based on assumptions that are generally violated for estimating selfing rates from F. The estimation of F itself is generally unbiased, but this quantity by itself is a facet of population structure and not the mating system (Ritland 1983; K. Ritland, *personal communication*). Also, F can change within generations, being reduced after each episode of selection against selfed individuals (Ritland 1990a), and near-neighbor matings can result in "effective selfing" (Ritland 1986). If adult tissues are sampled to calculate F, postzygotic selection, differential seed and seedling viability of inbred and outcross progeny are not accounted for, and the use of F as an indicator of the mating system may be biased downward (Mitton 1992).

Ritland (1990b) has extended the use of the F statistic by developing a procedure using changes in F between generations to measure inbreeding depression. The need for very large sample size, loci with several polymorphisms, and populations without extreme

Table 6–2. Probabilities of offspring genotypes given maternal genotypes

		Parental Genotype		
		AA	Aa	aa
Probability, given parental genotype, of progeny genotype	AA	$s + tp$	$s/4 + tp/2$	0
	Aa	tq	$s/2 + t/2$	tp
	aa	0	$s/4 + tq/2$	$s + tq$

Notes: t = frequency of outcrossing, s = frequency of selfing, p = frequency of A allele, q = frequency of a allele.
Source: Ritland 1983.

levels of outcrossing or inbreeding limits the application of the model (Ritland 1990a).

b. Mixed mating model

The mixed mating model is a robust model originally used for inbreeding crops. It assumes a mixture of outcrossing and selfing within a population. Outcrossing occurs at frequency t and selfing at frequency $s = 1 - t$. Expected genotype distributions for a single locus situation are present in Table 6–2. The model assumes (Clegg 1980, Brown et al. 1985, Mitton 1992):

- All maternal genotypes are exposed to the same pollen pool.
- Gametes or haplotypes in the pollen pool are constant over time and space.
- Outcrossing rates are the same for all maternal plants.
- No selection occurs between pollination and the time when seeds or seedlings are analyzed.
- Mendelian segregation of alleles in both pollen and ovules (e.g., no meiotic drive, no segregation distortion, no gametic selection).

The pattern of inheritance is determined by analyzing genotypes in progeny arrays, ideally those with codominant expression (as seen

with allozymes). This model can be applied where maternal genotypes are known, as in breeding experiments with alternate rows of
homozygotes planted in the field. This is the simplest form of an
estimator for t, and it can produce estimates of t for an individual
plant if p is known (K. Ritland, *personal communication*).

$$\hat{t} = \frac{\text{proportion of heterozygous offspring from aa mothers}}{\text{frequency of A in the pollen pool}} = \frac{H}{p}.$$

If heterozygous offspring of both AA and aa mothers are used,
the value of p for the maternal and paternal gene pool can be
estimated through an iterative process (Jain and Marshall 1967).
However, it usually is not necessary to determine parental genotypes
independently with electrophoresis, as maternal genotypes can be
inferred from progeny arrays. Although the inference should become less certain with small progeny arrays, the uncertainty is
incorporated into statistical procedures so that even progeny arrays
of size two can be used to estimate t and p if the progeny are half-sibs
(Ritland 1986, K. Ritland, *personal communication*). Combining
estimates of t across the two homozygous parental genotypes results
in a superior average estimate that is best produced by using
maximum likelihood procedures. If alleles are codominant, as in
most electrophoretic studies, progeny of heterozygous parents can
be used as well to develop a matrix of probabilities of offspring
frequencies, giving greater statistical power to the maximum likelihood estimate of outcrossing (t). Computer programs for analyzing
matrices can be extended to multiple loci (Ritland 1983).

Table 6–2 presents the probabilities of offspring types given the
maternal genotype. To obtain estimates of s and t, one numerically
finds the best fit of observed frequencies of transitions from parental
to progeny genotypes to the expected frequencies, as determined by
Table 6–2. Several statistical procedures (e.g., least squares, maximum likelihood) can be used for this.

As in any ecological model, some of the mixed mating model
assumptions are simplistic (Clegg 1980, Brown et al. 1985, Mitton
1992). All maternal plants are not exposed to the same pollen
genotype if pollen flow is limited and there is spatial heterogeneity
in genotypes. Assortative mating (e.g., based on pollinator preference for a floral morph, or time of flowering) results in variation in

the pollen available to different females. Outcrossing rates may differ among females and differential pollen fertility may vary outcrossing rates of males. Functional gender (relative contribution to the next generation through pollen or ovules) can also vary among individuals (Wyatt 1983).

If older seedlings are used to assess genotypes of the offspring, some of the assumptions of the model are violated and estimates of t may be biased, indicating more outcrossing than actually occurs. The effects produced by differential pollen fertility, postzygotic abortion, and differences in seed viability will not be taken into account (Mitton 1992). Segregation distortion and meiotic drive could also change the expected progeny genotype frequencies, although there has been little evidence of this to date (Mitton 1992). In addition to these violations of the assumptions, difficulties in interpretation arise when data from several different loci provide different estimates of outcrossing.

c. Biparental inbreeding and correlated mating models

New models are continually being developed to overcome some of the problems of the mixed mating model. Ritland's effective selfing model (1984, 1986) was designed to deal with biparental inbreeding (Uyenoyama 1986), or mating within local subpopulations of individuals with similar genetic makeup. Waller and Knight (1989) were able to distinguish between levels of inbreeding due to geitonogamous pollen transfer and assortative or biparental mating in *Impatiens* (Balsaminaceae). Their procedures are applicable to other species where multilocus data in progeny arrays can be obtained.

Correlated mating analyses (Schoen and Clegg 1984, Ritland 1989, Brown 1990) were developed to use on insect-pollinated species. The mixed mating model was originally developed for wind-pollinated species in which stigmas receive pollen from multiple sources and all progeny from one female consist of half-sibs. In entomophilous species, an insect often deposits lots of pollen from a single donor, resulting in seeds that are full sibs and all outcrossed. Using a mixed mating model on such a system results in overestimates of S. Correlated mating models allow progeny arrays to consist of full-sibs (Schoen and Clegg 1984) or mixtures of half- and full sibs (Ritland 1989, Brown 1990).

Correlated mating analysis examines the relationship among siblings in a cohort. For example, are they from the same paternal genotype and from the same type of mating (outcross or self), or are seeds within a fruit more likely to be full sibs? Studies of differential male fertility indicate that particular clones or certain age cohorts may have exceptional representation in the pollen pool, and pollen genotypes in the pollen pool may vary seasonally and among years (reviewed in Brown 1990, Mitton 1992).

d. Multilocus estimates of outcrossing

Multilocus estimates of t are less affected by nonrandom outcrossing and selection than single-locus models (Ritland and Jain 1981, Shaw et al. 1981, Brown et al. 1985). These models can at least partially separate biparental from uniparental inbreeding (Ritland 1990b). Data from three or four loci can be used to produce a t value with a variance approaching the minimum variance based on large numbers of loci (Brown et al. 1985). Because multilocus estimates reduce ambiguity in paternity and permit more unequivocal designations of outcrossing events (Ritland 1983, Mitton 1992), levels of inbreeding based on multilocus data tend to be lower.

Multilocus estimates are also valuable for estimating the outcrossing rate of an individual (Ritland 1990b). These estimates could be used by pollination biologists to correlate the rate of selfing with parameters such as pollen removal or insect visitation (K. Ritland, *personal communication*).

e. Paternity analysis

There has been much interest in recent years in trying to measure male reproductive success, about which little is known in comparison to female reproductive success. In the past, male reproductive success has typically been estimated indirectly, using indices such as pollen production, pollinator visitation, pollen or pollinia removal, or deposition on stigmas (see studies cited in Broyles and Wyatt 1990a). Another method of addressing this question is paternity analysis, which seeks to determine by genetic methods the actual contributions of individual males to the seeds collected during an experiment. Although studies of paternity analysis may involve some pollination techniques, most of the methods are analytical or statistical, so we will only outline them here. For details, see the references we have cited.

The genotype of each individual in a population is typically established using polymorphic loci determined by electrophoresis. Then the genotype of seeds collected from the plants is determined after germinating the seeds. Paternity analysis takes these progeny of (usually) known females and attempts to assign either a specific paternal individual or a paternal genotype. If the population is small and if the seeds are the result of controlled crosses, methods of paternity exclusion may work well. Unambiguous assignments of paternity are difficult under field conditions except possibly in small, very polymorphic populations (Brown 1990). For larger populations one must resort to statistical analysis of the likelihood of paternity, such as the highest ln of odds ratio (LOD) (Meagher 1986). Maximum likelihood models have been developed for assigning paternity, but homozygous genotypes may be correctly assigned more often (Devlin et al. 1988). Other models have been developed to assign paternity equally among all possible parental genotypes (Devlin et al. 1988) and to factor in the distance between maternal plants and their potential pollen donors (Adams and Birkes 1991). For information on these techniques and their use see Ellstrand 1984; Bertin 1990b; Hamrick and Murawski 1990; Broyles and Wyatt 1990a, 1990b.

Paternity analyses provide information on outcrossing rates of individuals, differences in functional gender, and assortative mating (reviewed in Brown 1990). They can also show mean and variance distance of pollen flow within a population and estimate the number of crossings from outside of the population (Hamrick and Schnabel 1985, Mitton 1992).

f. Computer programs for mating system analysis

A series of computer programs for estimating mating systems via the analyses of progeny arrays is available from K. Ritland, Department of Botany, University of Toronto, Toronto, Ontario M5S 1A1, Canada (or e-mail: Ritland@utcs.utoronto.ca). Please send a floppy disk along with your request. The programs employ maximum likelihood procedures and chi square to assess goodness of fit of the data to the model.

C. Analyzing Differential Reproductive Success

Individual plants may differ in their reproductive success. Lyons et al. (1989) divide the variation in success into three components: differences in male performance, differences in female performance, and differences due to interaction effects between male and female partners. The magnitude of the variance associated with the average performance in each component can be analyzed with complete or partial diallele crosses or factorial crosses (Lyons et al. 1989). Reproductive success of individuals can be measured at the pollination phase, the fertilization phase, and the seed maturation phase. Lyons et al. (1989) provide a thorough discussion of experimental design and analysis of experiments attempting to partition variation in plant reproductive success.

D. Neighborhood

Neighborhood, a concept introduced by Sewall Wright, refers to the area within a population where mating is effectively panmictic. The idea of neighborhood leads to the concept of "isolation by distance"; that is, where gene flow is restricted to small areas, local differentiation occurs and individuals separated by several neighborhoods are less similar genetically than those within a neighborhood. Seed and pollen dispersal distance affect neighborhood size. Therefore neighborhood is also affected by pollinator movements, pollen carryover, and differences in foraging behaviors when several pollinators visit a plant. Neighborhood area corresponds to a circle of radius 2σ, where σ is the standard deviation of gene dispersal. N_e is the number of individuals of an idealized population that would give rise to the observed rate of inbreeding (Wright 1943, 1946). The number of individuals in a neighborhood sets the upper limit for N_e. However, variations in fertility and population size reduce N_e in comparison to the actual census number of individuals in a neighborhood (Fenster, *personal communication*). Other factors that reduce N_e include selfing (pollen dispersal = 0), mating among closely spaced individuals, assortative mating, and large annual fluctuations in the number of breeding individuals (Levin and Kerster 1974).

For simplicity, let us initially assume that all gene dispersal is due to pollen movement from a source plant p. We assume that the population is stationary and that pollen moves equally in all directions so that mean pollen movement = 0. Because Wright's initial estimates of σ^2 were measured along a single axis (axial variance), variance determined with this procedure must be corrected to axial variance by dividing the radial variance by 2. The variance of pollen movement thus equals

$$\sigma^2_{(axial)} = \frac{\sum (p_i - (\bar{p}))^2}{n_p},$$

where n_p is the number of pollen donors and p_i is the distance of pollen movement of pollen grain i. (Note: From here on we will assume we are dealing with the corrected or axial variance). Because pollen is assumed to move equally in all directions ($\bar{p} = 0$) the equation reduces to

$$\sigma p^2 = \frac{\sum (p_i)^2}{n_p}.$$

Next, one must take into account gene dispersal by seed movement, which can be calculated in a similar manner. Measures of the total variance combine both pollen and seed dispersal variance (Crawford 1984):

$$N_e = 12.6 \left(\frac{\sigma_p^2}{2} + \sigma_s^2 \right) \cdot d.$$

The variance component from pollen is divided by 2 because dispersal through male function incorporates both pollen and seed dispersal. Because male gametes disperse and female gametes do not, the variance of pollen dispersal is halved. The variance component from pollen is multiplied by the outcrossing rate, as the dispersal of pollen from self-fertilization equals zero (Richards and Ibrahim 1978, Fenster 1991c).

Schmitt (1980) calculated variances in pollen movement caused by two different pollinators and multiplied each variance by the relative proportion of that pollinator's visits. Her work indicates that a small proportion of visits from a pollinator with long inter-plant flight distances can greatly increase neighborhood area.

Initial estimates of gene flow in populations indicated small neighborhoods and restricted dispersal (Ehrlich and Raven 1969, Levin and Kerster 1974). Some recent estimates (Schaal 1980, Levin 1983) suggest that gene flow in some species is more substantial than previous estimates indicated (see Hamrick 1987). Undetected pollen carryover and secondary seed movement can increase dispersal distances (Levin 1981, Hamrick 1987). Biased estimates also result if the following assumptions are violated (as they commonly are): (1) that pollinator flights are random in direction; (2) that pollen from donors at any distance from the maternal plant is equally compatible; (3) that seed viability is independent of the distance between parental plants.

Fenster (1991c) recently published results of experiments estimating neighborhood size of *Chamaecrista fasciculata* (Leguminosae). His study incorporated pollinator flight distances, carryover measures, and seed dispersal, as well as measures of dispersal of electrophoretic markers. In addition, he divided the flowering season into periods and factored in phenological variation in seed set, outcrossing rate, floral density, and pollinator activity. Neighborhood area estimates based on gene movement were greater than estimates based on pollinator flight distances, probably because of the effects of carryover. Yet the larger area estimate of a circle with radius 2.4 m still indicates restricted gene flow and potential for population subdivision. Subsequent research (Fenster 1991d) accounted for fitness disadvantages of progeny of individuals from the same neighborhood as compared with those from different neighborhoods. These data increased estimates of gene flow and neighborhood area to a circle with radius 3.0 m.

Procedures for estimating pollen movement are described in the Chapter 4, Section L. For additional information on genetic neighborhoods, you may wish to consult Levin and Kerster 1974 for a review of studies measuring pollen and seed movement or Roberds et al. 1991 for a model derived from Wright's neighborhood concept to quantify mating in clonal seed orchards of wind-pollinated species.

E. Carryover

Carryover is the movement of pollen from a pollen donor to flowers beyond the first pollen-receiving flower. Carryover means that all the pollen picked up by a vector is not deposited on the first flower that it visits. When carryover occurs, gene flow via pollen movement is underestimated by merely measuring pollinator flight distances between flowers (Levin and Kerster 1974). Carryover is often variable and difficult to measure, and studies of pollen carryover are relatively few. Differences or large variances in estimates for different plant-pollinator systems (Thomson and Plowright 1980, Campbell 1985), within a single species (Lertzman and Gass 1983, Waser and Price 1984, Geber 1985), or between morphs of a distylous species (Feinsinger and Busby 1987) make generalizations about carryover difficult (Zimmerman 1988b).

To estimate the number of flowers in a sequence that receive pollen from the initial donor, Hessing (1988) uses the point where pollen deposition from the initial flower is no longer assured (e.g., where the average number of grains received is less than one). When carryover is modeled as exponential decay, the curve does not intersect the x-axis and therefore overestimates the sequence number of the flower when pollen deposition is low (Hessing 1988). Also, if flowers are emasculated for the study (to avoid self-pollen on the stigma), pollinators may assume different positions and spend less time on flowers and thus change pollen deposition patterns. If plants need a minimum number of pollen grains to effect fertilization, one or two pollen grains that carried over long distances may not actually effect fertilization (Hessing 1988).

If pollen carryover is approximated by exponential decay, carryover loses much of its significance for gene flow between plants when multiple flowers are visited on a single plant or clone (Waddington 1981, Thomson et al. 1982, Galen and Plowright 1985). However, if optimal distances for pollen donation have an effect on which pollen grains fertilize ovules (Price and Waser 1979), pollen that has been carried over intermediate flowers may be the most successful (Waddington 1981).

Campbell (1985) describes carryover as the number of pollen grains deposited on a given flower in the sequence divided by the number deposited on the initial flower:

$$C_x = \frac{Y_x}{Y_1}$$

where x equals the sequence number of the flower. She compared empirical data from *Nomada* bees visiting *Stellaria* (Liliaceae) flowers to exponential and linear models. The fits were similar (r^2 = .51 and r^2 = .67, respectively). Carryover was different for bee flies and small bees.

Methods for measuring pollen movement are described extensively in the Chapter 4, Section L, and the effect of carryover on gene flow is described in Section C of this chapter.

F. Optimal Outcrossing Distance

Waser and Price (1983, 1985b, 1989, 1991a; Price and Waser 1979; Waser 1987; Waser et al. 1987) reasoned that limited gene flow in plant populations should result in genetic similarity of adjacent individuals. Mating between near neighbors could result in inbreeding depression, whereas mating between distant neighbors could disrupt local adaptions. Therefore, matings between individuals at some intermediate distances should result in the greatest fitness benefits. They coined this the optimal outcrossing distance. They have been able to demonstrate optimal outcrossing distances for *Delphinium nelsonii* (Ranunculaceae) and *Ipomopsis aggregata* (Polemoniaceae). For *D. nelsonii,* pollen quality was best, seed set was greatest, and offspring were more successful from 10-m crosses than from 1- or 100-m crosses. Similarly, 10-m crosses between *I. aggregata* plants resulted in more vigorous offspring than 1- or 100-m crosses.

Sobrevila (1988) reviews the studies that attempt to demonstrate inbreeding or outbreeding depression based on distance of pollen source. Ten of 14 species studied showed evidence of decreased reproductive success of crossings between plants in close proximity compared to distant plants. Evidence for outcrossing depression was found in all four of the species where very distant crosses were performed. Results of her own work on *Espeletia schultzii* (Compositae) is less straightforward. Early in the flowering season there were no distance effects on reproductive success, yet late in the season, when resources might be limiting, distance effects were apparent. Other attempts have failed to demonstrate optimal outcrossing distances (Carr 1990; B. Inouye, unpublished).

G. Self-Incompatibility

Self-incompatibility prevents the production of seed following self-pollination. It may result from a variety of mechanisms. Self-pollen either may not adhere to the stigma, adhere but not germinate, or germinate but be unable to penetrate or grow down the style (Richards 1986). Self-incompatibility is generally considered to be a prezygotic mechanism and does not take into account seed abortion or reduced viability of inbred offspring (de Nettancourt 1977). However, some authors use the term "postfertilization incompatibility" to describe systems like that found in *Asclepias syriaca* (Asclepiadaceae) where self-pollen tubes grow and penetrate ovules but the embryo does not divide and the fruit aborts (Kahn and Morse 1991).

Incompatibility systems are broadly classified as gametophytic or sporophytic, depending on whether the incompatibility reaction with the female tissue is mediated by the genotype of the haploid pollen grain or by the genotype of the diploid anther that produced the pollen. For a detailed examination of incompatibility systems, see de Nettancourt (1977). A more recent review is presented in Richards (1986).

1. Gametophytic Systems

The most common incompatibility system is the one-locus multiallelic gametophytic system (Richards 1986). Both alleles at what is termed the S locus are expressed by the female stigma and style, and one allele is expressed by the pollen grain. Large numbers of different S alleles may coexist in a population and are termed S_1,

S_2, S_3, etc. If the S allele in the pollen matches one of the S alleles in the female tissue, fertilization will not occur. Consequently, seeds will always be heterozygous.

Two parents can be fully compatible (e.g., female S_1S_2 x male S_3S_4) if neither haploid pollen grain shares the female alleles, semi-compatible (e.g., female S_1S_2 x male S_1S_3) if only half the pollen grains (S_3) can effect fertilization, or incompatible (e.g., female S_1S_2 x male S_1S_2) if each pollen grain shares one of the female alleles (Richards 1986). Single-locus gametophytic systems can be detected by the presence of semicompatibility and by crossing individuals and looking at the types of offspring produced (as determined by additional crosses of the offspring; Richards 1986). A maximum of four heterozygous genotypes can occur among the offspring if parents do not share any of the same S alleles. Fewer than four possible genotypes indicates a sporophytic system and more than four indicates a multilocus gametophytic system (Richards 1986).

Gametophytic systems typically have binucleate pollen grains with long viability and wet stigmatic papillae. The incompatibility reaction occurs as inhibition of the pollen tube within the style. The family Gramineae, a two-locus gametophytic system, is an exception to this generalization, exhibiting characteristics more like those found in sporophytic systems.

2. Sporophytic Systems

In sporophytic systems the anther confers substances responsible for the incompatibility reaction to the pollen grains. All pollen grains from the same flower exhibit the same phenotype (no semi-compatibility). Proteins from the anther are deposited in the exine, between the exine and intine, or in cavities in the exine, and may be held in place with the lipoprotein pollenkitt also deposited by the anther (Richards 1986). The incompatibility reaction can occur when incompatible pollen fails to adhere to the stigma, when pollen tubes are unable to penetrate the stigma, or when pollen tube growth in the style is inhibited (often in conjunction with the deposition of large amounts of callose). In general, sporophytic systems (except in heteromorphic sporophytic species) are characterized by trinucleate pollen with short viability, dry stigmas covered with a cuticle, and commonly inhibition at the level of the stigma (Richards 1986).

Diallelic sporophytic systems occur in 13 plant families. Multiallelic sporophytic systems are found only in the Asteraceae and Brassicaceae and all self-incompatibility in these two families is generally assumed to be under sporophytic control (Richards 1986).

3. Distinguishing Gametophytic and Sporophytic Self-Incompatibility

Richards (1986) shows that the type of self-incompatibility system can be identified from compatibility groupings at three levels:

> If the number of cross-compatible groupings among parents is one, suspect a diallelic sporophytic system. If it is three or more, suspect a multiallelic system.
>
> If all crosses are fully compatible or fully incompatible, it is a sporophytic system. If semicompatibility occurs, it is a gametophytic system.
>
> If siblings from a fully compatible cross are placed in cross-compatible groups, four groups indicate a one-locus gametophytic system, fewer than four groups indicate a sporophytic system, and more than four groups indicate a multilocus gametophytic system.

Morphological features that are highly correlated with gametophytic and sporophytic systems (listed in Sections 1 and 2 above) can be used to help confirm the designation.

4. Heteromorphic Systems

Heteromorphic species exhibit two or more floral morphs. Plants may produce long-styled, short-styled, or intermediate length–styled flowers. In dimorphic systems the term "pin" is often used to refer to flowers with long styles and short stamens and "thrum" to refer to flowers with short styles and long stamens. Fertilization may occur between morphs but not within morphs, although "illegitimate" pollinations do occur. There appears to be some "leakiness" in heteromorphic systems, with some bias in short-styled morphs toward self-fertilization. A suite of linked characters is associated with each morph, although every heteromorphic species may not exhibit all of the characters. The genetic basis for the incompatibility reaction is often a diallelic sporophytic system. Pin flowers typically produce small pollen grains and have long stigmatic papillae, and thrum flowers produce large pollen grains with short stigmatic papillae. There is evidence indicating that pin flowers are often more successful as females, receiving a great deal of legitimate and illegitimate pollen, perhaps because of the elevated position of the stigma, which is accessible to many types of pollinators. Thrum flowers may be more successful as males, requiring more specialized pollinators to transfer pollen to the recessed stigma. See Barrett 1985, 1990; Barrett and Glover 1985; and Barrett and Wolfe 1986 for further information on heterostyly.

Lord and Eckard (1984) describe a form of incompatibility between different types of flowers on the same plants of *Collomia grandiflora* (Polemoniaceae). This species has both chasmogamous and cleistogamous flowers that differ in style length, pollen tube growth rates, and stigmatic papillae. Crosses between chasmogamous and cleistogamous flowers are infertile. Thus each plant has a mixed mating system as chasmogamous flowers foster out-crossing and the cleistogamous flowers enforce inbreeding.

5. Cryptic Self-Fertility

Some species of plants appear self-sterile when pollinated with pure self-pollen (Bertin and Sullivan 1988). However, when mixed pollen loads are applied, some selfed-seeds are produced. This occurs in *Campsis radicans* (Bignoniaceae) and the selfed-seeds produced are lighter in weight and develop into smaller seedlings. In species like this, tests for self-fertility that apply pollen from only the selfing source will not conclusively demonstrate that no self-pollination occurs.

6. Overcoming Incompatibility Mechanisms

Incompatibility systems prevent self-pollination in many species, and in vivo pollination may be unsuccessful. Maintaining flowers under conditions of high humidity can sometimes overcome the incompatibility reaction. Carter and McNeilly (1975) enclosed flowers in polyethylene bags after self-pollination. Control flowers were enclosed in glassine bags that do not retain water. Pollen tube numbers and seed set after self-pollination were significantly increased in the humid environment provided by the polyethylene bags. Heat or CO_2 treatment can also break down incompatibility in some species (O'Neill et al. 1988; Leduc, Douglas, et al. 1990). Elevated CO_2 concentration of 3–5% v/v during the period of stigma–pollen tube interaction (i.e., after germination) masks the self-incompatibility reaction in several crucifers with sporophytic self-incompatibility (O'Neill et al. 1989). Keijzer et al. (1988) use the "cut-style" method to overcome interspecific incompatibility for lily breeding. Pollen is deposited on the top of a styleless ovary and pollen tubes are able to reach the ovules. Pollination of flowers in the bud stage is a method commonly used in horticulture to offset incompatibility mechanisms (Lawrence 1968). "Rescue pollination" is sometimes used to overcome compatibility barriers (Brown and Adiwilaga 1991). Rescue pollination involves a primary pollination with pollen bearing the desired but incompatible genotype (pioneer pollen) followed by a second pollination with a compatible pollen type (mentor pollen). The mentor pollen stimulates fruit and seed production, and some of the seeds produced bear the genome of the pioneer pollen.

H. Functional Gender

The majority of flowering plants are cosexual, having only one sex genotype such that individuals could theoretically function as both male and female. Only 15% of flowering plants have heteromorphic genes that designate plant sex (Schlessman 1987). Until the 1970s, plant gender classification was based solely on morphological characteristics such as the presence of hermaphroditic flowers or all pistillate flowers. However, researchers began to detect that hermaphroditic flowers did not always function as both males and females. For example, male and female plants of a dioecious species of *Solanum* (Solanaceae) were once classified as distinct species based on the different morphologies of apparently hermaphroditic flowers (Anderson and Levine 1982). Systems that appear to be androdioecious (with males and hermaphrodites) may exhibit cryptic dioecy because the hermaphrodites have nonfunctional pollen (Anderson and Symon 1989, Schlessman et al. 1990, Mayer and Charlesworth 1991). In dioecious plants, departures from strict maleness or femaleness are common (Schlessman 1986b). The distribution of staminate and pistillate flowers on ambisexual plants in a population may be either bimodal, with two basic sexes that sometimes have inconstant expression, or continuous, with individuals' male:female ratios falling along a broad range of values. In the past 15 years, more researchers have tried to quantify gender by measuring the relative investment of resources into male and female function or the number of offspring derived from each. Quantification permits description of plant gender along a continuum rather than as discrete categories and gives greater insight into the evolutionary processes responsible for sex allocation (Meagher 1988).

Functional gender of cosexual plants is based on the proportion of reproductive success derived from ovules or pollen (Lloyd 1979, Lloyd and Bawa 1984). The ultimate measure of male success for a plant is the number of offspring sired that reach reproductive maturity. For practical purposes, this is sometimes measured as the number of ovules fertilized by a male's pollen. The ultimate measure of success through female function is the number of viable offspring that survive to reproduction derived from ovules. Again, for practical purposes, this is often measured by collecting fruits just before the seeds are shed and counting viable seeds. It is often possible to differentiate viable seeds from undeveloped or aborted ovules on the basis of size, but to confirm viability it may be necessary to use other methods (see Chapter 2, Section F.2).

Functional gender measures the actual reproductive success derived from male versus female functions, whereas phenotypic gender is based on the relative investment in male versus female resources (Lloyd and Bawa 1984; this differs from Lloyd's 1980 definition, in which phenotypic gender is based on the individual's investment in male or female reproduction and functional gender relates the individual's investment to that of all plants in the population). Phenotypic gender is the relative production of female versus male products and equals gynoecial products divided by androecial products. Gynoecial products can be measured in terms of ovaries, ovules, fruits, or seeds and androecial products can be measured in terms of pollen, anthers, male flowers, and so on (Lloyd 1979). An index of the potential for seed production can be provided by counts of flowers, or ovules, although there may be environmental or physiological constraints on the development of all ovules into seeds.

Standardized phenotypic gender is an individual's expenditure on male or female products relative to the population investment (Lloyd and Bawa 1984).

l_i = male investment (pollen production, number of anthers, number of pollen grains, number of male flowers, etc.)

d_i = female investment (number of seeds, dry weight of fruit, etc.)

$E = \Sigma d_i / \Sigma l_i$ = the ratio of female to male investment for the entire population

Standardized phenotypic femaleness = $G_p = d_i / d_i + l_i\,E$

Standardized phenotypic maleness = $A_p = l_i\,E / d_i + l_i\,E$

Functional and phenotypic gender will not be equal if pollen or seeds from different sources differ in their success.

m_i = the actual fitness derived from female function

p_i = the actual fitness derived from male function

Measures of functional gender take this form:

$G_f = m_i / m_i + p_i$ (femaleness)

$A_i = p_i / m_i + p_i$ (maleness)

Broyles and Wyatt (1990a) calculated functional gender of *Asclepias exaltata* plants in order to test the pollen donation hypothesis. The number of seeds produced by each female was easy to determine. To determine male reproductive success, they used electrophoresis to classify individuals on the basis of seven polymorphic loci. Paternity analysis was employed to identify the unique pollen donor for each seed produced. They used the formula (terminology modified from that given above)

$$G_i = \frac{f_i}{f_i} + m_i$$

to calculate functional gender of each female (individual i), where f_i = the number of seeds produced by i, and m_i = the number of seeds sired as determined through paternity analysis. Subsequently, characters such as inflorescence size and flower number were related to functional gender.

Robbins and Travis (1986) define gender specialization as the enhanced ability to act as a pollen donor (or receiver) at the expense of the ability to act as a pollen receiver (or donor). This differs from functional gender because an individual's success as a male or female is based on fitness derived from that function and is dependent upon the plants to which the individual is mated. If an individual is a superior female, its pollen necessarily goes to inferior females which means that its success as a male is reduced, automatically creating an inverse correlation between male and female success.

Methods for quantifying sex ratios and sex allocation are discussed in Charnov 1982. Lloyd (1984) presents a simplified model for quantifying allocation:

$$w_i = m_i + \sum_{j=1}^{K} m_j e_i c_i ,$$

where

   w_i = combined male and female fitness of individual i;
   m_i = female fitness of individual i;
   p_i = male fitness = $\Sigma m_j e_i c_i$, that is, the sum of the male fitness
      over all females for individual i;
   e_i = eligible eggs of each mate (this can equal 1 if there is no
      incompatibility); and
   c_i = i's competitive share of the available eggs.
   K = number of mates.

If m_j is the same for all females, the equation reduces to:

$$w_i = m_i + K m_j c_i .$$

The model and its applications in evolutionary stable strategies maximizing fitness through sex allocation are further elaborated in Lloyd 1984.

Lloyd suggests that there are often upper limits to paternal fitness of animal-pollinated plant species that result in selection for reduced male allocation. These limits result from pollen losses exacted when pollinators eat, store, or brush pollen from a flower, and from pollinator damage to flowers that reduce attractiveness to further pollinators. In addition, multiple pollen grains taken from one flower may end up in competition with each other to fertilize a limited number of ovules on the same flower. Paternal fitness can theoretically be enhanced by increasing the number of pollen-producing flowers. Stanton et al. (1986) noted that resources often limit seed and fruit set, so increasing the number of flowers may function to augment the attractiveness to pollinators and thereby increase male fitness (pollen donation hypothesis). Their fieldwork with *Raphanus raphanistrum* (Brassicaceae) supports this idea.

Broyles and Wyatt (1990b) believe that nonfruiting flowers have not been selected solely to increase male reproductive fitness. They suggest that the pollen donation hypothesis is probably supported only in situations where reproduction is limited by maternal resources. By using paternity analysis and correlating the number of flowers with male and female reproductive output, they demonstrated that both males and females benefit from increased floral displays in *Asclepias exaltata*. A theoretical argument (Nakamura et al. 1989) against the pollen donation hypothesis suggests that large floral displays generally decrease male success because pollen is wasted by geitonogamy in self-incompatible species.

Pollinator behavior patterns may also influence realized reproductive performance and functional gender. For example, milkweed flowers (*Asclepias syriaca*) blossom for several days before their bumblebee pollinators become active (Morse 1987). Flowers remain open for 5 days, with little decrease in pollinia viability but with substantial loss of stigma receptivity. Thus flowers have relatively more male potential as they age. Flowers that bloom early in the season tend to function more as males. The same holds true for flowers in populations that are small or isolated (Morse 1987).

I. Gender Adjustment and Gender Diphasy

Gender adjustment involves a change in allotment of resources to male or female function in a system where investment ratios vary along a continuum. Gender diphasy (gender choice or sex change) is a switch in gender in a system where there is bimodal expression of gender. In dioecious plants, departures from strict maleness or femaleness are common (Bawa and Beach 1981). Ambisexual plants bear both staminate and pistillate flowers, but the distribution of flower types among plants in the population may be either bimodal or continuous. If the distribution is bimodal, ambisexuality could be merely "inconstant expression" of a gender phase (e.g., four female flowers on a plant with 100 male flowers). Without quantitative assessment of sex allocation, there may be confusion between gender modification and gender diphasy. Sex changes have often been attributed when previously pistillate plants bear some staminate flowers. However, many of these studies have not quantified the proportion of each type of flower on ambisexual plants, and conclusive evidence of gender diphasy is lacking. Sometimes several authors have examined the same data and reached different conclusions on the percentages of sex change and its significance in a species (Schlessman 1988). Schlessman (1988) reviews the evidence for gender diphasy and concludes that diphasy has been established for natural populations of *Arisaema* (Araceae), *Gurania* and *Psiguria* (Cucurbitaceae), *Panax trifolium* (Araliaceae), cultivated oil palms, *Elaeis guinensis* (Arecaceae), and Catasetinae orchids. Evidence is weak for diphasy in *Acer* (Aceraceae), *Atriplex* (Chenopodiaceae), and *Juniperus* (Pinaceae).

Theoretical models for sex change are based on the change in an individual's fitness derived from each sex by varying size or age (Schlessman 1988). In general, female plants have large resource requirements for fruit and seed production, whereas male investments may be less constrained by resources. Small plants may be able to derive more fitness through pollen than through ovules. Among diphasic plants, individuals in the male phase are generally smaller than those in the female phase, and the data indicate that there may be size thresholds for female expression (Schlessman 1988, 1991). Empirical data indicate that favorable conditions such as adequate moisture, rich soil, and sunlight favor females.

There are three requirements for gender modification (Lloyd and Bawa 1984, Schlessman 1988): individuals cannot control the conditions around them; all parts of the individual experience the same environment; and the individual can assess its fitness prospects (based on its size or environment) through male or female repro-duction. If fitness curves differ for males and females and these three conditions hold, there should be some type of gender modification (Schlessman 1988). Because most plants are simultaneous hermaph-rodites and can obtain some fitness through both male and female reproduction, gender adjustments are more common than gender changes.

Gender adjustment can occur within flowers by changing the number of ovules or stamens; by varying the number of male, female, or hermaphrodite flowers; or even by changing seed abor-tion levels (Lloyd and Bawa 1984). Although functional gender may vary because an individual's success is affected by the sexual behav-ior of other plants or by circumstances affecting the progeny, gender modification arises from an individual's own activities under given conditions.

Lloyd and Bawa (1984) suggest that researchers attempting to document gender modification should quantify the variation in all gender components, analyze the relationship between the variation and the plant's circumstances (i.e. size, environment), and provide theoretical and experimental analyses of how the modification results in fitness gains. Studies that attempt to quantify gender dynamics should be conducted for several growing seasons, as seed and fruit production in 1 year may have an effect on resource levels and reproduction in subsequent years (Schlessman 1986b).

An example of a study of gender modification that takes these factors into account is the work by Schlessman (1991) on dwarf ginseng (*Panax trifolium,* Araliaceae). Schlessman documented sex change in dwarf ginseng by marking and following individual plants at three sites for 4 years. Each spring he "sexed" plants by the size and number of floral buds (males produce 15–20+ flowers whereas hermaphrodites produce 6–8 flowers). Each year plants were reclas-sified as males, hermaphrodites, vegetative, or dead. He evaluated the overall frequency of gender change, correcting for differences in the male:hermaphrodite ratio of his sample population (roughly 1:1) and the natural sex ratio (male biased). He used log-linear models to test the influence of gender, site, and year on plant status in the following year. Schlessman employed G-tests to see whether changes to "vegetative" or "dead" were independent of gender, and Zar's method for testing differences among proportions was used to determine whether gender phases differed in their likelihood of switching states. To evaluate the effects of size and gender phase, Schlessman measured plants each year as a means of quantifying their resources. He used the sum of the lengths of the middle leaflets on each of the three leaves per plant as this measure was highly correlated with total leaflet length and dry root weight. Using one-tailed t-tests, he examined whether males would be larger than hermaphrodites and whether hermaphrodites would be larger than males. He tested for differences in the proportions of each gender that became larger or smaller. He also tested whether males that became larger would be more likely to change gender than those that became smaller or those that remained the same and whether hermaphrodites that became smaller were more likely to change than those that became larger or remained the same. Schlessman documented sex change in 83% of the individual plants and related sex expression to plant size.

J. Pollen-Ovule (P/O) Ratios

Cruden (1977) gathered evidence from 86 plant species that indicated that pollen-ovule ratios tend to reflect plant breeding systems. Pollen-ovule ratios were determined by estimating the number of pollen grains produced per flower and dividing this by the number of ovules per flower. Based on this ratio, Cruden classified plants into four categories: 0 = cleistogamous; 1 = obligately autogamous; 2 = facultative autogamous; 3 = facultative xenogamous; 4 = xenogamous. The general assumption behind this work was that plants that could self-fertilize did not need to produce so much pollen. Many authors have included P/O ratios in descriptions of plant breeding systems. P/O ratios are relatively easy to determine, using techniques for counting pollen described in Chapter 4, Section D, and by counting ovules per flower. Most authors count some subsample of the total amount of pollen, for example, using only two of five anthers. Koptur (1984) calculated P/O ratios for a plant with polyads in this way:

$$P/O = \frac{\overline{x}\ \#\ \text{stamens} \cdot \dfrac{8\ \text{polyads}}{\text{stamen}} \cdot \dfrac{\overline{x}\ \#\ \text{pollen grains}}{\text{polyad}}}{\overline{x}\ \#\ \text{ovules}} .$$

One should use caution in subsampling, as anthers at different levels in a flower may vary in pollen production. Individual species may deviate from the expected pattern because of variation in pollen size, seed size, or other measures of allocation.

Cruden's work on P/O ratios has been severely criticized for suggesting that pollen production is only to ensure ovule fertilization (Charnov 1979). However, theoretical evaluation of P/O ratios suggests that correlations with breeding systems fit evolutionary theory pertaining to local mate competition and to sex allocation in hermaphrodites (Charnov 1979, Queller 1984, Lovett Doust and Lovett Doust 1988, Bertin 1988). Even if his explanation was incorrect, Cruden's work is still valuable, as data from P/O ratio studies can be used to evaluate newer, evolutionary models (K. Ritland, *personal communication*).

K. Gametophytic Competition

Much of sexual selection theory revolves around the concepts of female choice and male-male competition. Female choice can take the form of stigma-pollen interaction and selective seed or fruit abortion based on pollen genotype. Some of the manifestations of male-male competition might be the production of large numbers of flowers that serve primarily as attractants for pollen vectors (Stanton et al. 1986), production of large quantities of pollen or of pollen packages (pollinia), nectar production patterns encouraging optimal forager movements, and morphologies that favor more precise, efficient pollen transfer (Stephenson and Bertin 1983, Willson and Burley 1983). In recent years evidence for male-male competition at the gametophyte level has been mounting (Snow and Spira 1991a, 1991b and references therein). Mulcahy (1979) hypothesized that pollen tube competition played a role in the rapid evolution of angiosperms, suggesting that rapid-growing pollen tubes sire more offspring, and their gametophyte superiority is reflected in the resulting offspring. Thus pollen "success" must have a genetic basis and be heritable, and there must be a positive correlation between pollen and sporophyte characters (Schlichting et al. 1990). The correlation between these characters has been documented in several studies (see references in Snow 1986), but the overall significance of pollen competition remains in question. For example, Mulcahy et al. (1978) reported that pollen tube competition resulted in more vigorous F_2 progeny in *Petunia hybrida* (Solanaceae). In contrast, Schlichting et al. (1990), working with zucchini, concluded that although pollen tube competition occurred, it played only a minor role in determining differences in progeny vigor, and that factors such as maternal effects, pollen-pistil interactions, and nonrandom abortion of seeds are also involved.

Until recently there has been little information on the role of pollen tube competition in natural populations. Snow (1986) evaluated natural rates of pollen deposition, pollen tube growth rates, and seed set of *Epilobium canum* (Onagraceae) to determine whether pollen tube competition could play a major role in determining paternity of seeds. Levels of competition varied greatly among styles. Snow and Spira's (1991a) research on natural populations of *Hibiscus moscheutos* (Malvaceae) demonstrated variation in pollen tube growth rate among pollen donors that was independent of maternal effects. This area of research promises to be fruitful in the next few years.

L. Pollinator Efficiency

The term "pollination efficiency" has been used in different ways by different researchers. For example, Cruden et al. (1990) define it as the number of pollen grains deposited per stigma divided by the number of grains produced per flower, or, in other words, an index of efficiency from the perspective of the plants producing and receiving the pollen. Alternatively, it is sometimes used as a variable related to the effectiveness of different pollinators, referring to the number of pollen grains transferred per visit of a particular flower visitor (Waser and Price 1990), or for a single plant species, pollen deposition and seed set effected by a species of pollinator (Waser and Price 1990). There is similar confusion in the use of terms such as pollination intensity (Snow 1982) or pollination effectiveness (Motten

1986, Galen and Newport 1987). Inouye et al. (1994) have proposed a lexicon for pollination biology to eliminate these overlapping definitions. We repeat some of their suggestions here, in addition to some other terms that may need unambiguous definitions:

total dispersal efficiency: the ratio of the number of pollen grains transferred to a vector visiting a flower, over the number of pollen grains produced by that flower.

partial dispersal efficiency: the ratio of the number of pollen grains transferred to a vector visiting a flower, over the number of pollen grains available on the flower at the time of the visit.

total acquisition efficiency: in a flower in which pollen is gathered by activities of the vector, the ratio of the number of pollen grains transferred to a vector visiting a flower, over the number of pollen grains produced by that flower.

partial acquisition efficiency: in a flower in which pollen is gathered by activities of the vector, the ratio of the number of pollen grains transferred to a vector visiting a flower, over the number of pollen grains available on the flower at the time of the visit.

total pollination effectiveness: the ratio of the number of pollen grains from a single flower deposited on receptive conspecific stigmas, over the total number of pollen grains produced by the original flower.

partial pollination effectiveness: the ratio of the number of pollen grains from a single flower deposited on receptive conspecific stigmas, over the number of pollen grains available on the original flower at the time of the visit.

pollination efficiency: the ratio of the number of pollen grains deposited on a receptive conspecific stigma in a single visit by a pollinator, over the number of pollen grains carried by the vector.

stigmatic fertilization success: the ratio of the number of pollen grains that succeed in fertilizing ovules, over the number of pollen grains deposited on the stigma.

fertilization potential: the ratio of the number of pollen grains that succeed in fertilizing ovules, over the number of pollen grains available at the time of the pollinator's visit (partial

fertilization potential) or the total number of grains produced by the source flower (total fertilization potential).

stigmatic seed set success: the ratio of the number of seeds
produced by a pollinator's visit, over the number of pollen
grains deposited on the stigma.

potential siring ability: the ratio of the number of seeds produced by a pollinator's visit, over the number of pollen
grains available at the time of the pollinator's visit (partial
potential siring ability) or the total number of grains produced by the source flower (total potential siring ability).

vector pollen load per source visit: the number of pollen grains
transferred to a vector during a single visit to a flower.

stigmatic pollen load per visit: the number of pollen grains
deposited on a stigma during a pollinator's visit.

stigmatic pollen–ovule ratio: the ratio of the number of pollen
grains deposited on a receptive conspecific stigma, over the
number of ovules the flower has.

We will discuss here techniques that might be used to assess some
of these measures.

At the simplest level, it may suffice simply to know whether the
visit of a particular flower visitor has transferred any pollen. This
observation typically is made on flowers that have not been visited
previously, such as flowers that were bagged before anthesis and
only unbagged while being observed, or flowers that are opening
for the first time and have been observed since they opened. Visual
inspections of stigmas after a visit may be sufficient if the pollen
grains are large enough, or it may be necessary to use techniques
such as those described in Chapter 4, Section D. In some cases there
may be a morphological indicator related to successful pollination,
such as retraction of the style (Neff and Simpson 1990) or tripping
of a flower with parts under tension (e.g., Schemske and Horvitz
1984). Schemske and Horvitz (1984) calculated different measures
of effectiveness as the percentage of flowers tripped per visit, fruit
set per flowers tripped, and fruit set per flower visit, for eight species
of Hymenoptera and Lepidoptera visiting a single species of flower.

This kind of observation for different kinds of flower visitors
allows the number of pollen grains that each species deposits during
a single visit or the number of seeds resulting from a visit to be

ascertained. For example, Kendall and Smith (1975, 1976), Parker (1981, 1982), and Tepedino (1981) compared seed set resulting from visits by two or more species of bees. Much of this research has involved commercially important crop plants being studied for the best way to ensure pollination.

Kislev et al. (1972) pointed out that there is a progressive reduction in the number of pollen grains as one proceeds from counts of pollen grains exposed by anthers, to counts of the grains picked up by an insect, to counts of the grains deposited on a stigma, and finally to counts of the number of pollen tubes reaching an ovule. They cited studies dating as far back as Joseph Kölreuter's in 1761 that have attempted to quantify these different stages, and they conducted their own study of pollen gathering by hawkmoths. The measure they were most concerned with was the number of grains picked up by a visit in comparison with the number deposited. They pointed out that it is easiest to measure this in cases (like hawkmoths) in which pollen is restricted to only part of the flower visitor's body.

To compare the significance of different pollinators, Lindsey (1984) used an index that ranged from 0 to 1.0 in intervals of 0.1, which incorporated analyses of insect size, specific foraging behavior relative to floral and inflorescence structure, and analyses of the consistency of movement within inflorescences, between inflorescences, and between plants. She also calculated pollination importance values for each species of flower visitor, which incorporated information on relative abundance of the species, its pollen-carrying capacity relative to other visitors in a population, and relative host plant fidelity (flower constancy) expressed as the average proportion of a species' pollen load that contained host pollen, and the index of pollination efficiency. Thus,

pollination importance value (PIV) = A x PCC x F x PE,

where A = abundance, PCC = pollen carrying capacity, F = fidelity, and PE = "pollination efficiency" — a qualitative measure that she determined from feeding behavior and contact with plant reproductive parts. For ease of expression and to illustrate their relative importance for pollination, each insect species was given a pollination importance index (PII) score that was calculated as the percentage of total pollination importance values of all visitor species collected in each yearly sample (i.e., PII = PIV/ΣPIV x 100 for all species). Although these indices may not work for comparisons among studies, they may prove useful within a single system, as Lindsey used them.

Thomson and Thomson (1992) stress that the relative efficiency of a single pollinator can only be judged when previous and subsequent visits by other types of pollinators are taken into account (see also Corbet et al. 1991). For example, a pollinator that removes lots of pollen but deposits little on stigmas may be valuable when it is the sole species. However, when a second pollinator that deposits much pollen on stigmas is also present, the first type of pollinator may actually be a liability, wasting much pollen that would effect fertilizations. Wilson and Thomson (1991) determined that pollen-collecting bees removed more pollen from *Impatiens capensis* (Caprifoliaceae) flowers than did nectar-collecting bees but also deposited less on stigmas. Because flowers received multiple insect visits, male reproductive success was dependent upon the sequence of the bee visitors.

Beattie (1976) used a measure called "pollinator service" to make comparisons among species. He counted the number of pollen grains in stigmatic cavities and then counted the number of ovules in corresponding ovaries. The percentage of ovules that could be matched to pollen grains was used as an index of pollinator service. As there is not necessarily a one-to-one correspondence between pollen grains and ovules that get fertilized, this is merely an index that can be useful for comparisons among species.

M. Pollen and Resource Limitation of Seed Production

Many plants produce far more flowers than fruits. These "excess" flowers could serve to increase male reproductive success by attracting vectors to disseminate pollen (pollen donation hypothesis; Stanton et al. 1986). Sutherland and Delph (Sutherland and Delph 1984, Sutherland 1986) examine fruit-flower ratios for multiple plant species and find significant differences among plant groups based on breeding systems, life forms, self-compatibility, and latitude of growth. Sutherland (1986) reviews the many hypotheses relating to "excess" flower production. Two of these hypotheses, sometimes considered as competing alternatives, have received much attention in recent literature. The first is that female reproductive success may be limited by inadequate resources (stored reserves, minerals, light, water) that prevent maturation of more fruits. The second is that pollinator service may be inadequate and fruit and seed production limited by a lack of pollen. The two types of limits to female reproductive success have different implications (Johnston 1991a). When resources are limiting, competition between males results in selection of characters that influence pollen dispersal. When pollen is limiting, selection favors those characters that promote pollen receipt. Because of these opposing selective regimes, theory predicts that plants should not be strictly pollen- or resource-limited (Haig and Westoby 1988). An evolutionarily stable strategy would result in a balance point between resource and pollen limitation, although any individual may fall somewhere short of the balance point in a given year, especially because plant stress and pollen availability cannot be accurately predicted for a given season (Haig and Westoby 1988).

However, empirical evidence for resource limitation of seed production is strong. Willson and Burley (1983) summarize studies that looked for resource limitation. The studies are categorized on the basis of the evidence for limitation: (1) Addition of fertilizer increases seed production. (2) Removal of competing plants increases seed production (Galen [1985] cautions that in exposed areas, removal of competing vegetation may also decrease soil moisture by removing ground cover). (3) Defoliation, xylem tapping, or low soil nutrients decrease seed production. (4) Good seed production in a given year depresses seed production in the subsequent year. Stephenson (1981) also reviews the evidence for resource limitation and selective abortion of pollinated fruits.

Many researchers have tried to demonstrate pollen limitation (Willson and Burley 1983, Zimmerman 1988b, summarized in Young and Young 1992), yet attempts to demonstrate pollen limitation have often been flawed and the results inconclusive. Addition of supplemental pollen to some flowers on a plant may result in seed set that is greater than that of control flowers. However, if resources are limiting, they may be differentially allocated to those fruits with the most or highest-quality seeds. The entire reproductive output of control and hand-pollinated plants may be similar, although the pattern of allocation among fruits differs. Zimmerman and Pyke (1988a) used two sets of control flowers. The first set of controls consisted of naturally pollinated flowers on experimental plants (those plants receiving supplemental pollen on some flowers), whereas the second set consisted of flowers on unmanipulated plants. Hand-pollinated flowers on experimental plants set more seeds than flowers on control plants, but control flowers on the experimental plants set fewer seeds than controls on the unmanipulated plants. This indicates that resources were shunted within plants. One means of dealing with this problem is to add supplemental pollen to all the flowers on treatment plants and compare seed production with control plants (Bierzychudek 1981, Stephenson 1984, Delph 1986, Johnston 1991a).

In a thorough, 4-year experiment using *Cypripedium acaule* (Orchidaceae), Primack and Hall (1990) included three reproductive treatments and three defoliation treatments: (1) hand-pollination of the single flower; (2) removal of flowering stalk and flower (3) control; and (1) removal of one leaf; (2) removal of both leaves (two leaves per plant); and (3) control.

By hand-pollinating, they were able to produce fruits in populations that rarely fruited naturally. They also demonstrated high costs for reproduction in the form of reduced growth and flowering of plants that matured fruits. Defoliated plants were capable of reproduction but had reduced probabilities of flowering subsequent years.

Even if all flowers on experimental plants receive supplemental pollen and total plant seed output exceeds that of controls, reproductive output the following year may be minimal. The high energy investment in maturing many fruits may deplete nutrient stores that would normally be available for future reproduction. Measuring the entire reproductive output is easiest for annuals, but by following the same plants for several seasons, the cost of maturing many fruits can be evaluated. Bierzychudek (1981) demonstrated that supplemental pollination of *Arisaema* (Araceae) increased seed set on experimental plants with no cost to the next year's reproduction. Additional thorough studies of this type are needed.

Even if plants are not generally pollen-limited, attracting more pollinator visits may have positive effects on male reproductive output and on the quality of female reproduction. High pollination intensity sets up the potential for pollen tube competition and selective abortion. The relative compatibility of the pollen source can influence which fruits mature. For example, the proportion of self-pollinated seeds can vary if resources or plant stress vary from year to year and those seeds are the first to be aborted (Lee 1988). Other factors, such as the order of flowering, can affect which fruits will be most likely to abort, and the total number of maturing fruits may affect the levels of fruit abortion (Stephenson 1981). Horvitz and Schemske (1988) demonstrated pollen limitation of fruit initiation but resource limitation of the number of fruits that reached maturity.

Willson and Burley (1983) note that pollen limitation has been well documented less frequently than resource limitation and question whether the presence of the introduced honeybee could suppress pollen limitation in many North American systems. However, for crop plants that often grow in large monocultures, it is a common practice to provide additional pollinators in the form of transportable bee colonies to maximize seed or fruit production. Corbet et al. (1991) designed a flow chart to help evaluate the adequacy of the pollinator community for maximizing seed production for crops and wildflowers in Europe (Figure 6–3).

Young and Young (1992) review 99 pollen supplementation studies conducted from 1979 to 1991. In 42% of the studies pollen supplementation increased female reproductive success, in 40% there was no effect, and in 17% supplementation decreased female fitness. The last effect may be an underestimate, as results of this sort are probably not published often (Young and Young 1992). They review 10 reasons that may account for this effect:

1. Pollen grains or pollen tubes may interfere with each other.
2. Pollen thieves are attracted to large pollen loads and damage tissues.
3. Pollinators remove pollen or damage tissue.
4. Hand-pollination damages female tissue.
5. Pollen from a single donor used for hand-pollinations may have less success than multiple donor pollen on naturally pollinated flowers.
6. Bagging of flowers has a detrimental effect.

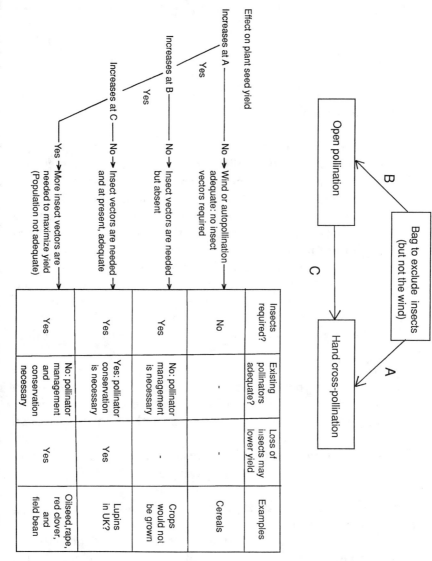

Figure 6–3. An experimental procedure to discover the extent to which a plant depends on insect pollination, and to investigate the adequacy of the existing pollinator community. From Corbet et al. 1991.

7. Stigmas that are hand-pollinated are not at peak receptivity.
8. Pollen used for hand-pollinations has low viability.
9. Pollen quantity used for hand-pollinations is insufficient.
10. Type I statistical errors are made (rejecting the null hypothesis when it is true).

Researchers attempting to demonstrate pollinator limitation should consider the following:

1. Flowers that bloom at different times in the season are subject to different pollination and weather regimes and may differ in seed output. It may be necessary to compare experimental and treatment flowers simultaneously.
2. High reproductive output one year may occur at the expense of the following year's output. Studies should be conducted for more than one flowering season.
3. If not all flowers are pollinated on experimental plants, resource shunting within the plant may increase the seed production of experimental flowers without increasing total plant output. This can occur through abortion of poorly pollinated flowers or through a decrease in flower production.
4. If additional pollen yields no increase in seed set beyond that of controls, then seed set is resource limited. However, when addition of pollen increases seed set, you have not demonstrated pollen limitation because of the above-mentioned reasons.
5. If supplemental pollination results in lower seed set than that of controls, the cause should be determined.

Schemske and Horvitz (1988) used path analysis to address questions of pollinator and resource limitation. Path analysis allows examination of both direct and indirect effects and permits the researcher to evaluate the magnitude of multiple variables affecting flower and fruit production. Schemske and Horvitz were able to discern changes in pollinator pool composition that were responsible for differences in fruit production among years. Path analysis can be conducted on natural systems (without experimental manipulation) and may help provide realistic assessments of the relationships among the variables in the system (Schemske and Horvitz 1988).

N. Phenotypic Selection

It is widely held that floral diversity is a function of coevolution between plants and their pollinators. Currently, researchers are attempting to quantify the selection potential of morphological characters by measuring the variance in those characters and are comparing the reproductive success of individuals with different morphological measurements (Galen 1989, Galen and Stanton 1989, Campbell 1989, Schemske and Horvitz 1989, Campbell et al. 1991). Some aspect of reproductive success, such as pollen export, visitation rate, pollen import, or seed set, is quantified for individuals in the population. One can examine the relative selection intensities imposed by male and female functions by relating male or female reproductive success to variation in a morphological trait or by identifying which traits are subject to selection through male or female components of reproductive success (Campbell 1989). In addition, the stage of pollination in which selection occurs can be identified (Campbell et al. 1991), or temporal variation in selection regimes can be evaluated (Schemske and Horvitz 1989). Although the statistical analyses of this type of research are beyond the scope of this book, the reader should be aware of the potential range of phenotypic selection analyses.

For example, using *Ipomopsis aggregata,* Campbell (1989) measured the amount of time in the pistillate phase, stigma exertion, flowering date, corolla length, corolla width, and distance from the base of the corolla to the highest and lowest anther. In this study, fitness effects were only measured during the pollination phase (not total fitness through both pollination and postpollination events). Campbell used fluorescent dyes as pollen analogues to quantify pollen donation and pollen receipt of plants with different morphological measurements. Because many of the morphological measurements were highly correlated, she used principal-components analysis and calculated selection gradients on principal-component scores. She was able to demonstrate that selection through male reproductive success favored long corolla tubes, late flowering, and a short pistillate phase. Selection through female function sometimes acted in opposing ways. Selection intensities were similar for male and female functions throughout the pollination phase but generally affected specific traits differently.

Galen and Stanton (1989) used phenotypic selection analysis to demonstrate that bumblebees exert strong selection pressures on flower size. Fitness traits measured included pollen remaining after a bee visit to a flower and pollen receipt by female-phase flowers. The number of visits each flower received and the length of each visit were also recorded. Morphological measurements included corolla flare, style length of male and female phase flowers, and pollen production. Selection through pollen export favored broad corollas. Selection through pollen receipt favored exerted stigmas. There was no evidence for opposing selection through male and female functions.

O. Isolation of Gametes

Sperm cells, egg cells, and embryo sacs can be isolated for manipulation for in vitro fertilization, genetic engineering, and study of the biochemistry and genetics of fertilization (Theunis et al. 1991). Techniques for isolation include enzymatic digestion of surrounding tissue, osmotic shock, microdissection, and grinding. Tests are available to evaluate the viability of the isolated material. Theunis et al. (1991) review about 90 publications on isolation of angiosperm gametes.

P. Suggestions for Planning Studies

1. **Variability in breeding systems.** Features of breeding systems such as self-incompatibility may not be absolute. Some individuals in typically self-incompatible species may set fruit when self-pollinated, and some individuals in typically dioecious species may bear both staminate and pistillate flowers (e.g., Bawa 1974, Schlessman et al. 1990). Sample

sizes for studies of breeding systems should be large enough to detect this sort of departure if it is significant to the nature of the study.

2. **Resource limitation of seed set or fruit production.** If you use seed set or fruit production as an index of successful pollination, beware that your results are not affected by resource limitation. This might be avoided by clipping styles on flowers that you are not using (e.g., Galen et al. 1985) or by watering and fertilizing plants to ensure that these resources are not limiting (e.g., Zimmerman 1983b, Jenner-sten et al. 1988). Bertin (1985) found that resource limita-tion was the likely cause of a temporal change in patterns of fruit set within a season.

3. **Multiple pollination mechanisms.** For example, Dafni and Dukas (1986) found evidence for insect-, wind-, and self-pollination in a single species.

4. **Variability in the degree of pollen limitation of seed set.** Johnston (1991a) found that female reproductive success was much less in one of three populations he studied.

5. **Variation in P/O ratio within a species.** Pollen-ovule ratios may vary with environmental factors. For example, Niesen-baum (1992) found that population P/O ratios were signifi-cantly greater in the shade than in the sun.

THE FAR SIDE

By GARY LARSON

"I don't have any hard evidence, Connie — but my intuition tells me that Ed's been cross-pollinating."

7. Animals

A. Collecting Insects

The U.S. Department of Agriculture, Agricultural Research Service, published a book that includes extensive information on equipment, collecting methods, and techniques for rearing, killing, and preserving insects (Steyskal et al. 1986; available on microfiche from the National Technical Information Service, 5285 Port Royal Road, Springfield, VA 22161. For more information, call NTIS order desk at (703) 487-4650. This book is out of print, but a revision is being prepared for the Entomological Society of America.). If you intend to send insects to a taxonomist, you should see a reference such as this one for the proper way to pin and label specimens. We will present a brief summary of some methods for capturing and preparing specimens. A general reference for techniques for trapping flying insects is Muirhead-Thomson (1991).

One of the advantages of working with insects is that, at least for now, there are none of the legal restrictions that one encounters if working with vertebrates. However, this does not relieve researchers of the obligation to treat them with respect. Try to minimize your collecting, particularly if the insects are not common.

1. Insect Nets
Aerial (lightweight fabric) nets especially designed for collecting butterflies and other flying insects work well for collecting pollinators. Most flower-visiting insects tend to fly up when disturbed, and you may find that it works better to lower a net slowly over an insect on a flower instead of sweeping at the flower (and usually picking it in the process). If you do sweep at a flying insect, swing the net

rapidly enough to force the insect to the bottom of the bag, and then turn the net so the bottom of the bag folds over the rim of the net.

BioQuip sells a nice range of insect nets, with different lengths of handles, sizes of nets, and types of net fabric. One model comes with a modular handle that permits addition of several extension sections for collecting insects in trees, and another comes without a handle (you can add your own) on a folding net frame so it will fit in your pocket (sometimes called a "National Park net" because it can be carried discreetly).

Once insects have been captured in a net, they can be transferred to a glass or plastic vial for inspection or to a killing jar. Hold the net bag in one hand and slip the other hand, with the open vial, into the bag. Once you have maneuvered the insect into the vial, use the net to keep it in until you can slip the top back on. Alternatively, if your killing jar is large enough, you can insert a portion of the bag containing the insect into the killing jar long enough to make the insect stop moving before you drop it into the jar. Butterflies or moths that are trying to beat their wings in the net (and losing scales) can be stopped by pinching their thorax through the net. Use the tip of your thumb to press the thorax against the flat part of your index finger. If you don't press hard enough you won't stop the insect, but if you press too hard you will rupture the thorax.

2. Killing Jars

There are basically two choices available when it comes to killing jars: cyanide or ethyl acetate. The trade-off here is between how quickly insects are killed (cyanide is faster) and how toxic the two are to humans. Commercially available killing jars usually have a lid modified with a small container suspended from it, in which to place the ethyl acetate (carbon tetrachloride is an alternative but is not recommended because it is a carcinogen and cumulative liver toxin) or cyanide. Sodium cyanide ($NaCN$), potassium cyanide (KCN), or even calcium cyanide ($Ca(CN)_2$) can be used. If you wish to make your own killing jars, they aren't much work. Ethyl acetate ($CH_3CO_2.C_2H_5$) is a liquid, and killing jars based on it should have some absorbent material such as plaster of paris in the bottom that will help retain it. A less elegant killing jar could be constructed with just a layer of paper towel to absorb the ethyl acetate; a jar like this could also function as a relaxing jar, keeping insects for a few days without their becoming brittle. A disadvantage of ethyl acetate is that it may discolor some specimens, especially Lepidoptera, confounding identifications (O. Pellmyr, *personal communication*). NOTE: **Try to avoid breathing the fumes!**

We have found that cyanide jars are easily made from glass "shell vials" (approximately 10 cm high, 2.5 cm diameter). Place about 1.5 cm of potassium cyanide (or sodium cyanide; it has the disadvantage of being hygroscopic, but Mitchell [1960] reports that it lasts longer) in the bottom of the vial. You can keep the cyanide in the bottom of the vial with a piece of cardboard; cut a 1.5-cm-wide strip from a noncorrugated piece of cardboard such as the backing on a pad of paper, and roll it tightly to a diameter that fits the vial snugly. Disks of blotting paper (Steyskal et al. 1986), a wad of cotton (Mitchell 1960), or a layer of plaster (O. Pellmyr, *personal communication*) will serve the same purpose. A cork makes a good lid for such a killing jar. Activate the killing jar by putting a few drops of water on the cardboard. If it loses its effectiveness after some time, add another couple drops of water. These jars will last for several years. CAUTION: **If you break such a cyanide jar and inject yourself with potassium cyanide by cutting yourself on the broken glass, you might receive a fatal dose.** To help prevent this possibility, wrap the bottom and partway up the sides of the vial with filament tape. Pellmyr (*personal communication*) suggests using Plexiglas vials, which would be less likely to shatter. LABEL THE VIAL AS POISONOUS so that if someone else finds it they will know what it is. Figure 7–1 illustrates the construction of some types of killing jars.

You will probably discover by trial and error how long to leave insects in the jar to kill them. If you don't leave them in long enough, you risk finding an insect wiggling around on the pin in your insect box. Some insects, especially beetles, can apparently "hold their breath" (by closing their spiracles) in a killing jar long enough that you think they are dead when they have not actually received a fatal dose. Insects that use cyanide as a defense (e.g., zygaenid moths; O. Pellmyr, *personal communication*) may also be resistant to cyanide killing jars. Most bees and flies are killed in a cyanide jar within a minute or two, whereas it may take considerably longer in an ethyl acetate killing jar. But don't leave insects in for more than a few hours. The fumes may change the colors of some specimens, and they may start to dry out and become brittle.

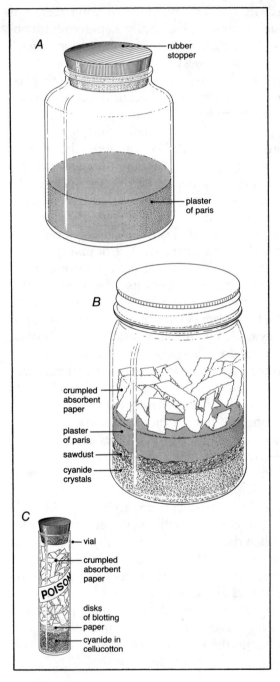

Figure 7–1. Widemouth killing jars for use with liquid (A) or solid (B) killing agents. Vial-type cyanide bottle for small insects with all-important "POISON" label (C). From Steyskal et al. 1986.

266

Insects can also be killed by putting them in vials in a freezer. Although some temperate-region insects can withstand freezing temperatures during the winter when they have produced the proper antifreezes, those caught during the summer can't. Or, if you are planning to preserve the insects in liquid anyway, you can drop them into the ethanol or other preservative in which they will be stored (see Section B below on preservatives).

For some insects it is easy to use killing jars to catch insects on flowers. If you approach a flower slowly so that you don't disturb the insects, or if you wait next to a plant for insects to arrive, you can often trap flower visitors by enclosing the flower and insects quickly in the killing jar. Empty vials can also serve for capturing insects, which can then be transferred to a killing jar.

3. CO_2

Carbon dioxide can also be used to anesthetize or to kill insects. Haslett (1989) used CO_2 from a pocket dispenser to anesthetize adult hoverflies (Syrphidae) once they had been transferred to vials from an insect net. (He then placed the vials on ice in a vacuum flask until they were transferred to a deep freeze in the lab, where they were stored for further analysis.) N. Gary (*personal communication*) has used portable CO_2 generators for anesthetizing bees in the field. These are constructed by putting dry ice (frozen CO_2) in an insulated container (e.g., a plastic canteen inside an insulated carrier on your belt), and putting a small hose into the container. As the dry ice evaporates, CO_2 gas is forced out through the hose, and because CO_2 is heavier than air it can be "poured" out of the hose onto insects on flowers, or into vials into which the insects are placed. If CO_2 is not generated quickly enough from warming of the dry ice, add water to the container to speed up the evaporation. Another technique for anesthesia is to drop insects into a chamber containing dry ice, but this method has the risk of accidental overexposure that may kill them (Gambino 1990).

If the CO_2 is being used only as an anesthetic (in which case the insects should only be subjected to it for a minute or two), you should be aware that it can have permanent effects on the foraging behavior of bees and perhaps other insects. Ribbands (1950) found that exposure for 2 minutes caused worker honeybees to change from hive to field activities at an earlier age than normal; this treatment also caused a marked reduction or elimination of pollen collecting by the bees and a significant reduction of life expectancy. Simpson (1954) also found a notable reduction of pollen gathering by bees that were treated with CO_2 for 10 minutes. Beckmann (1974) found that CO_2 narcosis of honeybees for 10 minutes caused retrograde amnesia, with about a 50% memory loss, which appeared to be permanent. Ebadi et al. (1980) found that short exposure (less than 2 minutes) to pure CO_2 did not alter the orientation of foraging honeybees as longer exposure (e.g., lasting hours) had been shown to do. They also found that although a treatment of 15 seconds did not affect survival or pollen foraging behavior, treatments of 30, 60, or 120 seconds reduced both survival and pollen-gathering behavior.

4. Sticky Traps and Water Traps

Sticky traps are probably not suitable for most pollination studies. They are likely to catch a wide variety of insects that don't visit flowers, and once the insects are caught, they are difficult to remove from the sticky surface without damage that will make them difficult to identify or preserve. However, sticky traps might be a useful means of comparing insect visitation to artificial flowers of different sizes, shapes, colors, or scents (e.g., see Finch and Collier 1989). There is at least one commercially available preparation for use in making sticky traps (Tangle-Trap). After the insects are caught, the sticky substance is usually dissolved with a solvent to release the insects, which can then be rinsed in Cellosolve followed by xylene (Steyskal et al. 1986).

House (1989), in a study of the pollination of rain forest trees, resorted to sticky traps as the only way to collect insects from the forest canopy flowers. There are several advantages to this method. It enables systematic sampling of a number of tree canopies over relatively short sampling intervals, the traps collect (without the need of artificial attractants) insects that visit flowering canopies, they keep trapped insects dry and isolated so that pollen loads can be retrieved from individual insects, and they provide a quantitative sample of insects arriving per unit area of crown per unit time. House made her traps from pieces of galvanized mesh (mesh size 12.5 x 12.5 mm, wire diameter 0.8 mm), 400 x 250 mm, so each trap had an area of 0.1 m^2. A curtain ring was attached to the midpoint on the top and bottom of the screen, and a short length of nylon cord was tied to each ring. A loop of fishing line (fired over a branch using a sinker and slingshot) was used to raise and lower the screen from the desired spot in the canopy. The traps were hung on the outer edge of the canopy close to flowering branches. Screens were changed every 3–7 days by unclipping the old one from the line and clipping on the new one. Sticky traps were carried from the field in a box fitted with dividers that kept screens from touching each other. Insects were removed by soaking the screens in kerosene for 10–20 minutes and filtering the kerosene through nylon mesh; they were then stored in 70% ethanol.

Water traps are simply water-filled bowls, which will drown insects that try to land in them. A few drops of a detergent may increase the capture success. Kirk (1984) used plastic dishes of 165-mm diameter and 60 mm depth, painted with a variety of colors to determine which insects were attracted to which colors. Pellmyr (1989) monitored both fly phenology and relative fly density by placing yellow bowls (20 x 15 x 8 cm; 8% reflectance for wavelengths <500 nm, 60% for wavelengths >550 nm, 35% at 530 nm, which was the midpoint of the rise in reflectance that occurred from the lower to higher wavelengths) containing water and detergent on pedestals at flower level in his study sites. He found that the traps measured movement as well as density, as few flies fell into the traps (or visited flowers) if movement rate was low (e.g., because of bad weather) even if fly density was high. These traps also caught bumblebee queens carrying *Calypso* orchid pollinia (O. Pellmyr, *personal communication*).

Yellow traps (either sticky or water traps) are particularly successful at attracting a variety of insects, although biting Diptera from at least six families land preferentially on darker colors (Kirk 1984). Sol (1966) found that for the three colors he tested, syrphid flies were caught most frequently in blue, then yellow, then white traps. Kirk found that flower-visiting thrips were attracted to white (without ultraviolet), blue, and yellow over other colors. Kearns (unpublished) used yellow plastic plates that she covered with Tanglefoot, but there are also commercially available yellow traps (e.g., Tangle-Trap).

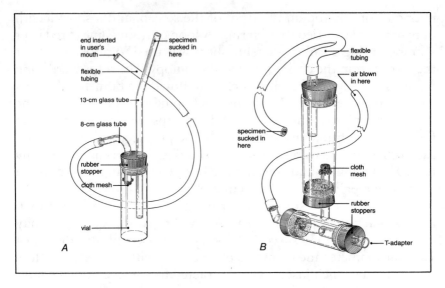

Figure 7–2. Vial aspirator (A); aspirator modified for blow type T-adapter (B). From Steyskal et al. 1986.

5. Vacuum Collectors

You may encounter situations in which insect nets or the other techniques described above are not adequate for collecting insects. One alternative is a vacuum collector. The ultimate in simplicity in a vacuum collector is probably an insect aspirator (Figure 7–2), which is a vial with two hoses entering it through a cork in the top. By sucking on one hose to create a vacuum, you can suck up insects with the other hose, catching them in the vial. Use a small piece of screen or mesh over the end of one the hoses inside the vial to prevent sucking the insect out of the vial and into your mouth. These are also available commercially (e.g., Carolina Biological Supply Co.). If you are concerned about coming into contact with allergens or toxins produced by the insects, you can modify the aspirator to work by blowing instead of sucking (Figure 7–2), or adapt it to work with a squeeze bulb.

Among the more exotic alternatives you might consider is the use of a battery-operated vacuum collector. Gary and Marston (1976) described a device they made based on a 12-volt automobile vacuum cleaner, powered by a car battery attached to a golf cart for

270

wheeling it around. They used a lightweight plastic pipe that could be extended with additional sections of pipe as the collecting end of the device, with a screen at the tip of the pipe to keep bees from being sucked into the vacuum cleaner (see their paper for construction details). Gary and Lorenzen (1987) described an improved device, also based on a 12-volt vacuum cleaner, modified to permit the use of small screen cages into which bees or other small insects can be collected. This modification helps to prevent buffeting and battering that might cause injury, the dislodging of pollen loads, or induce regurgitation of nectar. Perdew and Meek (1990) describe how to modify an inexpensive "cordless broom" to make a battery-powered aspirator for collecting adult mosquitoes and other small dipterans. Commercial vacuum collectors are also available (see Appendix 2).

6. Baits

If they were widely available, baits might be an attractive alternative for collecting insects, but there are only a few examples where they have been shown to work for pollinators. Sex pheromones, aggregation pheromones, or other chemical attractants might make it possible to collect insects that are normally inaccessible (e.g., those restricted to canopies of tropical rain forests), but may have the disadvantage of attracting insects that have not yet visited flowers (and may therefore not be carrying pollen). A disadvantage of sex pheromones might be that they only attract insects of one sex. In any case, the insects for which pheromonal lures have been developed do not appear to be important pollinators. Sigma Chemical Company sells pheromones for about 60 species of insects (primarily Lepidoptera, most economically important pest species, and some Diptera.)

Two examples, both from the tropics, demonstrate the use of components of floral scents to attract insects that pollinate flowers with those scents. Young (1989) used commercially available hydrocarbon and terpenoid compounds that are known components of floral fragrance oils from various species of *Theobroma* (Sterculiaceae) to attract midges that pollinate these flowers. Several researchers (e.g., Inouye 1975, Ackerman 1983, Roubik and Ackerman 1987) have used terpenes to attract male euglossine bees. These bees are pollinators of orchid flowers, as evidenced by the pollinia they carry. Much of the knowledge we have about these bees (e.g., their flight temperatures and the partitioning of orchid resources they visit) comes from collections made with cineole, methyl salicylate, eugenol, skatole, and other commercially available terpenes. These brightly metallic-colored bees are among the most spectacular of neotropical insects, and watching them buzz around a small sponge soaked with a terpene is a rewarding sight.

An additional "bait" that can be used to catch mosquitoes and other hematophagous Diptera is carbon dioxide, typically provided by dry ice, sometimes in combination with a small, "grain of wheat" light bulb. DeFoliart and Morris (1967) caught a variety of species of Culicidae (mosquitoes) and Tabanidae (horse-flies), as well as other blood-sucking Diptera that are not known as flower-visitors; their paper describes the design of the trap they used. A commercially produced trap is available (from BioQuip) that is battery-powered, and incorporates a small fan to draw mosquitoes into a mesh bag. Thien (1969) used a similar trap to catch mosquitoes for studies of a mosquito-pollinated orchid. We have had limited success with using a commercial trap to monitor the proportion of mosquitoes found carrying orchid pollinia at the Rocky Mountain Biological Laboratory. Careful examination of mosquitoes attracted to humans can also reveal the presence of a pollinium (small yellow dot on the head). Humans also make good bait for tabanid flies; males and some females feed on pollen and nectar (Borror et al. 1981), and can therefore be inferred to serve as potential pollinators.

Some Lepidoptera can be attracted with baits. Steyskal et al. (1986, p. 17) give a recipe for a bait based on fermented peaches and molasses (or brown sugar) that is attractive to noctuid moths (Noctuidae) and some butterflies. Utrio (1983) used a bait of brown sugar (100 g), beer (250 mL), and baker's yeast (1 g) that was fermented for 5 days in closed vessels, and then exposed in 25-mL vials with strips of filter paper to increase evaporation. He caught lots of noctuid moths and a smaller number of Geometridae. Pellmyr (*personal communication*) recommends a mixture of 10–15 ripe or overripe bananas, with 50 g of baker's yeast, left to ferment for at least 5–6 hours at about 40°C; although the container should have a lid, make sure it isn't on tight or you risk a messy explosion. A few grams of another ingredient, such as raspberry jam, may increase effectiveness of the bait. Some tropical Lepidoptera that

apparently use pyrrolizidine alkaloids as the basis of sexual pheromones (Pliske 1975b) are attracted as pollinators to plants containing those compounds (Pliske 1975c, DeVries and Stiles 1990) and
can be baited with dead stems, seeds, or foliage of those plants
(Pliske 1975a). Crude extracts of some of the alkaloids, placed on
absorbent pads, were attractive and could also be used as baits.

Gambino et al. (1990) used baited traps to monitor populations
of a Vespid wasp, the yellowjacket *Vespula pensylvanica*. Traps
consisted of a clear 1.5-L closed plastic cup containing 400 mL of
water and 5 mL of vegetable oil. A glass vial with a cotton wick was
attached to the lid, and 1 mL of a synthetic attractant, *n*-heptyl
butyrate, was added to the wick. Four 1-cm holes near the top of
the container walls served as entrances to the trap. Although these
wasps do not appear to collect or eat pollen, they do collect nectar
(P. Gambino, *personal communication*).

7. Malaise Traps and Window Traps

We have used Malaise traps in montane Colorado, and they
appear to be an efficient way to collect flies, and to a lesser extent,
other flying insects. Borror et al. (1981) mention that Malaise traps
often turn up rare or unusual species not collected by other means.
A Malaise trap (named after a Swedish entomologist, and not the
feeling you get when you see how many insects you have captured
and have to sort and count) consists of a barrier of mesh fabric with
a slanted fabric roof. Insects fly into the barrier, follow it up to the
roof, and then follow the slanted roof up to a collecting jar where
they are poisoned. We are not aware of any quantitative studies of
the efficiency of these traps in comparison to use of an insect net,
and it seems likely that there may be some insect species that are
not as susceptible to capture with the traps. Southwood (1978, pp.
243–244) reports that Malaise traps are apparently unbiased for
capturing larger Hymenoptera and some Diptera, but that the form
and color of the trap material will influence catches of Diptera.
Specimens captured with the traps may be useful for examination
of external pollen loads or of pollen in gut contents. We have also
found them helpful in monitoring changes in the abundance of flies
from year to year; the numbers of flies captured in weekly 36-hour
intervals appears to correlate well with visitation rates determined
by timed observations of flowers (Kearns 1990). Work by Hribar et

al. (1991) suggests that ultraviolet light reflectance might play a role in the effectiveness of these traps.

Another type of interception trap that has been used to monitor pollinators (Chaplin and Walker 1982) is a window trap. The general idea is to support a pane of glass (e.g., about 18 x 24 inches; pick a standard window size to get the cheapest price) in a wooden frame, with a water-filled trough below the glass on either side (Figure 7–3). Insects that don't see the glass will hit it, bounce off, and drop into the trough. Add a bit of a wetting agent, such as detergent, and ethanol as a preservative to the water in the trough. Chaplin and Walker (1982) found that a window trap was generally effective for heavy-bodied Hymenoptera, Coleoptera, Diptera, and some Lepidoptera (including both nocturnal and diurnal species).

8. Light Traps

At least some moths respond to light with a positive phototaxis, and this behavior can be used to capture them. The type of light bulb employed may affect the attractiveness of a light trap for different insect groups (Southwood 1978), but ultraviolet light is frequently used to attract insects instead of fluorescent or incandescent light sources. The lights can be shined onto a white sheet, and then moths can be picked off selectively, or the light can be incorporated into some sort of trap that will capture moths. For example, Kislev et al. (1972) presented an analysis of pollen loads of hawkmoths caught in light traps established for a study of Israeli moths. Kendall and Kevan (1981) used a portable blacklight (ultraviolet) trap to monitor noctural flight activity of alpine moths and discussed the significance of their results for noctural pollination. They used either a mild soapy solution of water and ethanol or an insecticide strip to kill the moths. Steyskal et al. (1986) describe several designs for light traps.

9. Fly Larvae

Some species of pollinators are predatory or parasitic; for example, some Syrphid fly larvae are predators on aphids, and others are found as scavengers in bumblebee nests. We have often found Syrphid larvae in bumblebee nest boxes at the end of the summer. Bombyliid fly larvae are often parasitic on bees, Tachinid

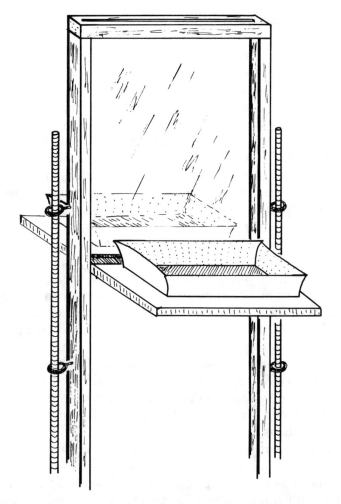

Figure 7–3. A window trap. Redrawn from a sketch by Chaplin (*personal communication*).

larvae are parasitic on Lepidoptera (and to a lesser degree Hemiptera, Coleoptera, Orthoptera), and some Calliphorid fly larvae can
be collected in carcasses. Information such as this may facilitate
gathering fresh specimens for collections or naive specimens for
controlled foraging experiments.

B. Preparing Collections

1. Pinning and Preserving Insects

Steyskal et al. (1986) present a thorough discussion of the techniques for mounting, labeling, and caring for insects. Some of this information is summarized here, but if you are not familiar with these techniques, you should see the full description.

In general, specimens should be "pinned" (mounted on insect pins) and labeled (see below). Stainless steel pins are available for use in tropical collections where standard pins might rust. Several sizes of insect pins are available for different sizes of insects. In general one should use the narrowest possible pin to minimize damage to the insect. However, if the pins are too thin to stick easily into the bottom of an insect box, use a double mount instead. Bees and flies large enough to be pinned should be pinned through the thorax, between or slightly below the base of the forewings, and slightly to the right of center, so you don't destroy the sculpturing of the central area of the scutum. Before specimens of bees have completely dried and hardened, it is desirable to pull the proboscis out into an extended position and to spread the mandibles apart so that their dentition and features of the labrum can be observed (Mitchell 1960). Straighten the legs, too, so they don't cover the lateral or ventral surfaces of the thorax. Identification of male bees may be facilitated by pulling the genital armature out into an exposed position.

Lepidoptera can be pinned — through the thorax at the thickest point or just behind the base of the forewings — on a spreading board, which has a groove for the thorax, and two boards along the side of the groove along which the wings can be spread while drying. Hold the wings in position with thin strips of paper pinned at the ends, to avoid putting pins through the wings themselves.

Insects too small or fragile for direct pinning should be double mounted on minuten needles (steel needles about 12 mm long and 0.1 mm thick, available from BioQuip; see Appendix 2) or carefully glued to "points," or triangular pieces of heavy paper (e.g., card stock). Points can be purchased pre-cut, or you can make your own with a point punch (a paper punch that makes triangles; available from BioQuip). Mount the point on a number 2 or 3 insect pin, and then glue the point to the right side of the specimen, taking care that the glue does not conceal critical characters (Figure 7–4). A. Nilsson (*personal communication*) suggests using transparent plastic (such as overhead transparency sheets) to make points so that one can still see the surface glued to the point. Specimens should be pinned while fresh (they will break if you try to pin them after they dry), but it is possible to store them in a freezer for a while before thawing and pinning them, or to relax them by putting them into a jar with a saturated paper towel for several hours (longer for larger specimens). If left in the relaxing jar for too long, the specimens may become moldy.

Pinned insects are typically stored in insect boxes or museum specimen drawers. Insect boxes are available in wood, fiberboard, or even cardboard, with foam bottoms that facilitate insertion of the insect pins. These boxes can be used to store, ship, or display insects. Smaller cardboard boxes (e.g., 10 x 10 cm and up) are also available for shipping insects.

Dried insects are susceptible to mechanical damage because of their fragility. To prevent damage, take care especially when shipping them. Put a sheet of corrugated cardboard across the top of the insect pins to prevent them from pulling out of the foam bottom during shipping. There should be room between the pins and the inside of an insect box or shipping box for this cardboard spacer. Put a flattened loop of tape on top of the cardboard to facilitate removing it. If necessary, additional space between the cardboard and the container lid can be filled up with a piece of bubble wrap, styrofoam sheet, or other packing material. You can also support large insects or double-mounted specimens with additional insect pins during shipping; place two pins at an angle to each other to form supports on either side of the mount.

We have had drying, pinned specimens reduced to a pile of wings overnight by mice that found them an easy source of protein. If your field site or laboratory has mice, be sure to put the insects somewhere inaccessible to them. The other pests of dried insects are dermestid beetles, whose larvae will burrow into specimens and usually destroy them. To prevent this kind of damage put moth balls or some other insecticide in the insect box, or store the boxes in an air-tight museum case that has an insecticide in it. Dermestids can get inside even what appear to be tightly closed insect boxes, so don't take chances with unprotected specimens.

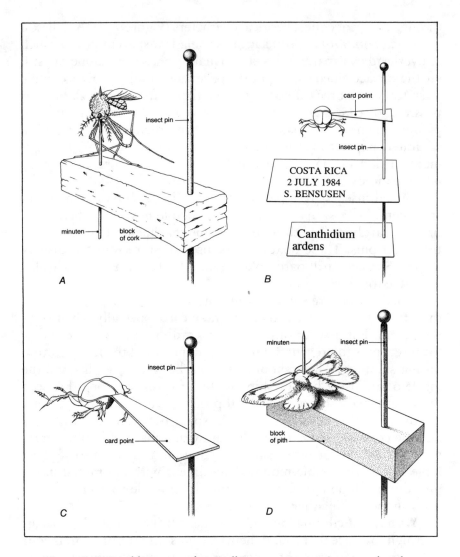

Figure 7–4. Double mounts for small insects: a mosquito pinned with a minuten needle to a block of cork on a regular insect pin (A); position of card point and lables on pin (B); attachment of card point to right side of specimen (C); a small moth pinned with a minuten needle to a block of pith on a regular insect pin (D). From Steyskal et al. 1986.

2. Preserving Insects in Alcohol

The following types of insects should be preserved in alcohol (70, 80, or 95% ethanol, not methanol or isopropanol) rather than pinned: ichneumonid wasps, mayflies, all soft-bodied insects (including all larvae and pupae) and most insects under 2 mm in length. Do not preserve adult bees, moths, true bugs (Hemiptera), or any flies other than minute Nematocera in alcohol. Place only one kind of insect in each vial. Use neoprene-, rubber-, or silicone-stoppered vials rather than screw-capped or shell vials (large mouthed, usually stoppered with corks; alcohol evaporates readily from these). Do not tape the vial caps; use Parafilm if necessary. Use clear glass vials of sufficient size to permit use of forceps or an eye dropper to remove specimens. To prevent dilution of the alcohol and subsequent decomposition of specimens, fresh alcohol should be placed in the vials within 24 hours after the initial immersion of specimens. All vials containing soft-bodied insects should be shipped with no air bubbles in the vials; to eliminate these, insert a paperclip or pin with the stopper (between the vial and the stopper), then remove the paperclip or pin after pushing the stopper in. Labels for specimens preserved in liquid should be inserted into the vial with the specimens and should be written with pencil or an ink that is not alcohol soluble.

Thrips (probably not pollinators in most cases, but perhaps important as pollen eaters in some systems) should be killed and preserved in alcohol-glycerin-acetic acid (AGA; 8 parts ethanol:5 parts distilled water:1 part glycerin:1 part glacial acetic acid; by volume).

3. Labeling Specimens

Whether they are pinned or preserved in liquid, the insects you collect should be labeled. At a minimum the labels should include information about the date and site of collection, and the name of the collector (Figure 7–4). Additional information, such as the plant on which the insect was collected, elevation, proboscis length, and a specimen number, can also be included on a second label if necessary. The labels should be on paper with a high linen content (preferably acid-free, as they may be in museum collections for a long time) and written with indelible ink.

Historically these labels were handwritten, but more recently entomologists have used commercial printers to generate labels (e.g., through BioQuip). One disadvantage of this method is that it usually entails a substantial waiting period between the time an insect is collected and the time labels are available. An alternative method that has been used is to generate text with a typewriter or letter-quality computer printer, which is then reduced either photographically or by using a reducing photocopier. Photographic reduction may be undesirable because of the darkroom work involved or the difficulty of finding appropriate paper, and photocopying has the disadvantage that the paper used in the copier may not be heavy enough, and the labels may have to be glued to a heavier stock.

The availability of personal computers equipped with laser printers offers an easier alternative for producing labels (Inouye 1991). Many word-processing and desktop publishing programs now offer the capability of specifying a wide range of point sizes for type. The size recommended for insect labels is 4 point, which permits about 5 lines of text on a label about 8 x 18 mm (Steyskal et al. 1986). At 300 dots per inch (the standard for most laser printers, although 400 dpi is also available on some models), 4-point type is very legible. You can use the "block copy" function of the word processor to generate multiple copies of the text quickly; type it once and then copy it. If you use a program that can support multiple columns of text on a page, you can place a large number of labels on a single page.

Steyskal et al. (1986) recommend using paper that is heavy enough so that the labels remain flat and do not rotate loosely on pins, such as 36-pound weight or two-ply, smooth-calendered bristol board (available from art-supply stores). They also recommend using linen ledger, 100% rag paper. Professional-grade herbarium sheets or archival-quality paper (e.g., acid-free paper available as expansion pages for use in albums; see Insect Labels in Appendix 2), are also suitable alternatives. Although these papers may not come in the standard 8.5 x 11–inch size that fits most laser printers, pieces this size can be cut from larger stock sheets with a paper cutter. These papers are more expensive than using regular laser printer paper but will also provide labels that are more suitable for long-term use in collections.

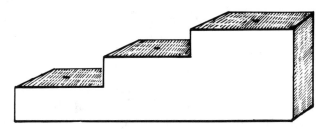

Figure 7–5. A pinning block, for adjusting insects and labels at a uniform height on a pin.

If you don't have your own (or a departmental) laser printer, you may find that you can have the printing done at a local computer service bureau by taking them an ASCII file with the text you want to use and suitable paper. They may charge a bit more than it would cost to have labels printed commercially by a printer, but you will probably be able to get them much faster. Sims (1989) found that a Mainframe Xerox 8700 laser printer produced better labels than a Hewlett Packard Laser Jet; the toner in the Xerox was more resistant when labels were submerged in an ultrasonic cleaner. This variation was a function of the differences in toners used in the two printers. Daly and Jordan (1989) found that labels printed on an Apple laser printer and labels photocopied on a Xerox copier were not as resistant to abrasion as handwritten labels following exposure to water, 70% ethanol, a humid chamber with phenol added, a hot plate, or paradichlorobenzene (PDB) crystals. However, it appears likely that laser-printed labels will prove satisfactory for the long term for pinned insects.

A simple tool called a stepping block or pinning block is a useful tool for someone who is preparing many specimens. It is a small block of wood with three steps for placing the specimen and labels at consistent heights on an insect pin. Each step has a small hole in it for inserting the insect pin (Figure 7–5).

C. Identifying Insects

If you are lucky, the pollinators you are working with may be easy to identify. For example field guides are usually available for hummingbirds and butterflies. If you are less lucky, you may find yourself working with minute flies, Microlepidoptera, parasitic wasps, or some other group for which there is no field guide, or perhaps not even any good taxonomic reference or expert. If you are working at a field station or near a museum, you may have access to reference collections or experts to help identify flower visitors.

The Systematic Entomology Laboratory of the Plant Sciences Institute of the U.S. Department of Agriculture's Agricultural Research Service is the primary U.S. federal agency responsible for providing identifications and other taxonomic services for insects and related organisms. They provide identifications of about 300,000 specimens per year. The Taxonomic Services Unit within the Systematic Entomology Laboratory receives, sorts, and distributes the specimens and reports the identifications (the service is free as of this writing). Although they are pleased to be of assistance, they ask that you determine whether sources of identifications are available in your own area or country before sending specimens to them. Material of agricultural interest is given the highest priority.

If you send specimens to the Systematic Entomology Laboratory, make sure that they are prepared and labeled carefully. For more information, request a copy of Form ARS-748A, Information for Identification Requests, and form ARS-748, the actual identification request form. The mailing address is:

Taxonomic Services Unit, Systematic Entomology Laboratory
USDA, ARS, BA, PSI
Building 046, Room 101A, BARC-West
Beltsville, MD 20705-2350 USA

The electronic mail address for the office is
MLACEY@ASRR.ARSUSDA.GOV; the telephone number
is (301) 344-3041.

The ability of this group to identify insects depends on the availability of experts for particular taxa. If no expert is available for a particular group, they may return your specimens without an identification. Alternatively, they may be able to recommend an expert elsewhere who would be willing to look at your specimens. The turnaround time is typically (from our experience) 6–18 months. If you include duplicate specimens, or some particularly interesting specimen, the taxonomists may ask for permission to keep samples for their museum or personal collections.

A general book with basic keys for all taxonomic groups of insect pollinators is: Ross, A. 1985. *American Insects: A Handbook of the Insects of America North of Mexico*. Van Nostrand Reinhold, New York, New York, USA. Here is a list of other references that you may find useful in trying to identify common North American insects:

Diptera

Cole, F. R. 1969. *Flies of Western North America*. University of California Press. (Out of print.)

Hull, F. M. 1973. *Bee Flies of the World. The Genera of the Family Bombyliidae*. Smithsonian Institution Press. Washington, D.C.

McAlpine, J. F., editor. 1987. *Manual of Nearctic Diptera*. Biosystematics Research Centre. Research Branch, Agriculture Canada. Monograph No. 28. (3 volumes).

Stone, A., C. W. Sabrosky, W. W. Willis, R. H. Foote, and J. R. Coulson. 1983. *A Catalog of the Diptera of America North of Mexico*. Smithsonian Institution Press. Washington, D.C.

Hymenoptera

Krombein, K. V., P. D. Hurd, D. R. Smith, and B. D. Burks. 1979. *Catalog of Hymenoptera North of Mexico*. Smithsonian Institution Press, Washington, D.C. (3 volumes). This book also lists host plants.

Laverty, T. M., and L. D. Harder. 1988. The bumble bees of eastern Canada. *Canadian Entomologist* 120:965–987.

Mitchell, T. B. 1960. *Bees of the Eastern United States*. 2 volumes. North Carolina Experiment Station, Technical Bulletin No. 141.

Stephen, W. P. 1957. *Bumble bees of Western America (Hymenoptera: Apoidea)*. Agricultural Experiment Station. Oregon State College (now University), Corvallis, Oregon 97331. Technical

Bulletin No. 40. Primarily a study of Pacific Northwest bumble bees (southern British Columbia, Washington, Idaho, Utah, Oregon, Nevada, and California), but we have found it useful for bumblebees in the Colorado Rocky Mountains, too.

Lepidoptera
Pyle, R. M. 1981. *The Audubon Society Field Guide to North American Butterflies.* Alfred A. Knopf, New York, New York.

Ferris, C. D., and F. M. Brown. 1981. *Butterflies of the Rocky Mountain States.* University of Oklahoma Press, Norman, Oklahoma.

D. Collecting Other Types of Pollinators

1. Birds
Permits (both state and federal) are required for capture, banding, maintenance in captivity, or marking of birds in North America. Federal Bird Marking and Salvage Permits are issued by the:

U.S. Department of the Interior
Fish and Wildlife Service
Bird Banding Laboratory
Office of Migratory Bird Management
Laurel, MD 20708
(301) 498-0205

Four types of federal permits are available to ornithologists in Canada and the United States: Master — Personal (for an individual permittee); Master — Station (for an organizational research program), and two types of subpermits (issued to persons who are not sufficiently experienced to qualify fully for a Master permit and to people assisting a master permittee with banding operations). Names and addresses of state permit offices are available from the Bird Banding Laboratory; they are typically run by a state's department of fish and game, division of wildlife, etc. (a list is available in the *North American Bird Banding Manual,* which is provided to federal bird-banding permit holders).

For information on how to use mist nets (the most common method) to capture birds, see Keyes and Grue (1982). The *North American Bird Banding Manual* includes this article as well as other information about banding techniques. Another reference for using bird traps or banding is: Bub, H. 1991. *Bird Trapping and Bird Banding: A Handbook for Trapping Methods All Over the World*. Translated by F. Hamerstrom and K. Wuertz-Schaefer. Cornell University Press, Ithaca, New York.

Hummingbirds, the most common avian pollinators in the New World, are often difficult to capture in mist nets because of their agility in flight. We have found it easier to use butterfly nets or traps to capture them. Once birds are accustomed to using a feeder, it is often possible to stand next to the feeder and catch them in an insect net (or even with bare hands if you want to impress an audience!). If the birds are marked, it is also possible to be selective about which ones are caught. Captured birds can then be banded, individually marked for later observation, and used to examine pollen loads or to study pollen carryover in flight cages (e.g., Waser 1988).

Traps allow some selectivity in capture of individual hummingbirds, although it may be more difficult to identify individuals when one is not standing right next to the feeder. Traps have the additional advantage of not requiring such quick reaction on the part of the trapper as the insect net technique does. One possible design for a trap (by Nancy Newfield, modified by Bob Sargent) is reproduced in Figure 7–6. The general idea is to build a wire cage large enough to hang a feeder in, with a trapdoor that can be lowered to capture the bird. If you suspend the feeder so that the feeding ports are slightly below the level of the trapdoor, it will be harder for the birds to escape before the door is closed. A hole covered with fabric will allow you to insert your hand to remove the bird from the trap. N. Newfield (*personal communication*) has found that trapping is far less traumatic to hummingbirds than capture in a mist net, but the birds must be accustomed to using a feeder as it does not usually work to put a flowering plant inside the trap.

To build the trap cut one piece of ½-inch hardware cloth 62 inches long and bend as shown in Figure 7–6. Overlap the 8-inch ends by 2 inches, and fasten them with wire. Cut out the 10 x 10–inch doorway. Cut three strips of foam padding 10 x 2 inches, fold them over the edges of the doorway, and glue them on with

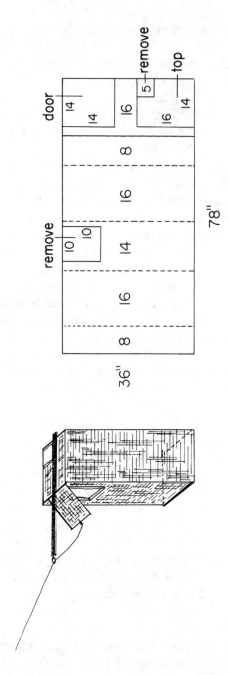

Figure 7–6. Pattern and finished design of a hummingbird trap made from a piece of $^1/_2$-inch hardware cloth (metal mesh), 36 inches wide, 78 inches long. Redrawn from descriptions by Newfield and Womack, Hummingbird Hotline #16, Sept. 1990.

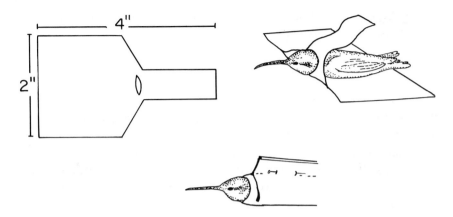

Figure 7–7. Pattern for flannel jackets for transporting hummingbirds. The sizes of the jackets will vary depending on the size of the birds to be transported. The size of the head hole is particularly important to prevent the bird from squirming out. Redrawn from Lasiewski 1962.

waterproof glue. Cut the top and door out; make the front edge of the top and the top edge of the door finished edges. Cut a 5 x 5–inch square out of the rear part of the top, and line it with foam. Attach the top to the body of the trap with wire, and attach the door to the top with two small S-hooks. Use a 15 x 15–inch piece of nylon net to cover the hole in the top, and attach it with binder clips; this provides access to the inside of the trap for removing birds. Wire a 1/2-inch square strip of wood across the top, and put an eyehook in the end. Tie a piece of string (or monofilament fishing line) to the door and run it through the eyehook and out to where you will be stationed while trapping. When you're not trapping, hold the door up by looping the string over a thumbtack or nail in the wood. To prevent birds from escaping if the door bounces open a bit when it drops, add a 2-inch-wide U of hardware cloth onto the door, so that it slips into the door opening when the door is closed.

Lasiewski (1962) described the construction of flannel restraining jackets (Figure 7–7) for holding and transporting hummingbirds, as well as a carrying case for handling large numbers of birds.

If you are particularly interested in working with hummingbirds, there is an informal newsletter (*Hummingbird Hotline*) distributed to those authorized to band them, edited by a volunteer. To ask to be added to the mailing list, write to: Ellie Womack, 1310 South Sycamore, Grove, OK 74344.

2. Bats and Other Mammals

Most researchers use mist nets to capture bats. An alternative might be to capture them at their roosts. Photography of bats on flowers may also be a means of gathering information about their foraging (e.g., Tuttle 1991).

Goldingay et al. (1991) trapped nonflying species of mammals by putting live traps baited with creamed honey on wooden stakes (1–2 m tall) near nectar-secreting inflorescences that they suspected were being visited by the mammals.

E. Collection of Pollen from Pollinators

1. Pollen Carried Externally

The pollen that is carried externally by flower visitors is of interest for several reasons. First, this pollen can effect pollination, and studies of the quantity of pollen picked up or deposited during floral visits are essential for an understanding of the dynamics of pollination. Second, it provides a record of the flower species (or in some cases morphs) that an individual has visited; the only caveat here is that grooming by pollinators may remove evidence of visitation. Third, to the degree that this pollen can be groomed off and consumed, it can also become a measure of the efficiency of pollen collection by pollen eaters.

Pollen carried on flower visitors can be removed in a couple of different ways that facilitate microscopic examination or quantification. The simplest situation occurs if you are working with plants that produce pollinia (Asclepiadaceae and Orchidaceae), as the pollinia are usually quite conspicuous and often easy to identify to species. Plants that produce loose pollen require more sophisticated techniques. If you are working with bees, it may be relatively easy to collect pollen that is already packed into corbiculae (pollen baskets) or abdominal scopae, but you may also want to collect pollen that adheres to other parts of the body. For Lepidoptera, Diptera, or other insects that do not collect pollen into some sort of external structure, then you may have to resort to a more specialized technique. Listed below are a variety of methods, classified by the type of animal from which the pollen is to be collected.

a. Insects

Beattie (1971a) described what has probably become the most widely used method for collecting pollen from insects. His technique (modified from a 1935 paper by Kisser) involves use of glycerin jelly containing a stain to make a semipermanent microscope slide. The ingredients are:

distilled water	175 cc
glycerin	150 cc
gelatin	50 g
crystalline phenol	5 g
crystalline basic fuchsin stain (also found in chemical supply catalogs as pararosaniline)	as desired

Phenol is added as a preservative, but it is also quite toxic to humans. T. Kelso (*personal communication*) has made this jelly without the phenol and found it worked well; it would develop mold after a month or two if not refrigerated. Phenol may be important in humid environments where fungi are more of a problem. If you make the jelly with phenol, avoid touching it with your hands or breathing the fumes.

To make the jelly add the gelatin to the distilled water in a beaker and heat until it is dissolved. Then add the glycerin and phenol and stir the mixture while warming it gently. Add basic fuchsin crystals to the mixture until you get the strength of color you desire. Beattie (1971a, p. 82) recommends "that the color of claret stains most pollen grains very clearly. Too dark a color may obscure morphological detail and too light a color may not highlight such detail sufficiently." Different species of pollen absorb different amounts of stain, so you may want to make batches of various shades. The mixture should be filtered through glass wool into sterile containers, which should then be sealed. Other stains can be substituted for the basic fuchsin, such as ruthenium red for staining intines, or a general stain such as fast green.

Take a small cube of the hardened jelly in some sterilized fine forceps or on the point of a dissecting needle. Use it to pick the pollen off of an insect, and then put it on a clean slide. If the pollen doesn't adhere well, warm the jelly slightly and try again. Warm the slide gently on a hot plate or over an alcohol lamp until the jelly melts; if you get it too hot, the jelly will denature and lose its consistency. Addition of a glass coverslip will make a semipermanent mount. This same technique works with whole insects, parts of flowers or insects, or intact stigmas.

Macior (1983) analyzed corbicular pollen loads by mounting them in glycerin jelly tinted with methyl green instead of basic fuchsin. Motten (1986) used a technique similar to Beattie's to remove pollen from insects. He dabbed cubes of agar (approximately 1 mm³) on the insect's body to pick up pollen grains. The agar was then dissolved in a drop of lactophenol with cotton blue stain on a microscope slide. Insects with very large pollen loads (especially in corbicular or scopal loads) could be subsampled with this technique. Pollen grains could then be counted or identified. In some situations, if you are familiar with the pollen collected by bees in your study area, you may be able to identify it by the color of the corbicular loads. If the corbicular load is multicolored, you can section the pellet of pollen to look at the changes in plants visited during the foraging trip.

If you don't need to keep the bee alive, you can remove the corbicular load after catching it in a killing jar. Lindsey (1984) removed pollen loads from freshly killed insects by using stiff-bristled brushes to brush the pollen into drops of cotton blue and lactophenol on microscope slides. She used the same brush for multiple insects but cleaned it carefully between samples to prevent contamination. Total counts of pollen grains were made if they did not exceed 1,000 grains; larger pollen loads were subsampled using an ocular grid. Buchmann and Shipman (1990) placed dead bees in snap-cap polyethylene vials with 10 mL of 90% ethanol, and then inserted the micro-tip of a Sonicator (Ultrasonics model W-220F) 1–2 cm into the vial for 15 seconds at a power setting of 5. This setting removed pollen from the bees without removing setae or other body parts. See Chapter 4 for additional related techniques for working with pollen.

The techniques described above are probably best for dead insects, or at least anesthetized ones. Parker (1981) described a method for removing pollen from live bees that might work for other insects as well. He caught a bee in a long tube, added distilled water, shook the vial up and down until no more pollen could be seen on the bee, and then released it. After the bees dried off, they flew off and were later observed foraging. The washed pollen was concentrated by centrifugation and then counted with a hemacytometer. Presumably the rinsing process could be repeated a second time to ensure more complete removal of the pollen. Primack and Silander (1975) used the same technique for bees and beetles but used alcohol, which might help rinse the pollen off better than water but would preclude releasing the insects unharmed.

Young (1988) scraped pollen from the bodies of flower-visiting beetles onto slides covered with fuchsin-stained gelatin, as she did not want to collect the beetles. J. Thomson (*personal communication*) designed a tool for removing corbicular loads from bumblebees as they returned to a nest box (Figure 7–8). The apparatus is inserted between the nest box and an entrance tunnel so that bees pass through it on the way to the nest box. When the bee is in the middle of the hardware cloth drum and underneath the block of foam that forms the roof of the middle section, push down on the foam to pin the bee against the mesh. Rotate the drum half a turn so that the trapped bee is now on the top, and remove the corbicular pellets through the mesh with forceps. By squeezing the bee harder, you can cause it to regurgitate a sample from its honey stomach, which can be collected in a capillary tube.

Olesen and Warncke (1989b) combined preservation of specimens (in 50% ethanol in individual vials) with pollen collection. To measure surface pollen loads, they washed the insects with ethanol from the vial, rinsed the vials again, and then concentrated the pollen by centrifugation. One advantage of this technique might be that pollen loss would be minimized during and after capture. In a study of pollen removal from anthers, Murcia (1990) kept bumblebees in vials for 2 days, during which time they groomed most of the pollen off their bodies. Then each bee was allowed to visit a single flower, after which the pollen was collected into a 3:1 mixture of absolute ethanol and acetic acid. To quantify the amount of

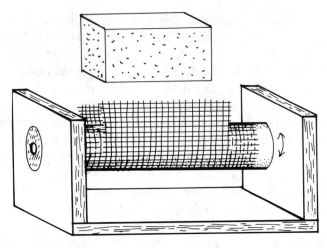

Figure 7–8. A tool for removing corbicular pollen loads from bees. Redrawn from a sketch by J. Thomson, *personal communication.*

pollen on the bee Murcia allowed the liquid to evaporate and counted pollen remaining on the bee or in the vial.

House (1989) wanted to collect and identify pollen from insects caught on Tangle-Trap sticky traps. A subsample of 10–50 insects was taken from each trap. These insects were stored individually in kerosene to dissolve the Tangle-Trap. The kerosene was then filtered through 13-mm diameter Sartorius cellulose nitrate membrane filters (type SM11301, pore size 8.0 μm), using a Millipore filtration apparatus and a hand-operated vacuum pump, to collect pollen carried externally by the insect. The apparatus was rinsed with clean kerosene between samples. House dried the filters in an oven at 60°C for 20 minutes, mounted them on glass sides, and applied nondrying immersion oil (n_D 23°C = 1.5.50) to clear them. A coverslip was placed over the filter to spread the oil and then sealed with clear nail polish. The slides could then be stored until examined with a microscope.

Goldblatt et al. also removed pollen from the body hairs and scopae of pinned insects by washing them with a few drops of absolute ethanol, and "teasing pollen loads off the scopae with a dissecting needle" (Goldblatt et al. 1989, p. 202). The ethanol was dropped onto a microscope slide, evaporated, and the residue was stained with Calberla's fluid and examined microscopically.

Kislev et al. (1972) examined pollen loads carried on the proboscides of hawkmoths captured in light traps. The proboscides
were removed with forceps and placed in drops of 5% KOH on a
slide placed on an electric hot plate at 90°C. This treatment relaxed
the proboscides and permitted the release of the pollen grains into
the solution. Each proboscis was transferred to another slide with
KOH to rinse it thoroughly. The pollen grains were cleared by
heating, with distilled water being added to prevent excessive
concentration of the KOH as it was heated. A drop of glycerin with
phenol (as a preservative) was added to the slide, which was then
sealed with paraffin. Despite these precautions, fungi eventually
ruined most of the slides.

Beattie (1971b) describes a detailed list of 25 parts of the insect
integument that he examined for pollen grains as part of a study of
the pollination biology of violets. He caught insects on flowers or
in flight near them and preserved them in an 8:2 mixture of 70%
ethanol:glycerol, which is viscous enough to minimize loss of pollen
from the integument. He examined the insects microscopically, and
for each region of the integument a score was determined:

0 — No pollen on region
1 — Very sparse pollen, fewer than 10 grains
2 — Patchy pollen cover, 10–40 grains
3 — General pollen cover, much of area covered
4 — Heavy pollen cover, grains may be several layers deep

The scores were summarized on a data sheet and graphed both
by region and on a diagrammatic representation of an insect (Figure
7–9). Beattie et al. (1973) and Green and Bohart (1975) used this
same list of tegument subdivisions to summarize the distribution of
pollen. The regions that were scored differed for Diptera and
Hymenoptera and are shown in Table 7–1.

Turnock et al. (1978) described a method of identifying pollen
grains on moths by using scanning electron microscopy (for more
detail, see Chapter 4, Section A.1.b). Because they found that most
pollen grains occurred on the proboscis of the moths they were
studying, they concentrated their attention on this part of the body.
Each moth's head was removed and glued to an aluminum stub. The
proboscis was then split carefully with a dissecting needle to expose

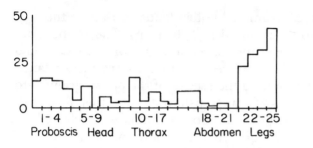 

Figure 7–9. Diagrams showing the distribution of pollen on flower visitors. The x-axis has code numbers for different parts of the integument, and the y-axis is a score for pollen coverage. On the diagrammatic insect, black indicates most dense coverage of pollen, hatched portions were small quantities, and white portions had no pollen. Redrawn from Beattie et al. 1973.

the inside of the proboscis tube. Some thoraxes, legs, and abdomens were also mounted on stubs to look for pollen. With this technique the researchers could identify pollen grains on and in the proboscis as well as on other body parts. In general, the pollen grains could only be identified to genus, subfamily, or family (Turnock et al. 1978). Although this method requires some preparation, as well as access to an SEM, it avoids the more tedious preparations required for acetolysis, other chemical fixing procedures, or critical-point drying. Bryant et al. (1991) used a similar technique for a study of pollen adhering to nectar-feeding noctuid moths and found that it took about 3 hours per moth for specimen preparation, photography, and pollen identification. Stelleman and Meeuse (1976) used SEM to identify and count pollen on syrphid flies.

Pollen from corbiculae can also be identified with SEM. Pellmyr (1985) took longitudinal cross-sections through one corbicula from each bumblebee that he collected and identified pollen with the help of a reference collection made at the study site. Bumblebees deposit pollen in almost vertical layers in the corbicula, so a cross-section will show the variety of pollen collected, unless some pollen is added laterally onto a very large corbicular load (Pellmyr 1985).

b. Birds and other vertebrates

Paton and Ford (1977) caught honeyeaters in mist nets and then collected pollen from them by holding them over a folded piece of paper and brushing pollen off their feathers and bills. The pollen

Table 7–1. Code numbers for parts of the insect integument

Body Region	Diptera	Hymenoptera
Proboscis	1. Anterior parts of proboscis: epipharynx, hypopharynx	1. Anterior parts of proboscis: epipharynx, maximillary stipes, galeae, laciniae, and palps
	2. Posterior parts of proboscis: labium	2. Posterior parts of proboscis: maxillary cardines, labial mentum, prementum, palps
	3. Lateral portions of above structures	3. Lateral portions of above structures
	4. Labellum	4. Paraglossae and glossae
Head	5. Vertex, ocellar triangle, dorsal occiput	5. Vertex, ocelli, dorsal occiput
	6. Vibrissae, folded mouthparts, proboscis fossa	6. Labrum, mandibles, folded mouth parts, proboscis fossa
	7. Eye	7. Eye
	8. Orbit, frons, parafacialia, face, buccae	8. Frons, genae, clypeus
	9. Annenal socket and antennae	9. Annenal socket and antennae
Thorax	10. Dorsal cervix, prescutum, humeri	10. Dorsal cervix, pronotum
	11. Ventral cervix, prosternal plates	11. Ventral cervix, prosternal plates
	12. Scutum	12. Mesonotal scutellum, prescutum (if present)
	13. Mesosternal plates	13. Mesosternal plates
	14. Scutellum	14. Mesonotal scutellum, metanotum (postscutellum), propodeal tergum
	15. Metasternal plates	15. Metasternal plates (if present), propodeal sternum
	16. Pleural plates, lateral parts of other thoracic plates, alulae, squamae	16. Pleural plates, lateral parts of other thoracic plates, alulae, squamae

Body Region	Diptera	Hymenoptera
	17. Halteres, pleural articulations of wings, tegulae	17. Pleural articulations of wings, tegulae
Abdomen	18. Terga of segments 1–3 or 4	18. Dorsal petiole, terga of segments 2–3
	19. Sterna of segments 1–3 or 4	19. Ventral petiole, sterna of segments 2–3
	20. Terga of remaining segments	20. Terga of remaining segments
	21. Sterna of remaining segments	21. Sterna of remaining segments
Legs	22. Coxae, trochanters, and femora of first pair	22. Coxae, trochanters, and femora of first and second pair
	23. Tibiae and tarsi of first pair	23. Tibiae and tarsi of first and second pair
	24. Coxae, trochanters, and femora of second and third pair	24. Coxae, trochanters, and femora of third pair
	25. Tibiae and tarsi of second and third pair	25. Tibiae and tarsi of third pair

Source: Beattie 1971b.

was then taken to the laboratory and sprinkled onto a warm drop of crystal violet in aniline oil on a microscope slide. Pollen was then identified to species by comparison with a reference collection. They found that contamination of the brush between samples was a problem. Although they cleaned the brush between uses by rinsing it in water or by brushing it vigorously against a cloth, several times it was found to have residual grains on it. However, because most pollen loads were hundreds of grains, the qualitative consequences of contamination were probably not significant.

Brown and Kodric-Brown (1979) studied the pollination ecology of a group of flowers visited by hummingbirds and looked at character displacement in pollen placement by collecting pollen samples from mist-netted birds. They pressed clear plastic tape (separate samples) against the bill, crown, sides (cheeks), and chin of each bird, and put the tape on a clean piece of white paper or a glass slide. These were then examined microscopically in the lab to determine the number of species represented and the number of grains of each species.

Feinsinger et al. (1987) worked with neotropical hummingbirds and used a similar technique to remove pollen from the bill, head plumage, and throat plumage of mist-netted birds. A piece of tape "was laid down dorsally from the bill tip to the nape of the neck, pressed down and gently rubbed, peeled off from the tip back, and mounted on a microscope slide." Then a second piece of tape was used to sample the underside of the bill and the throat. Pollen grains were then counted and identified by using a compound microscope modified with Hoffman Optics (see Chapter 4).

Wooler et al. (1983) used the glycerin-jelly technique described above to remove pollen from small marsupial pollinators. A small block (2 mm^3) of glycerin jelly was rubbed on appropriate areas of the animals' heads to pick up pollen, placed on a microscope slide, and then melted with a match or cigarette lighter. Goldingay et al. (1991) used a piece of tape (5 cm long), touching it three times to the snout of nonflying mammalian pollinators. The tape was placed on a microscope slide and pollen loads were scored by counting grains along three transects.

2. Gut Contents or Feces Containing Pollen

The exines of pollen grains are almost indestructible. Thus even after insects or other animals have ingested, digested, or excreted pollen, it can usually still be identified. For example, one assay of the diets of flower-visiting animals can be conducted simply by holding them and collecting feces, and then examining the contents microscopically. Bees, flies, butterflies, beetles, or moths captured on flowers and then held in vials for a few hours will probably defecate on the sides of the vials, and by examining the feces it should be obvious whether they have been eating pollen. A stain such as malachite green (which stains cellulose in pollen grains) might facilitate identification of pollen if it is present.

In addition to this nondestructive technique for sampling, it is possible to examine gut contents. For example, Pellmyr and Patt (1986) dissected beetles collected on flowers to confirm that they were eating pollen. Haslett (1983, 1989) carried out a study of the gut contents of syrphid flies, whose abdomens are often visibly yellow from pollen inside them. He dissected freshly killed flies in insect saline and observed pollen through the gut wall. He then sampled pollen from different regions of the gut, including the esophagus, crop, anterior and posterior limbs of the midgut, anterior hindgut, and rectum (Haslett 1983). These samples were then examined at a magnification of 400X. In his 1989 paper Haslett described a technique for preparing semipermanent stained slides of pollen sampled from the alimentary tract of flies, which should also work with other insects:

1. Dissect out the fly's gut (probably best done in insect saline solution).
2. Heat the isolated gut at 95°C for 3 hours in 5 mL of 10% KOH. This destroys the gut wall and cleans up the pollen grains enough that taxonomically important characters can be seen.
3. After the mixture has cooled, centrifuge it (3,000 rpm for 3 minutes). Decant off the KOH and rinse the pollen in distilled water. Centrifuge and decant again to leave a pellet of clean pollen.
4. Add a mountant such as glycerol jelly with a stain (Haslett used molten glycerol jelly with aqueous 0.5% safranine; see Chapter 4, Section A.1.a.ii) to form a suspension.
5. Use a prewarmed pipette to take up some of the suspension and spread it on a warm microscope slide. Add a coverslip and allow the slide to cool before examining it.

Olesen and Warncke (1989b) used a simpler technique for gut analyses. They mounted gut contents of flower-visiting flies in lactophenol-aniline blue for at least 8 hours before examining them. This stained the pollen blue and made it conspicuous. Kearns (1990) dissected intestines from muscoid flies, and stained them with basic fuchsin to aid in identification of the species of pollen. Pellmyr (1985) collected beetles from eight flower species and preserved them in alcohol. Subsequently he dissected their alimentary tracts, mounted the contents in Euparal, and identified the pollen he found with the help of a reference collection. He found it was possible to determine the sequence of pollen meals in these guts (*personal communication*).

It is also possible to dissect dried insects (e.g., those preserved on insect pins) to determine gut contents. Van der Goot and Grabandt (1970) did this with syrphid flies by softening them (by immersion for 10 minutes in 10% KOH at approximately 90°C) and then dissecting out the alimentary tract. They then pressed the pollen out of the tract and mounted it on a microscope slide. Leereveld (1982) found that 5 minutes in KOH was sufficient for softening and warned that excessive softening should be avoided to prevent the fly from coming apart with consequent loss of pollen from the digestive tract. He boiled the pinned insects while they were still on their pins to avoid loss of pollen through holes made by the pins. The exoskeleton of a softened insect can be removed with fine forceps or needles, and the digestive tract exposed.

Another way to prepare pollen from insect guts for examination, which would work with either fresh or dried specimens, is acetolysis. Benedict et al. (1991) removed the dorsal integument of boll weevils to expose the alimentary canal, which was removed and then processed with acetolysis to destroy organic detritus while leaving the pollen grains undamaged.

Wooller et al. (1988) examined feces of four flower-visiting bird species to determine whether they were eating pollen. Pollen grains in the feces were treated with acid fuchsin and malachite green to stain differentially grains containing protoplasm and empty grains (see Chapter 4). By confining them in a cage for a brief period after capture, it should be simple to collect feces from birds, bats, or other mammalian pollinators that are suspected of eating pollen.

Goldingay et al. (1991) trapped nonflying mammals in traps (baited with creamed honey) near nectar-secreting inflorescences. They examined feces deposited in the traps for the presence of pollen, as an indication of whether animals caught in the traps had been visiting inflorescences. A standard aliquot (2 mm^3) of fecal material was examined microscopically and the number of pollen grains was counted. Staining might facilitate observation of the grains.

3. Collecting Pollen from Insect Nests
Pollen is used as the primary larval food for many flower-visiting insects and can therefore be collected from nests. Honeybee pollen is often collected in commercial quantities by pollen traps. These

are basically wire mesh screens through which the workers must pass to get into the hive. The mesh size is adjusted so that the workers' bodies will fit through, but the lumps of pollen in their corbiculae are scraped off and drop below into a collecting bin. Pollen thus collected can serve as a food source for captive colonies of other types of bees and could be used to study resource utilization by the colony. If a pollen trap is not available, pollen could also be collected from the cells in which it is stored by the honeybees.

Although most other species of bees have much smaller colonies, it might be possible to use the same idea of scraping pollen loads off worker bumblebees or females of some solitary bees. Alternatively, pollen could be collected from the cells where it has been deposited for larval bees to feed on it. Raw (1974) collected samples from the cells of three solitary bee species (*Osmia*), stained them with basic fuchsin in 70% ethyl alcohol, and mounted them in glycerol jelly to identify them.

If the primary objective is simply to ascertain what species of pollen are being collected, it may be sufficient to wait until the colony or cell (of a solitary bee) is finished for the season, collect it, and then analyze the larval meconia (feces) that are ejected from bees before they pupate. The pollen grains should still be identifiable. Brian (1951) used this technique, collecting the hardened meconia found on the outside of pupal cocoons of bumblebees. Advantages of this technique are that it minimizes disturbance and doesn't remove food from the colony, analysis can be deferred until demands of the field season are over, and if necessary diets of individual larvae can be compared and the total pollen for a colony can be determined (Brian 1951). S. Buchmann (*personal communication*) has also used this technique with bumblebee colonies to compare pollen utilization among habitats by acetolyzing the remains of colonies.

F. Collecting Nectar from Pollinators

There are at least four reasons why one might want to collect nectar directly from flower visitors. One is to ascertain how much an individual forager is carrying in order to study the energetics of foraging. This permits estimation of the number of visits that a flower-visitor makes during a foraging bout. A second reason is to determine nectar characteristics, as in most cases the flower visitors are probably more efficient at extracting nectar than humans are (this assumes that the nectar collected from pollinators is not changed during or shortly after collection). This is particularly true for very small flowers, such as the individual florets of composite flower heads. A third reason is to look at pollen grains in the nectar loads as evidence of visits to particular species of plants. Fourth, if the nectar from a plant species has some unusual characteristic, such as the high concentration of potassium in *Allium cepa* (Liliaceae) nectar (Waller et al. 1972), analysis of nectar collected by a forager can show whether it has visited that plant species. Samples for any of these purposes are most easily collected from animals that have a crop for storing nectar. This technique is used most commonly on honeybees and, to a lesser extent, on bumblebees, both of which have honey stomachs or crops in which nectar is stored by foragers before they regurgitate it at the colony. It might also be useful for nectarivorous birds, which also have crops.

Methods used for studies of honeybee honey stomachs (reviewed by Gary and Lorenzen 1976) include dissecting the bee and removing the honey stomach to weigh it, and collecting the contents of a dissected stomach with a calibrated pipette. Less intrusive methods are also available, such as comparing the weights of outgoing and incoming foragers, comparing the weights of starved bees before and after feeding, or determining the mean weight of sugar syrup collected by groups of bees from the decrease in weight of a small dish of sugar syrup. Thomson et al. (1987) determined the weights of pollen and nectar separately for foraging trips by individually marked bumblebees by weighing the bee as it left the colony; weighing the bee when it returned; removing the corbicular pollen load and weighing the bee again; then waiting until the bee had regurgitated its nectar load into the honey pot in the nest box before a final weighing.

Gary and Lorenzen (1976) describe a technique that simplifies the collection, storage, and analysis of the contents of honeybee honey sacs. Although the contents can be expressed from a forager simply by squeezing the abdomen laterally and then collected at the bee's mouthparts (e.g., for carpenter bees, *Xylocopa,* see Wittmann and Scholz 1989), they wanted to develop a more reproducible technique. They mounted a pair of laboratory coverglass forceps on a base made from pieces of 4-mm-thick Plexiglas (Figure 7–10). They cut a V-shaped notch into a small rectangular piece of 28-gauge sheet metal, which they then added to the base near the tip of the forceps to hold the bee firmly in place during the collection procedure. They anesthetized bees (15 seconds of CO_2), placed them in the apparatus, and then squeezed with the forceps until the abdomen was flattened to force the nectar out of the bee. The nectar was collected in a microcapillary tube (1.0 mm internal diameter, 100 mm long) so it could be measured or saved (by sealing the ends of the tube with vinyl plastic putty) for chemical analysis.

Although Gary and Lorenzen's technique seems to work well for honeybees, it might be a bit harder to adopt for use with bumblebees because there is more variation in the size of bumblebee workers. Ranta and Lundberg (1981) obtained nectar samples from bumblebees by "pressing gently upon the tergites of the abdomen with a fingertip, and then the nectar contents were regurgitated on a slide." None of the pressure techniques will work with other insects that do not have a similar storage organ. The crop can also be dissected out of a bee, exposing it by removing the posterior portion of the abdomen (Heinrich 1972), but this is necessarily a destructive technique.

However the samples are collected, once available, they can be subjected to the same range of analyses as nectar collected directly from flowers. This assumes, however, that the honey sac samples are not changed significantly in their concentration or content during their residence in the bee. There do not appear to be many studies comparing the characteristics of nectar sampled directly from flowers and immediately after collection by a flower visitor. Park (1933) conducted a series of experiments, feeding bees a sugar syrup half a mile from the hive where he collected them. He found that nectar concentrations from honey stomachs were slightly more dilute than the syrup. The degree of dilution depended on the syrup concentration, ranging from -0.02% for 13% syrup to -1.80% for 64% syrup. This suggests that there is minimal alteration of nectar during collection and transport, at least by honeybees. In more recent work, however, Heinrich (1979a) found that at high ambient temperatures honeybees will regulate their head temperature by evaporative cooling using regurgitated nectar from their honeycrop.

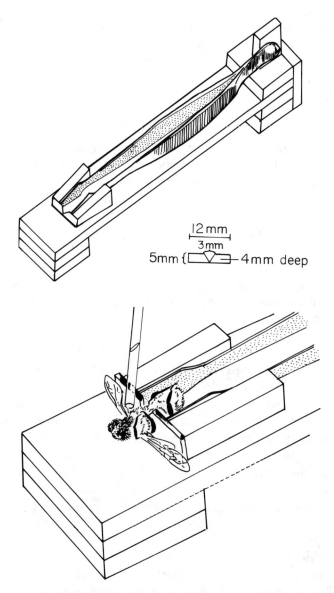

Figure 7–10. Apparatus for extracting contents of honeybee honey sacs. Redrawn from Gary and Lorenzen 1976.

Male bumblebees (*Bombus;* Bertsch 1984) and carpenter bees (*Xylocopa;* Wittmann and Scholz 1989) also dehydrate nectar, apparently as a way of increasing the efficiency of carrying resources for mating flights.

An alternative technique to squeezing out the contents of honey stomachs for determining the nectar loads of flower visitors was used by Pflumm (1977). He designed a way to measure how much of a nectar solution was consumed by a bee from a feeder. Nectar was supplied to the feeder (a microcapillary tube protruding from a glass plate) from a calibrated pipette with a syringe. As a nectar-collecting bee or wasp ingested nectar from the microcapillary tube feeder, he could replenish the supply from the pipette and measure the volumes being ingested. He used this device to determine that there were differences in the amount of nectar collected with changes in sucrose concentration, air temperature, and the shape of the feeder (capillary tube versus watch-glass–shaped feeders). The artificial flowers described below in section K.3 might also be used to determine some of this same information about nectar feeding animals.

Waller et al. (1976) and Waller and Martin (1978) were able to use two characteristics of onion (*Allium cepa* L.) nectar to identify foragers that had been collecting it. *Allium cepa* nectar has high levels of potassium, and atomic absorption spectrophotometry can be used to identify nectar-collecting bees that have been visiting it (from analysis of nectar in their honey stomachs). When exposed to long-wavelength ultraviolet light, onion nectar fluoresces brightly with a blue hue (Thorp et al. 1975). Waller and Martin (1978) also used this characteristic to determine whether the contents of honey stomachs, spotted onto filter paper, contained onion nectar.

A crude measure of crop contents in nectarivorous birds can be gained by gently blowing feathers covering the skin over the crop, to see how distended the crop is by its contents.

Roberts (1979) describes a method of assaying the sugar contents of flower visitors, which would also work for nectar samples removed from the animals. Whole insects can be mascerated in a vial with a known amount of distilled water, and the solution can then be analyzed as described in Chapter 5.

G. Morphological Measurements

Some of the most spectacular examples of evolution in flowers and pollinators are provided by their morphologies. The extremely long nectar spurs on some orchids and the extremely long proboscides that pollinators of those flowers possess are well-known examples. Darwin observed some of these orchids on Madagascar, and predicted the discovery of the long-tongued (25 cm!) sphingid moth that was discovered 40 years later and given the species name *predicta*. Figure 7–11 shows another spectacular example of a morphological correspondence between flower and pollinator. Proboscis length also appears to be an important morphological characteristic affecting resource partitioning and the organization of some bumblebee communities (Inouye 1977, 1978). This is not surprising, as characteristics of bills, tongues, or proboscides will influence the speed and efficiency of nectar or pollen collection.

Proboscis lengths of insects, or bill and tongue lengths of nectar-feeding birds, are probably the most significant morphological measurements for pollination studies. Wing length of different hummingbird species is also important, as it affects the aerodynamics of flight (e.g., Feinsinger et al. 1979). And of course morphology is also important as it relates to taxonomic characters. Thus pollination biologists may have to develop some familiarity with the terminology used to describe morphological characters in order to use keys to identify species. Figures 7–12 and 7–13 show the sorts of morphological characters that may be important to know for identification of Diptera and Hymenoptera.

1. Proboscis Length

Studies of resource partitioning or foraging often address the morphology of a foraging animal and its relationship to its prey or other food resources. In the case of flower-visiting insects the morphological measurement that is usually of most interest is proboscis or tongue length, sometimes in conjunction with head width. For example, in bumblebees resource partitioning is thought to depend in large part on the relationship between proboscis length and the corolla lengths of available flowers (e.g., Inouye 1978, 1980b; Ranta and Lundberg 1980; Ranta 1983). Even within a species a bumblebee's proboscis length may influence which flowers

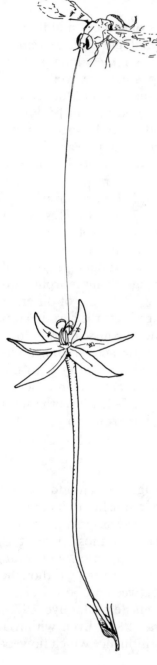

Figure 7–11. Long-tongued South African fly (*Megistorhynchus longirostris*, Nemestrinidae) and a flower that it visits (*Lapeyrousia fabricii*, Iridaceae). Redrawn from Vogel 1954.

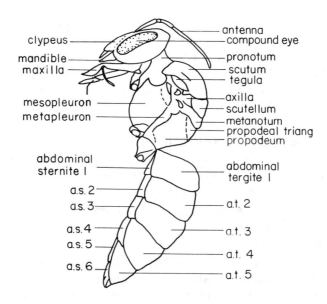

Figure 7–12. External morphology of a representative bee. Redrawn from Mitchell 1960.

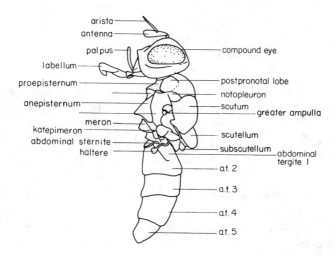

Figure 7–13. External morphology of a representative fly. Redrawn from McAlpine 1987.

it will visit (Inouye 1980b). The number of species of insects that use other body parts, such as legs, for collecting floral resources is quite small. Vogel and Michener (1985) and Steiner and Whitehead (1990, 1991b) describe oil-collecting bees in South Africa that use their forelegs to extract oil, and Inouye (Inouye and Pyke 1988) observed an alpine Australian dipteran species collecting nectar in a similar manner.

Proboscis length of an insect might seem like a relatively straight-forward measurement to make, but in fact there are a variety of ways in which it has been measured, which may result in significant differences reported for the same species. Part of the problem lies in the loss of flexibility of the mouthparts in preserved specimens, and part of it lies with the methodology used in measurement.

Another problem facing researchers is how to determine proboscis length for an individual forager under observation in the field. Pulling the proboscis to extend it for measurement may not work well on live specimens, and anesthesia by CO_2 may alter foraging behavior. Morse (1977a) suggests measuring insects immobilized by cooling to a torpid condition, at the same time that they are being given individual marks or tags. This problem has also been addressed successfully in bees by establishing the correlation between proboscis length and some other more readily determined measurement, such as one relating to wing length, in a series of freshly killed individuals, and then using this correlation to estimate proboscis length. Total wing length may not be satisfactory because of wing wear in older foragers, and radial cell length has been faulted (Morse 1977a) because it is subject to proportionately greater error. However, unless very old bees are being used, it is likely that at least one wing will still be intact, and if bees are being marked regularly in a study site then young ones would be marked as encountered (Morse 1977a). Morse found that the correlation coefficients between wing length and proboscis length ranged from r = 0.943 to 0.973 for four species of *Bombus*.

Harder (1982) suggests using the length between the proximal end of the wing and the distal end of the radial cell to predict proboscis length; he found that this measurement was highly correlated (r^2 = 0.97) with glossa length of bumblebees. He measured wing lengths by removing the wings, placing them in a microfiche reader, and reading the measurements from the screen of the reader.

To measure proboscis length of the bumblebees Harder removed the labium from the bee's head with a pair of forceps by grasping the prementum near its proximal end and pulling forward. If the labium had folded under the bee's head when it died, it was dissected out (severing the ligular arms and muscles that insert on the glossa by pulling them sharply in the direction of the labellum), and the glossa was inserted into a capillary tube to hold it straight. Before the glossa was measured, water was drawn up into the capillary tube so that the end of the glossa could be seen clearly.

There is also some possibility that the measurements recorded from the lab may not correlate precisely with function of the proboscis in the field. Thus Hawkins (1969) found that the effective length of a honeybee's proboscis was greater than the measured length, so that the bees could extract nectar from clover florets with long corollas because the flower yielded to the bee's pressure. Glossometers, such as capillary tubes from which bees or other insects can feed, may provide a more realistic measurement of proboscis length than measurements of the proboscis (Hawkins 1969). Harder (1982) attempted to develop a measurement technique that would provide an ecologically relevant measurement of proboscis length. He photographed bees feeding on a glass plate (16-mm cinefilm at 36 frames per second, camera placed below the tilted glass plate on which the bees stood; this plate also formed the lower side of a well from which the bees drank). These observations of the mechanics of the glossa during feeding were supplemented with field data on the depth to which bees of known glossa length had probed nectar spurs of *Impatiens biflora*. These data indicated that glossa length alone is a better measurement of functional proboscis length for bumblebees than glossa + prementum.

Not as much work has been done on morphological measurements of solitary bee proboscides, but a study by Shmida and Dukas (1990) found that body length and tongue length for solitary bees in a Mediterranean plant community in Israel were significantly correlated ($r = 0.832$, $p < 0.0001$).

Although it is easier to make measurements of insect mouthparts while they are still fresh and flexible, it is possible to use dried specimens. Medler (1962a) measured mouthparts of pinned specimens of cuckoo bumblebees (*Psithyrus*) by removing the heads and treating them with 10% KOH. The labium was removed, rinsed in water, and then the prementum and glossa were measured with an ocular scale in a binocular microscope. Macior (1983) also used dried specimens for proboscis measurements, soaking the heads first in 10% KOH for at least 12 hours, rinsing them in water, and then dissecting out the proboscides and mounting them in glycerin jelly.

Kislev et al. (1972) measured proboscis lengths of hawkmoths that had been killed in light traps. This was accomplished by detaching the proboscis from the head with forceps and placing it in a drop of 5% KOH on a microscope slide on an electric plate at 90°C. The proboscis then unrolled easily and could be measured. Although such treatment might produce a slightly different measurement than if it had been made while the proboscis was still fresh, this discrepancy was not determined.

Dipteran mouthparts appear to be stretchier than those of bees and Lepidoptera (Inouye, *personal observation*), and it is more difficult to obtain repeatable measurements. Gilbert (1981) measured the proboscis lengths of syrphid flies. The proboscis was immersed briefly in alcohol, the sclerotized parts were dissected off, and then a standardized procedure was used to make the measurement. He secured the head with a pin, grasped the labella with a pair of forceps, and in one movement pulled it "out to its furthest limit before the membranes split. The length was recorded as the distance from the lowest point of the face (a rounded projection of the lower mouth edge) to the tip of the labella, which pointed along the length of the proboscis" (Gilbert 1981, p. 246). He mounted mouthparts in Berlese fluid and used a camera lucida to make drawings of representative specimens. Despite their simple appearance, bee fly (Bombyliidae) mouthparts have an accordion-like section near the base of the external parts of the proboscis, allowing it to extend 1.6–1.8 times the "resting length" while they are visiting flowers (Pellmyr and Thomson, *unpublished observations*).

In addition to a microscope and optical micrometer, a vernier caliper can also be used to measure proboscis lengths if they are long enough. Calipers measuring to an accuracy of 0.02 mm are available and can be used even with relatively small bees or flies. Freshly killed specimens are the easiest to work with, as the mouthparts are still flexible for a few hours after death. However, fresh insects can be stored in a freezer and then thawed and measured at your convenience. Hold the insect between two fingers with its ventral surface up and use a pair of fine forceps to extend the mouthparts. While they are extended, use the calipers to measure them. If necessary, you may find you can hold the insect with the forceps by its mouthparts, letting its body provide some weight to help stretch the mouthparts while making the measurement. An illuminated magnifier or magnifying visor may facilitate the process.

2. Mass

Body mass is easily measured for studies of foraging behavior, thermoregulation, or time and energy budgets. If the flower visitors are being collected for other purposes, mass can be determined for freshly killed specimens. Bertsch (1984) took repeated measures of live bumblebees' weights by capturing them in a plastic petri dish and weighing them. Most bees quickly adapted to this interruption in their foraging on artificial flowers in a test chamber, and they resumed foraging after being released. Some electrobalances (e.g., Sartorius; J. Thomson, *personal communication*) will facilitate measurements of moving specimens by taking repeated measurements and averaging the readings. Unwin (1980) gives details for the construction of a simple microbalance that can be used in the field for insects for accuracies of about 1 mg. Although this balance doesn't provide the accuracy of an electrobalance, it can be constructed for much less than the cost of a commercial balance.

Once an avian pollinator has been caught, its mass can be determined easily. Hand-held spring balances (e.g., Pesola brand) are well suited for field use. Wrap the bird in a piece of fabric to immobilize it, weigh it quickly, and release it. Another elegant way to weigh the same individual repeatedly has been used in studies of hummingbirds by constructing a perch on top of a modified Pesola spring balance or an electrobalance and observing which marked birds were on the perch (Carpenter et al. 1983, Calder et al. 1990). Carpenter et al. (1983) modified a Pesola balance so that it functioned by being depressed from above and then placed a dowel perch on top of the balance so the weight of perched birds could be observed readily. Greater accuracy was obtained by using an electrobalance, replacing the pan assembly with a vertical rod and small horizontal perch. Such perches can be placed by a feeder or in a territory for use by the territorial male. The weights can then be recorded remotely without disturbing the bird. By weighing the same bird many times throughout the day or season, one can begin to consider the role that weight change plays in the physiology or migratory strategy of an individual (e.g., Carpenter et al. 1983, Calder et al. 1990).

J. Thomson (*personal communication*) points out that some laboratory models of nonelectronic analytic balances can be adapted for fieldwork even though they have a plug for line current. In some cases the only use of this electricity is to power a transformer to light a 6- or 12-volt DC light bulb for the display. If this is the case, you can cut the wires from the transformer and hook them instead to one or two lantern batteries. You may need to shade the display in order to read it in bright sunlight, and the balance will require a stable base.

3. Bird Bills and Tongues

There are five major groups of avian pollinators, each of which is restricted to one geographic area. Hummingbirds (Trochilidae) are found in the New World, honeyeaters (Meliphagidae) are restricted to Australia, honeycreepers (Drepanididae) are endemic to Hawaii, and sunbirds (Nectariniidae) and sugarbirds (Promeropidae) are found in Africa. Only rarely do other bird species serve as pollinators (e.g., Búrquez 1989). Paton and Collins (1989) present data on the masses and the bill lengths, curvatures, breadths, and depths of these nectar-feeding birds. Collins and Paton (1989) review data on body mass and wing and leg morphology for many of these same species. Ford and Paton (1977) also present data on beak morphology of honeyeaters. Most of these measurements are simple to make and unambiguous. One exception may be the measurement of bill length, which has been measured as the exposed culmen (not covered by feathers); the tip of the bill to the nares (nostrils); the tip to the junction with the skull (true culmen) (Paton and Collins 1989); the tip to the edge of feathers along the top of the upper mandible (Montgomerie 1984) (Figure 7–14). It is interesting that means and modes for sunbirds, hummingbirds, and honeyeaters are all centered on 20 mm (Paton and Collins 1989). This suggests that there may also be similarities in the flowers they visit.

There do not seem to have been many studies of the functional length of bills that might help determine which of these possible

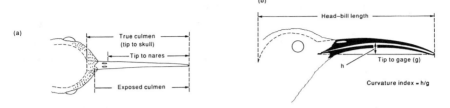

Figure 7–14. Measurements of bill morphology. Various indicators of bill length (a). Head-bill length and bill curvature, where h is the maximum height of the bottom edge of the upper mandible above the line from the tip of the bill to the gape (g), and a curvature index is defined as h/g (b). From Paton and Collins 1989.

measurements is the best to use, but as long as you indicate which measurement you are reporting, any of these may suffice. Montgomerie (1984) measured the maximum depth from which a bird could extract nectar by allowing it to forage from a tubular corolla of 6-mm internal diameter, filled with 20% sucrose solution (measured at 15-minute intervals until the distance did not increase over three consecutive measurements). This depth was a linear function of culmen (bill) length.

Paton and Collins (1989) point out that little detailed information is available on how nectar-feeding birds use their bills and tongues to collect nectar. Little is known about the depth to which bills are inserted, and the influence of bill length and shape upon the speed and efficiency of nectar collection. Paton and Collins point out the need for additional detailed comparisons of the flowers used by birds with different bill lengths and shapes, so that the adaptive values of particular tongue and bill morphologies can be determined. They suggest that the best method is to film the birds, because they move so quickly while collecting nectar. They also mention the need for studies that compare patterns of floral use by male and female nectarivores, or geographically isolated populations. Ewald and Williams (1982) studied the function of hummingbird bills and tongues by photographing them while they fed from a Plexiglas feeder.

4. Other Morphological Measurements

In addition to proboscis length, other morphological measurements may be relevant to pollination studies. For example, Hainsworth and Wolf (1972a) measured crop size in hummingbirds. To do this they fed birds a sugar solution containing a marker to make it X-ray dense, and then X-rayed the birds. In addition to this measurement, they measured the volume of sucrose solution that could be collected in 2–3 minutes following deprivation of food for an hour.

Dodson and Yeates (1990) measured body size of territorial flower-visiting flies, and confirmed earlier reports that wing length is a good estimator of thorax length (r = 0.81 for male *Comptosia* sp., and 0.97 for females; p<0.001, n = 20 for both cases). They found that it was easy to measure wing length quickly without damaging the flies, using a hand-held ruler and recording measurements to the nearest 0.5 mm.

An additional morphological measurement that has been studied in insects is reflectance, which is important in thermal balance. Willmer and Unwin (1981) describe the construction and use of a reflectometer they designed for field use with insects to determine the reflectance of the dorsal surface of the thorax, which overlies the flight muscles. Their tool measures light reflected from an insect's body onto the inside of a dome made from half of a table-tennis ball. They used a flashlight bulb as a light source and cadmium sulphide photoconductive cells as sensors connected to a meter circuit. In the laboratory, reflectance can be measured with a recording spectrophotometer with an integrating sphere attachment (e.g., Kirk 1984).

Although wing-beat frequency is not strictly a morphological measurement, it does depend on morphology and may be related to foraging behavior. Unwin and Ellington (1979) describe construction of an optical tachometer that can be used in conjunction with an oscilloscope to measure the wing-beat frequency of free-flying insects. Although the device is not available commercially, it should not be very difficult to construct from their description. Byrne et al. (1988) summarize data on wing-beat frequencies, mass, and wing loading for 160 species of insects, including some pollinators. To calculate wing surface areas for insects they projected images of the wings onto transparent acetate sheets (placing wings in a microfiche reader and tracing off of the display) and cut out the images. These were then weighed, and surface area was calculated using a linear regression relating template masses of known dimensions to their surface areas.

H. Pollinator Vision

The first experimental evidence that honeybees could see colors was provided in 1882 by Sir John Lubbock, who conducted experiments with colored papers (Grant 1950). Similar behavioral experiments by Knoll (1925) showed color discrimination by a moth. Since then trichromatic color vision, usually extending into the near ultraviolet, has been documented in many species of insect pollinators (Kay 1978, Menzel 1990), and tetrachromatic systems have been found for a few species of Hymenoptera and more commonly for Lepidoptera (Menzel and Backhaus 1991). Although the initial experiments were strictly behavioral, more recent work has relied on electrophysiology to investigate sensitivity to different wavelengths.

The human visual spectrum ranges from about 400 nm to about 700 nm and is trichromatic (primary colors blue, about 450 nm; green, about 520 nm; and red, about 700 nm). In contrast, bumblebees and honeybees also have trichromatic vision, but with primary colors of ultraviolet (336 nm), blue (432 nm), and yellow (532 nm peak absorbance) (Menzel and Backhaus 1991). Species of insects with tetrachromatic color vision have an additional, very long wavelength receptor system at about 600 nm (Menzel and Backhaus 1991). Color vision has also been proposed for other insects (reviewed in Kevan 1983), including wasps and flies (see Kugler 1951). Flies (the few species that have been tested) also have the ability to see ultraviolet; available evidence suggests that many insects have a separate ultraviolet-absorbing pigment that makes it possible to distinguish ultraviolet from longer wavelengths (Allan et al. 1991). All Lepidoptera that have been examined have color vision and the ability to see ultraviolet (Goldsmith and Bernard 1974), although there is great diversity in color-vision systems among butterfly species (some can see red, whereas others are about as red-blind as worker honeybees (G. Bernard, *personal communication*). Menzel

(1979), White (1985), and Menzel and Backhaus (1991) review insect visual pigments and color vision.

Hummingbirds and some other birds can also see into the near ultraviolet (Chen et al. 1984), and at least four mammal species (house mice, rats, gophers, and gerbils; Jacobs et al. 1991) have a retinal mechanism that is maximally sensitive to ultraviolet light (370 nm). Honeybees also have the ability to see polarized light (Menzel and Snyder 1974), which aids them in orientation on cloudy days. Some butterflies show innate preferences for particular colors (reviewed in Boggs 1986), whereas some moths apparently use only olfaction to locate the flowers they visit.

It is possible to create color photographs of flowers that show the ultraviolet reflectance patterns together with other floral colors. McCrea and Levy (1983) document a technique for doing this. For methods for visualizing ultraviolet reflectance patterns or quantifying reflectance at different wavelengths, see Chapter 3, Section C.1.

I. Physiological Measurements

For detailed studies of the energetics or physiological basis of plant-animal interactions, measurements of body temperatures, time-energy budgets, or even electrolyte balance may be appropriate.

1. Body or Flight Temperatures

One of the requisites for studies of the energetics of flight, or for general metabolic studies of pollinators, is body temperature. This is most readily studied in larger animals such as avian or mammalian flower visitors, because of the ease with which it can be measured and because they typically have a narrower range of temperatures than insects do. Even though hummingbirds can resort to torpor (Calder and Booser 1973), their metabolism is still restricted to basically only two normal temperatures. Electronic thermometers based on thermocouples or thermistors are available with a wide range of sensors, and one that can provide a rapid reading of rectal temperature is probably most suitable for birds or mammals. Calder (1971) and Calder and Booser (1973) were able to record body temperatures of incubating female hummingbirds by putting artificial eggs containing thermocouples inside their nests, but this technique doesn't help with free-flying birds.

A large number of studies have measured thoracic temperatures and thermal ecology of flower-visiting insects. In the context of pollination, this information is important because body temperature determines when insects can fly to visit flowers. Although it is much more difficult to calculate for insects than for homeotherms, because of the wide range of temperatures that most of them experience, it is also possible to use information on body temperature to help estimate the amount of nectar (or number of visits to flowers) that an insect must make to maintain a positive energy balance.

The technique used most commonly for measuring insect body temperatures in the field is the grab-and-stab technique (a variant of the "noose-'em-and-goose-'em" technique employed for lizard body temperatures), which involves inserting a fine thermocouple or thermistor probe into the thoracic flight muscles. If the insect is large enough, this can be a nondestructive process; large bees will fly away with no apparent injury (Inouye, *personal observation*). Measurements made in this way led to the discovery that some flower-visiting insects have very high thoracic flight temperatures (e.g., 40.7°C for male euglossine bees; Inouye 1975). This technique has been used for many bumblebee studies, for Diptera (Heinrich and Pantle 1975), and studies of Lepidoptera (e.g., Heinrich and Casey 1973). Heinrich (1972) grasped bumblebee workers as they were foraging on flowers and inserted a sub-mini-probe thermistor (Fenwal GC32SM2) from the ventral side to the approximate center of the thorax.

Body temperatures of flies can be measured in the same way as those of bees and butterflies. Inouye (unpublished) measured the body temperatures of flower-visiting calliphorid and tabanid flies in montane Australia with the grab-and-stab technique, using a thermocouple thermometer with a needle probe. At ambient air temperatures of 17°–22°C, 10 of 11 species had mean thoracic temperatures between 27.7°C and 32.6°C when foraging in the sun, and temperatures of 17.8°C to 26.4°C when foraging in the shade. In a similar study, Shelly (1984) and O'Neill et al. (1990) measured temperatures of robber flies, caught in flight, by inserting a fine thermocouple probe into the center of the thorax.

Several studies of insect flight temperatures have indicated that there can be significant cooling of the insect during the short interval between time of capture and the time the temperature is recorded. The cooling may occur because the insect is shaded from solar radiation, because it is immobilized (preventing muscular thermogenesis), or from heat loss from conduction. This effect can be minimized by measuring the temperature through the net (e.g., O'Neill et al. 1990), and by handling the insects with gloved hands (handball gloves work well for this; they are thick enough to prevent stinging by bees but flexible enough to facilitate handling small insects). Even though probes as fine as 0.33 mm in diameter are available, some flower-visiting insects are too small to be able to probe reliably. Shelly (1984) found that accurate readings were not possible for robber flies less than 20 mg in mass.

In addition to measuring temperatures of live insects, another technique can provide insight into the operative body temperatures of insects or those that an animal would achieve in a particular environment. O'Neill et al. (1990) inserted fine thermocouple wire into the bodies of freshly killed flies and then placed the flies at specific heights or particular angles with respect to the sun. The temperatures they recorded after equilibration were presumably similar to those that would have been measured from live flies.

Among the most thorough studies of insect heat budgets is one by Willmer and Unwin (1981). They made use of a number of specially constructed field instruments, such as a microbalance to weigh small insects, and a reflectometer for measuring the surface reflectance of the thorax (construction details are given in their paper). They also recorded heating and cooling rates (using freshly killed specimens with fine thermocouple wires implanted or attached to the thorax). These data permit estimates of the relative importance of size and color in determining rates of heat exchange and temperature, which are necessary for accurate predictions of heat budgets.

2. Energy Budgets

Data on the energy budgets of flower visitors can provide insight into the economics of plant-pollinator interactions. To construct an energy budget, one typically determines first a time budget, identifying what proportion of its time a pollinator spends in activity categories such as flight, perching, feeding, and so on. Then laboratory studies are required to measure the energetic cost of each of these activities. The resulting information can be used to determine, for example, how many flowers an animal must visit each day in order to maintain a positive energy balance. This information may prove instructive in interpreting patterns of pollinator behavior. For example, Wolf et al. (1975) calculated the "required foraging efficiency" (RFE; the efficiency required to produce a balanced energy budget over a 24-hour period) for nectarivorous birds as a way of exploring the general relationship between foraging efficiency and time budgeting.

Most studies of pollinator energy budgets have been of hummingbirds, which have a conveniently small number of activity levels (flight, perching, sleeping). The cost for each of these activities is calculated by multiplying the amount of time spent in each activity by the cost per unit time of that activity. For example, Wolf et al. (1976) calculated energy budgets for three species of alpine hummingbirds from Costa Rica. Wolf and Hainsworth (1971) describe the laboratory techniques necessary for measuring metabolic rates of flying and resting birds (a temperature-controlled metabolic chamber with an oxygen analyzer). Additional information about avian energy budgets is provided by Gill and Wolf (1975). The same concept can also be extended to the community level. Montgomerie and Gass (1981) compared the daily energy expenditure for a hummingbird community with an analysis of available energy and found that both temperate alpine and tropical lowland communities they studied were often near or at carrying capacity (all daily nectar production being used by birds).

Only a few studies have been made of insect energy budgets. These are more complicated because insects' body temperatures (and hence metabolic requirements) are much more variable than those of homeotherms. Thus, unless one is constantly monitoring body temperature (a possibility at least for tethered insects, by implanting fine temperature sensors), estimates of energy consumption will probably not be as precise as those for birds or mammalian pollinators. Pyke (1979, 1980) constructed energy budgets for bumblebee workers by assuming three sources of energetic cost: flight, thermoregulation, and nonflight (excluding thermoregulation). He calculated the total rate of energy costs (C)

to the bumblebee as $C = T + F + N$ calories/hour, where T = cost of thermoregulation, F = cost of flight, and N = cost of nonflight. After weighing bees and measuring the percentage of time (p) spent in each of these categories, he calculated that:

$$T = \max\{28.8 - 42.7p - 3.3(1-p),0\} \text{ calories/hour;}$$
$$F = 63.8p \text{ calories/hour;}$$
$$N = 4.9(1-p) \text{ calories/hour;}$$
$$C = T(p) + F(p) + N(p).$$

With these calculations he could then consider the question of optimal foraging in bumblebees (Pyke 1980).

3. Electrolyte Balance

Hiebert and Calder (1983) studied the sodium and potassium content of floral nectars visited by hummingbirds at the Rocky Mountain Biological Laboratory and the significance of these salts for electrolyte replacement in the hummingbirds (Calder and Hiebert 1983). Hummingbirds ingest large amounts of water from nectar, and when they excrete the excess they lose salts, too. Excretion of sodium and potassium accounted for one-half to one-third of the osmotic concentration of the urine, but these amounts could be replaced by the trace amounts in floral nectar. Nicholson (1990) conducted a similar analysis of osmoregulation in carpenter bees (*Xylocopa*). In this case nectar and pollen of the preferred food plants had very low levels of sodium, but the bees appear to be able to recycle almost all of the sodium (and potassium) that enters the rectum. Additional studies of osmoregulation by pollinators may shed light on the significance of ion concentrations in nectar and pollen.

Insect-pollinated plants can have very high levels of ions. Waller et al. (1972) reported levels of 3,600–13,000 parts per million of potassium in floral nectar of *Allium cepa* L. (cultivated onion; Liliaceae), which is about 10 times higher than levels in nectar of other species they examined. These high levels apparently deter visitation by honey bees, resulting in sporadic crop failures in the commercial production of onion seed.

J. Marking or Banding Animals

Studies of foraging behavior, population size, or migration may be conducted most easily on marked animals. Of the many different techniques available for marking animals, either individually or as a group, we review here some of those most appropriate for pollination studies. A general reference is Southwood 1978, and additional ideas may be found in Stonehouse 1978. Akey (1991) reviews a wide range of techniques for marking arthropods. Be forewarned that marking techniques all have the potential to affect the behavior, mortality, reproductive success, or other aspects of the biology of the marked animals (e.g., Harrison et al. 1991).

1. Insects
A variety of methods have been developed for marking insects. Because of their relatively short life span, techniques for marking adult insects do not have to be as permanent as those employed for avian or mammalian pollinators. However, some of the techniques produce lifelong marks. Techniques discussed here include external markers such as paints, fluorescent powders, tags, or mutilation of body parts, and internal marking using rare elements or dyes.

a. General techniques
To mark entire populations, trace elements such as cesium or rubidium can be introduced into larval diets of species in cultivation. This technique has been used for a variety of phytophagous insects, including Coleoptera, Diptera, Hemiptera, and Lepidoptera (reviewed in Pearson et al. 1989; see also Supplement 14 to *Southwestern Entomologist* 16[2], June 1991, which has 10 articles about elemental marking). The trace element can also be sprayed on (Pearson et al. 1989) or injected into (Fleischer et al. 1990) host plants to label them; the label will then be incorporated by herbivorous larvae that feed on the plant (e.g., Pearson et al. 1989). Hayes and Reed (1989) also suggested using artificial nectar to label adults and their eggs. Multiple trace element labels might increase the likelihood of detection (Hayes 1989). Labeled insects are detected by using atomic absorption spectrophotometry to identify the trace element. The detection threshold for rubidium found by Pearson et al. (1989) was 0.01 ng per sample, or about 0.1 ppm for the average

moth in their study, but the marking threshold will vary for every study depending on background levels of the marker element. A disadvantage of this technique is the expense involved in purchasing and detecting the trace element.

A simpler variation of this internal marking method is to use a dye as part of the larval diet and then to look for evidence of the dye in adults captured later. Showers et al. (1989) dissolved Calco Red N-1700 dye in corn oil and added it to a larval diet of black cutworm (*Agrostis ipsilon*) moths. Pheromone-baited traps caught moths at different distances from the site of release of the marked adults, which were then examined in the lab for marker dye in the abdomens for evidence of dispersal. Bell (1988) sprayed the same dye (9% crude cottonseed oil, 1% dye [w/v] and 90% water) at 2.8 kg/cm^2 onto wild geranium, an alternate host plant for tobacco budworms. They then used pheromone-baited traps to catch the adult noctuid moths in the area and examined them for the presence of dye. To detect levels of dye as low as 10^{-6} to 10^{-7} g, they rinsed individual moths with 1 mL of acetone, then mascerated moths in vials with 1 mL of acetone and chromatographed both mLs of acetone to look for the presence of dye. If they found dye in the mascerated moth but not in the rinse vial, they considered this a positive internal mark.

Strand et al. (1990) found that acridine orange worked well as an internal marker, fed to adult Diptera and Hymenoptera in water or honey. The optimal concentration differed among insect species, but concentrations of 0.001% to 1.0% marked the gut, usually the malpighian tubules, and sometimes the eggs of the experimental insects.

Dempster et al. (1986) investigated whether the elemental composition of adult Brimstone butterflies could be used as a chemoprint, or population marker. This technique relies on the existence of sufficient variation in elements such as K, Ca, Fe, Cu, Zn, Ti, P, Cl, and S among sites to differentiate populations. It also assumes that this pattern of representation of elements will be reflected in both larval and adult life stages, that it will not be disturbed by aging or feeding by adults, and that variation between populations will be greater than that among populations. Dempster et al. used wavelength dispersive X-ray fluorescence spectrometry to measure differences in elemental composition, and they found marked differences between the sexes and between individuals caught on different sites and in different years and seasons; K, Ca, P, Zn, Cl, and S were the most important elements for discriminating differences in principal component and canonical variate analyses. Their results indicate that there are indeed population differences resulting from variation in the composition of soil and food plants in different sites, but that as adults age and feed these differences are masked. They concluded that this technique has limited value in the study of dispersal, at least in that species of butterfly.

Kowalski et al. (1989) also attempted this method of chemoprinting with an anthomyiid fly. They used graphite furnace and flame atomic absorption spectroscopy for metallic cations, and concluded that the technique did not appear promising for studies of insect dispersal unless the weight of individual flies was considered.

Another mass-marking technique involves the use of micronized fluorescent dusts, of the same type used as a pollen analogue for studies of pollen flow. These pigment particles absorb radiant energy from ultraviolet light and convert it into longer wavelengths that we can see, emitting a light of a particular color. Thus it is easy to tell at night or in the laboratory whether a particular insect has been marked by examining it under ultraviolet light. Crumpacker (1974) described the advantages of this technique, which include rapid marking of many individuals without anesthesia, safety to users and marked individuals, persistence of the mark, and low cost. If the insects to be marked have many hairs and bristles, this will facilitate retention of the dust particles, but even insects with few hairs will usually retain enough particles for identification.

Molecular biology has also made possible a technique using an immunoassay. Hagler et al. (1992) describe the use of a solution of rabbit immunoglobulin G (IgG) as an external marker, which is then detected with an enzyme-linked immunosorbent assay (ELISA) using the complementary antibody (antirabbit IgG). They labeled hundreds of *Lygus* adults within minutes for a cost less than $100, and assayed more than 500 insect samples in only 4 hours. The specificity, sensitivity, and persistence of the marker were good in lab studies, but the technique does not appear to have been field tested yet.

b. Hymenoptera and Diptera

We have applied the fluorescent dusts with a small paintbrush directly to the bodies of insects on flowers, without causing them to leave the flowers (Kearns, *personal observation;* for a color photograph of this technique in use with honeybees at a feeder, see *National Geographic* 179[6]:96 [1991]). However, it is also possible to capture insects, put them in a vial containing the powder, and mark them heavily by shaking them around (Crumpacker 1974; Inouye, *personal observation*). Flies subjected to this kind of marking will usually spend some time grooming themselves when released, e.g., uncovering their eyes, but then seem to forage normally. Crumpacker found no evidence of any dust transfer during mating or any negative effect on mating behavior of marked *Drosophila pseudoobscura*. There was also no effect of heavy dusting on viability as long as the flies were not excessively moist at the time of dusting and they were transferred to a dry environment after marking. Heinrich (1981) cites a couple of other methods for mass marking, including use of an atomizer to spray foraging bees, and altering the entrance to a honeybee hive to make foraging bees pass over a block with fluorescent powder. One problem with using this method on bees is that they groom frequently, so that the marks may only be visible for a few hours to the unaided eye in daylight (but longer using an ultraviolet light and a microscope to look for evidence of marking).

Frankie (1973) used a slightly different technique to mark larger numbers of insects. Bees collected from flowers with an insect net were transferred to a 25 x 40–cm nylon net bag, which was tied off, and inserted into a 1.5-liter plastic container with a tight-fitting lid. A metal mesh floor kept the bag off the bottom of the container and out of a layer of fluorescent powder spread evenly on the container bottom. After the lid was closed, a bicycle-pump nozzle was inserted into a small hole at the base of the container and pumped about ten times to circulate the powder throughout the nylon bag. Insects were covered evenly with the powder, but were able to fly when released from the bag. Frankie (1973) found that bees with dye on them could be found a week later, although most of the powder was worn or groomed off, making detection difficult. There was only minimal contamination when marked and unmarked bees were collected at the same time in a net.

At the other end of the scale of marking studies, Fryer and Meek (1989) sprayed fluorescent pigment from an airplane to mark mosquitoes without having to capture them first. Pigment was transferred from the vegetation to adult mosquitoes for up to 4 days after application. If it is preferable to mark insects without capturing them, perhaps an atomizer or mister could be used as a small-scale alternative for marking insects that cannot be marked with a paintbrush.

Most insects can be marked individually with some sort of paint or marking pen. This method works well for foraging bees, which can often be marked with a small brush without disturbing them significantly as they feed on flowers (Inouye, *personal observation*). Peakall (1990) studied wasp movements by marking adults with various combinations of colored Liquid Paper to identify both marking day and the baiting station where the wasp was marked, and Kipp (1987) used a white, water-base marker (SNO-PAKE) to mark honeybees at a feeding station. One advantage of using typewriter correction fluid as these studies did is that it usually comes in small vials with brushes built into the lids, which work for marking the insects. Yeates and Dodson (1990) marked adult flies on the scutum or abdomen with unique marks to identify individuals. Wineriter and Walker (1984) compared the durability of 26 different marking materials on three species of insects and found that only one (Tech-Pen Ink, developed for marking laboratory glassware, available in 11 colors) adhered to and remained on all three species throughout their adult lives (in lab cultures).

Bees or flies can sometimes be marked while feeding on flowers, especially if they have their heads buried in a corolla so they can't see your movements. If it is not possible to mark the insects without catching them, you may be able to mark them through the mesh of the insect net in which you have caught them. This results in a colorful net after a while! Inexpensive small brushes appropriate for marking insects are usually available in the same hobby shops where modeling paints are sold. If you need individual markings and are working with an insect with a large enough thorax or abdomen, you may be able to use a spot code to mark a large number of individuals. A spot code is a pattern of colored dots or other marks placed on defined regions of an insect so that each pattern can be interpreted as equivalent to a number (Rooum 1989). Try to avoid putting paint on the insect's wings as this may interfere with flight (you can scrape off excess paint with fine forceps). J. Thomson (*personal communication*) prefers to use Flo-quil hobby paints instead of acrylic paints, and he suggests that because a brush can put on too much paint, alternatives such as a thin wire, grass stem, or toothpick will work better. In this way up to three color stripes can be put on the interalar region of even small bumblebees. If you put a white layer of Liquid Paper on the insect first (e.g., in the interalar area of bumblebees), then stripes of different colors of paint or ink may be easier to read (e.g., Thomson et al. 1982).

Bees can also be marked as they enter or leave a hive or nest. Free (1970a) describes methods for spraying honeybees with a stain at a hive entrance or forcing them to walk through a "marking block" at the entrance that marks them with fluorescent powders. The bees may be marked with enough powder that they will leave some on flowers, permitting determination of where they have been foraging. Although these methods were developed with honeybees, they could also be adapted for use with other species of bees.

A more elegant method for marking individuals employs numbered tags. A company in Germany makes small plastic tags (Opalith-Plättchen) in five different colors, with numbers from 0 to 99, for use in tagging queen honeybees to identify year-class or individuals. The tags come with a small bottle of glue and an applicator, and they work well on bumblebees (Inouye, *personal observation*; for a color photograph of honeybees marked with this type of tag see *National Geographic* 179[6]:96 [1991]) and other medium to large bees; the tags are about 2 mm in diameter. In addition, tags can be cut from a source such as an electronics parts catalog that has small print and glued onto bees (K. Rawson, *personal communication*). When gluing tags to the backs of bees, hold them for a few seconds while the glue dries, so they will not remove the tags by grooming when they are released; fast-drying glues work best. We find that handball gloves work well for holding bumblebees or other stinging insects. The leather used in their construction is light enough to provide the necessary flexibility in manipulating a live bee but usually thick enough to prevent stingers from penetrating to your fingers. Use the glove to reach into the insect net and remove the insect you wish to mark. Hold it by the legs or pinch the body gently between your fingers. An alternative to this type of handling is to catch the bees in (or transfer them from a net to) a vial, and put the vial in a refrigerator or on ice. For example, Morse (1981, 1982a) chilled bees to immobility before tagging them or painting marks on the pollinia they carried. He found no differences in the probabilities of recapture of bees that were chilled compared to unchilled bees, suggesting that this treatment has no significant effect.

Freilich (1989) describes a method for making your own tags by photoreducing a computer-generated spreadsheet filled with numbers. He had this done at a commercial print shop by producing a positive mechanical transfer, but you can accomplish the same via sequential reduction on a reducing copier, or use a laser printer to print small labels. He produced tags as small as 1 x 2 mm with three digits printed on them. Because he was using these tags on aquatic insects, he used plastic paper for the tags. By coating their top surfaces with clear epoxy, he eliminated a problem with surface abrasion. He glued the tags on with a cyanoacrylate glue (Superglue), and found that they would last for six months in a stream tank in the lab, and as long as 81 days in the field in a stream. This method should also work well for terrestrial insects.

Gary (1971) devised a capture-recapture method for tagging bees that semiautomated collection of data on bee movements. Small metal tags were cut from 0.25-mm shim steel with a metal punch and lined up on a piece of double-stick tape. They were then either painted or labeled with plastic bee tags (Opalith-Plättchen) that were glued on top. Foraging honeybees were captured, anesthetized for 15–20 seconds with CO_2, and then the metal tag was glued to the abdomen (instead of the thorax, to prevent fouling the wings with glue and because the abdomen is adapted for carrying loads). Beehives in the study area were equipped with a row of horseshoe magnets across the hive entrance, so returning bees had to pass through a "magnetic trap." Bees with metal tags hung up on the magnets until they could work their way free, leaving the tag behind. Thus the type and amount of glue used to apply the tag was important to keep it from coming off in the field yet allowing relatively easy removal at the hive. Recovery rates of tags were as high as 97.5%. Gary et al. (1976, 1977) and Roubik and Aluja (1983) report results of studies using this technique.

Heinrich (1981) cites a system employing electronic tags that has apparently only been used in one study. It involves putting a scanner at a hive entrance, to detect the individual resonating signature of each tag.

If you need to mark ants, use a fast-drying enamel spray paint. If you apply the spray from a distance of approximately 1 m, only the finest particles will mark them. This technique does not appear to contribute significantly to ant mortality, and the marks can last for several weeks (Paulson and Akre 1991).

c. Lepidoptera

Butterfly and moth wings present a large surface area for marking. Ehrlich and Davidson (1960) described a technique for marking Lepidoptera caught in butterfly nets. Specimens were marked on the edge of the undersurface of their wings with felt-tip markers; two different positions on each wing were used to indicate 1, 2, 4, 7 on one pair of wings and 10, 20, 40, 70 on the other pair. These numbers were chosen because they can be generate all others (e.g., positions 1 and 2 to represent 3, 1 and 4 to represent 5, etc.). Thus 154 butterflies could be individually marked; because they could easily differentiate sexes, up to 308 butterflies could be recognized individually. An index card slipped between the wings will facilitate marking. If larger butterflies or smaller marks are used, up to 1,000 butterflies can be marked with one color by using three positions on each of the four wings (e.g., Brussard 1971) (Figure 7–15). Brussard (1971) recommended Sharpie marking pens because they have a fine point, come in several colors, and the ink dries rapidly. He found that one person could catch and mark (or read the marks on) about 75 butterflies per hour in an area of about 1 hectare. Watt et al. (1977) modified the technique to include more positions; two on the margins of each wing, plus a central position to indicate 100, 200, 400, or 700 (other numbers are represented by combinations of these, e.g., 300 by marking both 100 and 200, 500 by marking 400 and 100, etc.), gave them the capability of marking up to 1,554 individuals. They used an acetone-based ink. Turner (1971) also used the 1, 2, 4, 7 code with a felt-tip marker, with marks in four positions on a white wing-band on the hind wings of a butterfly; hundreds were represented by a change in pen color.

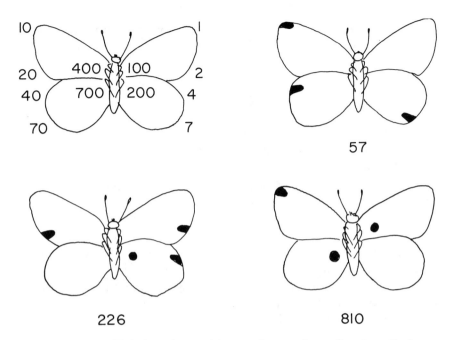

Figure 7–15. Technique for marking numbers on butterfly wings. Each wing has three possible positions, and a mark in that position indicates that number is part of the total for that individual. From Brussard 1971.

If the butterflies you are working with have a large enough wing area, another technique is possible. Paul Ehrlich's research assistants working at the Rocky Mountain Biological Laboratory monitor some butterflies by using a black marker to create a black patch on the wing, and then write a number on that with a silver metallic-color marker. This number can then be observed directly with binoculars while the insect is basking and has its wings spread open. This kind of mark is easier to read at a distance than the spot code (Brian Inouye, *personal communication*).

Rooum (1989) gives an interesting account of the history of using spot codes for marking bees. For example, Karl von Frisch was able to mark up to 601 individual honeybees by using combinations of five colors of paint and five different positions; e.g., the anterior and posterior margins of the thorax, with one or two dots in each position, and additional positions on the abdomen.

Another technique that has been used with Lepidoptera is to glue tags to the wings. Elger (1969) glued small pieces of colored paper to the wings of Arctiid moths, although these were not individual tags. Studies of migrating monarch butterflies have also relied on tags glued onto wings.

Singer and Wedlake (1981) pointed out a potential problem associated with capturing and marking butterflies. They had a 2% recovery rate of tropical papilionid butterflies captured, marked, and immediately released, and showed that the butterflies' behavior had been altered by being captured. By marking some butterflies without capturing them first (by crawling very slowly toward perched butterflies, with a marker in hand) they demonstrated that 21% of these butterflies were recaptured later. Thus the usual assumption that capture and marking do not alter subsequent behavior may not apply in some cases. In other studies of butterflies, however, neither Brussard (1971) or Watt et al. (1977) found evidence that the handling associated with capture and marking caused excessive dispersal.

d. Coleoptera

Goldwasser et al. (1992) conducted a study that required marking large numbers of flower-visiting neotropical scarab beetles. It is probably instructive to list the techniques that they tried that failed: painting the elytra with Liquid Paper, Flo-quil acrylic paint, or Testors enamel, applied with small brushes, toothpicks, or 1-cc syringes. The long drying times (due to very high relative humidity), sticky stigmatic exudates on the elytra, and burrowing activity by the beetles hindered application and retention of paint marks. As an alternative Goldwasser et al. developed a technique of cutting notches, either square or triangular, in the edge of the elytron using the field biologists' universal tool: scissors on a Swiss army knife (Figure 7–16). Each elytron was partitioned into four positions, and each of the eight resulting positions was assigned a value of 0 (no mark), 1 (a triangular notch), or 2 (a square). Using a base three system of counting, 6,560 individuals can be numbered uniquely. If three workers are marking beetles at the same time, one position can be reserved to indicate their identities, and each worker can mark 2,186 beetles without having to consult about the last number used by the others. The same technique could be used on smaller

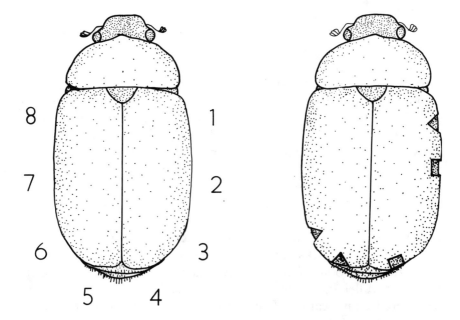

Figure 7–16. Positions and shapes of notches cut in beetle elytra in order to recognize individuals. Redrawn from Goldwasser et al. 1993.

beetles by reducing the number of notch positions. These marks were permanent, didn't appear to affect the beetles' longevity or ability to fly, and could often be interpreted even if the strip of elytron between adjacent notches was torn away, leaving only one side to each notch.

Bach (1990) marked adult beetles with different colors of Testors brand enamel paint. Lance and Elliott (1990) used Calco red dye and Calco oil blue V dye to label adult beetles feeding on an artificial diet. At concentrations of 1% or less, neither dye caused increased mortality, yet dye could be seen through most lightly pigmented portions of the cuticle. Disadvantages of this technique were that the blue marks appeared to fade after only several days and that detection of the marks by observers in blind evaluations was only 77% to 94% accurate.

2. Birds

Large nectarivorous birds such as sunbirds or honeyeaters can be marked using colored leg bands, using different colors and different legs to mark a large number of individuals. Hummingbirds, however, present a special problem because their small size makes impossible the use or observation of such leg bands. Several researchers have developed alternative techniques for marking hummingbirds to recognize individuals. It is possible to write numbers or symbols on the breasts of hummingbirds with marking pens, but we have found that these marks do not last for more than a few weeks, perhaps because the oil in the birds' plumage prevents the pigment from adhering well.

A more permanent marking can be achieved by using model paints, but be careful not to cover too much of the plumage, because this paint usually sticks the feathers together and may affect thermal regulation or grooming behavior. W. Calder (*personal communication*) and Miller and Inouye (1983) have used acrylic or other paints to color the tail feathers (e.g., three central retrices, or the three peripheral ones on each side with two different colors) of hummingbirds at the Rocky Mountain Biological Laboratory; this technique has the advantage that female birds can be identified while sitting on a nest because the tail feathers protrude over the edge. Calder, who probably holds the world's record for number of hummingbirds painted, is currently using fiber-tipped paint pens (with enamel paints and a mixing ball inside, e.g., Deco-color, or with somewhat less success, Faber-Castell or Testors; W. Calder, *personal communication*). These have the advantage of drying more quickly than model paint, but there are still problems with paint chipping, fading, or more rapid wearing of the ends of feathers.

Stiles and Wolf (1973) also used paint (Flo-Paque paint and Testors or Pactra airplane dope), putting one to three colored spots in a row across the upper back between the shoulders, while holding the bird "firmly against the index finger by the thumb and middle finger securing the wings below the body" (Stiles and Wolf 1973, p. 244) to prevent it from struggling. Make sure that the paint is fairly dry before releasing the bird or it will smear if the bird tries to preen itself soon after you release it. Thin the paint a bit first before applying it; although this kind of paint takes somewhat longer to put on, it also lasts a bit better (W. Calder, *personal communication*).

Miller and Inouye (1983), following a technique described by Baltosser (1978), glued small pieces of flagging, in different shapes and colors, to the backs of adult male Broad-tailed hummingbirds to identify individual territorial birds. Baltosser (1978) used a cork-borer to cut circular tags of 4.0–6.8 mm diameter, and pointed out that bicolored tags could be made by gluing different colors of flagging together. He suggested a design for a holding tube (constructed from the outside cardboard tube from Tampax tampons) to facilitate banding. Stiles and Wolf (1973) also used markers, but in this case made tags that attach to the legs. They modified an earlier unpublished technique by Ortiz-Crespo, who had used plastic leg streamers (which lasted up to 4 months). Instead, they used colored translucent acetate strips, with colored plastic tape to make stripes of different colors on the strips (Figure 7–17). The narrow part of the tag is crimped into a ring that will fit around the bird's leg and is then glued into place with a fast-drying glue that will dissolve acetate (they recommended Bond's Adhesive, but found that Duco Cement also worked). It is important to make sure that the tag is not too tight, or the bird may lose its foot because of loss of circulation! Waser and Calder (1975) stopped using this kind of tag at the Rocky Mountain Biological Laboratory because of concerns that they were impairing nest-building by females, but this decision was based on the observation of only two nests. Carpenter et al. (1991) made tags of vinyl flagging cut into 0.2 x 3–cm strips and attached them with Superglue to the bare skin at the base of the neck between two tracts of feathers. Up to three different colors were affixed to each bird. They found that these remained attached for the duration of a 6-month study.

Stiles and Wolf (1973) used both paint and leg tags on each bird to maximize their ability to identify them in the field. They also measured bill, wing, and tail lengths of each bird so that they might be able to identify them again if most of the markings were lost at the time of the next recapture. They cited evidence that these color-marking techniques did not alter the territorial or nesting behavior of birds, and it seems unlikely that they would alter foraging behavior either.

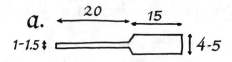

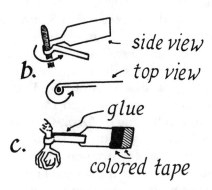

Figure 7–17. Plastic leg streamer used for marking hummingbirds. From Stiles and Wolf 1973.

Ewald and Carpenter (1978) used an ingenious technique to mark hummingbirds as they fed at feeders. They mounted a capillary pipette below the tip of the feeder and placed approximately 0.01 mL of thinned enamel paint in the tip of the pipette. A flexible tube ran from the other end of the pipette to the experimenter, who waited until a bird was feeding and then blew on the tube. The paint was then expelled on the bird's feathers, marking it between its upper chest and tail. By varying the color, amount, and viscosity of the paint, and the location of the mark, they could recognize many different individuals.

Don't forget that permits (both state and federal) are required for capture, banding, or marking of birds in the United States; see Section D.1 of this chapter. If the birds you are banding are also being used for studies of reproduction, beware that your bands aren't affecting mate choice (e.g., Burley et al. 1982).

3. Mammals

There is no federal permit for working with bats, but many states require a permit (from a fish and game department, division of wildlife, etc.). Go to a store that sells hunting or fishing licenses and ask for the name and address of the appropriate agency.

Mammalian pollinators can be tagged with individual tags, such as ear tags for marsupials or numbered leg bands for bats. For most species, fur dye may also work as a marking technique, with different patterns of dye used to identify individuals. If an appropriate dye or stain is used, such as a commercial fur dye (Fitzwater 1943; see Appendix 2), the marks will remain conspicuous until the animal molts (usually less than 6 months). If the animals can be induced to visit artificial feeders, or if they visit particular flowers often enough, it might be possible to use the technique described above for hummingbirds (Ewald and Carpenter 1978). Stebbings (1978) discusses additional techniques for marking bats, including fur clipping, tattooing, toe clipping, and chemoluminescent tags.

K. Foraging Behavior

Foraging behavior of flower visitors is important for many pollination studies, as it can directly influence many aspects of the interface between plants and their pollinators. The rates at which pollinators visit flowers and their patterns of movement among them can affect the success of pollination, gene flow, and the energy budgets of pollinators. Many aspects of foraging behavior are amenable to experimental studies, using techniques such as artificial flowers, or manipulating resources. Schmid-Hempel and Schmid-Hempel (1990, 1991) and Schmid-Hempel and Müller (1991) point out that one should be aware of the possible effects of parasites on the behavior of foraging pollinators; they present evidence that bumblebees parasitized by endoparasitic conopid flies forage in different ways than do unparasitized bees.

1. Visitation Rates
The rate of pollinator visits to flowers is important because when multiplied by the number of pollen grains transferred or the probability of pollination by a single visit, it can provide information about the likelihood that ovules will be fertilized. They are also a necessary component for calculations of net energy intake by pollinators and for calculations of time or energy budgets. Thus

visitation rates (in visits/unit time) can be measured either from the perspective of the flower or the flower visitor. Visitation rates can also serve as an index of the relative abundance of pollinators, either within or among years.

From a flower visitor's perspective, visitation rates are determined by three variables: (1) time/flower: the time expended to visit a single flower, which can be further subdivided into the amount of time required to land on and manipulate the flower (handling or access time) and the time required to extract the nectar it contains (extraction or ingestion time) (Harder 1983); (2) the spatial pattern of flowers: how far apart the flowers are, both within and among inflorescences; (3) traveling time: how fast the flower visitor can move within and among inflorescences.

Each of these variables can be influenced by a variety of factors, such as environmental variables, the temporal and spatial pattern of availability of pollen and nectar, the quality of the nectar and pollen, the presence of flowers competing for the services of pollinators, the number of flower visitors competing for floral resources, the visibility of nectar guides, and so on. Because it is not practical to control many of these variables in field studies, it may be necessary to design studies carefully in order to generate the desired information in a useful form.

There are some cases in which indirect measures of visits may prove informative. Flowers that are mechanically tripped or opened by pollinators, such as alfalfa flowers, the East African mistletoe, Broom, and Spanish Broom can provide this evidence. Table 7–2 lists a variety of such species. In straight-styled Australian *Banksia* (Proteaceae) species, anthesis is triggered by physical stimulation, such as a visit by a nocturnal mammal pollinator (but not by insects; however, untriggered flowers eventually open spontaneously) (Goldingay et al. 1991). In some of these species there may be a mechanism for resetting the movable parts in preparation for another visit; *Mahonia japonica* (Berberidaceae) flowers take 11–15 minutes to reset filaments following a visit, but some flowers (e.g., *Berberis* spp., Berberidaceae) can recover in 10–12 seconds (Percival 1965). Some visitors may also be able to extract nectar without consistently tripping flowers (D. Schemske, *personal communication*), and some tripping may occur without visits (Free 1970a). Aluri (1990) cites other examples in the Lamiaceae, Loranthaceae,

Table 7–2. Species showing evidence of an animal visit.

Flower Species	Common Name	Reference
Banksia spp. (Proteaceae)		Goldingay et al. 1991
Calathea spp. (Marantaceae)		Schemske and Horvitz 1984
Hyptis suaveolens (Lamiaceae)		Aluri 1990
Kalmia (Ericaceae)	laurel	Gleason and Crnquist 1963
Mahonia japonica (Berberidaceae)		Percival 1965
Medicago sativa (Fabaceae)	alfalfa	Free 1970a
Phragmanthera dshallensis (Viscaceae)	East African mistletoe	Gill and Wolf 1975
Sarothamnus scoparius (Fabaceae)	broom	Weberling 1989
Spartium junceum (Fabaceae)	Spanish broom	Weberling 1989
Stylidium graminifolium (Stylidiaceae)	grass trigger plant	Costin et al. 1979
Mimulus species (Scrophulariaceae)		Ritland and Ritland 1989
Thalia geniculata (Marantaceae)		Classenbockhoff 1991

Urticaceae, Fabaceae, Fumariaceae, Onagraceae, Rhizophoraceae, Cornaceae, Acanthaceae, and Marantaceae.

Flowers bearing pollinia, such as orchids or milkweeds, can be examined for the presence or absence of pollinia. Absence of the pollinia confirms a visit by a potential pollinator, although their presence does not confirm a lack of visitation, as nectar thieves may have visited the flowers. Obviously, deposition of pollinia also serves as confirmation of visitation, as does production of seed (at least in self-incompatible species). Nilsson and Rabakonandrianina (1988) recovered scales left by hawkmoths on orchid stigmas not only to confirm visitation but also to identify the moths involved (with a reference collection of scales from moth heads). Among flowers that showed any evidence of visitation, 47% had been pollinated and 60% had scales on their stigmas. It is also possible to use fluorescent or nonfluorescent pigments (e.g., Thomson 1982) or genetic markers (see Chapter 6) to provide evidence of visitation. Thomson used Helecon (non-Day-Glo, but ultraviolet-fluorescent) powders, which he applied to flowers as an aqueous suspension to eliminate wind dispersal of pigment during application. He used inconspicuous marks to identify focal flowers, and then applied pigment to the four nearest neighbors.

In studies of species visited by nectar or pollen robbers, a hole in the flower serves as confirmation of visitation. Unfortunately, nectar and pollen theft, which may also have a detrimental effect on pollination, do not usually leave any evidence behind. (See Inouye 1980a for a discussion of the terminology of floral larceny.) But in some instances other clues may be used as evidence of visitation. Craig (1989) reports that the long claws of foraging honeyeaters leave scar marks on the primary peduncle of inflorescences of *Phormium tenax* (Phormiaceae); he was able to use this as an indication of unobserved visitation. R. C. Plowright (*personal communication*) says that one of the best instances of a record of visitation is the pattern of marks left by bumblebee workers when they visit tomato flowers. These marks are used as an indirect index of pollination effectiveness in commercial efforts to utilize bumblebees as pollinators of greenhouse tomatoes. Neff and Simpson (1990) found that styles of sunflower florets retracted after successful pollination, providing a quick visual indicator that is faster than waiting for seed production.

Some flowers also show a marked but slower morphological response to pollination. This may include changes in petal color, as in *Lupinus* (Fabaceae) (Wainwright 1978, Gori 1989, Stead and Reid 1990), *Lantana* (Verbenaceae) (Mathur and Mohan Ram 1978), and *Cymbidium* (Orchidaceae) (Arditti et al. 1973); a change in stigma color, as in *Ourisia* (Scrophulariaceae) (Schlessman 1986a); wilting, as in carnation (Nichols 1977), *Petunia* (Solanaceae) (Gilissen 1977), and *Vanda* orchids (Burg and Dijkman 1967); or abscission of the corolla, as in *Digitalis* (Scrophulariaceae) (Stead and Moore 1977). These sorts of indicators might prove useful in some studies, but caution should be observed not to confuse them with age-related changes independent of pollination (e.g., age-dependent rather than pollination-induced color change in *Fuchsia excorticata* (Onagraceae) (Delph and Lively 1989). Gori (1983) reviews a variety of post-pollination changes in flowers.

Another indirect method of detecting visitation is possible for nonflying mammals. Cunningham (1991) put paint on the stems of *Banksia* inflorescences just before dusk and then noted the footprints of animals that had walked through the wet paint at night. He was then able to detect identifiable footprints around the inflorescences.

In most cases, however, visitation rates can only be quantified by direct observation, which becomes a tedious task if the rates are very low or a taxing one if they are high. Depending on the system, it may prove most efficient to follow an individual pollinator and record its activities. In other situations it will prove more profitable to watch a particular flower or patch of flowers and record visits to them. Which method to use will be determined in part by whether you are more interested in the flowers' or the pollinators' perspective. One potential problem with this variety of possible methods is that it may be difficult to compare data from different studies.

One high-tech (and probably expensive) alternative it to use a photographic technique. Goldingay et al. (1991) used automated photography to document nocturnal mammalian pollinators. Some super-8 movie cameras and video cameras have adjustable settings for exposure frequency, which facilitate making time-lapse movies of flowers; Fuhrman Diversified, Inc. makes a video camera system for fieldwork that has this capability (see Appendix 2, under Infrared Viewers). Shooting a frame every few seconds might be sufficient to record any visits to a flower; Washitani et al. (1991) photographed the flowers they were studying every 2 seconds with 8-mm film and found that visits averaged 10 seconds, so that they were almost assured of recording visits. Alternatively, a triggering device can be used to turn on a video or film camera when a flower visitor arrives. An infrared beam with a sensor can be deployed in front of the flower, so that whenever the beam is broken the camera is turned on. This technique has been used extensively for filming bowerbirds on their bowers (Borgia 1985), and experimentally for some pollination studies (D. Schemske, *personal communication*). If the camera is close enough to the flower, or has a telephoto lens, it may be possible to read tags on individual pollinators as they visit. Disadvantages of this technique are the cost of the equipment, the cost of developing film (not a problem if a video camera is used), and the need to protect the equipment from precipitation, sun, and theft. As equipment is miniaturized, it may become possible to use bar code readers and bar codes on flower visitors to record visits or activity at a hive.

Another method, which may work well only for buzz-pollinated flowers, from which pollen is collected by vibrating the anthers, is to use a tape recorder and microphone to record the floral handling sounds of pollen-foraging female bees. Cane and Payne (1988) used a modified Marantz PMD 200 cassette recorder with a high-gain preamplifier and Sennheiser ME88 spot head microphone to record foragers of seven species of bees foraging on *Vaccinium* (Ericaceae) flowers. They then used a 10-channel timer program to transcribe recorded foraging bouts in slow playback (accurate to 0.05 second), to determine mean floral handling durations, number of buzzes per flower, and interfloral travel rates for each species. Cane and Buchmann (1989) used the same equipment and analyzed data with computer-assisted, slow-playback transcription and event timings. Another method of detecting visitation in some buzz-pollinated flowers is to look for evidence of bees having handled the anthers. Buchmann (1992) reports that the enlarged anthers of buzz-pollinated plants that are grasped by bees with their mandibles and tarsal claws show small brown bruise marks (bee kisses) several hours after having been visited.

Nilsson (1979) used another method to trap visitors to *Herminium monorchis* (Orchidaceae), as part of a study of the influence of scent on visitation frequency. He enclosed inflorescences in plastic cylinders, which had tape lining the inside of the top of the chamber, with the sticky side out. Insects attracted to the inflorescence by the odor would apparently fly upward after visiting flowers and get stuck on the tape. Nilsson did not report on the efficiency of this method of trapping insects (i.e., whether it only trapped some types of visitors).

Thomson (1978) used a sweep net with a sharpened auxiliary rim to collect flowers and catch the insects on them at the same time. He then counted the number of flowers and insects in the bag to get an index of insects/flower. Although this destructive sampling may not be possible in some studies, the method might be useful for among-stand comparisons. One advantage is that samples can be stored in a bag and sorted later, if it is desirable to minimize field time.

Time per flower is usually measured with a stopwatch, which is started when the flower visitor first contacts a flower and stopped when contact is broken. For example, Gill and Wolf (1975) measured to the nearest 0.2 seconds how long a sunbird took to feed at a flower. They tried two different methods of data collection, which produced similar results. In one case they recorded times for individual visits and calculated a mean. In the other they accumulated times for multiple flower visits, and then calculated a regression of total time against the number of flowers visited; the means for individual visits fell within the confidence limits of the regression method. The advantage of recording individual times is that one can also then calculate a measure of variance. Photography can also be used to measure handling times. For example, Strickler (1979) filmed bees with a super-8 camera and then counted frames to determine how long bees spent on flowers. Video cameras could also be used for this.

If you are following individual pollinators to determine how many flowers they visit per unit time, you may find it necessary to have an assistant to help record data, as it is difficult to keep track of a flying insect or bird while also recording time per flower, time between flowers, or even just numbers of flowers visited. If you are satisfied with averages for time per flower or time spent in transit, you may be able to use two stopwatches simultaneously to record overall time and time on flowers and to calculate transit times by subtraction (e.g., Pleasants 1981). Or if the time per flower is too short for accurate timing with a stopwatch, you may have to record time per inflorescence (or capitulum) and divide by the number of flowers probed to derive an index of time per flower (e.g., Cartar 1991). An alternative to using stopwatches in the field is to use a tape recorder or video recorder and transcribe data later (e.g., Cartar 1991). Data collected on movie film or video can also be used to generate very accurate data on time per flower or transit times, by counting frames (e.g., Waser and Price 1985a). Appendix 4 gives a listing of a computer program that can be used with a laptop computer in the field to record behavioral observations; this may also facilitate collection of data while following a pollinator (see also Appendix 2, under Computer Programs). A programmable calculator may also prove useful in timing behavioral events (e.g., Pyke 1982b).

Arroyo et al. (1982, 1985), Inouye and Pyke (1988), Berry and Calvo (1989), Kearns (1990), and McCall and Primack (1992) have made an effort to use a single method in a variety of studies, in order to facilitate comparisons among them. This method involves watching a known number of flowers for 10-minute intervals and counting all visits by individual flower visitors during those intervals. The number of flowers being watched can be adjusted for the rate of visitation, with a larger number of flowers being watched if there are few visits. A wristwatch with a countdown timer is a convenient way to mark the 10-minute intervals. Measurements of air temperature, wind speed, light, and relative humidity are made before or after each observation period, and if enough observations are made, statistical relationships between these environmental variables and visitation rates can be calculated. (See Chapter 8 for more information about the instruments used for making these environmental measurements; see Appendix 2 for places to buy them.) Visitation data can be summarized as a number of visits per flower per minute for easy comparisons among studies (e.g., Arroyo et al. 1982, 1985; Inouye and Pyke 1988). A potential disadvantage of this method is that if visitation rates are very low, resulting in many measurements of 0 visits/minute, statistical analysis becomes difficult.

Visitation rates can be experimentally manipulated by a variety of techniques, including altering floral rewards (see Section K.6), preventing visitation (see Chapter 2, Section E), or removing species of pollinators (e.g., Inouye 1978). Experimental manipulations such as these can provide insight into the dynamics of plant-pollinator interactions. It is also possible to take advantage of the introduction of flower visitors to examine these interactions (e.g., the introduction of bumblebees into New Zealand, the introduction of honeybees into Australia, or the arrival of Africanized honeybees in the United States).

Be forewarned that visitation rate alone may not be a good index of resulting seed production. McDade and Davidar (1984) found a value of r^2 of only 0.065 for this relationship for *Pavonia dasypetala* (Malvaceae) flowers. This kind of discrepancy could be the result of self-incompatibility systems, ineffective pollinators, floral larceny, inbreeding depression, and so on.

Udovic (1981) developed a method for calculating an index of the expected number of visits by moth pollinators to *Yucca whipplei* (Agavaceae):

$$V_i = \sum_{j=1}^{n} M_i \, \frac{F_{ij}}{\sum_{k=1}^{m} F_{kj}} \cdot d_{j-1, j} \, ,$$

where

> V_i = the visitation index for plant i;
> M_j = the number of moths in the population at the jth census;
> F_{ij} = the number of open flowers on plant i at the jth census; and
> $d_{j-1, j}$ = the number of days between the jth census and the previous one.

The index assumes that for a given number of moths, the number of visits to a given plant during the time interval in question will be proportional to the fraction of open flowers in the population that belong to that plant (Udovic 1981).

2. Patterns of Movement

The patterns of movement of pollinators are important because — in combination with breeding system, pollen deposition, and carryover — they will determine an important component of gene flow (the other being movement of seeds). These movements may also be important in determining the population structure of the flower visitors themselves and have probably played a role in the evolution of floral morphology and displays. Differences between pollinator species in movements within or between inflorescences can also have an important influence on the amount of selfing versus outcrossing in self-compatible species. There has also been interest in determining whether the patterns observed in the field represent the optimal way for flower visitors to forage in order to maximize their collection of pollen and nectar.

Two parameters have been used to quantify flight patterns of flower visitors: flight distance, or the linear distance between two flowers visited in succession, and change in direction, or the difference in angular direction of a flight from one flower to the next in relation to the direction of the approach flight to the first flower (Waddington 1979a). The angular change is defined in degrees, with turns to the left being given a negative sign (range -1° to -180° and turns to the right a positive sign (range 1° to 180°); if there is no change in direction, the flight is recorded as 0°. Several studies of foraging by flower visitors have found that these quantitative descriptors of flight patterns change in relation to both flower density and the availability of floral resources, apparently in ways that maximize the net energy gain by the foragers (e.g., Waddington 1979a, Pyke 1978a). Of course these differences in flight patterns will have significance for pollen flow (e.g., Levin et al. 1971).

At least four methods have been used to investigate directionality in field studies. One (e.g., Woodell 1978) is to follow an individual flower visitor as it flies around, using a compass (either real or a mental one) to measure the direction of each flight and a cassette recorder to record the direction, distance, and perhaps time of each flight. Although one has to transcribe the data later, the recorder makes it possible to follow the movements without losing sight of the animal. Alternatively, one might use a portable computer as an event recorder to keep track of when each flower is visited and how long the transit time is between flowers (e.g., Turchin et al. 1991; see computer program listing in Appendix 4). In a variation of this method, one can map and label all the plants in a study site, follow animals as they visit plants and record which plants are visited (e.g., Thomson et al. 1987). A computer can then be used to make all the relevant calculations about flight patterns.

A third way to collect data on directionality was used by Lewis (1989) and Turchin et al. (1991). They placed wire stake flags along the flight path of insect visitors, with stakes next to each flower visited and sometimes additional stakes where turns were observed. If this method is used, it is helpful to have two people working, so one can keep track of the insect while the other is placing flags far enough behind that the insect is not disturbed. It is also easier to measure distances between flags with two people. Once the flags are in, the flight paths can be reconstructed after a certain number of flowers have been visited or after the insect is lost. Turchin et al.

(1991) discuss two methods for recording a path marked by numbered flags. One is to measure the spatial coordinates of each flag by triangulation with a measuring tape or compass, and the other is to measure the distance and direction of each move between two consecutive flags. They suggest that the first method is much more accurate and that triangulation with a compass is less accurate than use of a tape measure (because of the potential for accumulated errors). One person can measure the distance from a plant to each of two fixed stakes and then use trigonometry to calculate the location of the plants on an X-Y plane (a computer program for making these calculations is shown in Appendix 4).

The fourth method for monitoring movements among flowers is to map the study site. Heinrich (1979b) set up a twine grid with lines 15 cm apart, replicated the grid on a notepad, and drew in bees' patterns of movement to determine turning angles. Olesen and Warncke (1989c) tagged and mapped all flowers as they opened in their study site; the area was divided into 1-m^2 squares. The progress of flights was recorded, and later flight distances, angles of departure, and the percentage of insect moves to nearest neighbors were measured. Hodges and Miller (1981) didn't map plants for their study but did work within a grid and numbered plants within the grid, whereas Soltz (1986) did both and sketched the paths of foraging bees on graph paper over a map of the plot on which each plant of interest had been located. If you need to keep track of movement patterns in three dimensions, such as within the canopy of a tree, one possibility is to label all branches or flowers and then record the nearest reference point after each move the flower visitor makes (see Aluja et al. 1989 for a review of three-dimensional and other tracking techniques).

Another way that researchers have investigated movement patterns is with the help of artificial flowers, which are described in the next section. This method has been used with both bees (e.g., Waddington 1979a, 1979b, 1980) and hummingbirds (e.g., Pyke 1981a). Campbell (1985) used a similar technique by putting out potted plants in a 9 x 9–square grid of 81 plants. She used grids of a mixture of two species as well as a single-species grid to compare the effects of a competing resource on movements by both bee flies and solitary bees. For visits to one of eight nearest neighbors, she compared the probability of staying to visit multiple flowers on the same plant, the probabilities of moving different numbers of units on the grid between stops, and the probabilities of moving in particular directions relative to the last move. Differences between mixed and single-species plots in these behaviors would indicate either pollinator preference or pollinator constancy. Campbell used a randomized block design to test the effect of plot composition on the probability of an insect's remaining in the plot. She used three-way contingency analyses to evaluate the interactions between (1) plant composition, insect type, and flight distance, and (2) plant composition, insect type, and relative flight direction. And she used contingency analysis to determine whether the directions of consecutive moves were independent.

Woodell (1978) points out the distinction between overall directionality, or the tendency for the majority of flights to be in one direction, and sequential directionality, or the tendency of a flight to follow the direction of a previous flight. Pyke (1978b, 1981b), following bumblebees and hummingbirds as they visited flowers, recorded directions of movement (subdivided into four to eight compass directions) relative to the previous movement.

Several models of optimal foraging by flower visitors have been developed that consider both movements within an inflorescence (e.g., Pyke 1979, Best and Bierzychudek 1982) and movements among inflorescences. However, these studies are outside the scope of this book.

Anyone who has attempted to follow flower visitors for long periods of time has had the experience of losing sight of the animal after awhile. It may be leaving a patch of flowers to return to its colony, nest site, or territory, or it may be leaving for another patch of flowers. If you are studying flight distances, this introduces an inherent bias, in that you are not able to record long-distance flights. Although the frequency distributions of flight distances seem to be universally leptokurtic, long-distance flights are probably always somewhat underreported. A related topic that merits investigation is the distance that social or solitary bees will forage from a nest. Although there are some data for honeybees, little is known about bumblebees. Even less is known about the movements of some other groups of pollinators, such as Diptera. Because the foraging distance may influence dispersal distances for pollen, this may be a fruitful area for future research.

For additional information about this topic, see Handel (1983a) and Waddington (1983a, 1983b).

3. Artificial Flowers

Studies of foraging behavior can be confounded by a number of different variables, including wind speed, solar radiation, ambient temperature, availability of flowers, variation in the nectar or pollen content of flowers, etc. One way to control some of these variables is with artificial flowers. Many animals, including hummingbirds, bumblebees, and butterflies will forage readily from artificial flowers in the lab or in flight cages, and studies using these techniques have provided insight into the "rules" that some of these flower visitors use while foraging. The answers provided in this way are also valuable as input into the design of models of foraging behavior. Another reason for using artificial flowers is for choice experiments (see next section); different solutions or odors can be presented to foraging animals in order to discern their preferences or ability to discriminate among different diets or scents.

The use of artificial flowers for investigating animal behavior has a long history, largely ignored by more recent studies. For example, Plateau (1877; cited in Clements and Long 1923, p. 239) and several other researchers conducted extensive experiments over a hundred years ago on the responses of insect flower visitors to artificial flowers or real flowers that they mutilated in an imaginative variety of ways.

a. Making them

i. Insects. Most studies using artificial flowers have been done with bees, and many of these have involved wells drilled into a sheet of Plexiglas to hold an artificial nectar solution (usually a sucrose solution). The wells can then be refilled using a microliter syringe or repeating dispenser; sample well sizes that have been used are 1.0–2.5 mm in diameter, and 3.0–4.5 mm deep. In order to vary depth of the wells, Harder (1988) drilled a 1.6-mm-diameter hole, filled it with paraffin, and then reamed out the paraffin to a particular depth. Rather than use a whole sheet of Plexiglas you can cut out small pieces (e.g., 3 cm square) and deploy them as desired on a suitable background.

The wells can be highlighted for the bees by arranging "petals" around them. For instance, Heinrich et al. (1977) and Waddington et al. (1981) used a point punch to make blue and white 12-mm long "petals" from pieces of tape, and Dreisig (1989) used pieces of red tape to simulate petals. Studies that have used this sort of technique also include Cameron 1981 (Figure 7–18), and Cartar and Dill 1990. Alternatively, you can spray paint patterns by using a stencil cut in paper with a point punch. A similar but somewhat simpler technique is to put glass slides on top of small (2.5 cm diameter) "flower" shapes, with a drop of sugar solution on each slide. Manning (1957) used this technique to investigate the "flower constancy" of bumblebees to different shapes.

Some of the arrays used have been quite large. For example, Waddington (1979b) used a flower patch with 2,208 wells in 47 rows and columns on a sheet of Plexiglas 1.2 x 1.2 m; adjacent rows and columns were only 2.54 cm apart. Real (1981) used an array of 2,304 wells on a similar size sheet of Plexiglas. In contrast, Dreisig (1989) used an artificial inflorescence with only six flowers, and Waddington et al. used an array of only four flowers. If clear Plexiglas is used for the "flower patch," then wells can also be marked by designs on a piece of paper underneath the sheet of Plexiglas. These can be stamped with a rubber stamp, made with cut-paper shapes, or printed on a computer-generated printout (e.g., Waddington 1979b, 1980; Real 1981). To prevent scent cues from affecting foraging, the Plexiglas sheet can be washed carefully between runs.

Sometimes these Plexiglas sheets are put out inside flight cages (Cartar and Dill 1990), and sometimes they are incorporated into Plexiglas chambers to hold foraging bees (Heinrich et al. 1977, Cameron 1981, Waddington et al. 1981). Other researchers have taken advantage of captive colonies in the laboratory (e.g., Dreisig 1989) or put individual bees in screen cages with the artificial flowers (e.g., Waddington and Heinrich 1979). Real (1981) enclosed nests of wasps and bumblebees that were located in the field inside mesh flight cages. If ants are a problem (collecting "nectar" from your flowers), use Tanglefoot as a barrier or support the Plexiglas sheet on legs standing in cups of water.

More elegant feeders have also been designed, incorporating means to refill the flowers automatically. Heinrich et al. (1977),

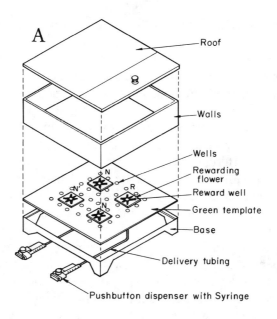

A

Roof

Walls

Wells

Rewarding flower

Reward well

Green template

Base

Delivery tubing

Pushbutton dispenser with Syringe

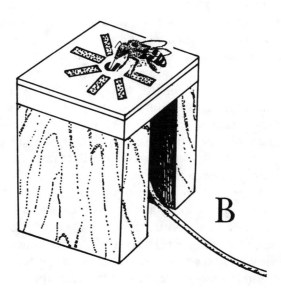

B

Figure 7–18. Artificial flower feeders as shown in Cameron 1981 (A) and Waddington 1982 (B). Figures reproduced from the original papers.

Waddington and Heinrich (1979), and Waddington et al. (1981) used the same type of feeder, also constructed from Plexiglas. A foraging surface was made from a single sheet of dark-green Plexiglas, laid over another sheet. The foraging surface had 2.5-mm diameter holes drilled through it, which aligned with small wells drilled into the lower sheet, where the nectar was placed. The holes in the upper sheet were made conspicuous by thin transparent Plexiglas squares with holes in the center, surrounded by artificial petals cut from different colors of tape with a point punch. Sucrose solution was fed into the wells in the lower sheet through polyethylene tubing leading to a repeating dispenser so that flowers could be refilled quickly. Waddington (1982) used a slightly different model (Figure 7–18), with individual flowers mounted on a block of wood, and a nectar well that could be refilled from a repeating syringe dispenser connected to the well by a piece of polyethylene tubing. Bertsch (1984) used an electric stepper motor to control a Hamilton syringe to refill artificial flowers. The amount needed to refill a flower after it was visited was determined electronically by optically monitoring the level in the feeder tube, and data on the number of the flower, time of visit, and amount of nectar collected were recorded automatically.

Free (1970b) placed a variety of two-dimensional model flowers between two sheets of glass for a study of the effect of flower shapes and nectar guides on the behavior of foraging honeybees. Drops of sugar syrup were placed on top of each flower and replenished as necessary. This arrangement protects the flowers, and allows the foraging surface to be cleaned quickly, but it would not work for testing responses of bees to ultraviolet reflectance patterns, as the glass is not transparent to them.

Brian (1957, p. 72) built artificial flowers out of funnels of colored paper, "waterproofed at the base to hold a drop of syrup and held on wire 'stalks' about six inches above the ground." This appears to be one of the first descriptions in the renaissance in use of artificial flowers. Her flowers, set out in a garden, were visited regularly by bumblebees once they were trained to use them ("by holding them over a 'flower' in a darkened tube until they began to drink" [Brian 1957, p. 72]). She found that bees visited them less frequently while the weather was sunny, apparently because the artificial flowers were not as attractive a resource as real ones. By

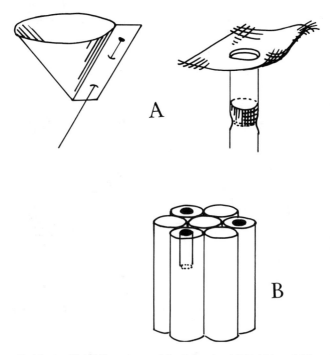

Figure 7–19. Artificial flowers used by Brantjes 1973 (A) and Kipp 1987 (B). Redrawn from the original papers.

varying the construction of the flowers Brian was able to investigate the effects of "flower" color and "corolla" depth. Brantjes (1973) used similar artificial flowers folded from colored paper, which were dipped into melted paraffin wax to waterproof them ("Trichter-blume" of Knoll 1921–1926) (Figure 7–19A).

Kipp (1987) made artificial inflorescences out of bundles of seven dowels, each 6 cm long. Some of the dowels had a paraffin-lined hole 6 x 22 mm, into which honeybees would crawl to collect the couple of µL of sugar solution that was placed inside (Figure 7–19B). In this study of directionality of foraging, the bees could be moved to the center of a hexaradial pattern of 37 inflorescences while they were involved in feeding.

Hartling and Plowright (1979) designed flowers that could be refilled remotely after having been emptied by a bee. They used a capillary tube for a corolla, a circular cardboard disk for a petal, a

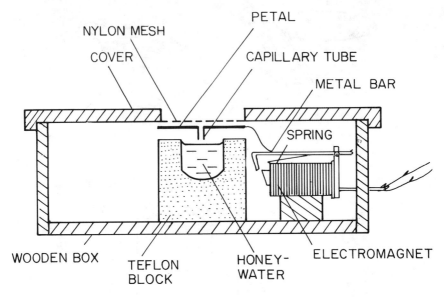

Figure 7–20. Artificial flower design. Redrawn from Hartling and Plowright 1979.

Teflon block to hold a reservoir of honey water, and a 12-V electromagnet. The capillary tube corolla was placed at the center of the cardboard disk petal, and the disk was mounted on a metal bar above the electromagnet. A black circular "nectar guide" was drawn on the petal around the capillary opening. The whole apparatus was housed in a wooden box, with an access hole in the side to facilitate cleaning and replenishment of the reservoir. The reservoir of honey-water solution was placed under the corolla, which was filled by capillary action when the electromagnet was activated, pulling the capillary tube down to touch the reservoir (Figure 7–20).

Marden (1984b) constructed artificial flowers of yet another design that also required bees to climb into them to reach the nectar reward. The flowers were made from 13 x 45–mm glass vials, with a nectar well (approximately 2.5 x 2.5 mm) drilled into a 4-mm-thick square piece of Plexiglas that was placed in the bottom of each vial. The vials were wrapped with blue tape, and 361 of them were arranged in a 19 x 19 grid on a 1.22 m^2 styrofoam board painted green. Each vial was set into a depression in the board that kept it in place. The flowers used by Marden and Waddington (1981) and

Waddington (1985) were somewhat similar. They were constructed of flat-bottomed plastic tubes (13 mm inside diameter; 13, 33, or 53 mm long) with a square piece of 3-mm-thick Plexiglas glued to the inside bottom of the tube. A hole drilled into the Plexiglas formed a well to hold a few microliters of a sugar solution. These wells were refilled with a repeating dispenser (Hamilton PB600-1) after honeybees visited them. The tubes were wrapped with different colors of plastic tape to provide different visual cues for the bees.

Harder (1986) investigated the relation of ingestion rate to flower depth and nectar concentration by using artificial flowers. Each feeder consisted of two nested tubes; the shorter outer one was placed with its end flush against the inner surface of a test chamber or thin paper card, and the longer inner tube with the sucrose solution (inside diameter 1.39 mm) was projected different distances into the outer tube to make bees reach to drink the nectar (their heads could not fit inside the outer tube). With this apparatus Harder could measure both lapping rates and the volume ingested per lap, with different concentrations of sucrose and different feeder depths. A video camera was used to film feeding, which was later observed in slow motion.

Waller (1972), Waller et al. (1972), and Inouye and Waller (1984) used artificial flowers for choice experiments designed to investigate the responses of honeybees to six different solutions presented simultaneously. Each flower consisted of seven small glass test tubes (0.9 mL, 0.25 dram), arranged as six around a central one, and held together with a rubber band. A blue Plexiglas disk, 3 mm thick and 3.6 cm in diameter, was placed on top of the glass tubes, and six microcapillary tubes (designed to hold 10 µL) were placed through holes in the Plexiglas and into the glass tubes for bees to use like drinking straws (Figure 7–21). A thin band of petroleum jelly on the rims of the test tubes prevented wicking of solution if the capillary tubes contacted the sides of the test tubes; a better solution might be to countersink holes in the bottom of the Plexiglas disk to hold the glass tubes centered below the holes in the disk.

Honeybees foraging on the flowers thus had a choice of six different capillary tubes, each protruding 1–2 mm above the blue disk, from which to drink. Bees were allowed to forage on the flowers until one of the glass tubes was emptied, and then the

Figure 7–21. Artificial flower feeder described in Waller 1972, Waller et al. 1972, and Inouye and Waller 1984.

amounts remaining in the other tubes were measured. The experiments were conducted in screen flight cages containing small beehives. If too many bees tried to forage simultaneously, wire mesh (hardware cloth) cones were placed over the flowers to restrict access to one or two, and in extreme cases a vacuum cleaner was used to suck up extra bees. These feeders also appear to work well with ants, although the amounts they drink are much smaller than those taken by bees (Inouye, *personal observation*).

For investigation of the potential effects of scent marking, it is convenient to use disposable artificial flowers to avoid having to clean flowers after each visit (P. Wetherwax, *personal communication*). Wetherwax uses plastic 0.3-mL sample cups for an autoanalyzer (Sarstedt, Inc.), with two small holes drilled in the top of the cap. One hole is in the center and has a piece of 10-μL capillary tube inserted through the hole so that the tube opening is flush with the top of the cap. The second hole, at the side of the cap, allows air to enter to replace nectar as it is removed. To make these flowers refillable, make a hole in the bottom, and attach a tube to the bottom of the capillary tube. Wetherwax attaches the tubing to a three-way stopcock that has a 5-mL syringe, for filling the system, and a 50-μL Hamilton syringe with a repeating dispenser (Hamilton PB600-1) so that the tube can be refilled with 1 μL of fluid after each visit.

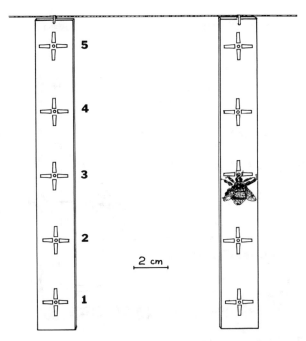

Figure 7–22. Artificial inflorescences. From Waddington and Heinrich 1979.

Although most of these artificial flowers were presented in a horizontal plane, many natural flowers are presented on vertical inflorescences. Waddington and Heinrich (1979) used artificial inflorescences with five "flowers" that could be presented vertically (Figure 7–22). These flowers used the same construction of a double layer of Plexiglas described above and had "petals" formed with tape. J. Thomson (*personal communication*) suggests using small square (e.g., 2.5 cm) Plexiglas flowers and arranging them as desired on a piece of tape. This method facilitates construction of arrangements with different spacing, or even different vertical orientations.

Alm et al. (1990) used a different design for their study of consumption of artificial nectar solutions by cabbage white butterflies (*Pieris rapae*) and honeybees. A graduated pipette on a ringstand served as a nectar reservoir, which was connected to a funnel (the feeder) with rubber tubing (Figure 7–23). A colored disk around the funnel served to differentiate solutions presented simultaneously. Consumption from the feeder was measured by the amount

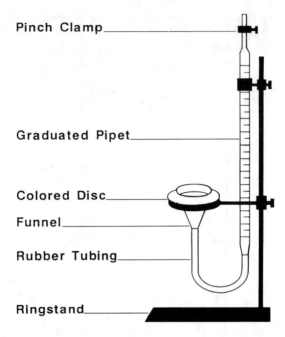

Pinch Clamp

Graduated Pipet

Colored Disc

Funnel

Rubber Tubing

Ringstand

Figure 7–23. Apparatus for measuring nectar consumption. From Alm et al. 1990.

released from the pipette to keep the feeder filled. J. Thomson (*personal communication*) suggested a somewhat similar feeder (untested) in which the nectar solution would be stored in a reservoir connected to the flowers with tubing, and the solution would be fed into the flowers (continuously or on a variable schedule) by raising the reservoir above the level of the feeder. A very slow motor, or a peristaltic pump might automate the refilling.

Erhardt (1991a) made butterfly feeders that are similar to those used by Waller for bees (see above). Test solutions were placed in 1-mL glass tubes, and each tube was capped with a piece of a Chore Boy cleaning pad (loose plastic loops) secured with a yellow rubber band (Figure 7–24). The meshwork of the cleaning pad provided a landing spot for the butterflies and allowed them to feed from the solution in the tubes. Three or four of these artificial flowers were mounted in a styrofoam holder (approximately 8 cm diameter) covered with green cloth, and a drop of lemon extract was added to the pad as an olfactory attractant. Consumption was determined at the end of a trial by removing and measuring the remaining

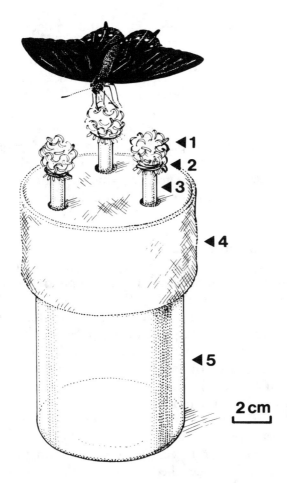

Figure 7–24. Butterfly feeder. From Erhardt 1991a.

solution with a syringe and correcting for evaporation by measuring control flowers covered with cheesecloth to prevent visitation.

Masters (1991) made artificial flower feeders for nectar preference tests with butterflies. He cut disks (1.8 cm diameter) from painted sheets of pressed foam mounting board (0.5 cm thick) and inserted into each disk a 1-cm section of hollow glass tubing (0.3 cm in diameter) so that an open end of the tube was flush with the flower surface. The other end of the tube was sealed with silicon aquarium sealant, and 0.5 μL of the artificial nectar was placed in the well. Two colors of flowers (yellow and orange) were arrayed on top of 49 evenly spaced dowels, 15 cm high and 0.8 cm wide, on a base of plywood (60 x 60 cm). To initiate a foraging bout a butterfly was placed on a flower, and its proboscis inserted into the well until it began to feed. Flowers were refilled after nectar had been removed (refilling did not inhibit further foraging). Boggs (1988) and Masters (1991) also induced butterflies to feed from 100-µL micropipettes by placing their proboscides on the micropipettes. Although these weren't used as artificial flowers, but for studies of rates of nectar feeding, this technique might permit investigation of some of the same variables for which artificial flowers are used.

Kugler (1956) built two kinds of artificial flowers for flies, as a way of elucidating their visual (for color and "nectar guides") and olfactory discrimination. One kind was made from colored paper circles, 20–30 mm in diameter, with an insect pin through the center so that it could only be reached by flying insects. The other model was a little square box, about 25 mm on a side and several mm deep, folded from a cross-shaped piece of paper. Flower parts could be added to the inside of the box as a way of conferring a floral odor. These, too, were placed on insect pins. A reward was offered on a smaller (6-mm diameter) piece of filter paper, soaked with 1–2 M sucrose, placed above the model on the insect pin.

Ants are more common as visitors to extrafloral nectaries than to floral nectaries, but they do pollinate some flowers. Lanza (1988) made artificial extrafloral nectaries for ants with the caps from microcentrifuge tubes. She filled one cap with the nectar solution and covered it partially with a second cap that had a hole in it. A thread was inserted through the hole and into the nectar solution to serve as a wick. This design could also be used as an artificial floral nectary.

ii. Hummingbirds. Hummingbirds are often already used to using feeders and seem quite adept at learning to feed from them. Thus it is relatively easy to conduct experiments with hummingbirds that require that they use feeders for choice experiments or to investigate feeding rates or movement patterns. There are several designs of feeders available commercially, although that they usually have several feeding tubes or access holes may be disadvantageous

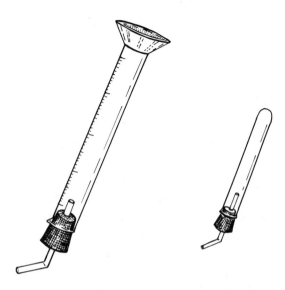

Figure 7–25. Design for a hummingbird feeder made from a test tube or graduated cylinder.

for some experiments. A simple feeder can be made by putting a rubber stopper with a glass tube through the center (from which the birds feed) into an inverted polystyrene graduated cylinder (Stiles 1976) or test tube (Figure 7–25). S. Hiebert (*personal communication*) suggests using a straight-sided bottle with a one-hole stopper and bent glass feeding tube; the end of the glass tube protruding through the stopper and into the bottle should be bent so it is not round, or else air bubbles rising into it tend to be held on the end of the tube instead of allowing more fluid into the tube.

In order to study the responses of hummingbirds to different arrays of artificial flowers, Pyke (1981a) designed a system that approximated an inflorescence. Individual flowers were constructed from the needles of disposable syringes, which have hollow plastic caps (18 mm long, 7 mm outside diameter, 4 mm inside diameter) into which the needles are molded. These needles were stuck into a styrofoam rod, the plastic caps serving as corollas to hold sugar water (e.g., 5 µL of 25% sucrose solution). The needles could be put in different arrays in the styrofoam rod to simulate different types of inflorescences. Pyke (1981c) used the same sort of feeder for honeyeaters too.

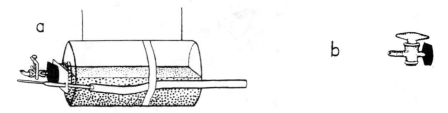

Figure 7–26. Hummingbird feeder. From Ewald and Carpenter 1978.

Hainsworth and Wolf (1976) and Montgomerie et al. (1984) used hummingbird feeders with different concentrations of sucrose and different "corolla" lengths to manipulate handling times and therefore foraging costs. Plastic tubing of different lengths was added to the end of the smaller-diameter tube that dispensed sucrose to simulate the effect of a corolla.

Montgomerie (1984) developed another type of artificial inflorescence for studying handling times of hummingbirds. Five tubular Plexiglas "florets," each with an internal diameter of 6 mm, were placed vertically 15 cm apart. Between trials he varied (one at a time) the nectar concentration, volume, corolla length, or corolla angle to examine their effects on handling time. The experiments were recorded on film and analyzed frame by frame to measure handling times.

Ewald and Carpenter (1978) used feeders with which they could regulate the amount of flow of a sugar solution. A 500-ml Nalgene container was closed with a cork with a short glass tube through it. A piece of flexible tubing was attached to the glass tube to form a feeding spout, and a hose clamp attached to the tubing was adjusted to regulate the flow rate. An air hole in the container allowed the flow to proceed evenly. By replacing the glass tube in the cork with a Pyrex stopcock that ended in a graduated tube, they could also release measured quantities of solution (Figure 7–26).

Hainsworth (1973) and Schlamowitz et al. (1976), in an attempt to duplicate the effect of a flower corolla on feeding rates of hummingbirds and sunbirds, added clear plastic tubing to the outside of a glass feeder spout. They varied the length of the tubing to simulate different flower morphologies, to investigate the effect of different morphologies on feeding rates.

iii. Manipulating real flowers. In some situations it may be possible or even preferable to use real flowers instead of artificial ones to generate a desired spatial pattern. For a study of bumblebee foraging, Manning (1956b) potted plants and moved them into the pattern he wanted. Schaal (1978) cleared all of the flowering *Liatris* (Asteraceae) plants from a subset of her study site and constructed in their place an experimental array by placing flower stalks (cut to identical heights) in water in bottles (obtained from the Lone Star Brewing Company). Bees foraged readily in the 10 x 10–square array consisting of 100 bottles placed in the prairie vegetation. Ginsberg (1986) also generated an artificial array, using soda bottles with knapweed stalks, to observe foraging by honeybees in a 5 x 5–square array with 50 cm between stalks. Waser and Price (1981) potted albino and normal *Delphinium nelsonii* (Ranunculaceae) plants and then created artificial populations with desired frequencies of albino plants. Feinsinger et al. (1991) manipulated densities of focal flowers and their neighbors by removing excess flowers or adding small branches with flowers. The cut branches were placed in containers filled with either half-strength Floralife solution (see Appendix 2) or a 50:50 mixture of decarbonated lemon-lime soda and water (which worked best; P. Feinsinger, *personal communication*). Both types of solution had 15 mg/mL $AgNO_3$ added to retard fungal growth. These solutions maintained inflorescences from 1 to 3 weeks, depending on the species. Nectar secretion continued in the flowers on cut branches but at significantly reduced levels.

Thomson (1988, 1989b) used pieces of plastic tubing, sealed at one end and filled with water, to hold freshly cut umbels. He created "inflorescences" by placing four umbel holders at the top of a short bamboo rod that he stuck into the ground; he then made an array of multiple rods. By replacing the flowers every 2–3 hours he ensured that there were adequate resources to attract bees; the bees foraged "normally" from these flowers. Waddington (1979a) placed a piece of pegboard Masonite (i.e., a flat surface, 0.61 x 1.22 m) with small holes drilled through it in a grid) over a water bath (a wooden frame lined with plastic), and then placed picked bindweed (*Convolvulus arvensis,* Convolvulaceae) flowers into holes in the pegboard so that their stems hung down into the water below. The flowers opened, shed pollen, and secreted nectar normally. Ashman and Stanton (1991) created a standardized inflorescence array as a way of estimating rates of pollinator visitation throughout the flowering season. The array was created by placing inflorescences in florists' water pics (vials for holding flowers) and placing the containers in a 2 x 2 arrangement adjacent to a natural population.

Vail (1983) also generated a desired spatial pattern of natural flowers, but without using picked flowers. Instead, he used a gasoline-powered string trimming tool to create patches of different densities of *Potentilla gracilis* (Rosaceae) in a natural meadow. Schmitt (1983c) picked inflorescences to create low-density plots of *Senecio integerrimus* (Asteraceae). Soltz (1986) manipulated density by removing or trampling plants to create desired densities in 8 x 23–m plots, and Heinrich (1979b) clipped off clover inflorescences to created desired flower densities. Wolfe and Barrett (1987) were even able to manipulate an array of an aquatic plant by placing inflorescences in jars containing water on a floating piece of plywood. Krannitz and Maun (1991) planted shrubs in groups of different sizes to determine the effects of floral display size on reproductive success.

Clements and Long (1923) replaced the corollas and calyxes of *Aconitum columbianum* (Ranunculaceae) flowers with crepe-paper floral envelopes, leaving the stamens, pistils, and nectaries intact. They also used this sort of manipulation on a variety of other species, and in some cases supplemented the normal flower with additional crepe-paper embellishments. In this way they could alter the outline and/or color of flowers to examine the responses of flower visitors. Nilsson (1988) constricted nectar spurs of an orchid to create the effect of flowers with shortened spurs, and showed that both pollinia removal and pollen receipt were reduced.

Bear in mind that manipulation of natural flowers may also have an effect on visitation by varying the attractiveness of an inflorescence or group of flowers. Stephenson (1979) found that inflorescences made smaller by the removal of flowers and buds had lower ratios of fruits to flowers, whereas flowers arranged into large inflorescences had a synergistic effect on the number of fruits set per inflorescence and per flower.

b. Training animals to use them

Honeybees, bumblebees, and hummingbirds are easily trained to use artificial flowers. Probably the best way is to use some sort of feeder that offers easy access to a sucrose solution, and then once the animals are used to feeding at that spot, substitute the artificial flowers. This process can often be accomplished in a few hours. An open dish of a scented sucrose solution or diluted honey placed near a honeybee hive will quickly attract bees, which can then be induced to feed on artificial flowers. Bumblebees constrained to a small foraging arena will often quickly find and use feeders, although in large screen flight cages they will often just fly against the screen instead of feeding.

Waddington and Heinrich (1979) familiarized bumblebees with their artificial flowers by placing a bee in an inverted 50-mL beaker over one of the flowers. After the bee accidentally discovered the sucrose reward, the beaker containing the bee was moved to other flowers. After several flowers had been visited in this fashion, the bees would forage spontaneously when the beaker was removed. Marden and Waddington (1981) and Waddington (1985) used a similar method to train honeybees, by holding them in the artificial flowers until they discovered the nectar. Soltz (1987) smeared honey onto artificial flowers until bumblebees were used to visiting them, and then in experimental studies filled only the wells with a standardized solution. J. Thomson (*personal communication*) suggests the following method that he has used with success on bumblebees; (1) chase the bee around the flight cage so that it will become hungry; (2) chill the bee by putting it in a vial (with air holes) in a refrigerator or ice chest; (3) after the bee is well cooled, take it out of the vial and put it on a flower. It will probably start feeding as soon as it is warm enough and then continue on to other flowers. Make sure the bee is completely warmed up and moving normally before you begin recording data.

There is some variation in the ability of individual Broad-tailed and Rufous Hummingbirds to figure out how to use feeders (Inouye, *personal observation*), but most seem to be quick to learn and will forage readily in flight cages. Waser and Price (1984) captured Rufous Hummingbirds at feeders, introduced them into outdoor flight cages, and trained them to visit individual flowers held in their hands. They have similarly trained birds to forage from feeders in the flight cages. Pyke (1981a) simply substituted his artificial inflorescence for a standard hummingbird feeder when he wanted birds to visit it.

We have not succeeded in getting muscoid flies to feed in flight cages or small screen cages, even on real flowers. The flies just fly against the screen and ignore the flowers. We have had slightly better success in the laboratory, by putting flies into large (1-gallon) glass jars, but the best technique we found was to put a fly in a small plastic vial with a flower. Although they would then visit the flower, their behavior did not appear completely normal. Even chilling the flies first for up to several hours did not help.

A common technique used for inducing butterflies to feed on artificial diets is to extend their proboscides with the aid of a pin until they contact the diet (e.g., Murphy et al. 1983). If the butterfly is hungry it will usually leave the proboscis extended and feed.

4. Choice Experiments

Choice experiments provide a means of testing the discriminatory abilities of flower visitors. This can include the ability to detect qualitative or quantitative differences in the sugars, amino acids, or other components of nectars, differences in the colors of flowers, flower size, flower volatiles, and so on. These differences can be significant because they can form the basis for reproductive isolation. Artificial flowers can be used for many of these tests, particularly those examining differences in nectar analogues. If differences in real or altered flowers are of interest, cafeteria experiments with these flowers are another alternative. Several studies (e.g., Thomson 1988, Thomson and Thomson 1989, Young and Stanton 1990b) have used florists' "aquapics," designed to hold cut flowers so that their stems are kept in water. By putting two or more of these together on a stick, the experimental flowers can be presented simultaneously to a foraging insect for its inspection (e.g., Thomson et al. 1982). As long as the flowers are moved slowly toward a foraging insect they don't seem to scare off potential visitors, which will often fly from natural flowers to the cut ones. This technique is also useful for experiments on pollen collection or deposition, as experimental flowers can be presented to pollinators rather than waiting for the pollinators to come to the flowers.

Another way of presenting flowers was used by Galen and Newport (1987), who put experimental flowers into plastic drinking straws attached to the end of a 1-m-long rod to present them to foraging bumblebees. Thomson (1988) used short pieces of plastic tubing at the top of a green bamboo wand to present cut umbels at the normal height and angle.

Choice experiments have also been used to investigate the preferences of hummingbirds for different sugars. Del Rio (1990) presented pairs of sugar solutions simultaneously in feeders made from glass tubes (inside diameter = 7.6 mm, with a 42-cm vertical section and lower 7-cm section bent upward at a 45° angle; the tip of the lower section was tapered to 2.5 mm in diameter). Birds were allowed to feed for 4 hours, after which the consumption of solutions was measured. Preferences were calculated as the ratio of consumption of one solution divided by the total consumption. Broom (1976) used a series of feeders, each with a different solution, to investigate the responses of hummingbirds to salt solutions. The birds' responses were measured either in terms of consumption or the amount of time spent feeding.

Waller et al. (1973) designed a method of testing the responses of honeybees to different flower odors and components of those odors. They used artificial flower feeders that provided a sucrose solution (Figure 7–27) to condition the bees to a particular odor and then tested their responses to the odors on flowers without a sucrose reward. The test arena was a 52-cm-diameter circular tray with receptacles for twelve 100-ml beakers, and a circular plastic disk that covered the tray. The beakers held the flowers or chemicals, and odors diffused up through thirteen 1.5-mm-diameter holes drilled in the disk above each beaker. The whole arena was placed on a table that rotated at 1/3 rpm to prevent a position bias. To prevent bees' footprint substance from affecting the results, a clean perforated metal screen was placed over each aroma site after a bee landed on it. This method worked well for honeybees and might be easily adapted for other flower visitors. Waller et al. (1974) used this technique to show that certain alfalfa clones could be reproductively isolated by the assortative pollination of honeybees responding to differences in flower scents.

Beker et al. (1989) constructed an apparatus similar to that of Waller et al. for an experimental study of associative conditioning using odors. They used a round polystyrene box (100 mm high, 230 mm diameter), placed on a rotating table. Eight glass vials (25 x 60 mm) were pushed into and through the upper lid of the box. Each vial contained both an aromatic material and a smaller vial (8 x 50

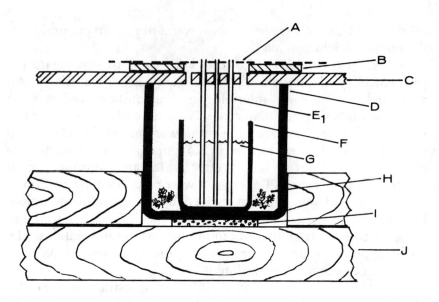

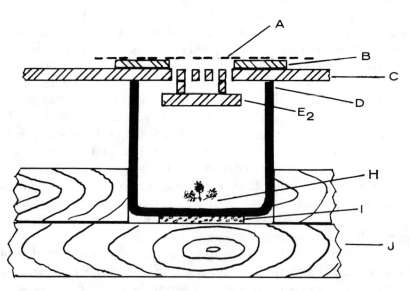

Figure 7–27. Apparatus for testing responses of bees to volatile compounds. From Waller et al. 1973. The upper figure shows how the feeder was used to condition bees by combining a sucrose reward with an odor source. The lower figure shows the apparatus as used to test responses to volatile compounds.

mm) containing 70% sucrose. A piece of perforated plastic net was placed over the exposed mouth of the vial, so that a bee had to land on the net and insert its proboscis through the net in order to obtain the sucrose solution. The net was replaced after each visit to prevent contamination by footprint substance.

Dobson (1987) used a similar test apparatus. To test whole flowers she filled a 5 x 5–cm glass container with flowers, and concealed them under 8-ply cheesecloth. To test odors from whole pollen, 8–12 mg of pollen was placed in a 2-cm-diameter aluminum dish, covered with wire screen. Pollenkitt lipids and odors from internal pollen lipids were extracted from 8 mg of pollen and placed on filter papers under wire screen. The dishes or glass containers, containing samples from each of four plant species, were put inside a circular wire-screen cage (22 cm diameter), and the responses of bees introduced into the cage were scored. Responses were classified as feeding (landing on the samples and probing with mouthparts) or nonfeeding (landing but not probing).

Another method for choice experiments involving odors is the use of wind tunnels. Drost et al. (1986) describe construction of a small flight tunnel. Lewis and Takasu (1990) put different odors onto filter paper and then used a wind tunnel to compare responses to them. Although they were not working with floral odors, the methods they used could be easily adapted for them.

Cock (1978) reviews eight different methods of analysis for assessment of preference. Although most of these methods are in the context of predator-prey relationships, they could probably also be used for experiments with flower choice.

5. Color Preferences
One of the striking features of many flower-filled meadows is the diversity of colors represented. It is generally assumed that these colors, as well as the patterns of colors in individual flowers, serve as cues for pollinators. For example, work by Waser and Price (e.g., 1985a) demonstrated the significance of differences in responses by pollinators to white versus blue flowers of *Delphinium nelsonii*. Many of the designs of artificial flowers described above could be used either to test the ability of pollinators to discriminate different colors or to test color preferences. For example, Brian (1957) used her paper flowers to look at responses to six different colors, and

Manning (1957) used his artificial flowers to look at visitation to two different colors. Knoll (1925) put samples of different colors of paper in a regular array between two sheets of glass. When moths tried to feed from the paper, as if they were flowers, they left marks from their proboscides on the glass. This allowed Knoll to investigate responses to both color and brightness.

Haslett (1989) investigated the color responses of hoverflies (Syrphidae) to five different colors by offering an array of painted plastic disks set out on a background of artificial grass. There was no nutritional award associated with these disks, and he just recorded the numbers of individuals of each species that "visited" the different disks. Pointing out the dangers of using human descriptions of colors, as insect color vision can be quite different from human color vision, Haslett suggested using a reflectance spectrophotometer to define colors by their reflectance spectra.

Dafni et al. (1990) studied the responses of beetles to different colors by using colored plastic cups similar in size to flowers that the beetles visited. The cups were mounted on sticks 15–25 cm long to match the height of the real flowers, and the number of beetles landing in each color of cup was monitored. They used a spectrophotometer to match the paint used on the plastic cups with the colors of real flowers. To test the responses of beetles to the presence or absence of another beetle (a dead beetle or a black paint spot), they used water-filled petri dishes with detergent (a wetting agent) in the water. These water traps, or bowls filled with water and a drop of detergent and maybe some formalin as a preservative, also work as a way to investigate responses to color by other flower visitors. Southwood (1978) reviews several studies that have used water traps for this purpose. One advantage of water traps is that captured insects can be removed without damage and are therefore easy to identify.

It isn't difficult to use hummingbird feeders to examine color preferences. The simplest technique is probably to color the sugar solution in clear feeders designed so that birds can see the solution (e.g., Bené 1941, Collias and Collias 1968, Miller and Miller 1971). Stiles (1976) put colored disks (approximately 2 cm in diameter) around the feeder tubes on his hummingbird feeders. He mixed acrylic paints to obtain the desired colors and tried to match brightness, which he measured using a spectrophotometer fitted with a reflectance unit. Miller et al. (1985) used a similar technique to look at color responses by hummingbirds. They fitted feeders with 1.5-cm red or yellow cardboard discs placed over the end of the glass sipper tube on the feeder.

Goldsmith and Goldsmith (1979) worked with hummingbird feeders equipped with tungsten lamps and interference filters in order to test color discrimination. They presented four different feeders and attempted to train the birds to particular wavelengths in order to examine their ability to discriminate. The birds selected feeders on the basis of hue, not brightness.

Miller et al. (1985) developed a sophisticated laboratory system to record the time and duration of each visit to a set of six feeders, as well as use of perches. It used photocells on the back of each feeder and a microswitch on each perch; a computer recorded the data for each visit or perching event. Each feeder was made from a 7.5-cm square metal plate with a 2.5-mm hole; behind the hole was a modified pipette containing a sucrose solution. They used this system to test the significance of color and position as cues for orientation and compared their results with a field study using feeders.

In the near future it may be possible to use arrays of mutations of flower color to investigate the responses of flower visitors to variation in color. For example, by inserting transposons at key sites in the genes of *Antirrhinum majus* (Scrophulariaceae), it is possible to produce a range of flowers with color mutations of the red wild type, including white albino, ivory, mixed ivory and red, mixed white and ivory, and reddish-brown (Luo et al. 1991; or see photograph in *Nature* 353(6339):20). There is also existing variation in some crop plants that might be used for experimental studies. For example, alfalfa (*Medicago sativa*) plants can have white, yellow, blue, or purple flowers (which can be used for studies of crossing and selfing; Steiner et al. 1992).

6. Manipulating Resources

Experimental manipulations of nectar and/or pollen can provide insight into the foraging behavior of flower visitors and can be relatively simple to implement. For example, Kodric-Brown and Brown (1978) reduced nectar availability for flower-visiting hummingbirds by cutting down inflorescences of flowers that the hummingbirds usually visit. Although planting more flowers would be a long-term experiment, Gass and Sutherland (1985) enriched patches of flower with sucrose solution (35%, w/w, close to the mean measured for nectar of the manipulated flowers). They used Hamilton (PB600-1) repeating dispensers with lengths of tubing (intravenous medical tubing) attached to facilitate adding 5-µL units of solution to *Aquilegia* (Ranunculaceae) and *Castilleja* (Scrophulariaceae) flowers at night. They added 7–18 times as much solution as would be found in unenriched flowers. Thomson (1989) added 35% sucrose solution to flowers and found that bumblebees developed preferences for the enriched flowers, which lasted even after the nectar additions were discontinued. Pyke et al. (1988) also added nectar to bird-pollinated (honeyeaters) flowers to investigate the effects on pollen removal and seed set. Although they started out by adding 46% (w/w) sucrose solution, as this matched the mean concentration of natural nectar at their site, they reduced this by half after noticing that the solution added to previously treated flowers was becoming extremely viscous.

Preventing visits to flowers for a while is another way of increasing nectar availability. Cibula and Zimmerman (1987) bagged plants for 40–50 hours in order to look at the responses of foraging bumblebees to increased nectar volumes. Watering plants may also significantly alter nectar production. Zimmerman (1983b) was able to effect a 75% increase in nectar availability in experimental plants by watering them, and Zimmerman and Pyke (1988d) found under some circumstances (dry conditions at the end of flowering) that control plants had only 13% of the nectar production per flower of watered plants.

It is not coincidental that bird-pollinated flowers are used commonly for nectar addition experiments. These flowers are often large and tubular, offering easy access to the nectary compared to smaller insect-pollinated flowers. Another method for adding resources that may work in some experiments with hummingbirds is to add feeders. Kodric-Brown and Brown (1978) used this approach to study territoriality.

Although it may be difficult to add pollen to a flower, it is possible to cover some flowers to avoid pollen removal and then make these flowers available after pollen has been collected from control flowers. It is relatively easy to remove anthers from flowers by picking them with fine forceps, although this may affect the time visitors spend per flowers (Kearns, *unpublished observation*). Thomson (1989b) varied pollen supply by emasculating some plants and leaving others intact. On a more limited scale, it may also be possible to prevent anthers from dehiscing by gluing them; Buchmann and Cane (1989) used this technique to plug up the pores of anthers of buzz-pollinated flowers. In some flowers it may be possible to remove pollen simply by shaking the flowers; H. Dobson (*personal communication*) uses this method on *Rosa* (Rosaceae) flowers.

Another way to manipulate resources is to move (or remove) whole plants. This kind of manipulation can provide insight into competitive interactions between plants for the services of pollinators. Campbell and Motten (1985) used both of these types of manipulations, removing flowers of one species from a mixture of two and creating desired combinations of flowers of both species by using potted plants. Pollinators, too, can be manipulated in a similar fashion. Inouye (1978) temporarily removed foraging bumblebees of one of two species in a meadow in order to look at the effect on resource partitioning by the bees. Individual bees were caught in small plastic cups and stored in a cool dark place for a day or two; mortality was minimal, and the bees were released and resumed foraging.

7. Pollen-Harvesting Rates

The rate at which pollen is harvested by flower visitors is an important component of their foraging behavior because it affects their efficiency in gathering food. It is also important from the plants' perspective, as it influences the dynamics of pollen transfer (although there may be no correlation between pollen removal and subsequent deposition; Wilson and Thomson 1991). Amounts of pollen removed by a flower visitor are usually measured by comparing amounts of pollen in anthers before and after visits. These data, in combination with measurements of time spent visiting a flower, provide information on rates of pollen harvesting. To estimate the amount of pollen in an anther before it is visited, you might be able to rely on a regression of number of pollen grains on undehisced anther length (see Chapter 4).

Strickler (1979) observed several species of bees foraging for pollen from *Echium vulgare* (Boraginaceae) and measured their efficiency in removing pollen. She estimated the average amount of pollen removed per anther by comparing dry weights (measured to ± 0.01 mg) of a sample of anthers from which pollen had been removed and a sample of unvisited anthers. She found that typically 61% to 84% of the available pollen was removed. Foraging rates were then calculated by dividing the mean weight of pollen collected per flower by the mean time required to visit a flower.

Snow and Roubik (1987) collected anthers from bagged, unvisited flowers and from flowers that had been visited just once. They stored the anthers dry for several months and then removed the pollen by vibrating the anthers with a 440 Hz tuning fork. The pollen was then weighed to the nearest 0.1 mg. They used a particle counter to determine the number of pollen grains per mg. Wilson and Thomson (1991) used a particle counter to count pollen grains after single or multiple visits to previously bagged flowers. To count pollen not removed, the androecium was put into a microcentrifuge tube, air-dried, preserved in 70% ethanol, and later sonicated for 30 minutes to remove the pollen grains. The sample was diluted with 1% NaCl to a volume of 200 mL, and subsamples were counted with a Coulter electronic particle counter. Schmid-Hempel and Speiser (1988) used a less quantitative approach, making a visual assessment of the pollen content of anthers (0, 10, 25, 50, 75, or 100% of maximum pollen content). They compared morning and afternoon censuses to calculate the percentage of available pollen that was removed for an average male flower (which they used as an estimate of male reproductive success).

Wolfe and Barrett (1989) collected anthers from flowers that had received a single visit from a bumblebee, and undehisced anthers from unvisited flowers. Because certain anthers were difficult to reach, some of the collections included part of the perianth tube. They placed the anthers in 70% ethanol and later subjected them to acetolysis, which digested the anther sacs and perianth tubes. Samples of the acetolyzed pollen were placed on a hemacytometer, and the pollen grains were counted with a compound microscope.

8. Nectar-Harvesting Rates

The amount of nectar collected by each lick of a nectarivore's tongue may also be of interest for some studies. This can be determined by observations of birds at feeders (measuring the number of licks and the amount of nectar collected), or estimated from preserved specimens. Paton and Collins (1989) preserved tongues of freshly killed honeyeaters in 4% formalin for 48 hours and then transferred them to 70% ethanol for permanent storage. To measure nectar loads, tongues were removed from alcohol, hydrated for 10 minutes in water, dried with tissue paper and weighed, and then dipped into a sucrose solution and reweighed. Kingsolver and Daniel (1983) developed a model that uses data on the role of tongue morphology as a mechanical determinant of nectar-feeding strategy in hummingbirds; it predicts the optimal sucrose concentrations for different nectar volumes, given optimal licking behavior.

Boggs (1988) measured the rate of nectar feeding in butterflies as one of the important components of nectar foraging. Modifying a method from May (1985), she used a 100-μL micropipette partially filled with sucrose solution as a feeder. The micropipette was placed horizontally inside a temperature-controlled box.

Another perspective on the morphology of nectar feeding is an engineering perspective on fluid feeding. Kingsolver and Daniel (1983) and Daniel et al. (1989) adopt this perspective to consider the mechanics of fluid flow through mouthparts of hummingbirds and butterflies. They use their mechanistic models to predict the optimal range of nectar concentrations (maximizing energy intake) for flowers visited by these pollinators. Their results might be useful in designing and interpreting experiments with artificial nectar solutions.

9. Pollen Deposition and Carryover

Pollinators vary in the efficiency with which they deposit pollen (i.e., the amount of pollen obtained versus amount deposited), and in the dynamics of pollen deposition (e.g., how much from one visit is deposited at the next flower). These aspects of pollination are important in determining the population genetics of plant populations, and are therefore of both practical and theoretical interest. Despite their significance, there have been very few studies to date of the dynamics of pollen transmission. For examples of techniques for studies of this topic, see Chapter 4, Section H.

10. Flower Constancy

Flower constancy, or the degree to which animal flower visitors restrict consecutive visits to a single species, is significant for studies of both flower visitors and the plants they visit. It is an important aspect of the foraging behavior of flower visitors, which can affect both the rate at which they discover new species as they come into flower and the efficiency with which they collect floral rewards. Flower constancy will also affect the efficiency with which pollen is transferred from one plant or flower to another, as well as stigma clogging and gene flow. The terminology and theory of this aspect of pollination biology are still somewhat unsettled. A variety of terms have been proposed to replace the general term "flower constancy," in an attempt to distinguish among different mechanisms that might produce the same behavior.

Heinrich (1976) introduced the qualitative terms "major" and "minor" in reference to the foraging specializations of bumblebees. Waddington (1983a) reviewed studies of this topic and presented reasons for using the more general term "floral-visitation-sequences" (F-V-S). He suggested using a transition probability matrix as a means of quantifying the behaviors represented by this term. Waser (1986) made distinctions among a "fixed preference," as may be seen in oligotropic or oligolectic solitary bees; a "labile preference," which probably reflects choices made on the basis of floral morphology and rewards; and constancy, which derives from intrinsic limitations in the nervous system of a pollinator. In this view, constancy shares some characteristics with fixed preferences but differs in that the specialization on particular flower species "should differ among conspecific individuals depending on the sequence of acceptable flowers each has encountered" (Waser 1986, p. 594). Lewis (1989, pp. 5–6), in considering "flower visit consistency" of a butterfly species, distinguished among fixed preference and two types of adaptable preference, including constancy ("when the insect continues to visit the flowers of the first species it encounters that is above [a reward] threshold") and labile preference ("a pattern of visits in which the insect visits more than one flower species, eventually confining its visits to the most rewarding"). Lewis described a model that suggests that constancy and labile preference may result from a common mechanism. Experimental work by Lewis (1986) and Waser (1986) suggests that both constancy and labile preference exist and may co-occur.

Array experiments are one way to investigate flower constancy. For example, Waser (1986) put flowers in vases in a regular array and observed bees foraging on them. He suggested using a constancy index proposed by Bateman (1951) for analyzing transitions between two types of flowers. The transitions are noted as:

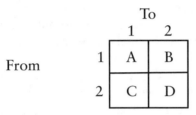

The index is:

$$CONS = [(AD)^{1/2} - (BC)^{1/2}] / [(AD)^{1/2} + (BC)^{1/2}]$$

and ranges from -1 (complete inconstancy) to 0 (random transitions) to +1 (complete constancy).

Lewis (1989) and Lewis and Lipani (1990) tested for visit consistency by comparing the likelihood that an inflorescence that a butterfly encounters (i.e., one on or near the flight path) would be visited, when the last inflorescence visited was of the same or a different species. To do this she used a logit analysis that "breaks down the odds of visiting a flower species into prior odds and odds multipliers due to history, species and the interaction of the two" (Lewis 1989, p. 3). This analysis, Waser's use of Bateman's index, and Waddington's suggestion of a transition matrix appear to be the only methods proposed so far for quantitative analysis of flower constancy, and their use has been quite limited to date.

The most direct method for determining flower constancy is to monitor individual flower visitors as they move from plant to plant, recording the species that they visit. For example, Lewis (1989) followed butterflies and marked each inflorescence they visited with a numbered flag. She later reconstructed the flight paths and noted the numbers and identities of flowers they visited (as well as those they did not visit that were within 10-cm of the flight path). She then used a logit analysis to test for effects of history, plant species, and their interaction on the odds of visiting an inflorescence. Thomson (1981b) points out that the method of following an individual flower visitor in the field, however, may confound constancy by choice with passive constancy that may result when an animal that might otherwise make random choices enters a monospecific patch of flowers. To circumvent this potential problem, and to facilitate data collection, Thomson presented bees with experimental bouquets of flowers in small vials of water at the end of a meter-long rod. The flowers could be presented to bees as they foraged naturally, generating large sample sizes in short order.

Direct observation may not be practical or possible in all cases, however, so indirect measures may also be required. One of these is to look at the pollen loads carried on the flower visitor's body, either as loose pollen or in a structure such as a corbicula where it is collected by the insect as it forages. Cane and Payne (1988) collected scopal pollen loads from bees, mounted them in gelatin with Alexander's stain, and then identified pollen along two perpendicular 100-grain transects. Alternatively, pollen in the insect's alimentary tract may also shed light on what flowers it has been visiting (see Section E.2 above for techniques). Problems with this method are that it may not reflect some species that are visited if the insect doesn't eat that type of pollen, or that it may not provide an accurate quantitative record of visitation. It is probably wise not to equate the presence of pollen in the alimentary tract with effectiveness as a pollinator; Leereveld et al. (1981) found pollen from reputedly anemophilous flowers on and inside syrphid flies but did not have any conclusive evidence that the flies were effective pollinators.

Cock (1978) discusses how to assess preference, reviewing eight methods that have been used in the literature and pointing out major drawbacks in some. Although his discussion is in the context of predator-prey relationships, it also has applications to flower choice by flower visitors.

11. Photographic Studies of Behavior

The speed with which some flower visitors work makes it difficult or impossible to observe in detail many aspects of their behavior. In order to study them it may be necessary to use either still or motion photography. Although still photography may suffice for some types of studies or for documenting visits to flowers, motion pictures, either on film or videotape, are much more useful. For instance, Macior (1964) used 16-mm motion-picture film to record the behavior of bumblebees foraging in a glass-enclosed flight cage. The bees were photographed at 32 and 64 frames per second, as well as by close-up stereophotography. These photographic records then allowed a careful analysis of the bees' behavior on the flowers. Video recorders with macro lenses would permit the same sort of analysis, but without the expense of film required for cinematography. In another study, Laverty (1980) used videotapes made with a macro lens in the field to analyze foraging behavior. The videotapes were played back at one-eighth of the recorded speed in order to analyze the images carefully. Harder (1983) used a video camera to record bees at artificial flowers, and then timed feeding behavior to the nearest 1/70 second by viewing the images in slow motion and timing with a stopwatch. He also used videotape to look at the grooming behavior of bumblebees after they visited flowers with varying amounts of pollen (Harder 1990). Macior (1968) combined photography with tape recordings of bees. The tape recordings were analyzed with an oscilloscope and used to determine the frequency of wing vibration (pollen is removed from the flowers they were visiting by vibration of the wings).

Photography also works well with avian flower visitors. Hainsworth (1973) and Schlamowitz et al. (1976) filmed birds (hummingbirds and sunbirds) at 18 frames per second at a feeder in order to count licks per second. A frame-by-frame analysis of the film facilitated careful measurements. Hainsworth and Wolf (1972b) filmed foraging by hummingbirds to determine foraging time per flower to the nearest 0.05 second. Waser and Price (1985a) also counted frames to determine handling times per flower for both hummingbirds and bumblebees.

If you do resort to still photography, use a lens that has a long enough working distance (between the surface of the lens and the subject) that you won't scare off the flower visitor. For most insects pollinators, a distance of about 30–40 cm is sufficient. You may also need to use either very fast film or an electronic flash (which doesn't seem to bother insects) to get good depth of field and a fast enough shutter speed to prevent blurring. A ring flash works well for macrophotography, as it provides an evenly distributed light source.

12. Manipulating Cues for Pollinators

In order to determine whether visual or scent cues are used by flower visitors, it is possible to manipulate them. Experiments like these may also make it possible to discover which cues are used for long-distance attraction versus short-distance orientation. The simplest experiments are probably those in which scent cues are isolated from visual cues. One way to do this is by placing glass chimneys over inflorescences. Knoll (1921–1926) used this technique to investigate the importance of color versus odor in the attraction of a bombyliid fly (*Bombylius fuliginosus*) and hummingbird hawkmoth (*Macroglossum stellatarum*) to flowers. If these chimneys are open at the top, then presumably insects following a scent cue would approach the inflorescences from above, whereas those using visual cues would approach them directly and run into the glass. Knoll (1925) used another technique to investigate foraging by a hawkmoth. He put flowers between two pieces of glass (10 x 15 cm, with a 2-mm gap) sitting in a groove in a wooden block. This separated the visual and olfactory cues from the flower. The moths tried to feed from these flowers and in the process left marks from their proboscides on the glass that showed where they were trying to find nectar.

Nilsson (1979) isolated visual and olfactory cues in an orchid species by covering inflorescences in plastic cylinders, which were supported by attaching them to the ground with a stiff wire. The cylinders were adjusted so that the lower edge was well below the height of the lowest flower. The upper end of each cylinder was covered with tape on its interior so that the sticky side faced out. This permitted Nilsson to trap insect visitors and to compare visitation to inflorescences with and without flowers (or floral odors). Pellmyr and Patt (1986) isolated visual and olfactory cues by putting leaves, a spathe, a spadix, or combinations of these inflorescence parts of *Lysichiton americanum* (Araceae) into petri dishes. Some dishes were sealed, providing only visual cues, and some had 100 small perforations allowing odors to pass through. The responses of the beetle pollinators indicated that olfactory cues initially attracted beetles and that odor then elicited a search behavior for the appropriate yellow color typical of the spathe. Pellmyr (1986b) was able to find odorless morphs of *Cimicifuga simplex* (Ranunculaceae), and by putting inflorescences in vases he could display them next to naturally fragrant flowers to compare the responses of pollinators. Because he was able to identify the two major components of the floral odor, he was also able to add these chemicals (isoeugenol and methyl anthranylate) to naturally scentless flowers and then compare their attractiveness against the naturally scented ones. Beaman et al. (1988) excluded visual cues by covering *Rafflesia* (Rafflesiaceae) flowers with cloth and then leaves, allowing a hole for scent cues, and excluded scent cues by wrapping flowers in polyethylene film or by filling the cup-shaped flowers with water. Scent cues could also be manipulated by adding artificial scents.

It is also possible to alter the visual appearance of flowers by mutilation, painting, or other alterations. Stuessey et al. (1986) picked the ray corollas from *Helianthus* (Asteraceae) flowers to investigate their adaptive significance. Many flowers will twist on their pedicels to regain their normal orientation if the flower stalk is bent, and this reorientation might facilitate experiments on the responses of pollinators to altered orientation. For example, if you bend a stalk over, wait a few days until the flowers have reoriented on the stalk, and then return the stalk to its original position the flowers would then be presented to visitors with this altered orientation (at least until they again change their orientation). Heinrich (1979b) used this technique for experiments with bumblebee foraging.

Alterations of flower color are also possible. Jones and Buchmann (1974) carried out manipulations of flowers in the field to test the behavioral responses of bees to ultraviolet cues. Their manipulations included twisting flowers 180° and 90°, removing anthers, gluing ultraviolet-reflective lateral petals over the normally absorbing banner (using Elmer's Glue-All), and covering reflective parts with a yellow felt-tip marker to make them absorbing. Waser and Price (1985a) painted flowers with acrylic paints to examine the effect of flower color on pollinator behavior. UV Killer (see Appendix 2; Hribar et al. 1991) might also work for manipulations of ultraviolet reflectance patterns, as might the ultraviolet brighteners that are used in most laundry detergents.

Although it does not appear to have been used for field studies, another technique for manipulating flower color is possible in anthocyanin-containing flowers. Anthocyanins tend to turn red in acid conditions and blue in alkaline conditions. When exposed to fumes of ammonia, some blue flowers (e.g., *Delphinium, Clematis* [Ranunculaceae], and pansies) will turn red, and some red flowers (e.g., geraniums and peonies) will turn blue-green. This color change is easily demonstrated with picked flowers put in a container with a dish of ammonia or glacial acetic acid, and it could probably be adapted for use on unpicked flowers by enclosing them in plastic bags with vials or wicks with the appropriate solution. It may take up to a couple of hours for the color change to be completed, posing the possibility of creating a whole range of color shades. Prolonged exposure of red geraniums to ammonia will result in a second color change to light yellow. Many white flowers will change color to yellow when exposed to ammonia vapors; this change is a consequence of the presence of flavones. (This information on color changes is from a class laboratory manual by Bastiaan J. D. Meeuse; see Meeuse 1982.) Red roses, purple irises, orange gladioli, and blue bachelor's buttons will turn white within minutes when exposed to sulphur dioxide gas, and in some of these the color is restored after the flower is dipped in phosphoric acid or hydrochloric acid (Nygaard 1992, cited in *Science News* 141(16):255). One way to accomplish this color change is to put sodium sulfite in a jar with vinegar, seal the jar, and put it in warm water for 15 minutes; these chemicals react to produce sulphur dioxide.

Scora (1964) produced flowers of *Monarda punctata* var. *arkansana* (Lamiaceae), which normally have spots on the lower lip that appear to serve as a nectar guide, without spots. He did this by inducing teratologies with colchicine. He experimented on the significance of the spots for pollinator behavior by putting out unspotted flowers either in a mixture with control flowers or in a monoculture. In a mixture, bees and wasps ignored the unspotted flowers, whereas in a monoculture they landed on the flowers but left without feeding. This kind of manipulation as well as mutations in flower color and shape may be used frequently in the future to investigate plant-pollinator interactions at the behavioral level.

It is also possible to use artificial flowers to investigate pollinators' responses to visual or olfactory cues. Lunau (1991) used this technique to identify the significance of stamens as an optical signal for bumblebees. With the artificial flowers, he could isolate the optical and olfactory cues provided by anthers and thereby demonstrated an innate releasing mechanism in bumblebees that responds to the optical signals of anthers.

L. Estimating Pollinator Population Size

Monitoring pollinator populations can be difficult. Although it is easy to measure visitation rates to flowers, measures of abundance independent of visits to plants are much more difficult to obtain. Perhaps for this reason, there do not appear to be many data available on long-term variation in pollinator populations. Hummingbirds are perhaps the best-studied avian pollinator because they are easy to catch and band. For example, Calder et al. (1983) and Inouye et al. (1991) have monitored hummingbird populations at the Rocky Mountain Biological Laboratory for almost two decades. However, the number of birds caught at feeders is an inverse function of the number of flowers available in the surrounding meadows (Inouye et al. 1991). When flower densities are low, the birds appear to be more dependent on feeders as a source of food. Pyke (1988) has monitored honeyeater abundance over a period of several years by counting the number of birds detected in an instantaneous scan of a circular plot of radius 20 m.

Some of the problems associated with counting flower visitors, given this variation in numbers of flowers, might be avoided by using a standardized array of flowers. For example, Ashman and Stanton (1991) put out a 2 x 2–array of inflorescences in water pics and monitored insect activity regularly throughout the flowering period of their study species. This method might also work for between-year comparisons, although one then encounters the problem of variation between years in the densities of other species of plants that may be competing for the same pollinators. Pettersson (1991) monitored the relative abundance of flower-visiting moths over two years by catching all visitors to a particular clump of plants every night between 11 P.M. and 12. Fenster (1991c) measured bee-flower ratios throughout a flowering season as an index of pollinator activity. During each of the five time periods constituting the flowering season, he counted the number of bees in permanent plots at 15-minute intervals and repeated this an average of 20 times.

If the pollinators in question can be detected by sound, a regular census of insects or birds heard along a standard route may suffice. For instance, Waser (1979) listened for the mechanical flight noise made by male Broad-tailed Hummingbirds flying nearby. A visual census may be required for insects that cannot be detected by listening. Poulson (1973) established plots of 2.2 m^2 in a field of flowers and made counts at intervals from 2 hours to twice daily. Ranta et al. (1981) walked a line transect at least twice each day to count bumblebees, or else conducted 10-minute counts in 5 x 5–m quadrats. Bowers (1986) censused bumblebees at least three times a week in randomly placed circular plots of 5-m radius; each count lasted 10 minutes. Pyke (1982b) walked a regular route up an altitudinal transect and counted bees he encountered. Tepedino (1981) estimated the size of a bee population at weekly intervals using the Manly-Parr mark-recapture method; bees captured on flowers were marked with paint. This type of study should provide the most reliable data, but it may still be dependent on the abundance of flowers where the insects are caught. For additional information on mark-recapture methods, see Southwood (1978).

Some of the other techniques discussed under the section on collecting insects may provide good alternatives to counts on flowers. For example, data on fly abundance obtained from Malaise traps parallel trends in data on visitation rates of flies to flowers at the Rocky Mountain Biological Laboratory (Kearns 1990). Kuno (1991) reviews some aspects of sampling and analyzing insect population data.

Monitoring population sizes and activity at a single beehive or nest is much simpler. If the bees are in a hive or nest box, it is usually easy to open it to observe the bees inside. If this is not desirable, it is also possible to count the bees as they enter or leave the hive. Gary (1967) describes a model for beehives: a cone made of 8-mesh galvanized wire cloth that is large enough to cover the hive entrance so that bees have to fly through a 1.5 x 10–cm opening. This facilitates visual counts of bees leaving the hive. He restricted use of the cone to 30-second observation periods so as not to upset the flight behavior of incoming foragers. This technique might work well for ground-nesting bees as well, to slow down their arrivals and departures to facilitate observations.

Spangler (1969) describes a simple photoelectric counter for use with small honeybee colonies, which could also be used for other types of counting or monitoring tasks. Liu et al. (1990) describe a similar but computer-based system for counting bees as they move in or out of a hive. It is based on infrared light-emitting diodes (LEDs) that shine onto phototransistors; bees walking through the infrared beam cause an electrical pulse in the phototransistor that can then be counted by the computer. The same system also measured temperature inside and outside of the hive, solar radiation, relative humidity, wind speed, and the weight of the hive. S. Boyd (*personal communication*) developed a technique for bumblebees in a nest box. She added a short section of clear plastic tubing to the entrance hole in the nest box and created a landing platform by cutting off the top half of a short section of the tube on the end farthest from the nest box. By forcing bees to enter or leave the colony through this tubing, you can slow them down sufficiently to facilitate recognition of tagged or marked individuals.

M. Trap-Nesting Pollinators in the Field

Some insects, including both social bees like bumblebees and solitary species such as megachilid (leaf cutter) bees, can be induced to nest in artificial nest boxes, holes drilled in blocks of wood, or in specially prepared plots of ground. Access to nests such as these can facilitate collection of nectar or pollen samples, marking of foragers, or just provide an interesting way to observe pollinators. If you succeed in getting bees to use the nests you provide, you may also be able to maintain them in captivity (see the next section).

Several studies were conducted in the 1960s to determine whether it was feasible to domesticate bumblebees by inducing them to nest in boxes (e.g., Hobbs et al. 1960, Hobbs 1967). Although the efforts to manage bumblebees like honeybees were not successful, much was learned about the best way to get them to nest in boxes. The basic technique is to build a wooden box (e.g., a 15-cm cube with an overhanging hinged roof), line it with upholsterer's cotton (available from upholstery shops), and put them out in the spring as the queens emerge from hibernation. Richards (1978, 1987) reviews the various techniques that can be used to entice species that normally nest underground. About 25% to 30% of boxes are colonized (Inouye, *personal observation;* Richards 1987, Gamboa et al. 1987).

Solitary species can also be induced to nest in trap nests. Maeta et al. (1985) describe several different kinds of observation nests used in laboratory studies of a xylocopine bee. One involved putting a length of pith in a glass tube, which was covered with black paper except during observations. Much of the work on trap-nesting of bees has been done with species that are important as pollinators of crops, for example *Osmia* (Megachilidae) pollinators of alfalfa. Torchio (1972, 1976, 1982a, 1982b, 1984a, 1984b, 1989a, 1989b) discusses a variety of possible techniques. The observation box he describes in his 1972 paper has attracted at least 20 species of bees and greatly facilitated our understanding of their biology (P. Torchio, *personal communication*). Most of these techniques involve drilling holes into blocks of wood or styrofoam and then lining the holes with paper straws or glass tubes. Another technique for creating holes is to cut grooves in flat boards and then to stack boards to close the open sides of the grooves. Fricke (1991) describes a method of constructing "pre-split" nests that facilitate subsequent observation of the contents. He cut wood blocks into two sections, routed a drill-guide channel in one section, bound the two sections back together with masking tape, and then drilled the desired hole size and depth. The nests were then dipped in paraffin to make a watertight seal between the two pieces of the block and redrilled to remove paraffin.

The phenology of insects that can be induced to nest in artificial domiciles could be manipulated for experimental studies. Torchio (1981) stored cocoons of solitary bees (*Osmia*) at 4°C, and later incubated them at 27°C for several days until the adults emerged. Bumblebee colonies could also be manipulated by temperature to experiment with the effects of forager size (because workers early in colony development are smaller than those produced later), for example.

N. Maintaining Pollinators in Captivity

Many experimental pollination studies are better conducted in the laboratory, or perhaps in flight cages, so that experimental variables can be more easily controlled. If it is desirable to use the same individual over an extended period of time, you may have to figure out how to do so responsibly so that the animal remains in good health.

Honeybee and bumblebee colonies can be collected or purchased (see Appendix 2), and they are easy to maintain in captivity. They can either be fed sugar, nectar, or a sugar solution and given access to pollen, or they can be allowed to forage outside. Bumblebee queens can be captured in the spring, after having emerged following hibernation, or in the fall after mating and prior to hibernation. Because they do not spend much time foraging between mating and hibernation, it is much easier to find them in the spring while they are searching for nest sites. The queens can be kept in the lab and will rear colonies there (Plowright and Jay 1966), which can then be used for studies of foraging.

A successful design described by Plowright and Jay (1966) is a small wooden box with an annex for nesting (Figure 7–28). The larger box (17.5 x 12.5 x 10 cm internal dimensions) contains a feeding tube, has screened ventilation holes, is lined with corrugated cardboard, and has a glass roof. The smaller chamber (7.5 x 7.5 x 5 cm internal dimensions) is lined with upholsterer's cotton (available from upholstery stores) and is supplied with a lump (0.75 g) of fresh pollen. This design permits observation of the queen when she is not brooding and, if the roof of the nesting annex is removed, monitoring of the brood mass or larval bees. See Plowright and Jay (1966) for additional details. Free and Butler's (1959) book on bumblebees has information on the collection and study of bumblebee colonies and on starting colonies in captivity, and Holm 1966 has a section on propagating natural populations (by putting out domiciles). This work is also reviewed by Morse (1982b). Heinrich (1979c), Laverty (1980), and Mjelde (1983) describe work conducted with captive colonies of bumblebees.

Bumblebees are also reared commercially, primarily for pollination of tomatoes in greenhouse cultivation. Heemert et al. (1990) describe their technique, which begins with field-collected queens.

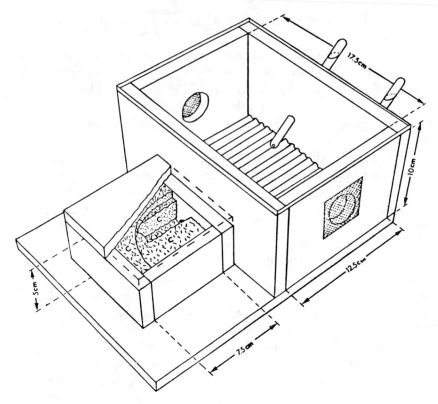

Figure 7–28. A bumblebee nest box. From Plowright and Jay 1966.

If it is necessary to break diapause in the queens, the technique of Röseler (1985) is used (CO_2-narcosis). The queens are put individually into small wooden cages at 29°C and 50% to 60% relative humidity in the dark. Every 2–3 days sugar-water (50%) and pollen are put into the box, and four to five young honeybees are added to assist the queen (it is not yet known how they assist the queen). When the box (or subsequent colony) is handled, a red light source (invisible to bees) is used. When the first bumblebee workers emerge, the colony is transferred to a bigger box with two compartments. The big compartment, where the bees nest, is 19 x 19 x 17 cm. The small compartment (19 x 9.5 x 17 cm) serves as an area where the bees can defecate. Ventilation is provided by making a number of holes covered with metal gauze. Heemert et al. (1990) provide additional information about maintaining continuous cultures of bumblebees.

Solitary bees can also be maintained in observation nests. Michener and Brothers (1971) describe an observation nest for use with burrowing bees, such as halictines. It consists of two pieces of glass, separated by a layer of fine soil. Burrows in these nests are visible from either side of the nest.

Free (1970a) describes the history of use of blowflies (Calliphoridae) for pollination. The incentive for this work was that these flies were found to be more suitable than honeybees for cross-pollination of crops such as *Allium cepa* (onion; Liliaceae), *Daucus carota* (carrot; Apiaceae), and *Brassica oleracea* (brussels sprouts, kale, and cabbage; Brassicaceae), particularly in plant breeding studies. Blowflies can be trapped or reared. Adult flies can be trapped using animal refuse such as fish heads or road kills, placed under a screen cage. Free describes a couple of methods of rearing flies, using resources such as cattle lungs or moist dog biscuits. Blowflies are apparently the only type of Diptera that has been reared successfully in large numbers for pollination studies.

Lepidoptera can be reared and maintained in the laboratory as well. The principal problem is finding an appropriate larval diet for the species you want to use. If you can work with a species for which a diet has already been determined, you may save yourself a lot of trouble. Some species and their diets are available commercially (see Appendix 2). In some cases it may be easiest to grow the host plant (or a closely related agricultural plant if one is available), whereas in other cases you may be able to mix an artificial diet. If you want to maintain a continuous culture, you may also have to find the best way to get individuals to mate in the laboratory. Often males must be 2–10 days old, whereas females must be just emerged. If they won't mate in the laboratory, try putting them in a cage outdoors (W. Watt, *personal communication*).

Avian pollinators can be maintained in captivity, but a permit may be required (as it is in the United States). Wooller et al. (1988) kept honeyeaters in an outdoor aviary and fed them a mixture of honey, water, and milk-based invalid health food, provided in bird feeders fixed to perches. (S. Hiebert, personal communication, suggests that it is probably not a good idea to feed any bird milk products.) Several researchers have maintained hummingbirds in captivity. Stiles (1976) housed them in holding cages 50 x 50 x 100 cm in size. They were given sugar water in feeders, a protein-vitamin mixture once a week (Lasiewski 1962), and access to *Drosophila* daily. Montgomerie et al. (1984), del Rio (1990), and Tiebout (1991) used similar methods. Scheithauer (1967) also reviews methods of feeding captive hummingbirds. S. Hiebert (*personal communication*) suggests that older avicultural work with hummingbirds has been superseded by better techniques. Commercially prepared diets are now available for hummingbirds; to prepare them all you have to do is add water. The major danger with this food source is that protein components spoil rapidly, so you should change food frequently (at least daily), and use sterile techniques when preparing it (e.g., soak and wash feeders in a solution of chlorine bleach, 5 ml/L of water, for at least an hour). Hiebert also recommends that captive hummingbirds not be given sugar water as a regular food item and that molting birds be provided with fruit flies. She found that Perky Pet brand feeders (models 214 or 215) worked particularly well for her captive birds. See the entry on hummingbirds in Appendix 2 for sources of more information about maintaining hummingbirds.

O. Methods for Nocturnal Pollination Studies

Nocturnal pollination is probably restricted to three groups of animals: moths, bats, and marsupials (e.g., Turner 1982). The normal difficulties of working with pollinators are compounded by the lack of light for making observations, and methods specific to nocturnal studies all relate to trying to compensate for this lack of light. Alternatives include using a light source that the animals can't see (e.g., infrared) or amplifying the available light. A couple of companies (see Appendix 2) manufacture infrared viewing systems.

The best (but most expensive) way to compensate for the absence of light is to work with a night-vision scope, which uses electronic light amplification. Even weak moonlight can provide sufficient light for an image intensifier to provide a reasonable view of flowers or pollinators, but a flashlight beam can help to ensure that sufficient light is available. Farnsworth and Cox (1988) describe how to make a low-powered laser illuminator designed to provide supplementary light for night-vision devices. Monocular or binocular models of night-vision scopes are available, and a variety of lenses can be used to provide different magnifications. DWI had limited success using a night-vision scope to observe foraging sphinx moths on *Ipomopsis* (Polemoniaceae); the moths are such swift flyers that it is difficult to keep them in view. A flashlight helped to locate moths by the reflecting eye-shine. Dafni et al. (1987) used an Intrametrix IRTV 445 Forward-looking infrared device for nocturnal observations of hawkmoths.

Although some studies of moths indicate that they may have tetrachromatic vision, with receptors in the range of red light, it may still be appropriate to use red lights to observe them at night. Sheck and Inouye (unpublished) used flashlights covered by red photographic filters to make observations of moths foraging at night on *Frasera speciosa* (Gentianaceae) at the Rocky Mountain Biological Laboratory. Willson and Bertin (1979) made night observations with lanterns covered with red cellophane. Erhardt (1991b) used a strong flashlight to observe moths on flowers but found that he had to avoid shining the full beam of light on them or they would fly off; the weaker light from the edge of the beam did not deter them. Such observations of moths on flowers often yield species that rarely come to light traps.

Goldingay (1990) studied marsupial yellow-bellied gliders (*Petaurus australis*) in Australia. This arboreal nocturnal species feeds extensively on nectar. He observed them at night with a 100-W 12-V spotlight and a pair of binoculars. (Portable spotlights like this that plug into the cigarette lighter of an automobile are available at most automobile parts stores.) The gliders became habituated to the spotlight and could be observed without apparent disruption to normal behavior, but they were also observed at times with a 55-W red spotlight.

If you need to mark plants or a path in order to find them at night, try using reflective tacks (for putting on trees) or stick-on reflectors (see Appendix 2), which will reflect light from a flashlight. Stick-on reflective material could be used for tagging plants as described in Chapter 2, Section B, and would have the advantage of being visible both night and day.

If you are simply trying to confirm the activity of flower visitors at night, you may be able to use photographic methods. Goldingay et al. (1991) used this technique to record nonflying mammals visiting inflorescences. An infrared triggering mechanism in combination with a camera and an electronic flash would be a relatively inexpensive way to accomplish this. Another method to confirm nocturnal pollination is an experiment with bags, covering some flowers only at night and others only during the day. Goldingay et al. also used this technique to confirm the significance of nocturnal pollination for *Banksia* species.

A recent book on ecological and behavioral methods for the study of bats (Kunz 1988) summarizes a variety of techniques that might be useful for studies of bat pollination, including techniques for capturing, marking, observing, surveying, detecting, maintaining, and photographing them.

P. Suggestions for Planning Studies

1. Potential for both nocturnal and diurnal pollination. Some flowers may be visited by both diurnal and nocturnal pollinators (e.g. Bertin and Willson 1980, Jennersten 1988), which may differ in their effects as pollinators (e.g., Jennersten and Morse 1991). Observations at several times throughout the day and night might confirm this, or a nonobservational technique such as bagging some flowers during the day and others at night might also suffice.

2. Effects of body size of worker bumblebees on flower visitation. Visitation patterns of worker bumblebees, which are probably more variable in size within a species than most flower visitors, can be affected by body size. Smaller individuals may visit different species of flowers than do larger individuals (e.g., Morse 1978, Inouye 1980b). The time required for a bee to visit a flower can also be affected by its body weight and glossa length (Harder 1983).

3. Effects of competitors on a pollinator's resource utilization. Experimental removals of one of the species of bumblebees foraging in an area (e.g., Morse 1977b, 1982c; Inouye 1978) have shown that the remaining species may alter their behavior. If you are interested in patterns of resource utilization, from either the pollinator or the flower's point of view, remember that they may vary in time or space depending on the composition of the pollinator or plant community.

4. Seasonal changes in pollinators and seed set. If a plant species is in bloom for long enough, the pollinator fauna may change (e.g., Wolfe and Barrett 1988, Ashman and Stanton 1991). Or the other species of plants in bloom in the area may change, affecting visitation patterns even if the pollinator fauna doesn't change (e.g., Jennersten et al. 1988). Either of these kinds of changes may result in seasonal differences in seed set (e.g., Jennersten et al. 1988). Schmitt (1983a) found that individuals flowering for the first time during the middle of the flowering period had a greater probability of setting seed, and a stronger relationship between seed number and flower number, than plants flowering for the first time early or late in the season.

5. Differences among pollinator types. A wide range of pollinators may visit a single plant species (e.g., Alcorn et al. 1961). Different pollinator species visiting the same plant species may, because of variation in their size and foraging behavior, remove and deposit different amounts of pollen (e.g., review in Thomson and Thomson 1992, Ashman and Stanton 1991, Kearns 1990). This variation means that pollinators can differ in their effects on neighborhood size (e.g., Schmitt 1980). It may also be important to know whether the flower visitors you are watching are traplining or area-restricted. If they are, and you are watching a single plant or small area, you may be collecting data from only one or a few animals, whose behavior might be idiosyncratic. Marking flower visitors so you can recognize them can help avoid this potential problem. Even within a species, foragers collecting nectar versus pollen may either collect or deposit different amounts of pollen (e.g., Wilson and Thomson 1991).

6. Effects of microclimate on flower visitation. Although some flower visitors (e.g., birds, mammals, some insects) are homeothermic or capable of endothermic temperature regulation, others (e.g., many Diptera, solitary bees) are dependent on solar radiation to achieve the body temperatures required for flight. For such species, microclimates have an important effect, and in shady areas there can be an important effect of sun flecks on visitation rates (e.g., Beattie 1971c). Careful observation and measurement may be required to elucidate the effects of microclimate on visitation.

7. Differences in pollinators in different habitats. If a flower species has a wide distribution, encompassing different habitats, the pollinators that visit it may differ in each habitat. This may occur most commonly regionally or along elevational gradients, but sunny versus shady or disturbed versus unperturbed habitats may also exhibit such differences. For example, Pellmyr (1986b) found three different pollinator guilds along the altitudinal range of an herbaceous species of Ranunculaceae in Japan.

8. Effects of marking flowers or flower visitors. Some pollinators are attracted to particular colors. For example, hummingbirds will fly up to red flags to investigate them. If there is a chance of such an effect in your study, don't use colors to differentiate treatments within an experiment. There is also the potential for some effect of marking flower visitors, although this does not seem to have been demonstrated yet.

9. Effects of magnet species on visitation rates. Species with flowers that are highly attractive to pollinators (magnet species; Laverty 1992) may produce increased visitation rates to co-flowering species that normally are rarely visited. This may produce unusual differences in visitation rates between populations, depending on whether magnet species are present or absent.

8. Environmental Measurements for Pollination Studies

Both plants and pollinators are affected by environmental variables, and for many types of studies it may be necessary to quantify one or more parameters. Scale, both temporal and spatial, is an important issue in this regard. In some cases the data available from the nearest weather station (e.g., an airport, an official U.S. weather station, or field station) may suffice, if only a general description of the climate is necessary. At the other extreme, it may be necessary to use micrometeorological techniques in order to monitor the microclimate inside an individual flower, as this can influence the characteristics of nectar (e.g., Corbet and Delfosse 1984). Most pollination studies, however, will probably require some intermediate scale of resolution, such as in the general vicinity of the flowers.

Richard Primack (Department of Biology, Boston University, 5 Cummington St., Boston, MA 02215) has initiated a small-scale international effort to standardize collection of meteorological data in conjunction with data on visitation rates by flower visitors. This method has now been used to quantify the effects of environmental variables on visitation in several different habitats (Arroyo et al. 1985, Inouye and Pyke 1988, Berry and Calvo 1989, Kearns 1990, McCall and Primack 1992). The instrumentation is relatively simple and not prohibitively expensive. It includes a hand-held sling psychrometer to measure ambient temperature and relative humidity (calculated from the difference in temperature between wet- and dry-bulb thermometers), an anemometer for measuring wind speed, and a photographic light meter for measuring light levels. These instruments are read before or after every 10-minute observation period during which insect visits are counted to a known number of flowers. The data on numbers of visits per flower per 10 minutes (see Chapter 7, Section K.1, for more information on quantifying visitation rates) can then be correlated with the meteorological data to determine the significance of temperature, wind speed, and light levels for visitation.

One alternative to use of these hand-held instruments is an automated system involving portable sensors, which can be set up next to a patch of flowers while it is being observed. We have used a Campbell Scientific data logger with sensors for light levels and temperature, in conjunction with a hand-held anemometer and sling psychrometer, but all of these measurements could also be made by appropriate sensors attached to a data logger. Advantages of this system are that more frequent measurements can be made if they are required (sampling frequency can be set as high as every several seconds), and the data can be transferred directly to a computer for analysis.

Unwin (1980) and Unwin and Corbet (1991) discuss a variety of aspects of microclimate, with particular reference to how it affects plant-animal interactions. They include a variety of techniques for how best to measure it, and if microclimate is an important aspect of your study you should refer to these publications.

A. Air Temperature

Air temperature can affect pollination studies via its effects on both plants and animals. Flower development and opening, nectar secretion, anther dehiscence, and seed development are all likely to be dependent on ambient temperature. Similarly, air temperature affects the activity of flower-visiting insects. For example, del Rio and Búrquez (1986) found that for pollination of *Mirabilis jalapa* L. (Nyctaginaceae) to occur, ambient temperature had to be above 13°C; below that temperature the hawkmoth pollinators didn't fly. Between 13° and 21°C (air temperature measured at dusk) pollination efficiency increased linearly. Pollination was therefore mostly autogamous on cool nights, and allogamous on warm nights. Visitation rates, if not mating systems, are probably always affected to some degree by air temperature, either through its effect on the plants, on the animals, or their interaction.

The sling psychrometer mentioned above may suffice for measuring air temperature and has the advantage of also providing information about relative humidity (see Section D). The next level of sophistication, and cost, is an electronic thermometer. A thermocouple or thermistor sensor attached to a meter calibrated to read it is an obvious possibility. If the sensor is very small, you may get numerous fluctuations in the meter readings, reflecting microclimatic differences. Advantages of electronic sensors include that they are less likely to break than glass thermometers, they give almost instantaneous readings, and they can be attached to a recorder.

B. Flower Temperature

Meeuse (1973) used an interesting technique to measure flower temperature in heat-producing flowers. He painted a thin film of "liquid crystals" on the flowers and then used the color changes in the film to determine the temperature distribution in the flowers. Unfortunately, this technique only worked well on dark flowers or flower parts, as the color changes were difficult to observe on light backgrounds. The recipe for the liquid crystal solution is:

Cholesteryl proprionate, 1.2 g
Cholesteryl decanoate, 1.5 g
Cholesteryl oleate, 7.3 g

Heat the materials together on a 70°C waterbath. This provides a "basic mixture" that can be modified to manipulate the temperature range over which the solution will change color. The basic mixture changes from copper red to yellow to metallic green to peacock blue between 41.2° and 42.2°C. Adding more cholesteryl proprionate lowers the indicator range; an extra 1.10 g changes the range to 25.9°–29.2°C, which is more appropriate for most flowers.

C. Solar Radiation

Solar radiation is important for pollination studies because it affects both plants and pollinators (as well as pollination biologists).

It affects plants generally because of its consequences for photosynthesis, temperature, and water relations, and specifically with regard to pollination biology because of the consequences for nectar production (e.g., Pleasants 1983). It is no less important for flower-visiting animals, as many insects are dependent on solar radiation to achieve the temperatures required for activity and flight.

There are many ways to measure solar radiation. The appropriate method will be determined by your budget, your requirements for accuracy, and whether you need continuous data or occasional readings. If a constant record is needed — for example, to quantify day length or the number of hours of sun or shade — a mechanical pyranometer (solarimeter) with a chart recording may suffice. This instrument measures sunlight by the difference in temperature between black and white plates, covered with a hemispherical glass dome. Some models are mechanical in nature whereas others (probably more accurate) rely on thermocouples to measure the temperature difference. A more sophisticated solution is an electronic instrument to measure solar irradiance (in $MJ \cdot m^{-2} \cdot d^{-1}$) and quantum flux (in μm) (e.g., Lambda Instruments model LI-185A light meter equipped with pyranometric and quantum sensors; Young and Smith 1980). Or a meter that measures only photosynthetically active radiation (PAR; usually measured in Photosynthetic Photon Flux Density, or the number of photons of wavelength 400–700 nm incident per second per m^2) may be appropriate for some studies.

Perhaps the least expensive solution for instantaneous measurement of light levels is use of a light meter, which is in effect a solarimeter with a spectral response limited to the range visible to humans. For instance, photographic light meters, which use a photocell to measure light levels, can be very sensitive and provide sufficient information about illumination. The sensor Primack recommends for use in quantifying environmental parameters while recording visitation rates of pollinators to flowers is a Seiko photographic light meter (Model L-248, Multi-lumi), although any light meter that can provide readings in the appropriate units of luminance will work. Unwin (1980) includes tables that convert information from reflected light readings taken with a photographic light meter into values of luminance (candela per square meter or square foot, or foot-lamberts), and levels of incident light readings into values of lux (lm/m^2), milliphots, or foot-candles. The primary disadvantage of photographic light meters is that they are not measuring the same wavelengths of light that insects see or that are used for photosynthesis by plants. Thus they are inappropriate for some types of studies.

If you can be satisfied with a relative index of light availability, for example for a comparison of sites, the Diazo paper technique may suffice. This technique was described by Friend (1961), and modified by Lubbers and Christensen (1986), who used it as a measure of light penetration through a forest canopy to the under-story plant they studied. The Diazo paper reacts to ultraviolet radiation, and because the proportion of ultraviolet light absorbed by leaves is similar to the absorption of photosynthetically active radiation, this method provides an index of percent light penetration (Lubbers and Christensen 1986).

The method is relatively simple. Make a 1-inch square stack of Diazo paper (you need about 15 sheets) and top it with a cardboard or black paper square with a few holes punched in it (with a hole punch for notebook paper). Do this work in a dimly lit area to reduce exposure of the paper to light. Insert the stack into a clear plastic bag (e.g., coin envelope) or plastic petri dish and keep it in the dark until you want to begin measuring light. For example, put an envelope out at the corner of a quadrat at dusk and then collect it 24 hours later. After collection, remove the stack from the envelope and expose it to an ammonia-saturated atmosphere for 10–20 minutes. Diazo paper is sensitive to ultraviolet wavelengths and after exposure to ammonia the layers of paper through which UV has penetrated will turn white, whereas the rest of the paper turns blue. The more light there is, the greater the number of layers will turn white. Under full sunlight during the summer, Friend (1961) found that 11–12 pages in the stack would show bleaching. The average (for holes within stacks) number of white layers per stack can be compared among stacks (or sites).

D. Relative Humidity

Relative humidity is most likely to influence pollination through its effects on plants; as long as nectar or other water sources are available, humidity by itself is not likely to have an impact on the activity of flower-visiting animals. Humidity affects nectar concentration (Corbet et al. 1979) and pollen presentation via anther dehiscence. Although there are few data available, in habitats with relatively low humidity there is probably a steep gradient of relative humidity between the nectary of a plant and the ambient environment (e.g., Corbet and Delfosse 1984).

The least expensive way to measure ambient relative humidity is with a sling psychrometer. This instrument consists of a pair of thermometers, one of which has a wet wick on its bulb and the other of which measures the air temperature. The psychrometer is swung around to create evaporative cooling with the wet wick, and the relative humidity is calculated from the difference (wet-bulb depression) in temperature between the wet- and dry-bulb thermometers. The accuracy of this technique depends on the reliability of the thermometers and the success of the evaporative cooling; if the thermometers are not swung around for long enough, the potential degree of evaporative cooling will not be achieved and the relative humidity will be overestimated. This potential shortcoming may be overcome by using mechanical psychrometers with fans to create the evaporative cooling.

Unwin (1980) describes the construction of a miniature electric psychrometer, which uses small thermocouples housed in pieces of polyethylene tubing and read by a thermocouple indicator as temperature sensors. He gives details for construction of an indicator designed for use with this psychrometer. The advantages of this psychrometer are that it requires sampling only a small volume of air in order to determine the humidity, and it has no moving parts.

Humidity can also be measured electrically; this is the technique used for automated measurements. The best of these methods is probably the capacitance hygrometer, which measures the change in capacitance caused by absorption of water into a dielectric compound (usually aluminum oxide). Relatively inexpensive digital electronic meters that measure humidity, often in combination with temperature, are available (see Appendix 2).

None of the techniques described above will work for measuring the humidity inside a flower because the thermometers or sensors are too large to fit in the flowers. The only technique that seems possible on this scale is described by Unwin (1980) and was used by Corbet and Delfosse (1984). It depends on the fact that the concentration of an aqueous solution in equilibrium with air is related to the relative humidity. Thus very small droplets of a suitable solution can be exposed to the air in a flower and allowed to equilibrate there before their concentration is determined with a refractometer. Unwin (1980) recommends two different solutions for different ranges of humidities: sucrose for humidities over 90% and potassium acetate ($KC_2H_3O_2$) for the range 40% to 100%. A sampling device for holding the solutions can be made with epoxy-coated copper wire (size 00, or 0.4 mm diameter, 36 swg), which is bent to form small loops. Unwin (1980) describes a jig for making the loops. To make the jig, insert insect pins (size 00, 0.4 mm diameter) into a piece of wood at 5-mm intervals and cut them off at a height of about 5 mm. Then wind a length of wire around the sequence of pins, forming a series of loops in the wire. Alternatively, a strip of nylon mesh could be used, with solution placed into the holes in the mesh (e.g., insect netting).

To measure the humidity in a flower, put a drop of solution in each loop, insert the wire with its series of loops into the flower, and let the solution equilibrate for about 30 minutes. Then remove the wire and measure the refractive index of each drop. The equilibration time will depend on how closely you have matched the solution you started with to the equilibrium solution. A couple of trial runs will help you determine the appropriate starting solution. If you are too far off from the equilibrium concentration, your droplet may either evaporate completely or get so big that it falls off the loop (Corbet, *personal communication*). Corbet and Willmer (1981) used the potassium acetate method to measure relative humidity inside two species of tropical ornithophilous flowers. Although small droplets typically equilibrate in less than half an hour in free air, Corbet and Willmer found that they took much longer (e.g., 3–7 hours) in the corolla tube of one of the species. The range of relative humidity measurements was about 66% to 98% over the length of the flowers. Droplets in the middle region of the corolla gained water more slowly than those at either end, suggesting that the nectar in the base of the corolla and ambient air at its mouth were providing sources of water vapor. Evaporation from the surface of the nectar pool was probably responsible for the concentration differences (<1.5% as sucrose) they found at different levels of the nectar.

Table 8–1. Constant Humidity with Sulfuric Acid Solutions

Density of acid solution	%	Relative H_2SO_4 Humidity
1.00	0	100.0
1.10	15	93.9
1.20	28	80.5
1.25	34	70.4
1.30	40	58.3
1.35	45	47.2
1.40	51	37.1
1.50	60	18.8
1.60	69	8.5
1.70	78	3.2

Source: Weast 1972 (page E-40). Reprinted by permission of CRC Press.

Unwin and Corbet (1991) present a figure showing the approximate relationship between concentration of sucrose and potassium acetate solutions and relative humidity. However, Corbet (*personal communication*) reports that another researcher found a big discrepancy between their curve and hers, and suggests that researchers determine their own calibrations. This can be done by putting the droplets in a well-ventilated area with a constant humidity and checking the equilibrium concentrations against the results of a sling psychrometer.

For some experiments it may also be desirable to maintain a constant relative humidity. Environmental chambers make this possible on a relatively large spatial scale, but for small chambers (e.g., a petri dish or other small-volume container) another technique may be appropriate. It is possible to maintain a constant humidity by including a sulfuric acid solution of a particular density inside the chamber. Table 8–1 lists the density (D_4^{20}, kg/L) of the acid solution required to maintain a particular relative humidity.

E. Wind Speed

Wind speed can have an effect on visitation rates by pollinators to flowers, with extreme wind speeds preventing any visitation. It can be quantified in different ways depending on the temporal and spatial scale required. The most traditional means of quantifying wind speed is an anemometer, typically with three cups to catch the wind and turn a shaft. Digital and analogue hand-held models are available (see Appendix 2), as are models that can be attached to a data logger. Vane-type anemometers are also available for field use. The most sensitive sensors, and the ones most suitable for microscale measurements, are hot-wire anemometers. They rely on the rate of cooling of a heated wire or thermistor to quantify wind speed, and are much smaller and more fragile than mechanical anemometers. Young and Smith (1980) used a Thermonetics model HWA-103 hot-wire anemometer to measure wind speed. In either case, placement of the anemometer at the same height as the flowers being studied will provide the most relevant data. Grace (1989) gives a thorough discussion of the techniques and measurements of wind speed for plant physiological ecology.

F. Soil Moisture

There are only a few quantitative or experimental studies of the effects of soil moisture on pollination biology, which could include effects on flowering, nectar production, or seed production. Precipitation (probably through its effect on soil moisture) is the proximal cue for flowering for both some tropical (e.g., Augspurger 1980) and temperate (e.g., desert) species. Flowering in these species can be manipulated to examine the consequences of flowering in or out of synchrony with conspecifics (e.g., Augspurger 1980). Watering during a dry year may also prolong the flowering period of some species. If the species you are working with preforms flower buds, the response (if any) to increased soil moisture may not be immediate. For *Frasera speciosa* (Gentianaceae) in the Colorado Rocky Mountains, there is a two-year lag between the time of watering and the time flowers appear (Inouye, unpublished). For these and other reasons you may want to monitor soil moisture.

Soil moisture can be measured in several ways. The simplest is probably gravimetric: dig up a sample, weigh it, dry it (e.g., in an oven for 24 hours), and reweigh it. The difference in weight can be attributed to the water the sample originally contained. The disadvantages of this method are that it is destructive, at least of the sampling site, and that water removed by drying the soil may not reflect accurately the amount of water available to plants. Another common method of measuring soil moisture is electrically with buried gypsum blocks containing a pair of electrodes. The blocks are buried at the desired depth in the soil. The resistance between the electrodes (measured with a meter) is inversely proportional to the amount of moisture in the gypsum block, which is at equilibrium with the soil. These blocks are relatively inexpensive (see Appendix 2), but may only last a few field seasons. The meters used for measuring resistance in the blocks are often calibrated in percentage available moisture, which may suffice if you only require a relative index, but these units don't mean much to a plant physiologist (or perhaps the plant either).

A more sophisticated method for measuring soil moisture is with a tensiometer, a water-filled tube with a porous tip on the lower end and a sealed vacuum gauge on the upper end. It is installed vertically in the soil with the tip at the depth you wish to monitor. As soil moisture is depleted by evaporation, by plants, or by drainage, water is sucked from the tensiometer, creating a vacuum, which is measured by the gauge. This vacuum is reduced when precipitation adds water to the soil and water is drawn back into the tube through the porous tip. A high-tech alternative is a neutron probe, which measures the density of hydrogen nuclei in the soil. The most sophisticated method is probably time-domain reflectometry, which measures the propagation time of a waveform traveling over a fixed-length probe buried in the soil; the signal travels more slowly in the presence of water. The cost of this type of system (over $8,000) probably places it beyond the budgets of most pollination studies. Rundel and Jarrell (1989) review a variety of methods of measuring soil moisture and discuss some of their advantages and disadvantages.

It may also be possible to use an indirect measure of plant water deficits instead of monitoring soil moisture. Búrquez (1987) found that leaf thickness, as measured with a micrometer, was strongly correlated with relative water content. He marked four areas on a leaf with ink, away from any major veins, and used the mean of these four measurements. Although he suggested that it is necessary to calibrate this technique for each population, or perhaps even each individual, this measurement provides a nondestructive field technique.

G. Reporting Time of Day

For behavioral studies it may be important to know what time of day observations were made, as day length may affect either the plants or the animals. To facilitate reconstruction of when during the day you made your observations, it is useful if you report not only the time but also the relationship of that time zone to Greenwich Mean Time, and the time of sunrise and sunset on that day. For example, Corbet (1978a, p. 26) reported: "Times are given as British Summer Time, 1 h ahead of Greenwich Mean Time. On the dates in July 1971–76 when series of records were made the time of sunrise was between 04.44 and 5.06 hours BST; solar noon was between 13.04 and 13.06 hours; and sunset was between 21.06 and 21.23 hours BST." Hoopingarner and Adey (1985) also pointed out that day length varies with both season and latitude and that time schedules may change with designations such as Standard or Daylight Time. They suggested that when reporting data on insect foraging, the time of observation should be given as the hour-after-dawn and the day length: "For example, an observation made at 12.00 EDT (Eastern Daylight Time) in East Lansing, Michigan, on 1 June would be reported as 6/15 or 6.15, since it was the 6th hour of a 15-hour photophase. With a standardized time system, such as we suggest, it would no longer be necessary to look up solar tables for the different locations under comparison." They showed examples of figures with axis labels using this method of reporting times.

H. Suggestions for Planning Studies

1. Many environmental variables can affect pollination. Beattie (1976) found, in studies of *Viola,* that daily duration of sunlight, direction of slope of the habitat, and number of days of sunshine preceding sampling all affected percent pollination. Preliminary studies may be required to determine what environmental variables you should measure for your study.

9. Experimental Considerations

A. Documenting Studies

Of course the study you are conducting is so thorough that there are no unanswered questions left! But perhaps some day in the future someone will wish they had access to your study site, original data, or confirmation of identifications of the plants you worked with or the flower visitors you studied. If you have properly documented your study, it may not be difficult for them to reconstruct what you did. Such documentation should include the deposition of representative plants and flower visitors in museums or other permanent collections, and sufficient descriptions in the materials and methods section of publications. However, providing access to original data is not as simple, nor is documentation of the exact location of study sites.

Journals are reluctant to include extensive tables of data, which raises the issue of how best to make original data available to future researchers who might be interested in them. Several recent trends may help to facilitate this more thorough documentation of studies. One is the efforts being made by the National Science Foundation to require mechanisms for database generation and maintenance at field stations and by some research projects that it supports (e.g., the Long-Term Ecological Research program). Accompanied by what seems to be a growing appreciation for the value of carefully collected and documented data sets, this trend may encourage the extra effort it takes to provide the requisite documentation. If data are being collected with or stored on a microcomputer, then making them available to others may simply require adequate documentation of the format, row and column contents in a spreadsheet, or entries in a database.

The availability of video recorders should make it relatively simple to document the location of study sites. By videotaping the areas where you have worked, or even just photographing them, and depositing this information in an appropriate place, you can facilitate relocation of the sites. Field stations or museums you are associated with may provide appropriate repositories for such information.

B. Sample Sizes

It takes time to collect data, and taking too many samples is a waste of time and resources and may be detrimental to a plant or pollinator population. Yet taking too few samples may lead to a meaningless study or errors in interpretation (Eckblad 1991). If the purpose of sampling is to determine a mean, and you know how accurate you wish to be in your estimate of the mean, you can use a statistical formula to determine the appropriate sample size. This requires that you have some information (usually from a prestudy sample) of the mean and variance that you are likely to encounter in your study. Eckblad (1991) presents the following sample size equations, where "accuracy" refers to the percentage within which you want to estimate the mean (e.g., ± 10%, represented in the formula as 0.1):

$$\text{sample size} \approx \frac{(t\text{–value})^2 \, (\text{sample variance})}{(\text{accuracy} \times \text{sample mean})^2}$$

The t-value would typically be the value for 0.05 level of significance. If you have already calculated the coefficient of variation (standard deviation divided by the sample mean), an equivalent equation is:

$$\text{sample size} \approx \left[\frac{(\text{coefficient of variation} \times t\text{–value})}{\text{accuracy}}\right]^2$$

It may be instructive to plot the number of samples needed (y-axis) to achieve a particular accuracy for an estimate of the mean (x-axis), as a way of deciding whether it is realistic to try to achieve a particular level of accuracy. Although these formulae assume a normally distributed metric character, they are moderately robust with other distributions (Eckblad 1991). You might want to use one of them with an initial small sample and then recalculate the required sample size again once you have taken a larger sample. Krebs (1989) also discusses their use and recommends that they be used conservatively. Eckblad will provide spreadsheet models for several different programs to facilitate the required calculations. (Send $23 to Oakleaf Systems, P. O. Box 472, Decorah, Iowa 52101; specify Macintosh or IBM-PC version; call [319] 382-4320 for information.)

The sample-size equations can also be rearranged to calculate accuracy for a given sample size:

$$\text{accuracy} \approx \frac{(\text{coefficient of variation})\,(t\text{--value})}{\sqrt{\text{sample size}}}$$

If you are trying to plan a study without access to preliminary data, Tables 9–1 and 9–2 showing data from the literature might help you to estimate required sample sizes. Bear in mind that there are two components to variation in these data, intra- and interplant differences. Most of these data appear to include both components.

C. Natural Systems with Characteristics Good for Certain Types of Work

Certain types of natural systems are particularly well suited for pollination studies. Although it may be luck that leads you to such a system, these suggestions might help in a more directed search.

- **plants with genetic markers** — e.g., wild radish, which has a petal-color polymorphism, white and yellow flowers (Stanton et al. 1986), and crop plants, such as *Cucumis melo* (Cucurbitaceae) (Handel 1982).
- **plants with pollen color polymorphisms,** for studies of pollen carryover — e.g., *Erythronium americanum* (Thomson and

**Table 9–1. Means and standard deviations for floral characters
(measurement in mm unless otherwise indicated; SE = standard error)**

Character	Mean	s.d.	N	Reference
Corolla width	3.9	0.38	30	Waser and Price 1984
Corolla width	3.64	0.28	39	Campbell 1989
Corolla width	25.7	1.37	30	Waser and Price 1984
Corolla width	1.36	0.31	21	Inouye and Pyke 1988
Corolla width	5.58	0.38	30	Inouye and Pyke 1988
Spur length	10.21	0.53	27	Steiner and Whitehead 1991b
Anther length	24.1	1.31	30	Waser and Price 1984
Style length	24.0	5.54	14	Waser and Price 1984
Pollinium length	1,173 μm	48.0	70	Morse and Fritz 1985
Stigma length	16.0	16.0 (SE)	10	Ritland and Ritland 1989
Ovary length	5.8	0.3	10	Ritland and Ritland 1989
Nectary-stigma	32.8	1.53	133	Johnston 1991b
Petaloid sepal length	13.2	1.32	45	Ågren and Schemske 1991
Flower orientation	13.1°	1.3 (SE)	30	Schlessman 1986a

Table 9–2. Means and standard deviations for proboscis or beak
lengths; measurements in mm

Species	Mean	s.d.	N	Reference
Apis mellifera honeybee	5.77	0.11(SE)	12	Rinderer et al. 1981
Bombus flavifrons (glossa) bumblebee workers	6.25	0.28	12	Inouye 1980b
Bombus vagans (glossa) bumblebee workers	5.37	0.68	42	Medler 1962b
Bombus ruderatus (glossa) bumblebee queens	10.80	0.39	8	Medler 1962b
Meliphaga cratitia (beak) Purple-gaped Honeyeater	14.4	1.03	65	Ford and Paton 1976

Plowright 1980); *Erythronium grandiflorum* (Thomson et al. 1986). These species have a natural pollen dichromism.
- **plants with short generation times** — e.g., rapid-cycling *Brassica,* which can complete a generation in a matter of weeks (Williams and Hill 1986, Tomkins and Williams 1990; see Appendix 2).
- **plants with sporophytic incompatibility systems** (so flowers don't have to be emasculated to prevent self-pollination) — e.g., *Raphanus sativus* (Marshall 1991).
- **plants with many flowers,** so that manipulations can be repeated — e.g., *Raphanus sativus* (Marshall 1991).
- **plants with large seeds** so that seed paternity can be assessed easily with electrophoresis — e.g., *Raphanus sativus* (Marshall 1991).
- **plants with long styles** that facilitate pollen-tube growth studies, as do pistils with multiple branches, so that pollination treatments can be applied to separate stigmas of the same flower (eliminating among-flower variation) — e.g., *Hibiscus moscheutos* (Snow and Spira 1991b).
- **heterostylous plants** (with different sizes of pollen) for studies of pollen movement.

Appendix 1.
References to the Pollen Literature

General References

Brooks, J. B., P. R. Grant, M. Muir, P. Van Gijzel, and G. Shaw, editors. 1971. Sporopollenin. Academic Press, London, England.

Echlin, P. 1968. Pollen. Scientific American 218:80–90.

Erdtman, G.1969. Handbook of palynology. Hafner, New York, New York, USA.

Faegri, K., P. E. Kaland, and K. Krzywinski. 1989. Textbook of pollen analysis. Fourth edition. John Wiley and Sons, Inc. New York, New York, USA.

Kremp, G.O.W. 1965. Morphologic encyclopedia of palynology. University of Arizona Press, Tucson, Arizona, USA.

Kummel, B. G. and D. M. Raup, editors. 1965. Handbook of paleontological techniques. W. H. Freeman, San Francisco, California, USA.

Moore, P. D., and J. A. Webb. 1978. An illustrated guide to pollen analysis. John Wiley and Sons, New York, New York, USA.

Shivanna, K. R., and B. M. Johri. 1985. The angiosperm pollen. John Wiley and Sons, New York, New York, USA.

Taxonomic Groups

Identification guides for specific taxa are also available:

Gymnosperms

Birks, H.J.B. 1978. Geographic variation of *Picea abies* (L.) Karsten pollen in Europe. Grana 17:149–160.

Birks, H.J.B., and S. M. Peglar. 1980. Identification of *Picea* pollen of late Quaternary age in eastern North America: a numerical approach. Canadian Journal of Botany 5l8:2043–2058.

Millay, M. A., and T. N. Taylor. 1976. Evolutionary trends in fossil gymnosperm pollen. Review of Palaeobotany and Palynology 21:65–91.

We thank Dr. Susan Short, palynologist at the University of Colorado, for compiling most of this list of important references.

411

Appendix 1

Aceraceae

Helmich, D. E. 1963. Pollen morphology in the maples (*Acer* L.). Papers of the Michigan Academy of Science, Arts and Letters 48:151–164.

Amaranthaceae

Tsukada, M. 1967. Chenopod and amaranth pollen: Electron microscopic determination. Science 157:811.

Asteraceae

Bolick, M. R., J. J. Skvarla, B. L. Turner, V. A. Patel, and A. S. Tomb. 1984. On cavities in spines of Compositae pollen — a taxonomic perspective. Taxon 33:289–293.

Feuer, S., and A. S. Tomb. 1977. Pollen morphology and detailed structure of family Compositae, tribe Cichorieae. II. Subtribe Microseridinae. American Journal of Botany 64:230–245.

Tomb, A. S. 1975. Pollen morphology in the tribe Lactuceae (Compositae). Grana 15:79–89.

Betulaceae

Leopold, E. B. 1956. Pollen size-frequency in New England species of the genus *Betula*. Grana Palynologica 1:140–147.

Cactaceae

Tsukada, M. 1963. Pollen morphology and identification. II. Cactaceae. Pollen et Spores 6:45–85.

Caryophyllaceae

Chanda, S. 1962. On the pollen morphology of some Scandinavian Caryophyllaceae. Grana Palynologica 3:67–89.

Chenopodiaceae

McAndrews, J. H., and A. R. Swanson. 1967. The pore number of periporate pollen with special references to *Chenopodium*. Review of Palaeobotany and Palynology 3:105–117.

Tsukada, M. 1967. Chenopod and Amaranth pollen: Electron microscopic determination. Science 157:811.

Fagaceae

Smit, A. 1973. A scanning electron microscopical study of the pollen morphology in the genus *Quercus*. Acta Botanica Neerlandica 22:655–665.

Gramineae

Andersen, S. T., and F. Bertelson. 1972. Scanning electron microscope studies of pollen of cereals and other grasses. Grana 13:79–86.

Köhler, E., and E. Lange. 1979. A contribution to distinguishing cereal from wild grass pollen grains by LM and SEM. Grana 18:1113–1140.

Juglandaceae

Whitehead, D. R. 1963. Pollen morphology in the Juglandaceae. I. Pollen size and pore number variation. Journal of the Arnold Arboretum, Harvard University 44:101–110.

Whitehead, D. R. 1963. Pollen morphology in the Juglandaceae. II. Survey of the family. Journal of the Arnold Arboretum, Harvard University 46:369–410.

Lythraceae

Graham, A., S. A. Graham, J. W. Nowicke, V. Patel, and S. Lee. 1985. Palynology and systematics of the Lythraceae. I. Introduction and genera *Adenaria* through *Ginoria*. American Journal of Botany 72:1012–1031.

Graham, A., S. A. Graham, J. W. Nowicke, V. Patel, and S. Lee. 1985. Palynology and systematics of the Lythraceae. II. Genera *Haitia* through *Peplis*. American Journal of Botany 74:829–850.

Graham, A., S. A. Graham, J. W. Nowicke, V. Patel, and S. Lee. 1985. Palynology and systematics of the Lythraceae. III. Genera *Physocalymma* through *Woodfordia, Addenda,* and conclusions. American Journal of Botany 77:159–177. Includes key to pollen of Lythraceae.

Onagraceae

Ting, W. S. 1966. Pollen morphology of Onagraceae. Pollen et Spores 8:9–36.

Rosaceae

Reitsma, T. 1966. Pollen morphology of some European Rosaceae. Acta Botanica Neerlandica 15:290–307.

Saxifragaceae

Ferguson, I. K., and D. A. Webb. 1970. Pollen morphology in the genus *Saxifraga* and its taxonomic significance. Botanical Journal (Linnean Society of London) 63:295–312.

Appendix 1

Major Regional Groups

Local reference collections of the common flowers of an area are easily prepared by collecting, staining, and mounting pollen on microscope slides. Regional reference collections include:

Adams, R. J., and J. K. Morton. 1972, 1974, 1976. An atlas of pollen of the trees and shrubs of eastern Canada and the adjacent United States. Parts 1–3. University of Waterloo Biology Series no. 8. Waterloo, Ontario, Canada.

Bassett, I. J., C. W. Crompton, and J. A. Parmelee. 1978. An atlas of airborne pollen grains and common fungus spores of Canada. Research Branch, Canada Department of Agriculture Monograph No. 18. 321 pp.

Heusser, C. J. 1977. A survey of pollen types of North America. American Association of Stratigraphic Palynoligists Contribution Series 5A, 111–129.

Jasonius, J. 1978. A key to the genera of fossil angiosperm pollen. Review of Palaeobotany and Palynology 26:143–172.

Kapp, R. O. 1969. How to know pollen and spores. Wm. C. Brown, Dubuque, Iowa, USA.

Lewis, W. H., P. Vinay, and V. E. Zenger. 1983. Airborne and allergenic pollen of North America. The John Hopkins University Press, Baltimore, Maryland, USA.

Lieux, M. H. 1980–1982. An atlas of pollen of trees, shrubs, and woody vines of Louisiana and other southeastern states. Parts 1–3. Pollen et Spores 22:17–57, 191–243; 24:21–64.

McAndrews, J. H., and J. E. King. 1976. Pollen of the North American Quaternary: The top twenty. Geoscience and Man 15:41–49.

McAndrews, J. H., A. A. Berti, and G. Norris. 1973. Key to the Quaternary pollen and spores of the Great Lakes region. Life Science Miscellaneous Publication, Royal Ontario Museum.

Markgraf, V., and H. L. D'Antoni. 1978. Pollen flora of Argentina. The University of Arizona Press, Tucson, Arizona, USA.

Martin, P. S. 1969. Pollen analysis and the scanning electron microscope. Scanning Electron Microscopy. Proceedings of the second annual Scanning Electron Microscope Symposium, April 1969. pp. 89–103. (Southwestern U.S. pollen taxa).

Moriya, K. 1976. Flora and palynomorphs of Alaska. Kodansha, Tokyo, Japan.

Punt, W., editor. 1976. The north-west European pollen flora. I. Elsevier, Amsterdam, Holland.

Richard, P. 1970. Atlas pollenique des arbres et de quelques arbustes indigènes du Québec. I. Introduction générale. II. Gymnospermes. III. Angiospermes (Saliacées, Myricacées, Juglandacées, Corylacées, Fagacées, Ulmacées). IV. Angiospermes (Rosacées, Anacardiacées, Acéracees, Rhamnacées, Tiliacées,

Appendix 1

Cornacées, Oléacées, Caprifoliacées). Le Naturaliste Canadien 97:1–34, 97–161, 241–306.

Wingenroth, M., and C. J. Heusser. 1983. Pollen of the high Andean flora. Instituto Argentino de Nivologia y Glaciologia, Mendoza, Argentina.

Journals

American Journal of Botany

Grana — An international journal of Palynology (previously Grana Palynologica)

Japanese Journal of Palynology (in Japanese)

Journal of Allergy

Journal of Palynology

New Phytologist

Palaeobotanist

Palynology

Palaeogeography, Palaeoclimatology, Palaeoecology

Pollen et Spores

Quaternary Research

Review of Palaeobotany and Palynology

Appendix 2.
Sources of Equipment and Supplies

If a supplier or manufacturer is only referred to once in this list, the address and telephone number are listed in that spot. If they are referred to more than once, look for this information in the alphabetical list of suppliers and manufacturers at the end of the appendix.

Adsorbent, for collecting floral odors.
> Carbotrap and Porapak Q
>> Supelco, Inc.
>> Supelco Park
>> Bellefonte, PA 16823-0048
>> (800) 359-3041; (814) 359-3441

> Tenax TA
>> Alltech Associates (see Appendix 3)
> Chromosorb G (see Appendix 3)

Air samplers, for windborne pollen (See also Pollen samplers).
Burkard Manufacturing Co. makes several types of volumetric air sampling devices. SKC, Inc. makes a portable, rechargeable vacuum pump that might work well for sampling known volumes of air and trapping pollen on a filter (see Vacuum pumps).

Anemometers.
Simerl Instruments makes a couple of different models of hand-held anemometers. One of them (model BTC) is part of the suite of instruments recommended by the International Pollination Project for recording environmental parameters in conjunction with visitation rates. Model DIC is a hand-held digital read-out electronic anemometer, $295. Model BTC is a hand-held analogue anemometer, $125.
> Simerl Instruments
> 528 Epping Forest Road
> Annapolis, MD 21401
> (301) 849-8667

> Vector Instruments (make integrating cup anemometers; S. Corbet,
>> *personal communication*)
> 115 Marsh Road
> Rhyl, Clwyd, LL18 2AB
> United Kingdom
> 9745 50700

Aquapics, for holding cut flowers. Check with your local florist.

416

Appendix 2

Ascorbic acid test paper, for testing for the presence of ascorbic acid. VWR carries EM Quant Test Strips (EM-10023-1, $28.20/100) and J. T. Baker Testrips (JT4409-1, $27.75/100).

Atomizers, for dispensing fluorescent powder dyes (See Powder blower).

Balances, spring for weighing birds in the field.
Pesola scales, a brand of spring scales, are available in several weight ranges. We have used them for weighing hummingbirds, and they should work for most avian or mammalian pollinators. They are available from Avinet and other sources.

Bee tags.
The most commonly used bee tags seem to be Opalith-Plättchen, which are plastic tags developed for use in apiculture (tagging queens to make them easier to identify by year-class or individual). The tags are available in 5 colors, in numbers 1–99, and are sold in a kit that includes glue and a small brush. There does not appear to be a distributor for these in the United States, but they are readily available from several sources in Europe. (If you only need a small number of the tags, contact David Inouye.)
The tags are available from the manufacturer:

Chr. Graze KG
Fabrik für Bienenzuchtgeräte
7056 Weinstadt-Endersbach
Bei Stuttgart
Germany
telephone: (0 71 51) 6 11 47; telex 7 262 213 apig d; fax 7151-609239

(Price as of fall 1991 was DM 22.50 for a kit with 500 tags, plus DM 5 for airmail shipping.)
They are also available from retailers, including:

Gustav Nenninger
8741 Saal A. Saale
Germany

Bienen Mathys AG
3762 Erlenbach i. S.
Switzerland

Bird-banding supplies.
Avinet carries mist nets, poles, spring balances, etc.

Black light insect trap, for attracting noctural and crepuscular insects.
BioQuip Products

O. B. Enterprises, Inc.
4585 Schneider Drive
Oregon, WI 53575
(608) 835-9416

Butterfly larval diet.
Carolina Biological Supply carries larval diets and larvae of several Lepidop-
teran species.

Bumblebee colonies.
At least one North American company sells established bumblebee colonies
for use in commercial hothouse pollination (especially of tomatoes):
Bees-under-Glass Pollination Services, Inc.
Box 310, RR 1
Cantley, Quebec J0X 1L0
Canada

Cages, screen, for use as flight cages to enclose or exclude insects.
Available in a variety of sizes, with different mesh sizes. A 6 x 6 x 6-foot cage
(without frame) costs $103–$173 (depending on mesh size) and weighs about
7 pounds. (BioQuip Products also carries cages.) You may also be able to use
a screen tent designed for camping. Tetko carries a wide range of woven
screening media. Sullivan and Brampton sells several different kinds of nylon
netting and shade fabric if you want to make your own cages.
Lumite
P. O. Box 977
Gainesville, GA 30503
(800) 241-7566; fax (404) 531-1347

Calipers, for measuring flowers or proboscides.
Several models of digital calipers are available, e.g., Brown and Sharpe
Digit-Cal, Fowler Sylvac, Ultra-Cal, Max-Cal, and Mitutoyo. DataQ is a
computer program designed to interface digital calipers and programs such
as spreadsheets, database managers, or statistical packages. For more information, contact:
David L. Schultz
909 Valley Road
Aiken, SC 29801
(803) 649-0411

For about $350 an electronic caliper with RS232 interface is available from
Fred V. Fowler Co.
66 Rowe St.
P. O. Box 299
Newton, MA 02166
(617) 332-7004; fax (617) 333-4137

Appendix 2

Digi-matic caliper with LCD display, data processor
  Forestry Suppliers

  Camlab
  Nuffield Road
  Cambridge CB4 1TH
  England

Several models of digital calipers and software are available from
  Leslie F. Marcus
  Planetarium Station
  P. O. Box 524
  New York, NY 10024
  (212) 769-5721 or (212) 595-3089

Capillary tubes, micro, for collecting nectar from flowers.
  Microcap brand from Fisher Scientific is available in sizes including 1, 2, 3,
  4, 5, and 10 µl, $7.35/100.
    Shandon Scientific
    112 Chadwick Road
    Astmore Industrial Estate
    Runcorn, Cheshire WA7 1 PR
    England

Climatological data.
  This is the title of a publication available for each state in the United States,
  published monthly by the National Climatic Data Center. Daily records and
  monthly summaries are provided for precipitation and several temperature
  variables for a large number of weather stations. You may find copies of this
  in a large library; subscriptions are also available. National Climatic Data
  Center, Asheville, NC 28801.

Color charts for describing flower color (see Tucker et al. 1991 for a review of
  available charts). Royal Horticultural Society colour charts (Catalog #526)
  are available for £30 (+ £6 postage).
    Mail Order Department
    RHS Enterprises Ltd.
    RHS Garden
    Wisley, Woking
    Surrey GH23 6B
    England

Another color guide is: Smithe, F. B. 1975. Naturalist's color guide. American
Museum of Natural History. ISBN 0-913424-03-X, $21.50. (Look under
QL767.S63 in your library.) Send orders to:

Appendix 2

American Museum of Natural History
Members' Book Program
Central Park West at 79th
New York, NY 10017
(212) 769-5500

Computer programs, for behavioral observations with laptop computers.
The Observer
Noldus Information Technology bv
e-mail address: NOLDUS@RCL.WAU.NL
available from:
Fernon Electronic Imaging
P.O. Box 948
St. Michaels, MD 21663
(410) 745-2274; fax (410) 745-9399

Crit-O-Seal, for sealing microcapillary tubes containing nectar samples.
Available from most scientific supply companies. Sigillum, an alternative (S. Corbett, *personal communication*), is manufactured by:
Modulohm I/S
Vasekaer 6-8
DK2730 Herlev, Copenhagen
Denmark

Data loggers, for recording meteorological (or other data).
Campbell Scientific, Inc. (also has offices in Canada, England)

LI-COR, Inc.

Grant Instruments (Cambridge) Ltd.

Densitometer, for measuring density of photographs (slides or negatives) of flowers, to determine reflectance in comparison to a scale included in the photograph. Ask someone who works with thin-layer chromatography, X-ray film, or DNA sequencing; they may have one you can use.
Biomed Instruments, Inc.
1020 S. Raymond Ave.
Suite B
Fullerton, CA 92631
(800) 927-2290; (714) 870-0290; fax (714) 879-6385

Dialysis tubing, for covering flowers.
Spectropor dialysis membrane 045336 is made by Spectrum Medical Industries, Inc., of Los Angeles, and is distributed by many sources, including Fisher Scientific and Carolina Biological Supply.

Difco Laboratories, for bacteria, yeasts, and molds for biological assays for vitamins or amino acids.
P. O. Box 331058
Detroit, MI 48232-7058
(800) 521-0851; (313) 462-8500

Electronic components, for making your own instruments.
Radio Shack

Electromail
P. O. Box 33
Corby Northamptonshire NN17 9EL
England

Euparal, neutral mounting medium (e.g., Ramanna 1973).
Carolina Biological Supply (catalog number 86-1890, 50 ml, $18.47)

Field osmometer (Unwin and Willmer 1978).
See instructions in this reference for building a field osmometer.

Field stations.
For a copy of the Organization of Biological Field Stations' directory of (primarily North American) field stations, contact
Dr. Richard Coles
OBFS
Tyson Research Center
Washington University
P. O. Box 258
Eureka, MO 63025

Flagging, brightly colored vinyl or polyethylene tape, for marking plants or study sites.
Forestry Suppliers, Inc.

Ben Meadows

Flight cages or greenhouses.
Several researchers at the Rocky Mountain Biological Laboratory use collapsible Quonset hut–like structures with either screen or fabric covers as flight cages or greenhouses. They are available in a variety of sizes from:
Hansen Weather-port Corporation
300 S. 14th St.
Gunnison, CO 81230
(303) 641-0480

BioQuip Products also sells a screen cage that works well as a flight cage. It is supported by metal poles (e.g., electrical conduit). Park Seed Wholesale has greenhouse equipment and supplies for greenhouses.
See also Cages, screen.

Appendix 2

Floralife, for extending the life of cut flowers. Try your local florist, or:
Carlsted's
2252 Dennis St
Jacksonville, FL 32203
(904) 354-4634

Forceps.
BioQuip Products

Edmund Scientific

Small Parts, Inc.

Rio Grand Albuquerque (a supplier of jeweler's tools)
6901 Washington N.E.
Albuquerque, NM 87109
(800) 545-6566; (505) 345-8511

Funding for pollination studies.
There do not appear to be any sources of funding dedicated solely to pollination studies, as there are for some disciplines. However, if you are working with orchids, or with birds, there are some potential sources of funding to which you can turn, e.g.:
American Herb Society

American Horticulture Society

Orchid Society

Sigma Xi

The American Ornithologists' Union publishes a booklet on "Grants, Awards and Prizes in Ornithology" that may be of interest to researchers working on ornithophily. Send $7.00 for the booklet, handling and shipping (U.S. dollars only, payable in advance) to:
Max C. Thompson
Assistant to the AOU Treasurer
Dept. of Biology
Southwestern College
100 College St.
Winfield, KS 67156-2499.

Another source of research funding for advanced predoctoral candidates in the fields of zoology and paleontology is Frank M. Chapman fellowships in ornithology and Theodore Roosevelt fellowships for study of North American fauna. Request information booklet and application forms from:

Appendix 2

The Office of Grants and Fellowships
Dept. H
American Museum of Natural History
Central Park West at 79th
Street New York, NY 10024.

For behavioral studies, try the Animal Behavior Society Research Grants program. Maximum award is $1,000. All ABS members are eligible, although funding priority is given to graduate students and recent Ph.D.'s in their first year of postdoctoral training. See an issue of the society's newsletter for current information on where to get application forms.

Fur dye, for marking mammalian pollinators.
Nyanzol D fur dye ($30/pound) can be ordered from:
Belmar, Inc.
P. O. Box 145
100 Belmont St.
N. Andover, MA 01845
(508) 683-8726

Glassine envelopes, for collecting fruits or seeds.
BioQuip Products

E. F. Fullam, Inc.
900 Albany Shaker Rd.
Latham, NY 12110
(518) 785-5533; fax (518) 785-8647
3 sizes available, 3.25 x 4" ($29.95/1,000), 4 x 5" ($52), 6.5 x 9 cm ($34)

Glucose test paper, for checking nectar samples for the presence of glucose.
Boehringer Mannheim Chemstrip GP test strips
Fisher Scientific BC-200743, 100 for $18.10

Uristix, available from:
Ames Division
Miles Laboratories Ltd.
Stoke Poges, Slough SL2 4 LY
England

Herbarium supplies, for preparing plant specimens.
Carolina Biological Supply Co.

University Products, Inc.

Hoffman Modulation Contrast System.
The Hoffman Modulation Contrast System consists of a filter (the modulator) in the objective lens of a microscope, and a special aperture in front of the condenser. It is available as new standard equipment or for upgrading compound microscopes. This system converts phase gradients into intensity variations, giving a three-dimensional appearance to objects like pollen grains.

It can be used in combination with an incident light fluorescence method. It may offer some benefits over phase contrast viewing systems. Objectives range in price from $600 to $2,370; condensers are $500–$1,245, and control polarizers are $65–$125.

Modulation Optics Inc.
100 Forest Drive at East Hills
Greenvale, NY 11548
(516) 484-8882

HPLC equipment and supplies, for nectar analyses.
Rainin Instrument Co.

Hummingbird food, for keeping hummingbirds in captivity.
Nektar-Plus
Nekton U.S.A., Inc.
1917 Tyrone Blvd.
St. Petersburg, FL 33710
(813) 381-5800

Bio-Nektar
Biotropic U.S.A. (also has books and a quarterly journal about caring for and raising hummingbirds)
P. O. Box 50636
Santa Barbara, CA 93150
(805) 969-9377

Image analysis software, for morphological studies.
MorphoSys captures images from a specimen viewed with a video camera and permits measurements of the image such as distance, angles, and area. Available from:
Exeter Software
100 North Country Road
Setauket, NY 11733
(800) 842-5892; (516) 689-7838; fax (516) 751-3435

Infrared viewers, for making observations of nocturnal pollinators.
Edmund Scientific

A video camera system for fieldwork with an infrared light source is made by
Fuhrman Diversified, Inc.
905 South 8th Street
La Porte, TX 77571
(713) 470-8397

Herbach & Rademan (Model ELC or ELT, battery-operated, about $450)
401 E. Erie Ave.
Philadelphia, PA 19134-1187
(800) 848-8001; fax (215) 425-8870

Appendix 2

Insect boxes.
> BioQuip Products

> University Products (has boxes for mailing specimens, too)

> Watkins and Doncaster

Insect labels.
> BioQuip Products will print custom labels.

> University Products has acid-free papers for making your own.

> Zellerbach Corp. makes Dulcet 70, an acid-free paper that works well in laser printers and comes in 8.5 x 11–inch sheets.

Insect media, for rearing insects in the lab.
> Bio-Serv
> P. O. Box 450
> Frenchtown, NJ 08825
> (800) 234-7378; (201) 996-2155; fax (201) 996-4123

Insect mounting boards, for pinning Lepidoptera.
> BioQuip Products

> Carolina Biological Supply Co.

> Watkins and Doncaster

Insect nets, for catching insect pollinators.
> BioQuip Products

Lovins Micro-slide Field Finder.
> This is a microscope slide that is divided into a grid of 1-mm squares, with 0.1-mm intervals marked. Each square is coded with a number and letter as a guide to locating a particular position on a microscope slide (find the field of view you want, replace the slide with the field finder slide, and record the position). Cruden et al. 1990 used the slide itself as a way to facilitate counting of pollen grains. The slides are manufactured by Teledyne Gurley. One distributor is:
> > Fisher Scientific, catalog number 12-454, $133.50.

Magnifiers.
> **Comparator,** magnifier mounted on a transparent base, with a measuring reticle. Available in 6X, 9X, 12X, with a variety of reticle patterns. Edmund Scientific Co.
> **Binocular.** The OptiVisor is worn on your head and can be tilted up out of the way when not needed. Magnifications from 1.5X to 3.5X are available, for about $27 each.

Donegan Optical Company, Inc.
15549 West 108th Street
Lenexa, KS 66219

(Also available from Carolina Biological Supply Co., Edmund Scientific, Forestry Suppliers.)

Malaise traps, for monitoring insect populations.
BioQuip Products

Golden Owl Publishers; $217 for portable, Townes-design Malaise traps.
182 Chestnut Road
Lexington Park, MD 20653
(301) 863-9253

If you want to make your own, netting is available from Sullivan and Brampton.

Meteorological instruments.
Campbell Scientific

Edmund Scientific

LI-COR, Inc.

Delta-T Devices Ltd.

Grant Instruments (Cambridge) Ltd.

Microbalance, suitable for weighing small insects.
Cahn electronic analytical balances are one example; available from Fisher Scientific, etc.

Micropipettes, for sampling nectar.
Drummond Microcaps, available in sizes from 0.25 to 200 µL. Other brands are available, too.
Cole Parmer Instrument Company

Fisher Scientific

VWR Scientific

Microscope, field.
Discovery Scope
P. O. Box 136759
Fort Worth, TX 76136
(817) 237-1681

Mist nets and other supplies, for work with birds.

      AVINET, Inc., carries mist nets, net poles, spring scales, calipers, head-lamps, etc.

      National Band and Tag Co., carries a large line, including bird bands.

Museum cases, for storing herbarium sheets, dried insects, or other preserved pollinators.

      Interior Steel Equipment Co.
      2352 E 69th St.
      Cleveland, OH 44104
      (216) 881-0100

      Lane Science Equipment Corporation
      225 West 34th Street
      New York, NY 10122

Netting, for making pollinator exclusion cages.

      Sullivan and Brampton sells olive drab nylon marquisette netting that works well for this.

Night vision devices.

      Edmund Scientific sells a camera-adaptable model

      Hamamatsu Photonics (image intensifiers, e.g., C2100 night viewer)
      360 Foothill Road
      Box 6910
      Bridgewater, NJ 08807-0910
      (201) 231-0960

      ITT Defense Technology Corporation (a night-vision binocular)
      Electro-Optical Products Division
      7635 Plantation Road
      Roanoke, VA 24019
      (703) 563-0371

Particle counters.

      Coulter Corporation
      P. O. Box 2145
      Hialeah, FL 33012-0145
      (800) 526-6932; (305) 885-0131

      Particle Data
      115 Hahn Street
      Elmhurst, IL 60126
      (800) 323-6140; (708) 832-5653

      HIAC-ROYCO (distributed by Pacific Scientific, Inc.) particle size analyzer.

Appendix 2

Pinning and collecting supplies, for work with insects.
BioQuip Products carries a wide assortment of field equipment, laboratory equipment, supplies, and storage systems. Carolina Biological Supply also has a wide variety of supplies for entomological studies.

Pipettors, e.g., for filling artificial flowers, or real ones.
A variety of brands (e.g., Oxford, Rainin, Eppendorf) of pipettors are available, including some with fixed capacities from 1 μL on up, some adjustable ones, and some repeating microdispensers that might facilitate repetitive operations. Expect to pay $150 to $300. Thomson (*personal communication*) recommends a battery-powered, repeat-dispensing pipettor from Rainin Instrument Co. for filling artificial or real flowers with nectar. Rainin also makes a manually operated model (Pipetman Precision Microliter Pipette). Eppendorf makes a repeating pipettor that dispenses up to 48 samples of 10–50 μL at 1-second intervals.

Planimeter, for measuring the area of flowers or leaves.
Try a surface-area meter or computerized method of measuring area (e.g., SigmaScan software) from:
Jandel Scientific
65 Koch Road
Corte Madera, CA 94925
(415) 924-8640

Jandel Scientific
Schimmelbuschstrasse 25
D-4006 Erkrath 2
Germany
0 2104 / 3 60 98

Pollen diluent for diluting pollen for hand pollination.
Agtech Developments (NZ) Ltd.
25 McPherson Street
Richmond 6805/23
New Zealand

Pollen, for feeding captive bees.
Try your local beekeepers to see if they have a pollen trap on a hive. Several different types of live pollen (e.g., from fruit trees, shipped frozen for pollination) are available from:
FirmYield Pollen Services, Inc.
2326 Rudkin Road
Yakima, WA 98903
(509) 452-1495

CC Pollen Co. (in 55-pound drums, $2/pound)
3627 E. Indian School Road, #209
Phoenix, AZ 85018-5126
(800) 875-0096

Appendix 2

Pollen sampler.

Kramer-Collins 7-day drum sampler, designed for sampling pollen grains, fungal spores, or other airborne particles, which impact on double-coated cellophane tape applied to a 15.24-cm diameter drum. Can be modified to sample a range from 1 hour to 30 days.

G-R Electric Manufacturing Co.
1317 Collins Lane
Manhattan, KS 66502
(913) 537-2260

Sampler is $300, optional timer is $250, vacuum pumps are $250–$300 (prices as of summer 1991). For pictures and sample data, see: Kramer, C. L., M. G. Eversmeyer, and T. I. Collins. 1976. A new 7-day spore sampler. Phytopathology 66(1):60–61. Eversmeyer, M. G., C. L. Kramer, and T. I. Collins. 1976. Three suction-type spore samplers compared. Phytopathology 66(1):62–64.

A Rotorod sampler takes short-term continuous air samples of pollen or mold spores with 12-V battery power. It collects pollen (or other airborne particles like spores) with a spinning collecting rod covered with silicone grease. Model 92 with a retracting head costs $250. Contact:

Sampling Technologies, Inc.
10801 Wayzata Boulevard, Suite 340
Minnetonka, MN 55305
(612) 544-1588

Pollen traps, for honeybee hives.

Most suppliers of equipment for honeybees carry pollen traps. Two sources (for others, look at advertisements in *American Bee Journal*) include:

Western Bee Supplies, Inc. (catalog #WB679, $16.47)
P.O. Box 171
Polson, MT 59860
(800) 548-8440

Brushy Mountain Bee Farm, Inc.
Rt. 1, Box 135
Moravian Falls, NC 28654
(800) 233-7929; (919) 921-3640; fax (919) 921-2681

Chris Plowright (*personal communication*) recommends the Ontario Agricultural College (OAC) model, available from:

F. W. Jones and Sons (catalog #BT955, $39.95)
68 Tycos Drive
Toronto, Ontario M6B 1V9
Canada
(416) 783-2818

Appendix 2

Pollination bags (see also Dialysis Tubing.)
Applied Extrusion Technologies makes bags of several sizes from Delnet nonwoven polyolefin plastic resin for seed hybridization research.
P. O. Box 852
Middletown, DE 19709
(800) 521-6713; (302) 378-8888; fax (302) 378-4482

Lawson Bag sells a variety of pollination bags, for crops such as sorghum, corn, wheat, and sunflowers. The bags are available in glassine, parchment, or polyethylene.
P. O. Box 8577
Northfield, IL 60093
(800) 451-1495; (708) 446-8812; fax (708) 446-6152

Note: Corn States and Carpenter Paper Co., cited in some publications as sources of bags, no longer make them.
  For covering large plants, try material designed for covering agricultural crops, sometimes called crop cover or floating row cover (because it's light enough that plants lift it up as they grow). Gardening centers or seed companies often carry it. Two such sources include:
Park Seed Wholesale

Gardener's Supply Company
128 Intervale Road
Burlington, VT 05401

Powder blower, for applying fluorescent dusts to insects or flowers (e.g., Stockhouse 1976). Model 175 powder blower, for even diffusion of powders, $31.50.
DeVilbiss Health Care, Inc.
2575 Asbourne Drive
Lawrenceville, GA 30243
(800) 338-1988; (404) 339-6665

Protein test strips, for testing nectar for the presence of protein.
Albustix (for testing for protein in urine, based on tetrabromphenol blue, but also used for testing nectar, Willmer 1980). Available from Fisher Scientific, AM28-70, $20.00/100.

Radiation sensors for measuring light levels.
Decagon Devices, Inc.

LI-COR, Inc. has a variety of radiation sensors.

Skye Instruments Ltd.
Unit 5
Dole Industrial Estate
Llandrinodod Wells
Powys LD1 6DF
United Kingdom
0597 824811

Rapid-cycling *Brassica.*
Crucifer Genetics Cooperative (CrGC)
Department of Plant Pathology
1630 Linden Drive
University of Wisconsin
Madison, WI 53706
(608) 262-8638
BITNET address: crgc@wisc.macc.bitnet
INTERNET address: crgc@vms.macc.wisc.edu

The CrGC was established to develop, acquire, maintain, and distribute information about seed stocks of various crucifers and crucifer-specific symbionts (pathogens). They also distribute seeds of the Rapid-Cycling Brassicas developed by Paul Williams, which have a 35-day life cycle. Sample prices are 25 or 100 seeds for $3.00. Tomkins and Williams (1990) describe the potential for using rapid-cycling *Brassica campestris* (*rapa*) L. for classroom studies of plant development and pollination.

Carolina Biological Supply Co. carries these plants, as well as dried bees for making "beesticks" (Williams 1980) for pollinating flowers.

MacIntyre Fast Plants Kits are available from:
Mottingham Garden Centre
Mottingham Lane
London SE12 9AW
United Kingdom
081 857 2425

Reflectors, for marking plants or paths for nocturnal studies.
Light-reflecting tacks can be ordered from Cabela's. Stick-on reflective strips in four colors are available from:
Road Runner Sports
6310 Nancy Ridge Road
Suite 107
San Diego, CA 92121-3266
(800) 551-5558

Refractometers, for measuring refractive index of nectar samples. Many different brands are available, including:
Atago: available from Cole-Parmer Instrument Co. (e.g., Brink and deWet 1980)

Bausch and Lomb (e.g., Pleasants and Chaplin 1983; Southwick et al. 1981 ground the cover plate for use with 1-μL samples)

Bellingham and Stanley manufactures a series of refractometers. Some of these can be specially modified (by adjusting the prism head setting) to work with microliter quantities. Contact the manufacturer with details of the samples anticipated; a "nominal charge" is made for this service. Model 12-01, with a range of 0–50% sugar, is about $300.
 Bellingham and Stanley, Inc.
 5815 Live Oak Parkway
 Suite 2C
 Norcross, GA 30093
 (404) 662-8573

 Bellingham and Stanley Limited
 Longfield Rd.
 North Farm Industrial Estate
 Tunbridge Wells, Kent TN2 3EY
 England
 0892 36444

Fisher #5242 temperature-compensated (e.g., Reader 1977)

Erma hand refractometer (e.g., Toledo and Hernández 1979)

Kyowa HR-5, 0-90%, #11654 (300 g, under 12 cm in length; e.g., Wright 1985). Available from
 Wood's Photo Supplies
 P. O. Box 98845
 TST Kowloon
 Hong Kong
 3-663229; telex 37844 WPS HX

Leica (was American Optical) 10431 temperature-compensated hand refractometer (e.g., Neill 1987): available through VWR, Fisher. For other distributors call Leica at (716) 891-3000.

Fisher Scientific carries at least two brands of hand refractometers designed for calculating dissolved solids, calculated in units of Brix (percent dissolved sucrose). Cole-Parmer Instrument Co. also carries a variety of types of refractometers.

Relative humidity meter.
 For those with moderate budgets who would rather not use a sling psychrometer for measuring relative humidity, there is an electronic hygrometer available. The meter also measures air temperature and air velocity with the right probes and can serve as a thermocouple thermometer. Cole-Parmer Instrument Co. carries it, for about $400 for the meter, $350 for each probe. As it

is battery-operated, it should work well in the field. Other models that measure only humidity or humidity and temperature range from $130 to $775. Another model of hand-held, battery-operated, combination hygrometer/thermometer is available from both Fisher Scientific (catalog number 11-661-8) and some other suppliers, for about $184. A pocket-size relative humidity "pen" is also available from many major scientific instrument suppliers (e.g., VWR, Fisher) for about $159.

Kane-May Ltd.

Vaisala (UK) Ltd.
11 Billing Road
Northampton NN1 5 AW
England
0604 22415

Screen, fine mesh (wire cloth), for straining mascerated anthers to purify a suspension of pollen grains.
Small Parts, Inc.

Sealant, for sealing microcapillary tubes containing nectar samples.
Fisher Scientific (Critoseal or Fisher brand sealants are examples)

Seed counter, electronic, for counting seeds.
Old Mill Company has a large variety of counters and accessories for them.
12011 Guildford Road
Suite 102
Annapolis Junction, MD 20701
(301) 725-8181

Sling psychrometer, for measuring relative humidity and air temperature by using both wet- and dry-bulb thermometers.
Cole-Parmer Instrument Co. carries 4 kinds.

Bacharach Instrument Co. makes models with Celsius and Fahrenheit mercury thermometers.

Fisher Scientific carries the Bacharach Instrument model (catalog number 11-664, $54.95) used by several groups of pollination biologists.
625 Alpha Drive
Pittsburgh, PA 15238
(412) 782-3500

Soil moisture blocks and meters, for measuring soil moisture.
Forestry Suppliers, Inc.

SoilMoisture Equipment Corporation
P. O. Box 30025
Santa Barbara, CA 93105
(805) 964-3525

Soil sample bags, for bagging flowers.
Forestry Suppliers, Inc. carries bags made with unbleached muslin or tyvek (with drawcords around the top) or polyethylene (with wire and tape around the top).

Spectroradiometer, for measuring spectral radiation.
LI-COR, Inc.

Spring balances, for weighing birds.
Avinet

Stake wire flags.
Forestry Suppliers, Inc.

Ben Meadows

String, for delimiting perimeters of plots.
Large cones of cotton string are often used in mail rooms for bundling mail, and you may be able to purchase one from them.

Surface-area meter, both portable and conveyor-belt models.
LI-COR, Inc.

Decagon Devices, Inc.

Delta-T Devices Ltd.

CID, Inc.
P. O. Box 9008
Moscow, ID 83843
(800) 767-0119; (208) 882-0119

Syringes, microliter, for sampling nectar, filling flowers, etc. Hamilton microliter syringes work have often been used for this purpose. They are available in 10, 25, 50, 100, and 250-μL sizes, with cemented or removable (more expensive) tips. About $27–$35, available from Cole-Parmer Instrument Co. We recommend repeating pipettors for filling flowers, as they can usually hold larger volumes and require less maintenance. Hamilton makes a repeating unit to attach to their syringes, model PB600.

Tags, for plants (a variety of plastic and metal seals and tags).
National Band and Tag Co.

Seton Name Plate Company
P. O. Box GH-1331
New Haven, CT 06505
(800) 243-6624; (203) 488-8059

Also available from Seton Ltd, Banbury, UK; Seton GmbH, Langen, Germany; and Seton Inc., Toronto, Canada.

Soodhalter Plastics, Inc. (a source for plastic drink stirrers)
1327 E. 18th St.
Los Angeles, CA 90021
(213) 747-0231

Their Stir Stix are 15 cm long, available in 12 different colors, and sturdy enough for long-term marking. $21/thousand. Other smaller plastic toothpicks, etc., are available, too.

Tangle-Trap, for making sticky traps for insects.
This is an adhesive made from polybutenes, designed to be applied to a trap surface. It will remain tacky until covered with insects or foreign matter. It is available as a paste, spray, or brushable formula, and flat yellow or spherical red traps are available (designed for whiteflies and apple maggots, respectively). A product information sheet is available from the Tanglefoot Co. Other sticky materials that may work include petroleum jelly or high vacuum grease. Commercially prepared sticky traps are also available from Park Seed Wholesale.

Tanglefoot, a tree pest barrier.
This is a mixture of castor oil, waxes, and resins that can be applied to a tree trunk or plant stem as a barrier to crawling insects (e.g., nectar-thieving ants). It comes as an aerosol spray, in tubes, squeeze bags, caulking tubes, or tubs (a product information sheet is available from the Tanglefoot Co.). The product is available from garden shops; contact the manufacturer for a list of dealers if you can't find one. One source is listed here:
Carolina Biological Supply Co. 65-4120, 1 lb. $7.86, also available as an aerosol in a spray dispenser, 65-4121, 5 g $5.48.

Tech-Pen, marking pens, available in 12 colors.
Mark-Tex Corporation
161 Coolidge Avenue
Englewood, NJ 07631
(800) 222-0876; (201) 567-4111

$20 for a set including pen and refill, refills, and points (also available separately). These are also stocked by a variety of suppliers, including:
Lux Scientific Instrument Corp.
2460 N. Dodge Blvd.
Tucson, AZ 85716

Appendix 2

Tensiometers, for measuring soil moisture.
Several models are available from Forestry Suppliers, for about $40–$50 each.

Thermometers, thermocouple, for measuring insect body temperatures.
Kane-May Ltd.

Physitemp (formerly Bailey Instruments); their BAT-12 microprobe
thermometer works well for field studies of insect temperatures. It
is battery-powered, accepts a variety of sensors, and is very portable.
154 Huron Avenue
Clifton, NJ 07013
(201) 779-5577; telex 238790

Thin-layer chromatography sheets, for analysis of nectar contents.
Eastman-Kodak TLC plates, coated with silica gel (Corbet et al. 1979) (have
a polyester film backing, so they can be cut into smaller sizes. There are also
other brands with polyester backing.
VWR, Fisher Scientific, etc.

Time-domain reflectometry, for measuring soil water content.
Campbell Scientific TDR soil moisture measurement system (about $8,000),
Campbell Scientific, Inc.

UV fluorescent powders.
Day-Glo Color Corp. Day-Glo brand thermoplastic fluorescent pigments.
Available in 11 different colors, $6.40–$7.00/pound for quantities less than
150 pounds; standard packaging is 50-lb. bags, but you may be able to get
smaller (e.g., 1-lb) quantities upon request. Average particle size is about 5
microns.

Radiant Color Corp. produces organic daylight fluorescent pigments in 10
different colors. Particle size is about 6 microns for their Series R-105
pigments, pricing for quantities of 5 pounds or less is $12/pound.

Shannon Luminescence. Carries 5 colors of inorganic "invisible" pigments,
white in ambient light and fluorescing under ultraviolet. Also six colors of
organic pigments that are colored in both ambient and ultraviolet light.
Particle sizes are about 10 microns, prices are $40/ounce, $55/4 ounces,
$95/pound.

USR Optonix Inc. manufactures five colors of fluorescent luminescent
pigments.

Note: We have not been able to communicate with U.S. Radium Corp.
(Luminescent Chemicals Division, 170 E. Hanover Ave. Morristown, NJ
07960), which was a supplier for some studies in the past. If you can't locate
a source of these powders, look for fluorescent chalk, and crush it to make
your own.

UV insect trap.
Carolina Biological Supply Co. carries a portable AC-powered one that fits
into a plastic bucket for storage or carrying.

Appendix 2

UV Killer.
A compound for treating fabric to eliminate UV reflection but that may work with other materials as well. 18-oz spray bottle. Try your local sporting goods store or:
    Gander Mountain, Inc. (catalog #199G7813, $7.99)
    Box 248
    Highway W
    Wilmot, WI 53192
    (800) 558-9410; (414) 785-4500

Also available from the manufacturer together with a special detergent for washing treated fabric, for $14.90 and $3.00 shipping,
    Atsko
    2530 Russell S.E.
    Orangeburg, SC 29115
    (803) 531-1820

UV lights, portable.
BlackEye Fluorescent light. Operates from 12-V cigarette lighter or 6 D-size batteries; has both UV and standard fluorescent white bulbs. $19.95 from Cabela's. A larger model is also available for 12V power only.
    Battery-operated UV hand lamps. Several models are available from UVP (distributed by VWR Scientific, Sargent Welch Scientific, American Scientific, Fisher Scientific, and many others).
    Ultraviolet Products
    5100 Walnut Grove Avenue
    P. O. Box 1501
    San Gabriel, CA 91778
    (800) 452-6788; (818) 285-3123; fax (818) 285-2940

    LUX Scientific Instrument Corp.
    2460 North Dodge Boulevard
    Tucson, AZ 85716
    (602) 327-4848; fax (602) 327-5102

Shannon Luminous also has some portable UV lights.

Vacuum collectors, for sampling insect density or extracting insects from vegetation.

    Burkard Manufacturing Co. Ltd. makes a couple of different models, including one mounted on a backpack for field use.

Vacuum pumps, for collecting floral odors.
SKC Air-Check Personal Air Sampler, model 22-4, portable and rechargeable, about $400.00.
    SKC, Inc.
    334 Valley View Road
    Eighty Four, PA 15330-9614
    (800) 242-1276

Appendix 2

Videotapes of pollination.
Plant Reproduction, and Sexual Encounters of the Floral Kind, both available from:
Insight Media
121 West 85th St.
New York, NY 10024
(212) 721-6316

Waterproof paper and notebooks.
Rite-in-the-Rain field books, available from Forestry Suppliers

Weather stations.
Campbell Scientific sells equipment for setting up a weather station.

Edmund Scientific

LI-COR, Inc. sells a meteorological data logging system

Radio Shack

Sources of information

Bee-L is an electronic bulletin board, or research communication network accessible by computer. The file server, or mainframe computer that maintains the mailing list and forwards messages submitted to the bulletin board, is accessible from both BITNET and INTERNET, as well as at least one commercial network (Compuserve, via INTERNET). Southwick (1989) describes how it works. In brief: (1) Send the message SUBSCRIBE BEE-L xxx, where xxx is your full name, to the e-mail address: LISTSERV@ALBNYVM1.BITNET; (2) to send a message to other subscribers, send it to BEE-L@ALBNYVM1.BITNET.

Bumble Bee Quest Newsletter, a quarterly publication begin in June 1991, to stimulate interaction and communication among bumblebee researchers. Year-round rearing and commercial use in pollination are particular emphases. Edited by Margriet Wyborn, under the auspices of the International Commission for Plant-Bee Relationships. To receive the newsletter, send your name and address to:
Dr. D. T. Fairey
Research Station
Box 29
Beaverlodge T0H 0C0
Canada

Melissa, *The Melittologist's Newsletter,* is a regular newsletter for the dissemination of information on the study of solitary and social bees.

438

Appendix 2

Editor
Dr. Ronald J. McGinley
Department of Entomology
Smithsonian Institution NHB-105
Washington, D.C. 20560

International Bee Research Association
18 North Road
Cardiff CF1 3DY
United Kingdom
(0222) 372409/372450; telex: 23152 monref G8390
E-mail: MUNNPA@CARDIFF.AC.UK

According to their literature, "IBRA is an educational and scientific charitable trust, funded annually by subscriptions from Members and income from sale of publications and other services. IBRA provides the world's most comprehensive information and advisory service on all aspects of bees and beekeeping — thereby helping to promote bee-keeping and bee research worldwide." IBRA publishes *Bee World, Apicultural Abstracts,* and the *Journal of Apicultural Research.*

International Commission for Plant-Bee Relationships
The ICPBR is a section of the International Union of Biological Sciences, founded in 1950. Some of its "working groups" will be of interest to pollination biologists, including the pollination group, the nectar group, and the honey and pollen group. The ICPBR and its working groups sponsor symposia and circulars (newsletters). Its official organ is the journal *Bee World.* For membership information, contact:
J. N. Tasei
Secretary ICPBR
Laboratoire de Zoologie
INRA
86600 Lusignan
France
(33)49-55-60-90; fax: (33)49-55-60-88

International Society of Hymenopterists
James M. Carpenter, Secretary
Museum of Comparative Zoology
Harvard University
Cambridge, MA 02138

Dues are $5.00/year; send to the treasurer
Gary A. P. Gibson
Biosystematics Research Centre
Agriculture Canada
K. W. Neqtby Bldg.
Ottawa, Ontario K1A 0C6
Canada

The society has an annual meeting, usually with the Entomological Society of America, and is spearheading development of a new journal, *Journal of Hymenoptera Research*.

Wildflowers

Information about particular wildflower species might be obtained through local botanical clubs and wildflower societies. A list of the groups in the United States is available from

New England Wildflower Society, Inc.
Garden in the Woods
Hemenway Road
Framingham, MA 01701

Addresses and phone numbers for suppliers:

If they are available, we have listed toll-free (area code 800) telephone numbers. Because these are not available for people calling from outside the United States, we have also listed regular telephone numbers. If you are trying to get information from outside the country, have access to electronic mail, and know someone in the United States, you might send them an electronic mail message (usually at no cost, and almost instantaneous) and ask them to make a toll-free call for you.

Aldrich Chemical Company
940 West Saint Paul Avenue
Milwaukee, WI 53233
(800) 558-9160; (414) 273-3850

Avinet
P. O. Box 1103
Dryden, NY 13053-1103
(607) 844-3277

Ben Meadows Company
3589 Broad Street
P. O. Box 80549
Atlanta (Chamblee), GA 30366
(800) 241-6401

BioQuip Products, Inc.
17803 LaSalle Ave.
Gardena, CA 90248
(213) 324-0620; fax (213) 324-7931

Appendix 2

Burkard Manufacturing Co. Ltd.
Woodcock Hill Industrial Estate
Rickmansworth, Hertfordshire WD3 1PJ
England
0923 773134

Cabela's
812 13th Ave.
Sydney, Nebraska 69160
(800) 237-4444; fax (308) 254-2200

Campbell Scientific, Inc. (also has offices in Canada, England)
P. O. Box 551
815 West 1800 North
Logan, UT 84321
(801) 753-2342; fax (801) 752-3268

Carolina Biological Supply Company
2700 York Road
Burlington, NC 27215
(800) 334-5551; (919) 584-0381

Customers in the Western United States, order from:
Box 187
Gladstone, OR 97027
(800) 547-1733; (503) 656-1641

Overseas customers call (919) 584-5555, Export Department.

Cole-Parmer Instrument Company
7425 North Oak Park Avenue
Chicago, IL 60648
(800) 323-4340; (708) 647-7600

Day-Glo Color Corporation
4515 St. Clair Avenue
Cleveland, OH 44103
(216) 391-7070

They also have sales representatives in Newark, Chicago, Los Angeles, Charlotte, and in the Netherlands, Germany, and Canada.

Delta-T Devices Ltd.
128 Lower Road, Burwell
Cambridge CB5 0EJ
England
0638 742922

Appendix 2

Edmund Scientific
 101 E. Gloucester Pike
 Barrington, NJ 08007-1380
 (609) 573-6250

Fisher Scientific (one of the larger scientific supply companies, has many branch offices)
 711 Forbes Avenue
 Pittsburgh, PA 15219
 (412) 562-8300 (there are several regional toll-free numbers)

Forestry Suppliers, Inc.
 P. O. Box 8397
 Jackson, MS 39284-8397
 (800) 647-5368; (601) 354-3565

Grant Instruments (Cambridge) Ltd.
 Barrington
 Cambridge DB2 5QZ
 England
 0763 260811

Kane-May Ltd.
 Burrowfield
 Welwyn Garden City
 Hertfordshire AL7 4TU
 Great Britain

Kontes Scientific Glassware
 Spruce Street
 P. O. Box 729
 Vineland, NJ 08360
 (609) 692-8500

Lawson Bag
 P. O. Box 8577
 Northfield, IL 60093
 (800) 451-1495; (708) 446-8812

LI-COR, Inc.
 P. O. Box 4425
 Lincoln, NE 68504
 (800) 447-3576; (402) 467-3576

National Band and Tag Co.
 721 York Street
 Box 430
 Newport, KY 41072
 (606) 261-2035

Park Seed Wholesale
Cokesbury Road
Greenwood, SC 29647-0001
(800) 845-3366

Radiant Color
2800 Radiant Avenue
Richmond, CA 94804
(415) 233-9119; (800) 777-2968

Rainin Instrument Co. Inc.
Mack Road
Woburn, MA 01801
(800) 472-4646; (617) 935-3050 (California office [510] 654-9142)

Shannon Luminous Materials, Inc.
304A North Townsend Street
Santa Ana, CA
92703
(714) 550-9931

Small Parts, Inc.
6901 N.E. Third Avenue
P. O. Box 381736
Miami, FL 33138
(305) 751-0856

Sullivan and Brampton Company
251 South Van Ness Ave.
San Francisco, CA 94103
(415) 861-4455; fax (415) 861-8075

Tanglefoot Company
314 Straight Avenue, SW
Grand Rapids, MI 49504
(616) 459-4139

Tetko, Inc.
P. O. Box 346
111 Calumet St.
Lancaster, NY 14086
(716) 683-4050; fax (716) 683-4053

University Products, Inc.
P. O. Box 101
Holyoke, MA 01041-0101
(800) 628-1912; in MA (800) 336-4847; (413) 532-3372

USR Optonix Inc.
Kings Highway
Box 409
Hackettstown, NJ 07840
(908) 850-1500; fax (908) 850-5756

VWR Has many regional offices, including:
P. O. Box 3551
Seattle, WA 98124
(800) 333-6336; (206) 575-1500

P. O. Box 626
Bridgeport, NJ 08014
(800) 234-9300

Watkins and Doncaster
P. O. Box 5
Cranbrook
Kent TN18 5EZ
England

Appendix 3.
Chemicals and Stains

This appendix contains information for ordering most of the chemicals and microscopy supplies that are mentioned in the book. A number of these chemicals are toxic or dangerous. Please use caution in working with them, and dispose of them properly. If you don't know how, you can get information from the company that supplies them, or from environmental safety offices or chemistry departments on most university or college campuses. We have listed for each chemical some of the typical uses for pollination studies and citations to some studies that have used it. We have also included a source (use of catalog numbers from a particular company is not meant as an endorsement of that company) and representative sample sizes and prices (1991 or 1992 catalogs). Some of these chemicals are available in different grades, and you may be able to use a lower-cost grade than we have listed here. For example, some of the stains are available in a lower-cost grade or certified by the Biological Stain Commission (and more expensive); the lower-cost grade would probably work just as well for some application such as dyeing pollen grains for studies of pollen flow. Similarly, some biochemicals are available in lower-cost grades or tested for plant cell culture.

Note about **radioactive isotopes**: Their use is regulated in most countries in the world. In the United States, information about federal and state licensing is available from:

U.S. Nuclear Regulatory Commission
Division of Licensing and Regulations
Washington, D.C. 20555

You can probably get this same information, which is specific for some states, from radiation safety officers on most university campuses or colleges.

Addresses for ordering chemicals:

Aldrich Chemical Company
940 West Saint Paul Avenue
Milwaukee, WI 53233
(800) 558-9160; (414) 273-3850

Alltech Associates, Inc. (has overseas offices, too)
2051 Waukegon Road
Deerfield, IL 60015-1899
(800) 255-8324; (708) 948-8600

Eastman Fine Chemicals (Kodak chemicals are also distributed by Fisher
  and VWR)
(800) 225-5352; (716) 588-4817

Fisher Scientific (has many branch offices as well)
711 Forbes Avenue
Pittsburgh, PA 15219
(412) 562-8300; there are several regional toll-free numbers

Fluka Chemical Corporation
980 South Second St.
Ronkonkoma, NY 11779
(800) 358-5287

Passaic Color and Chemical
2836 Patterson St.
Patterson, NJ 07501
(201) 465-3932

Ted Pella, Inc. (electron microscopy supplies)
P. O. Box 492477
Redding, CA 96049-2477
(800) 237-3526; (916) 243-2200

Pfaltz and Bauer, Inc.
172 East Aurora Street
Waterbury, CT 06708
(800) 225-5172; (203) 574-0075; fax (203) 575-3181

Pylam
1001 Stewart Ave.
Garden City, NY 11530
(800) 645-6096

Sigma Chemical Company
P. O. Box 14508
St. Louis, MO 63178-9916
(800) 325-3010

   Sigma Chemical Co., Ltd.
   Fancy Road
   Poole, Dorset BH17 7NH
   England 0202 733114 (charges reversed)

Appendix 3

Sigma Chemie GmbH
Grünwalder Weg 30
W-8024 Deisenhofen
Germany
089 613 01-0

Sigma Chimie S.a.r.l.
L'Isle d'Abeau Chesnes
B.P. 701 38070
St. Quentin Fallavier Cedex
France
74 82 28 00

Chemicals

Acetic acid, glacial ($C_2H_4O_2$) (for acetolysis; for chromatography solvent for sugar analyses, Grant and Beggs 1989; for PABA stain for nectar, Baker and Baker 1982)
Sigma Chemical Co. A 6283, 100 mL $8.90

Acetic anhydride ($C_4H_6O_3$) (for acetolysis)
Sigma Chemical Co. A 6404, 200 mL $7.00

Acetone (for chromatography solvent for detecting sugars, Magnarelli et al. 1979)
Fisher Scientific A11-1, 1 L $23.75

Acetonitrile (C_2H_3N; methyl cyanide) (for solvent for sugar analyses by HPLC, Southwick et al. 1981)
Sigma Chemical Co. A 3396, 500 mL $11.50

Acid fuchsin ($C_{20}H_{17}N_3O_9S_3Ca$) (for Alexander's stain, Alexander 1980)
Sigma Chemical Co. A 3908, 25 g $18.75

Acridine orange ($C_{17}H_{19}N_3$) (used as an internal marker for insects, Strand et al. 1990; a stain for visualizing cytological features of pollen tubes, Alves et al. 1968)
Sigma Chemical Co. A 6529, 5 g $7.00

Aerosol-OT (a biological detergent, for preparing pollen for SEM, Lynch and Webster 1975; used as a surfactant for staining lipids on flowers, Slater and Calder 1988)
Sigma Chemical Co. A 6627, 10% (w/v) solution, 100 mL $9.75

Agar (for picking pollen off of insects, Motten 1986)
Sigma Chemical Co. A 6549, 100 g $12.90

Albustix (for testing for protein in urine, based on tetrabromphenol blue, but also used for testing nectar, Willmer 1980)
Fisher Scientific AM28-70, 100 for $25.35

447

Alcian blue (for staining pollen tube walls, Hewitt et al. 1985; for staining pectinaceous material in fixed pistils, Ciampolini et al. 1990)
Sigma Chemical Co. A 3157, 25 g $36.00

Amido black (see Naphthol blue black)

p-Aminobenzoic acid ($C_7H_7NO_2$; PABA) (for detecting sugar concentrations, Saini 1966, Baker and Baker 1982, Grant and Beggs 1989; for visualizing chromatograms, Saini 1966)
Sigma Chemical Co. A 0129, 100 g $14.05

Ammonium hydroxide (NH_4OH) (for detecting sugars on paper chromatograms, Trevelyan et al. 1950; for gas chromatography of amino acids, Gilbert et al. 1981; for analyses of alkaloids in nectar, Deinzer et al. 1977)
Sigma Chemical Co. A 6899, 500 mL $6.55

Ammonium nitrate (NH_4NO_3) (for pollen germination medium; Leduc, Monnier, and Douglas 1990)
Sigma Chemical Co. A 9642, 500 g $13.35

Amyl acetate (for preparing pollen for SEM, Lynch and Webster 1975)
Aldrich Chemical Co. 22,747-1, 100 g $7.30

Aniline (C_6H_7N) (for making aniline-oxalate spray for visualizing sugars on chromatograms, Partridge 1951)
Sigma Chemical Co. A 9880, 20 mL $4.10

Aniline blue (for use as a pollen stain, Cruzan 1989, Ramming and Hinrichs 1973, Mulcahy and Mulcahy 1982; for staining callose in fixed pistils, Ciampolini et al. 1990) Methyl blue is sometimes used as a substitute.
Aldrich Chemical Co. 86,102-2, 25 g $11.30

Aniline diphenylamine spray reagent (for detecting reducing sugars on chromatography plates, Watt et al. 1974, Magnarelli and Anderson 1977)
Sigma Chemical Co. A 8142, 100 mL $14.75

8-anilino-naphthalenesulfonic acid ($C_{16}H_{13}NO_3S$) (for staining nectar or stigmatic exudates for the presence of protein, Heslop-Harrison et al. 1974 [referred to as 1-ANS])
Sigma Chemical Co. A 1028, 5 g $10.50

p-Anisidine hydrochloride ($C_7H_9NO \cdot HCl$) (for visualizing sugars after paper chromatography, Percival 1961, Baskin and Bliss 1969)
Sigma Chemical Co. A 1006, 10 g $16.20

Anthrone ($C_{14}H_{10}O$) (for determining carbohydrate content of nectar, McKenna and Thomson 1988; for testing nectar or mosquitoes for presence of fructose, Van Handel 1967, 1972; Magnarelli and Anderson 1977; Magnarelli et al. 1979)
Sigma Chemical Co. A 1631, 10 g $8.25

Arabic gum (for Hoyer's medium, Radford et al. 1974)
Sigma Chemical Co. G 9752 (gum arabic), 100 g $5.00

Appendix 3

Araldite 6005 resin (for embedding pollen grains for sectioning prior to TEM, Feuer 1990)
Ted Pella, Inc. 18056, 450 g $10.70

Auramine O ($C_{17}H_{22}N_3 \cdot Cl$; Color Index 41000) (for detecting lipids in stigmatic secretions, Ciampolini et al. 1990, or in fixed stigmas, Lord and Kohorn 1986)
Sigma Chemical Co. A 9655, 25 g $20.00

Barium carbonate ($Ba^{13}CO_3$) (for generating labeled CO_2, Pleasants et al. 1990)
Sigma Chemical Co. 27,719-3, 1 g $21.10

Basic fuchsin ($C_{20}H_{20}ClN_3$) (see also Pararosaniline) (for basic fuchsin jelly, Beattie 1971a)
Eastman Fine Chemicals, Kodak #01762C 25 g $21.20

Benzidine ($C_{12}H_{12}N_2$) (for testing for peroxidase enzymes, Galen et al. 1985; for solvent for TLC for nectar sugars, Jeffrey et al. 1969) CAUTION: **This is a carcinogen.**
Sigma Chemical Co. B 3503, 1 g $21.95

bisBenzimide ($C_{25}H_{24}N_6O \cdot 3HCl$; Hoechst No. 33258) (a fluorescent DNA probe used for staining pollen tubes, Hough et al. 1985)
Sigma Chemical Co. B 2883, 100 mg $18.75

BHT ($C_{15}H_{24}O$; butylated hydroxytoluene) (for amino acid analyses of honey, Gilbert et al. 1981)
Sigma Chemical Co. B 1378, 100 g $6.80

Bismarck brown ($C_{18}H_{18}N_8 \cdot 2HCl$; Color Index 21000) (used as a pollen stain, Peakall 1989)
Sigma Chemical Co. B 5263, 25 g $28.00

Bismuth nitrate ($Bi(NO_3)_3$) (to test for the presence of alkaloids in nectar, Baker and Baker 1975)
Sigma Chemical Co. B 9383, 25 g $3.85 (see also Dragendorff's reagent)

Borax ($Na_2B_4O_7$; sodium tetraborate) (for making solution for visualizing chromatograms, Block et al. 1958)
Sigma Chemical Co. B 0127, 100 g $8.50

Boric acid (H_3BO_3) (for Brewbaker-Kwack medium, Brewbaker and Kwack 1963; for making bromocresol purple spray for visualizing sugars on chromatograms, Block et al. 1958; for making TLC medium, Jeffrey et al. 1969)
Sigma Chemical Co. B 0252, 100 g $6.50

Boron trifluoride-methanol (BF_3) (for testing for alkaloids in nectar, Mattocks, 1967; for testing for lipids in nectars, Buchmann and Buchmann 1981)
Sigma Chemical Co. B 1252, 14% solution, 100 mL $9.10

Brilliant green ($C_{27}H_{34}N_2O_4S$; Color Index 42040) (used as a pollen stain, Peakall 1989)
Sigma Chemical Co. B 4014, 25 g $29.00

Appendix 3

Bromcresol purple ($C_{21}H_{16}Br_2O_5S$) (used to visualize sugars on chromatograms, Block et al. 1958, Watt et al. 1974)
Sigma Chemical Co. B 5880, 5 g $7.50

Bromphenol blue ($C_{19}H_{10}Br_4O_5S$) (used as a stain for the presence of protein in nectar, Baker and Baker 1973–doesn't specify whether they used sulfone form or sodium salt)
Sigma Chemical Co. B 0126, sulfone form, 5 g $6.10

n-Butanol ($C_4H_{10}O$) (for chromatography for testing for sugars in mosquitoes, Magnarelli and Anderson 1977, Magnarelli et al. 1979; for chromatography solvent for sugar analyses, Hough et al. 1950, Saini 1966, Watt et al. 1974, Grant and Beggs 1989)
Sigma Chemical Co. BT-105, 500 mL $10.35

tert-Butanol ($C_4H_{10}O$) (for chromatography of sugars, Saini 1966)
Sigma Chemical Co. B 2138, 500 mL $12.15

Cacodylate buffer (for fixing flowers for location of osmophores, Stern et al. 1986)
Ted Pella, Inc. 18851, 100 g $48.50

Calcium chloride ($CaCl_2$) (for use as a drying agent in analysis of lipids in floral oils, Buchmann and Buchmann 1981; or for drying pollen, Rouse 1985)
Sigma Chemical Co. C 1016, 20 mesh anhydrous, 100 g $6.10

Calcium chloride, dihydrate ($CaCl_2 \cdot 2H_2O$) (for pollen germination medium, Leduc, Monnier, and Douglas 1990)
Sigma Chemical Co. C 3881, 500 g $15.50

Calcium nitrate ($Ca[NO_3]_2 \cdot 4H_2O$) (for Brewbaker-Kwack medium, Brewbaker and Kwack 1963) (for semi-vivo culture of pollen tubes, Rao and Kristen 1990)
Sigma Chemical Co. C 1396, 500 g $16.20

Calco Red N-1700 (dye used in insect diet to produce marked adults, Bell 1988, Showers et al. 1989, Lance and Elliott 1990). Calco was a brand name for a company that is no longer in business, but the dye is still available from Pylam. A suitable replacement may be Passaic Color and Chemical Co. oil red 2144, also available from Pylam. 1-pound jar, $45 for either product.

Calco oil blue V (dye used in beetle diet to produce marked adults, Lance and Elliott 1990)
Pylam, 1-pound jar, $80.

Calcofluor white (see Fluorescent brightener)

Camphor (for testing for the presence of surfactants on nectar, Corbet and Willmer 1981)
Sigma Chemical Co. C 9380, 25 g $7.00

Carbon disulfide (a solvent for extracting floral odors, Bergström et al. 1980)
Aldrich 27,066-0, 500 mL $12.60

Carmine (Color Index 75470) (for making acetocarmine stain, Cruzan 1989, Radford et al. 1974)
Sigma Chemical Co. C 6152, 25 g $95.00

Chloral hydrate (for Hoyer's medium; or for clearing ovules, Palser et al. 1989)
Sigma Chemical Co. C 8383, 100 g $11.10 (a class IV DEA controlled substance)

Chloroform ($CHCl_3$) (for testing mosquitoes for presence of fructose, Van Handel 1972; solvent for PABA for detecting sugars, Baker and Baker 1982; used in analysis of alkaloids in honey, Deinzer et al. 1977; for eluting lipids from floral oils, Buchmann and Buchmann 1981)
Fisher Scientific C294-1, 1 L $25.80

Cholesteryl decanoate ($C_{37}H_{64}O_2$; cholesteryl n-decylate) (for making liquid crystals for visualizing temperature of flowers, Meeuse 1973)
Sigma Chemical Co. C 4633, 1 g $6.30

Cholesteryl oleate ($C_{45}H_{78}O_2$) (for making liquid crystals for visualizing temperature of flowers, Meeuse 1973)
Sigma Chemical Co. C 9253, 250 mg $16.80

Cholesteryl propionate ($C_{31}H_{52}O_3$) (for making liquid crystals for visualizing temperature of flowers, Meeuse 1973)
Sigma Chemical Co. C 7133, 5 g $13.10

Chromosorb G (porous polymer support for gas chromatography, for collecting floral odors, Williams 1983)
Sigma Chemical Co. C 5389, 100 g $38.70

Cobalt chloride ($CoCl_2$) (for making dip sticks for measuring nectar depth in flowers, see Chapter 5; for pollen germination medium; Leduc, Monnier, and Douglas 1990)
Sigma Chemical Co. C 3169, 100 g $19.90

Coomassie brilliant blue ($C_{47}H_{48}N_3O_7S_2Na$) (for staining for proteins in nectar, Bradford 1976, Scogin 1979; to stain for protein in stigmatic solutions, Lord and Kohorn 1986) This is a trademarked name for a stain that is also available as Brilliant blue G:
Sigma Chemical Co. B 5133, 5 mg $8.30

Cotton blue (see Aniline blue)

Crystal violet ($C_{25}H_{30}N_3Cl$; Color Index 42555) (for staining pollen, Paton and Ford 1977; R. L. Tremblay (*personal communication*) found that it killed orchid flowers when injected into pollinia, and Peakall (1989) also found that it killed flowers)
Sigma Chemical Co. C 0775, 25 g $18.00

Cupric sulfate, pentahydrate ($CuSO_4 \cdot 5H_2O$) (for pollen germination medium, Leduc, Monnier, and Douglas 1990)
Sigma Chemical Co. C 6283, 250 g $16.20

Cyclohexane (C_6H_{12}) (for collecting floral odors, Williams 1983)
Sigma Chemical Co. C 8517, 100 mL $13.00

Dansyl chloride ($C_{12}H_{12}ClNO_2S$) (for dansylating amino acids in nectar samples to make them UV fluorescent, for identification on chromatograms, Baker and Baker 1976a)
Sigma Chemical Co. D 2625, 500 mg $7.50

DAPI (See 4′,6-Diamidino-2-phenylindole)

o-Dianisidine, tetrazotized ($C_{14}H_{12}N_4O_2Cl_2\cdot ZnCl_2$; also known as Diazo blue B, and Fast blue salt) (used for stigmatic esterase assay, Lord and Kohorn 1986)
Sigma Chemical Co. D 3502, 10 g $5.05

2,6-Dichlorophenol-indophenol ($C_{12}H_6Cl_2NO_2Na$) (used as a reagent to test for acid compounds in nectar, Buchmann and Buchmann 1981; to test for antioxidants in stigmatic solutions, Baker et al. 1973)
Sigma Chemical Co. D 1878, 1 g $6.30

Diethylamine ($C_4H_{11}N$) (used in chromatography solvent for identifying honey contents, Willmer 1980)
Sigma Chemical Co. D 3131, 250 mL $6.50

Diethyl ether (used as a solvent for extracting floral odors, Bergström et al. 1980)
Aldrich 34,436-2, 950 mL $19.00

Diglime (see bis-2-Methoxyethyl ether)

4′,6-Diamidino-2-phenylindole ($C_{16}H_{15}N_5\cdot 2HCl$) (DAPI; used to stain generative cells of pollen tubes, Shivanna et al. 1991a)
Sigma Chemical Co. D 1388, 5 mg $21.55

2,2-Dimethoxypropane ($C_5H_{12}O_2$) (for dehydrating fixed gynoecia for histological studies, Lord and Kohorn 1986)
Sigma Chemical Co. D 8761, 100 mL $8.15

p-Dimethylaminobenzaldehyde ($C_9H_{11}NO$) (for testing for alkaloids in nectar, Mattocks, 1967)
Sigma Chemical Co. D 2004, 25 g $8.35

Dimethyl-dichlorosilane ($C_2H_6Cl_2Si$) (for silanizing glass for use in fragrance analyses)
Sigma Chemical Co. D 3879, 50 mL $5.55

2,4-Dinitrophenylhydrazine ($C_6H_6N_4O_4$) (for a colorimetric assay for vitamin C in nectar, Lüttge 1962)
Sigma Chemical Co. D 2630, 10 g $6.60

3,5-Dinitrosalicylic acid ($C_7H_4N_2O_7$) (for making reagent for photometric nectar analysis, Kapyla 1978)
Sigma Chemical Co. D 1510, 25 g $6.95

Appendix 3

Diphenylamine ($C_{12}H_{11}$) (for making acidified aniline-diphenylamine [ADPA] reagent for revealing sugars on chromatograms, Watt et al. 1974)
    Sigma Chemical Co. D 2385, 5 g $11.75

Dragendorff's spray reagent (0.11 M potassium iodide and 0.6 mM bismuth subnitrate in 3.5 M acetic acid, for the detection of alkaloids and quarternary nitrogen compounds in nectar, Baker and Baker 1975)
    Sigma Chemical Co. D 7518, 100 ml $12.00

EDTA ($C_{10}H_{16}N_2O_8$; ethylenediaminetetraacetic acid) (used as a buffer for neutron activation analysis, Gaudreau and Hardin 1974). See also the entry for ethylenediaminetetraacetic acid.
    Sigma Chemical Co. ED, free acid, 100 g $6.50

Ethidium bromide ($C_{21}H_{20}N_3Br$) (for counterstaining stigmas, Waser and Fugate 1986, Waser et al. 1987; used as a fluorescent DNA probe for pollen tubes by Hough et al. 1985; enhances fluorescence of pollen grains attached to a stigma, Dumas and Knox 1983). **CAUTION: This is mutagenic.**
    Sigma Chemical Co. E 8751, 1 g $10.75

Ethyl acetate (for chromatography solvent for sugar analyses, Percival 1961, Baskin and Bliss 1969, Grant and Beggs 1989)
    Fisher Scientific E145-1, 1 L $39.00

Ethylenediaminetetraacetic acid (EDTA, sodium salt; Na_2EDTA) (for pollen germination medium; Leduc, Monnier, and Douglas 1990). See also the entry for EDTA.
    Sigma Chemical Co. E 6635, 100 g $14.85

Ethyl ether (for chromatography solvent for sugar analyses, Magnarelli and Anderson 1977)
    Fisher Scientific E134-1, 1 L $35.80

Europium oxide (Eu_2O_3) (used as a marker for neutron activation analysis; see references in Gaudreau and Hardin 1974)
    Sigma Chemical Co. E 5626, 1 g $16.25

Evans blue ($C_{34}H_{24}N_6O_{14}S_4Na_4$) (used as a pollen stain, Linhart 1973). CAUTION: **This is a carcinogen.**
    Sigma Chemical Co. E 2129, 10 g $15.75

Fast blue B (for use with α-naphthyl acetate to test for stigmatic esterase activity, Mattson et al. 1974)
    See o-Dianisidine.

Fast green ($C_{37}H_{34}N_2O_{10}S_3Na_2$; Color Index 42053) (for staining pollen tubes, Levin 1990)
    Sigma Chemical Co. F 7252, 5 g $7.90

Ferric acetate ($Fe(OH)(CH_3CO_2)_2$) (for making acetocarmine jelly, Radford et al. 1974)
    Pfaltz and Bauer, Inc. F00610, 100g $23.00

Appendix 3

Ferrous sulfate, heptahydrate ($FeSO_4 \cdot 7H_2O$) (for pollen germination medium; Leduc, Monnier, and Douglas 1990)
> Sigma Chemical Co. F 7002, 250 g $9.90

Fluorescein diacetate ($C_{24}H_{16}O_7$) (for staining pollen tubes, Shivanna and Heslop-Harrison 1981)
> Sigma Chemical Co. F 7378, 10 g $9.95

Fluorescent brightener ($C_{40}H_{42}N_{12}O_{10}S_2Na_2$; also known as Calcofluor white) (for increasing the fluorescence of pollen tubes, Jefferies and Belcher 1974)
> Sigma Chemical Co. F 6259, 1 g $12.70

Folin and Ciocalteu's phenol reagent (for protein determination using the method of Lowry et al. 1956)
> Sigma Chemical Co. F 9252, 100 ml $17.40

Formaldehyde (CH_2O) (for making FAA preservative)
> Sigma Chemical Co. F 1635, 500 ml $9.45

Formic acid (CH_2O_2) (for chromatographic analyses of amino acids, Baker and Baker 1977)
> Sigma Chemical Co. F 0507, 100 ml $11.55

Fructose ($C_6H_{12}O_6$) (for a standard for nectar analyses)
> Sigma Chemical Co. F 0127, 100 g $4.95

Galactose ($C_6H_{12}O_6$) (for a standard for nectar analyses)
> Sigma Chemical Co. G 0750, 10 g $9.20

Gelatin (for basic fuchsin jelly, Beattie 1971). Available in a variety of gel strengths (depends on the animal source of the gelatin). A representative one is listed here.
> Sigma Chemical Co. G 9382, 100 g $7.00

Glucose ($C_6H_{12}O_6$) (for a standard for nectar analyses)
> Sigma Chemical Co. G 8270, 100 g $8.25

Glutaraldehyde ($C_5H_8O_2$) (for fixing gynoecia for histological studies, Lord and Kohorn 1986, Ciampolini et al. 1990)
> Sigma Chemical Co. G 5882, 10 mL $36.80

Glycerin (see Glycerol)

Glycerol, or glycerin ($C_3H_8O_3$) (for Alexander's stain, Alexander 1980; for making basic fuchsin jelly, Beattie 1971a; for making lactophenol, Radford et al. 1974)
> Sigma Chemical Co. G 7757, 500 mL $8.90

Glycol methacrylate (for embedding fixed gynoecia for histological studies, Lord and Kohorn 1986)
> Ted Pella, Inc. 18350 (GMA kit), $26.70

454

Hematoxylin solution, Mayer's (for stain-clearing for observations of ovules, Stelly et al. 1984)
Sigma Chemical Co. MHS-1, 100 mL $10.50

Heptafluorobutyric anhydride ($C_8F_{14}O_3$) (for amino acid analyses of honey, Gilbert et al. 1981)
Sigma Chemical Co. H 1006, 5 mL $22.00

Heptane (C_7H_{16}) (for extracting floral odors, Brantjes 1978)
Sigma Chemical Co. H 9629, 25 mL $3.50

Hepes buffer ($C_8H_{18}N_2O_4S$) (for pollen germination, Peakall and Beattie 1989)
Sigma Chemical Co. H 3375, 10 g $9.90

Hexamethyldisilazane (a reagent for preparing trimethylsilyl derivatives for gas chromatography, Sweeley et al. 1963, Baskin and Bliss 1969)
Sigma Chemical Co. H 4875, 25 mL $7.40

n-Hexane (C_6H_{14}) (a solvent for extracting floral odors, Bergström et al. 1980)
Sigma Chemical Co. H 9379, 100 mL $16.20

Histidine ($C_6H_9N_3O_2$) (for constructing a histidine scale for scoring concentrations of amino acids in nectars, Baker and Baker 1973)
Sigma Chemical Co. H 8000, 5 g $5.25

Hydrochloric acid (for the pararosanilin method of determining stigmatic receptivity, Bancroft 1975)
Sigma Chemical Co. 251-2, 2.0 N, 50 mL $9.75

Hydrogen peroxide (H_2O_2) (for testing for alkaloids in nectar, Mattocks 1967; for testing for peroxidase enzymes on stigmas, Galen et al. 1985)
Sigma Chemical Co. H 1009, 30% solution, 100 mL $14.05

Hydrogen sulfide (for visualizing sugars on paper chromatograms, Trevelyan et al. 1950)
Fisher Scientific 10-599L, 5.6 ft^3 $170.00

Iodoplatinate spray reagent (used as a reagent to test for presence of alkaloids in nectar, Buchmann and Buchmann 1981)
Sigma Chemical Co. I 0256, 100 mL $14.75

Isopropanol (used in analysis of honey, White and Rudyj 1978)
Sigma Chemical Co. 405-7, 480 mL $10.50

Lacmoid (for staining callose, Ramming et al. 1973)
Sigma Chemical Co. L 7512, 5 g $9.50

Lactic acid ($C_3H_6O_3$) (for Alexander's stain, Alexander 1980; for polyvinyl lactophenol mounting medium, I. Baker, *personal communication*).
Sigma Chemical Co. L 1875, free acid, 30% solution by weight, 50 mL $18.55

Lactophenol (see Chapter 4, Section E.4.b for a recipe)

Lanthanum oxide (La_2O_3) (a possible choice for use as a marker for neutron activation analysis)
  Sigma Chemical Co. L 4000, 10 g $4.10

Lead citrate (($C_6H_5O_7$)$_2Pb_3 \cdot 3H_2O$) (stain for preparing pollen or pistils for electron microscopy, Feuer 1990, Ciampolini et al. 1990, Rao and Kristen 1990)
  Fluka Chemical Corp. #15326, 25 g $11.20

Lugol solution (I/KI solution, for staining pollen grains for the presence of starch, Olesen and Warncke 1989a)
  Sigma Chemical Co. L 6146, 100 mL $6.90
    (Or make your own by adding 80 g KI to 10 g I in 100 mL distilled water; Olesen, *personal communication*. For use, take 2 mL of this solution and add up to 100 mL of distilled water.)

Magnesium sulfate ($MnSO_4$) (for pollen germination medium; Leduc, Monnier, and Douglas 1990)
  Sigma Chemical Co. M 7506, 500g $26.15

Magnesium sulfate heptahydrate ($MgSO_4 \cdot 7H_2O$) (for Brewbaker-Kwack medium, Brewbaker and Kwack 1963)
  Sigma Chemical Co. M 7774 (tested for plant cell culture), 500 g $15.75

Malachite green oxalate (($C_{23}H_{25}N_2$)$_2 \cdot 3C_2H_2O_4$; Color Index 42000) (for Alexander's stain, Alexander 1980)
  Sigma Chemical Co. M 9015, 25 g $16.75

Methanol (CH_3OH) (for testing mosquitoes for presence of fructose, Van Handel 1972, Magnarelli and Anderson 1977; for eluting lipids from floral oils, Buchmann and Buchmann 1981; a solvent for extracting floral odors, Bergström et al. 1980; for PABA stain test for nectar, Baker 1982)
  Sigma Chemical Co. M 3641, 500 mL $9.25

bis-2-Methoxyethyl ether ($C_6H_{14}O_3$; Diglyme) (for testing for alkaloids in nectar, Mattocks, 1967)
  Sigma Chemical Co. M 8132, 100 mL $6.30

Methyl blue ($C_{37}H_{27}N_3O_9S_3Na_2$) (sometimes used as a substitute for Aniline blue)
  Sigma Chemical Co. M 9015, 25 g $26.00

Methylene blue ($C_{16}H_{18}ClN_3S$) (used as a pollen dye, Stephens and Finkner 1953, Stelleman 1978, Handel 1983b)
  Sigma Chemical Co. M 9140, 25 g $18.50

Methyl ethyl ketone (C_4H_8O) (for solvent for TLC of nectar sugars, Jeffrey et al. 1969)
  Sigma Chemical Co. M 2886, 500 ml $10.95

Methyl green ($C_{27}H_{35}BrClN_3 \cdot ZnCl_2$) (used as a pollen stain, Macior 1983)
  Sigma Chemical Co. M 5015, 25 g $40.00

Microstik (adhesive for attaching pollen grains to specimen stubs for SEM, Lynch and Webster 1975)
Ted Pella Inc., 16033, 14 ml $4.95

α-Naphthol ($C_{10}H_8O$) (for detecting ketoses and ketose-containing oligosaccharides in chromatographed nectar samples, Grant and Beggs 1989)
Sigma Chemical Co. N 1000, 10 g $8.65

Naphthol blue black ($C_{22}H_{14}N_6O_9S_2Na_2$; "Amido-black") (used as a stain for the presence of protein in nectar, Baker and Baker 1973)
Sigma Chemical Co. N 9002, 50% dye content, 100 g $10.30

Napthoresorcinol ($C_{10}H_8O_2$) (for making a spray to visualize sugars on chromatograms, Partridge and Westhall 1948)
Sigma Chemical Co. N 6250, 1 g $10.50

α-naphthyl acetate ($C_{12}H_{10}O_2$) (for detecting stigmatic esterase activity, Slater and Calder 1988)
Sigma Chemical Co. N 6875, 5 g $8.50

Neutral red ($C_{15}H_{17}N_4Cl$; Color Index 50040) (used as a pollen dye, Linhart 1973, Stelleman 1978; for staining for the presence of osmophores in fresh flowers, Knudsen and Tollsten 1991)
Sigma Chemical Co. N 6634, 25 g $22.00

Nile blue chloride ($C_{20}H_{20}N_3OCl$) (for screening for lipids in nectar, Baker and Baker 1975)
Sigma Chemical Co. N 5383, 5 g $13.70

Ninhydrin fixer spray reagent ($C_9H_6O_4$) (for visualizing amino acids on chromatograms, Baker and Baker 1973)
Sigma Chemical Co. N 0507, 100 mL $12.00

p-Nitroaniline ($C_6H_6N_2O_2$) (used as a reagent to test for phenolic compounds in nectar, Buchmann and Buchmann 1981)
Sigma Chemical Co. N 2128, 50 g $4.35

Nitro blue tetrazolium ($C_{40}H_{30}Cl_2N_{10}O_6$) (to test for pollen enzyme activity, Hauser and Morrison 1964, Mayer 1991)
Sigma Chemical Co. N 6876, 50 mg $9.65

Orange G ($C_{16}H_{10}N_2O_7S_2Na_2$; Color Index 16230) (used as a pollen stain, Peakall 1989)
Sigma Chemical Co. O 7252, 25 g $17.25

Orcein (used as a pollen stain, Goldblatt and Bernhardt 1990)
Sigma Chemical Co. O 7505, 5 g $25.00

Osmium tetroxide (OsO_4) (for fixing pistils or pollen for electron microscopy, Ciampolini et al. 1990, Miller and Nowicke 1990; to stain for unsaturated lipids in stigma secretions, Baker et al. 1973; for fixing flowers for location of osmophores, Stern et al. 1986) CAUTION: **Vapor causes blindness.**
Sigma Chemical Co. O 5500, 250 mg, $28.30

Appendix 3

Oxalic acid ($C_2H_2O_4$) (for detecting the presence of sugars, Baker and Baker 1982; for making PABA stain for estimating nectar concentration, Baker 1979)
Sigma Chemical Co. O 0376, 100 g $7.40

Paraformaldehyde (for fixing flowers for location of osmophores, Stern et al. 1986)
Sigma Chemical Co. P 6148, 500 g $9.95

Pararosaniline ($C_{19}H_{19}N_3O$) (the principal component of mixtures commonly referred to as basic fuchsin; could probably be substituted for it in Beattie's 1971a recipe)
Sigma Chemical Co. P 7632, 1 g $5.35 (also available as certified stain, P 1528)

Pararosaniline chloride ($C_{19}H_{18}N_3Cl$) (used hexazotised with α-naphthyl acetate to detect stigmatic esterase activity, Slater and Calder 1988)
Sigma Chemical Co. P 7632, 1 g $5.35; they can provide instructions for preparing hexazonium salt, which is unstable for prolonged storage.

n-Pentane (for a solvent for HPLC analysis of nectar, Buchmann and Buchmann 1981; a solvent for extracting floral odors, Bergström et al. 1980)
Fisher Scientific P399-1, 1 L $30.00

Periodic acid (H_5IO_6) (for use with Schiff's Reagent to detect insoluble polysaccharides in sectioned gynoecia, Lord and Kohorn 1986, Ciampolini et al. 1990).
Sigma Chemical Co. P 7875, 25 g $8.20

Peroxidase indicator (to test for peroxidase activity on stigmas; not cited by any pollination reference, but might work)
Sigma Chemical Co. 95-6, 1 mL $9.75

Phenol (C_6H_6O) (used as a preservative, Beattie 1971a; used in chromatography solvent for amino acid analyses, Baker and Baker 1973; for making lactophenol, Radford et al. 1974) CAUTION: **Phenol is toxic.**
Sigma Chemical Co. P 3653, 500 g $40.55

Phenylhydrazine hydrochloride ($C_6H_8N_2 \cdot HCl$) (for making phenylhydrazine reagent for nectar sugar analysis; Morrow and Sandstrom 1935)
Sigma Chemical Co. P 6926 10 g $7.50

Phloxin (Color Index 45405) (used with methyl green to stain cytoplasm and cellulose in pollen grains, Stanley and Linskens 1974)
Sigma Chemical Co. P 8894, inquire for price.

Phloxine B ($C_{20}H_2Br_4Cl_4O_5Na_2$; Color Index 45410) (used as a pollen stain, Peakall 1989)
Sigma Chemical Co. P 4030, 25 g $37.00

Phosphate buffer (used as a pre-treatment for pollen grains for triphenyltetrazolium chloride test, Diaconu 1968 — who did not specify what pH was used (cited in Stanley and Linskens 1974); for fixing gynoecia for histological studies, Lord and Kohorn 1986)

Sigma Chemical Co. P 3288 (pH 7.2 at 25°C), 12 vials, enough for 45.6 L $31.00

Phosphoric acid (H_3PO_4) (ortho-) (for adjusting pH of solvent for HPLC analysis of sugars, Southwick et al. 1981; for sugar staining reagent, Saini 1966; for reagent for testing for presence of protein in nectar, Bradford 1976; for reagent for visualizing chromatograms, Block et al. 1958)
Fisher Scientific A242-500, 500 mL $37.10

Phosphorus pentoxide (P_2O_5) (for use in a desiccator, Gilbert et al. 1981)
Sigma Chemical Co. P 0679, 250 g $13.60

Phthalic acid ($C_8H_6O_4$) (for making aniline-phthalate reagent for visualizing chromatograms, Partridge 1949)
Sigma Chemical Co. P 8657, 10 g $22.50

Platinic chloride (H_2PtCl_6) (for making iodoplatinate reagent for visualizing alkaloids on chromatograms, Smith 1969)
Sigma Chemical Co. P 5775, 1 g $30.45

Polyvinyl alcohol (for making polyvinyl lactophenol mounting medium)
Eastman Fine Chemicals, Kodak #153 9709, 1 kg $41.70

Porapak Q (porous polymer support for gas chromatography, for collecting floral odors, Williams 1983)
Alltech Associates, available as a prepacked column.

Potassium acetate ($KC_2H_3O_2$) (for a solution for measuring relative humidity inside flowers, Unwin 1980)
Sigma Chemical Co. P 1147, 500 g $11.10

Potassium carbonate (K_2CO_3) (for use as an alternative to acetolysis, Erdtman 1969)
Sigma Chemical Co. P 4379, 100g $4.80

Potassium chloride (KCl) (for pollen germination medium; Leduc, Monnier, and Douglas 1990)
Sigma Chemical Co. P 3911, 500 g $15.75

Potassium hydroxide (KOH) (for clearing insect bodies or plant parts, Kislev et al. 1972; for acetolysis)
Sigma Chemical Co. P 1767, pellets, 250 g $6.40

Potassium iodide (KI) (stains starch in pollen grains, Stanley and Linskens 1974)
Sigma Chemical Co. P 8256, 100 g $13.80

Potassium nitrate (KNO_3) (for Brewbaker-Kwack medium, Brewbaker and Kwack 1963)
Sigma Chemical Co. P 8291 (tested for plant cell culture), 500 g $15.50

Potassium permanganate ($KMnO_4$) (for testing for antioxidants in nectar, Baker and Baker 1975)
Sigma Chemical Co. P 6142, 500 g $15.85

Appendix 3

Potassium phosphate, monobasic (KH_2PO_4; anhydrous; monopotassium phosphate) (for staining styles with aniline blue, Weller and Ornduff 1989; for pollen germination medium; Leduc, Monnier, and Douglas 1990; for making decolorized aniline blue, Currier 1957)
Sigma Chemical Co. P 5379, 100g $7.70

Potassium phosphate, tribasic (K_3PO_4) (for a brightener for staining pollen tubes, Jefferies and Belcher 1974)
Sigma Chemical Co. P 5629, 500 g $13.80

1-Propanol (for chromatography solvent for sugar analyses, Grant and Beggs 1989)
Fisher Scientific A414-500, 500 mL $14.75

Pyridine (C_5H_5N) (for chromatography solvent for sugar analyses, Sweeley et al. 1963, Grant and Beggs 1989)
Sigma Chemical Co. P 3776, 100 mL $10.30

Rubidium chloride (RbCl) (for marking herbivorous larvae by spraying it on their host plants, Pearson et al. 1989)
Sigma Chemical Co. R 2252, 5 g $12.15

Ruthenium red (for staining pectinaceous material in fixed pistils, Ciampolini et al. 1990)
Sigma Chemical Co. R 2751, 25 mg $9.00

Safranin O ($C_{20}H_{19}N_4Cl$; Color Index 50240) (used as a pollen stain, Peakall 1989)
Sigma Chemical Co. S 8884, 25 g $27.00

Samarium oxide (Sm_2O_3) (used as a marker for neutron activation analysis, Gaudreau and Hardin 1974, Handel 1976)
Sigma Chemical Co. S 1125, 10 g $17.50

Schiff's Reagent (for use with periodic acid to detect insoluble polysaccharides in sectioned gynoecia, Lord and Kohorn 1986, Ciampolini et al. 1990).
Sigma Chemical Co. 395-2-016, 500 mL $36.00

Silver nitrate ($AgNO_3$) (for detecting reducing sugars, Trevelyan et al. 1950, Grant and Beggs 1989)
Sigma Chemical Co. S 6506, 5 g $10.20; 925-30 (0.5 N solution) 1 gal $6.75

Silver protein (for electron microscopy of pollen tubes, Rao and Kristen 1990)
Sigma Chemical Co. S 6767, 10 g $19.90

Silyating agents (for producing trimethylsilyl derivatives of sugars for gas chromatography, Loper et al. 1976, Elmqvist et al. 1988); see Hexamethyldisilazane.

Sodium acetate ($NaC_2H_3O_2$) (for making phenylhydrazine reagent for nectar sugar analysis; Morrow and Sandstrom 1935)
Sigma Chemical Co. S 8750 250 g $7.30

Appendix 3

Sodium amytal (for Nitro blue tetrazolium test for pollen viability, Hauser and Morrison 1964)
> Sigma Chemical Co. A 4430, 100 mg $5.45 (Amobarbital, now a controlled substance)

Sodium carbonate (for use as an alternative to acetolysis, Erdtman 1969)
> Sigma Chemical Co. S 2127, 500 g $13.95

Sodium chlorate (for bleaching pollen grains after acetolysis, Erdtman 1960)
> Aldrich 24,414-7, 100 g $8.00

Sodium hydroxide (NaOH) (used in detection of sugars on paper chromatograms, Trevelyan et al. 1950)
> Sigma Chemical Co. S 5881, 500 g $15.05

Sodium molybdate, dihydrate ($Na_2MoO_4 \cdot 2H_2O$) (for pollen germination medium; Leduc, Monnier, and Douglas 1990)
> Sigma Chemical Co. S 6646, 100 g $20.30

Sodium nitrite ($NaNO_2$) (used for the pararosanilin method for stigma receptivity, Bancroft 1975)
> Sigma Chemical Co. S 2252, 500 g $11.10

Sodium potassium tartrate ($C_4H_4KNaO_6 \cdot 4H_2O$) (for making reagent for photometric nectar analysis, Käpylä 1978)
> Sigma Chemical Co. S 2377 500 g $12.60

Sodium pyrophosphate ($Na_4P_2O_7$) (for stabilizing hydrogen peroxide solution for alkaloid analysis of nectars, Mattocks 1967)
> Sigma Chemical Co. S 9515, 100 g $6.80

Sodium succinate ($C_4H_4O_4Na_2$) (for Nitro blue tetrazolium test for pollen viability, Hauser and Morrison 1964)
> Sigma Chemical Co. S 2378, 100g $8.65 (Succinic acid, disodium salt, hexahydrate)
> Fluka has the anhydrous form. #14160, 100 g $10.40

Sodium sulphite (Na_2SO_3) (for softening plant tissues, Jeffries and Belcher 1974, Palser et al. 1989)
> Sigma Chemical Co. S 0505, 250 g $7.75

Spurr's resin (for embedding fixed pistils for electron microscopy, Ciampolini et al. 1990)
> Ted Pella, Inc. 18300, $33.00

Sucrose ($C_{12}H_{22}O_{11}$) (for a standard for nectar analyses)
> Sigma Chemical Co. S 9378, 500 g $13.95

Sudan III ($C_{22}H_{16}N_4O$) (for screening for lipids in nectar, Baker and Baker 1975)
> Sigma Chemical Co. S 4131, 25 g $7.15

Sudan IV ($C_{24}H_{20}N_4O$) (for screening for lipids in nectar, Baker and Baker 1975, Kevan et al. 1983)
    Sigma Chemical Co. S 4261, 25 g $28.00

Sudan black ($C_{29}H_{24}N_6$) (for screening for lipids in stigmatic secretions, Ciampolini et al. 1990; for screening for osmophores in epoxy-embedded sectioned floral material, Stern et al. 1986)
    Sigma Chemical Co. S 0395, 25 g $18.50

Sulfuric acid (H_2SO_4) (for acetolysis, Erdtman 1969; for anthrone test for carbohydrates, Van Handel 1967; for analysis of sugar contents of nectar, Roberts 1979, Schemske et al. 1978)
    Sigma Chemical Co. 680-2 (2/3 N), 100 mL $7.00

Tenax GC (porous polymer support for gas chromatography, for collecting floral odors, Williams 1983)
    Alltech Associates, 15g $130.00

Thiocarbohydrazide (CH_6N_4S) (for electron microscopy of pollen tubes, Rao and Kristen 1990)
    Sigma Chemical Co. T 2137, 5 g $8.70

Thiourea (CH_4N_2S) (for anthrone reagent for assaying for sugars, Cresswell 1989)
    Sigma Chemical Co. T 7875, 100 g $7.65

Tin (Sn) (powder, for labeling pollen for detection by backscatter SEM and X-ray microanalysis, Wolfe et al. 1991)
    Aldrich 26563-2 (-325 mesh, 44), 500 g $53.10

Toluidine blue o ($C_{15}H_{16}ClN_3S$; Color Index 52040) (used as a pollen stain, Peakall 1989)
    Sigma Chemical Co. T 0394, 25 g $33.00

Trichloroacetic acid ($C_2HCl_3O_2$) (for making a spray to visualize sugars on chromatograms, Partridge and Westhall 1948; for electrophoresis of proteins in nectar, Chrambach et al. 1967)
    Sigma Chemical Co. T 4885, 100 g $10.90

Trimethylchlorosilane (for gas chromatography, Sweeley et al. 1963, Baskin and Bliss, 1969)
    Sigma Chemical Co. T 4252, 25 mL $5.25

2,3,5-Triphenyltetrazolium chloride ($C_{19}H_{15}N_4Cl$) (for evaluating pollen quality, Cook and Stanley 1960, Heslop-Harrison, Heslop-Harrison, and Shivanna 1984; for visualizing reducing sugars on chromatograms, Trevelyand 1950; also for testing seeds for viability, Moore 1962)
    Sigma Chemical Co. T 8877, 10 g $9.05

Tris-HCl ($C_4H_{12}ClNO_3$; Tris[hydroxymethyl]-aminomethane hydrochloride) (for buffering 2,3,5-Triphenyltetrazolium chloride dye, Stanley and Linskens 1974)
    Sigma Chemical Co. T 3253 (TRIZMA Hydrochloride), 100 g $16.20

Appendix 3

Trypan blue ($C_{34}H_{24}N_6O_{14}S_4Na_4$) (used as an addition to aniline blue stain for pollen tubes, Kambal et al. 1976)
  Sigma Chemical Co. T 0776, 5 g $6.70

Tween 20 (polyoxyethylene [20] sorbitan monolaurate) (used as a wetting agent for neutron activation analysis, Handel 1976)
  Sigma Chemical Co. P 1379, 100 mL $8.85

Uranyl acetate (for staining fixed pistils for transmission electron microscopy, Ciampolini et al. 1990).
  Ted Pella, Inc. 19481, 30 g $17.50

Xylene (for cleaning paraffin from pollen slides) Fisher Scientific X4-4, 4L $46.95

Zinc (Zn) (powder for labeling pollen for detection by backscatter SEM and X-ray microanalysis, Wolfe et al. 1991)
  Aldrich 32493-0 (-100 mesh, 149, but with a broad size distribution), 10 g $41.00

Zinc sulfate, heptahydrate ($ZnSO_4 \cdot 7H_2O$) (for pollen germination medium; Leduc, Monnier, and Douglas 1990)
  Sigma Chemical Co. Z 0501, 500 g $18.80

Appendix 4.
Computer Programs

We have included several programs here that may be of use for pollination studies. They include:

1. a BASIC program for using a laptop computer as an event recorder;
2. another BASIC program for the same purpose, designed for following butterflies;
3. pseudocode for a program that calculates coordinates of successive positions from measuring tape triangulation fixes;
4. pseudocode for a program that calculates coordinates of successive positions from compass fixes;
5. pseudocode for calculation of move lengths and turning angles from spatial coordinates;
6. an algorithm that combines displacements measured at regular time intervals into moves; and
7. a FORTRAN program for mapping plants from measurements taken from two stakes.

The first program comes from Petersen (1990), the last comes from Barbara Thomson (*personal communication*), and the others are from Turchin (*personal communication*; they were used for Turchin et al. 1991).

I. A BASIC program for using a computer as an event recorder

The program is copied (with a few corrections) from Petersen 1990. It doesn't have the fancy touches that commercial programs do, but on a portable laptop computer it may suffice for simple observations of flower visitors in the field.

During an observation period, the program will record the type, duration, and exact time of each behavior. The following data can be displayed on the screen, stored in a data file, and (optionally) printed:

1. a behavior code (any single alphanumeric character);
2. the duration of the behavior (in seconds);
3. the start time of the behavior (in seconds since midnight);
4. the ending time of the behavior (in seconds since midnight); and
5. the "Current Behavior," or last code entered, so that you can verify the behavior that is currently being recorded.

To use this program on a DOS-based computer, enter the listing below, including the line numbers, into an ASCII file and save it with the filename EVENT.BAS (or contact one of the authors of this book by e-mail and we'll send it to you so you

464

won't have to type this yourself). To execute the program, type BASIC EVENT, while you are logged into a directory that has a copy of BASIC (or a directory listed in the PATH statement in your AUTOEXEC.BAT file). You will then be prompted to enter an experiment name, experiment description, and output file name, and asked whether you want the data sent to a printer as well. To start recording a series of behaviors, press the code for the first behavior (and then the ENTER key). When the next behavior begins, press the key coding for it (but it is not necessary at this point to press the ENTER key). When you are finished collecting data, press the X key. To quit the program and return to DOS, type SYSTEM.

```
10 '=============BEHAVIORAL EVENT RECORDER==============
20 DIM C$(500),D(500),S(500),E(500)        '500 behaviors maximum per
                                            expt.
30 GOSUB 400                               'Get Experiment Information,
                                            open files, date
40 CLS : LOCATE 5,10 : PRINT "During the experiment, Press X or x to EXIT."
50 LOCATE 7,10 : PRINT "PRESS CODE FOR FIRST BEHAVIOR AND
    RETURN:"
60 INPUT CODE$
100 '********* BEGIN MAIN PROGRAM ***********************
110 N=1                                    'Counter for number of events
120 CLS : PRINT EXPTN$                     'Print Experiment Name, etc.
130 PRINT EXPTDES$ : PRINT DATE$
140 PRINT HDR$
150 START=TIMER                            'Start the timer
160 NXT$=""                                'Next behavior not yet re-
                                            corded
170 WHILE NXT$<>"X" AND NXT$<>"x"          'Continue until eXit code is
                                            pressed
180 IF NXT$<>"" THEN GOSUB 300             'Output to screen and file
190 NXT$=INKEY$                            'Check keyboard buffer for in-
                                            put
200 WEND                                   'EXit code encountered - stop
210 GOSUB 300 : C$ (N) = "X"               'Output final event data
220 FOR I=1 TO N-1                         'Print event information to
                                            file & screen
230 PRINT #2, USING OUTIMG$ ; C$ (I) ; D(I) ; S(I) ; E(I) ; C$(I + 1)
240 IF OUTX = 2 THEN PRINT #1, USING OUTIMG$ ; C$ (I); D(I); S(I); E(I);
    C$(I+1)
250 NEXT
260 END '**************END MAIN PROGRAM*********************
300 '--------COMPUTE EVENT DURATION AND OUTPUT--------
```

```
310 ENDEV=TIMER                                   'Record end of event
320 DUR=ENDEV-START                               'Event duration
330 STEMP=START : START=TIMER                     'Restart the timer
340 C$ (N)=CODE$ : D(N)=DUR                       'Assign values for later output
350 S(N)=STEMP : E(N)=ENDEV
360 PRINT #3, USING OUTIMG$; CODE$; DUR; STEMP; ENDEV; NXT$
370 N = N + 1                                     'Next behavior counted
380 CODE$ = NXT$ : NXT$=""                        'Set CODE$ to current behav-
                                                   ior
390 RETURN
400 '———————INITIALIZING SUBROUTINE———————
410 'Constants, open appropriate files, input experiment information
420 HDR$= "Behavior Duration Start End Current Behavior"
430 OUTIMG$=" \ \ ######.##   ######.##   ######.## \ \"
440 CLS : LOCATE 5,10 : PRINT "BEHAVIORAL EVENT RECORDER (500
    Max)"
450 LOCATE 7,10 : PRINT "Experiment Name: " : INPUT EXPTN$
460 LOCATE 9,10 : PRINT "Experiment Description (P characters):"
470 LOCATE 11,1 : INPUT EXPTDES$
480 LOCATE 13,10 : PRINT "Output file path\name: " : INPUT FILEN$
490 LOCATE 15,10 : PRINT "Input Number for desired output option :"
500 LOCATE 17,15 : PRINT "1- Disk file only"
510 LOCATE 18,15 : PRINT "2- Disk and hardcopy (be sure printer is ON)"
520 LOCATE 20,15 : INPUT "Choice :", OUTX
530 OPEN "LPT1:" FOR OUTPUT AS #1        'Open files
540 OPEN "O", #2, FILEN$
550 OPEN "SCRN:" FOR OUTPUT AS #3
560 IF OUTX<>2 THEN RETURN               'No hardcopy printout
570 PRINT #1, "Experiment Name: ", EXPTN$
580 PRINT #1, "Experiment Description: ", EXPTDES$
590 PRINT #1, "Date: ", DATE$
600 PRINT #1, "Output filename: ", FILEN$ : PRINT #1, " "
610 PRINT #1, HDR$ : PRINT #1, " "
620 RETURN
```

Appendix 4

The following programs are provided courtesy of Peter Turchin. The first is the actual Basic program for a TRS-100. The other listings are in pseudocode. These are the programs referred to in Turchin, Odendaal, and Rausher 1991.

II. The BASIC program for recording behavioral observations in the field.

This program was used by Odendaal et al. (1989) for following *Euphydryas* butterfly females. They recorded the following behavioral events: beginning of flight ("fly"), termination of flight ("land"), oviposition ("ovipos"), nectaring ("nectar"), and harassment by conspecific males ("chased"). The letter codes are used to control the BASIC program. They indicate when observations were initiated and terminated, and allow the user to handle errors, to enter comments, and to leave the program.

Type C to display Codes. A record is stored in file RECORD.DO.

The BASIC program	Explanation
10 OPEN "RECORD.DO" FOR APPEND AS 1	Open the data file
20 PRINT #1,DATE$;" ";TIME$	Write to file date and time
30 DEFSTR A-C	Character strings begin with A,B,C
40 CLS	Clear screen
50 PRINT "PRESS B TO BEGIN OBS"	Prompt user
60 A=INKEY$	Get user input
70 IF A="" THEN 60	Cycle to 60 until a key is struck
80 NUM=VAL(A)	Translate character A to numeric
90 IF NUM=0 THEN 140	If A is 0 or character, go to 140
100 PRINT #1,A;" ";TIME$	Write to file the code and time
110 PRINT A	Print the code on screen
120 SOUND (10000-1000*NUM),5	Make sound, pitch depends on A
130 GOTO 60	Back to waiting for user input
140 JUMP=INSTR("BCESW",A)	Prepare to handle letter codes
150 IF JUMP<> 0 THEN 210	If code is B,C,E,S,W, go to 210
160 IF A="X" THEN 510	If code is X go to exit

170 SOUND 5000,2	Invalid code, make irritating sounds
180 SOUND 5000,2	"
190 PRINT "INVALID CODE ENTERED:";" ";A	Alert user about invalid entry
200 GOTO 60	Go back to wait for user input
210 ON JUMP GOSUB 230,270,350,440,480	Depending on letter code, go to . . .
220 GOTO 60	After finishing task, go to input
230 LINE INPUT "INSECT NUMBER? ";AINS	Begin obs, get insect number
250 PRINT #1,A;" ";TIME$;" INSECT ";AINS	Record it in the file
260 RETURN	Back to user input
270 CLS	Clear screen for codes to follow
280 PRINT "CODES:"	Start showing codes
290 PRINT "1:FLY 2:LAND 3:OVIPOS"	"
300 PRINT "4:NECTAR 5:BASK 6:CHASED"	"
310 PRINT " "	"
320 PRINT "B:BEGIN OBS C:CODES E:ERROR"	"
330 PRINT "S:STOP OBS W:COMMENT X:EXIT"	"
340 RETURN	Back to user input
350 BEEP	Error handling starts
360 PRINT "ERROR! REPLACE CODE WITH:"	"
370 A=INKEY$	Get the correct code
380 IF A="" THEN 370	Cycle back until a key is struck
390 PRINT A	Show correct code on screen
400 PRINT #1,"ERROR: REPLACE ABOVE WITH ";A	Write it to the file
410 RETURN	Back to user input
440 PRINT #1,A;" ";TIME$	Write to file S for stop
460 PRINT "OBSERVATION TERMINATED."	Print it to screen
470 RETURN	Back to user input
480 LINE INPUT "COMMENT: ";BC	Get the comment from user
490 PRINT #1,"C: ";BC	Write it to the file
500 RETURN	Back to user input
510 PRINT #1,A;" ";TIME$	Write the time of exit to file

520 CLOSE 1   Close the file and leave pro-
                                         gram

III. Pseudocode for a program that calculates coordinates of successive positions from the measuring tape triangulation fixes (a double asterix denotes power).

INPUT F: a N x 2 matrix of fixes for N subsequent positions of the followed
   animal.
   1st column distance AC, 2nd — distance ACB
   AB: length of the baseline.
OUTPUT C: a N x 2 matrix of spatial coordinates for N positions.
Step 1. For i = 1, 2, . . . N do steps 2-4.
Step 2. FIX1 = F[i;1]
   FIX2 = F[i;2]-F[i;1]
Step 3. Calculate x and y coordinates (assuming that the origin is at A, and x-axis
   is AB):
   X = (AB**2 + FIX1**2 - FIX2**2)/(2AB)
   Y = sqrt(AC**2-X**2)
Step 4. Output X and Y to C:
   C[i;1] = X
   C[i;2] = Y
Step 5. Stop

IV. Pseudocode for a program that calculates coordinates of successive positions from the compass fixes (a single asterix denotes multiplication).

INPUT F: a N x 2 matrix of fixes for N subsequent positions of the followed
   animal.
   1st column direction from A to C, 2nd — from C to B
   AB: length of the baseline.
   DIRAB: direction from A to B.
OUTPUT C: a N x 2 matrix of spatial coordinates for N positions.
Step 1. For i = 1, 2, . . . N do steps 2-5.
Step 2. FIX1 = 180 - F[i;1]
   FIX2 = 180 - F[i;2]
Step 3. FIX1 = FIX1 - DIRAB
   FIX2 = FIX2 - DIRAB

Step 4. Calculate x and y coordinates (assuming that the origin is at A, and x-axis
is AB):
X = AB/(1 + (tan FIX1)/(tan FIX2))
Y = X*tan FIX1

Step 5. Output X and Y to C:
C[i;1] = X
C[i;2] = Y

Step 6. Stop

V. Pseudocode for calculation of move lengths and turning angles from spatial coordinates.

INPUT C: a N x 2 matrix of spatial coordinates for N successive positions.

OUTPUT M: a (N-1) x 1 array of move lengths
D: a (N-1) x 1 array of move directions
A: a (N-2) x 1 array of turning angles

Step 1. For i = 1, 2, . . . (N-1) do steps 2-4.

Step 2. Calculate the coordinate change for each displacement:
DISPLX = C[i+1;1] - C[i;1]
DISPLY = C[i+1;2] - C[i;2]

Step 3. Calculate move length and direction of each move:
MOV = sqrt(DISPLX**2 + DISPLY**2)
DIR = arcsin(DISPLX/MOV)

Step 4. Output MOV and DIR:
M[i] = MOV
D[i] = DIR

Step 5. For i = 1, 2, . . . (N-2) do steps 6-8.

Step 6. Calculate turning angles (assuming turns left are negative, and turns right
are positive angles)
TURN = D[i] - D[i+1]

Step 7. Adjust turning angles to lie within [-180, 180]
If TURN -180, then TURN = TURN + 360
If TURN 180, then TURN = TURN - 360

Step 8. Output turning angles
T[i] = TURN

Step 9. End

Appendix 4

VI. An algorithm that combines displacements measured at regular time intervals into moves.

INPUT C: a N x 2 matrix of coordinates of spatial positions.
   MAXDEV: the limit of allowed deviation, set by the user.

OUTPUT DELIM: numbers of spatial positions that delimit moves
   R: reduced matrix of coordinates in which only those spatial positions are kept that delimit moves

Step 1. Set i = 1. Initialize DELIM as an empty array, and append i to it.

Step 2. While i<N do steps 2-11

Step 3. Set j = 1.

Step 4. Increment j by 1. If (i+j)>N, then go to step 12. Otherwise continue with step 5.

Step 5. Calculate the direction from postion i to (i+j)
   DISPLX = C[i+j;1] - C[i;1]
   DISPLY = C[i+j;2] - C[i;2]
   DISTj = sqrt(DISPLX**2 + DISPLY**2)
   DIRj = arcsin(DISPLX/DISTj)

Step 6. Initialize DEV: a j x 1 array of deviations.

Step 7. For k = 1, . . . j do steps 8-9

Step 8. Calculate distance and direction between positions i and (i+k)
   DISPLX = C[i+k;1] - C[i;1]
   DISPLY = C[i+k;2] - C[i;2]
   DISTk = sqrt(DISPLX**2 + DISPLY**2)
   DIRk = arcsin(DISPLX/DISTk)

Step 9. Calculate the distance from position (i+k) to the line connecting i to (i+j), and assign the value to DEV
   DEV[k] = absval(DISTk x sin(DIRj-DIRk))

Step 10. If any of the deviations in DEV are larger than MAXDEV, then do step 11, otherwise go to step 4.

Step 11. Append (i+j-1) to DELIM. This is a move break point.
   Set i = i+j-1
   Go to step 2.

Step 12. Append N to DELIM.

Step 13. Set M = the size of DELIM. Initialize R as a M x 2 matrix.

Step 14. For i = 1, 2, . . . M do step 15.

Step 15. R[i;1] = C[DELIM[i];1]
   R[i;2] = C[DELIM[i];2]

Step 16. Stop.

VII. A FORTRAN program for generating a map of points measured to two different fixed stakes (from Barbara Thomson).

```
C       AB and BC are the distances to B, the point being mapped, from the stakes
C       at A and C, the endpoints of baseline AC. X and Y are the coordinates
C       of point B in a Cartesian system with point A as the origin and segment
C       AC lying on the X-axis. The program expects to read two distances for
C       N points B. All of these points should be on the same side of line AC.
        read(2,*) N
        read(2,*) AC
        do 100 i = 1,N
          read(2,*) AB, BC
          ABBC = AB + BC
          if (ABBC.lt.AC) then
              write (6,*) 'Problem with plant ', i
          else
              S = (AB + BC + AC) /2.
              THETA = 2. * ASIN(SQRT(((S-AB)*(S-AC)) / (AB*AC)))
              X = AB * cos*(THETA)
              Y = AB * sin (THETA)
              write(6,*) X, Y
          endif
100     continue
        stop
        end
```

Appendix 5.
Glossary

A glossary for the study of pollination biology could potentially rival the length of this book, as there are so many different disciplines that contribute to the field, and each has its own vocabulary. We have tried to include terms particularly relevant to field studies of pollination.

abscission: the normal separation of leaves or fruits from the plant body by disintegration of specialized parenchyma cells.

abscission zone: the area at the base of a leaf, fruit, or other organ where a layer of specialized parenchyma cells forms and eventually disintegrates, resulting in the loss of the organ.

acetolysis: a process for preparing pollen for preservation that removes tissue from pollen grains, leaving the exine intact.

acropetally (of flowering sequence): flowers on an inflorescence opening from base to apex.

actinomorphic (of flower shape): radially symmetrical.

agamospermy: the production of seeds without sexual reproduction through sporophyte budding or from a diploid embryo sac.

allelopathy (of pollen): inhibition of pollen germination in sympatric species.

allogamy: fertilization resulting from pollen of a different flower.

allophilic: flowers that lack morphological adaptations to direct pollinators and can be used by short-tongued insects.

allopolyploidy: having more than a diploid sets of chromosomes, including some nonhomologous chromosomes, derived from two or more parents.

allotropous (uncommonly used): flower-visiting insects that are only slightly adapted for pollination, lacking both structural adaptations and flower constancy.

ament: a catkin; an elongate axis bearing apetalous, unisexual flowers.

ambisexual: bearing male and female parts.

aminoid: a description of the odor of fly-pollinated flowers that smell like sweat, feces, urine, or rotten meat.

Andrenidae: a family of solitary or communal ground-nesting bees.

androdioecious: a group of plants with both androecious and hermaphroditic individuals.

androecious: possessing only staminate flowers.

androecium: a collective name for all of the stamens in a flower.

andromonoecious: cosexual, but bearing both hermaphroditic and staminate flowers.

anemophily (anemogamy): wind pollination.

angiosperms: flowering plants; plants characterized by ovules surrounded by tissue, embryo sacs, double fertilization, and usually highly developed vascular systems.

antepetalous stamens: those opposite the petals.

antesepalous stamens: those opposite the sepals.

anthecology: refers to "the study of all aspects of the interactions between flower-visiting (anthophilous) animals and the flowers they visit, as well as to the pollination biology of those flowers that are pollinated by wind or water" (Baker and Baker 1973, p. 243); pollination biology.

anther: the floral organ that forms male spores (microspores), or pollen; normally consists of two lobes (thecae), each with two pollen sacs (microsporangia) in which pollen development takes place.

anthesis: the time when pollen is shed or stigma becomes receptive.

Anthomyiidae: a family of muscoid flies; with dark coloration, often very hairy; larval habits varied, but most are plant (often root) feeders.

anthophilous: flower-loving; refers to animals that visit flowers.

apocarpous: bearing separate (unfused) carpels.

Apoidea: the superfamily of bees; Hymenoptera, whose larvae are nourished with pollen and nectar; also Hymenoptera, bearing at least some branched hairs (wasps do not).

apomixis: reproduction without sex either through agamospermy or vegetative reproduction.

apomorphous (of flower colors): advanced.

aposporous: exhibiting a type of agamospermy in which the embryo sac is usually derived from the nucellus instead of the normal archesporium.

apparent reabsorption rate: "difference between apparent secretion rate in undisturbed flowers and gross secretion rate" (see reabsorption). Apparent reabsorption rate should approach influx rate (Búrquez and Corbet 1991, p. 370).

apparent secretion rate: "rate of change of solute content of nectar in undisturbed, unvisited flowers" (see gross secretion rate) (Búrquez and Corbet 1991, p. 370).

approach herkogamy: a hypothesized stage in the evolution of reciprocal herkogamy (Barrett 1990), during which stigmas are positioned above and separated from anthers.

archegonium: the female reproductive organ derived from a single gametophyte cell and ultimately bearing the egg nucleus in gymnosperms, Bryophytes, and Pteridophytes.

Appendix 5

archesporium: the tissue of the nucellus that produces the cell that undergoes meiosis in the formation of the embryo sac.

assortative pollination: mating among similar individuals (e.g., plants with the same flower color, height, or time of flowering).

autofertility: self-pollination within a flower (e.g., by corolla abscission, curling of the stigma into the anthers, etc.).

autogamy: self-fertilization.

automimicry: nectarless plants mimic rewarding flowers on different plants of the same species (e.g., in dioecious plants where flowers are similar but only one sex offers a reward; populations or individuals of nectarless varieties) (Dafni 1984).

autopollination: self-pollination within a flower resulting from the position of anthers and stigmas, or the change in position of these structures as the flower ages.

autopolyploid: bearing more than two sets of homologous chromosomes.

Baker's law: colonizing species are self-fertile.

balanced breeding system: a combination of outcrossing and selfing such that outcrossing maintains genetic variability and selfing produces local adaptations.

banner petal: the upper petals of a papilionaceous flower.

basipetally (of flowering): flowers on an inflorescence opening from apex to base.

bee fly: a fly of the family Bombyliidae.

bifid (of stigmas): two-pronged.

bilabiate (of a corolla): two-lipped, as in Scrophulariaceae and Lamiaceae.

binucleate: a pollen grain containing a tube nucleus and a generative nucleus.

Bombyliidae: a large family of flies known as bee flies that are parasitic or hyperparasitic as larvae and flower-visiting as adults.

bract: a reduced, modified leaf found on the inflorescence.

breeding system: "all aspects of sex expression in plants that affect the relative genetic contributions to the next generation of individuals within a species" (Wyatt 1983, p. 55); mating system.

Brix scale: a scale for measuring sugar concentration; some refractometers are calibrated in Brix % scale, which shows the number of grams of sucrose contained in 100 grams of solution.

bumblebee: a member of the genus *Bombus;* large, hairy social bees found almost worldwide, especially important in high-altitude and high-latitude pollinator communities.

buzz pollination: a pollination mechanism involving vibration of flowers (usually with poricidal anthers) by bees as a way of obtaining pollen; the vibrations are produced by "shivering" of the flight muscles and are often audible to an observer.

Appendix 5

Calliphoridae: a family of Diptera called blowflies; in the muscoid group.

callose: β 1,3,glucan; a complex polysaccharide found in sieve elements of phloem, pollen mother cell walls, germinating pollen grains, and within pollen tubes in the form of localized deposits known as callose plugs.

calyx (plural calyxes, calyces): all the sepals of a flower.

campanulate: bell-shaped (refers to corolla of a flower).

cantharophily (cantharogamy): the floral syndrome involving pollination by beetles.

capitulum: an inflorescence of flowers or florets crowded together on a receptacle; e.g., flower heads of Asteraceae.

carina: the keel (two united lower petals) of a papilionaceous flower.

carinate (of flowers): like a keel; bearing a ridge.

carpel: stigma, style, and ovary; female floral organ.

carpenter bees: subfamily of bees (Xylocopinae); bees that nest in wood or plant stems by excavating tunnels; similar in appearance to bumblebees.

catkin: see ament.

cauliflory: production of flowers directly on the stem from older parts of the trunk of a tree. Best known in cocoa trees but also exhibited by a variety of tropical woody plants.

centripetal (of flowering): flowers opening from the outside to the inside of an inflorescence.

certation: 1. differential growth rates of pollen tubes in a stigma when they bear different but compatible S alleles; 2. a phenomenon in dioecious species with heteromorphic sex chromosomes, in which the sex ratio of progeny depends on the number of pollen grains deposited on the stigma. An excess of female offspring when many pollen grains are deposited has been attributed to the competitive superiority of X-bearing pollen.

chasmogamy: opening of the perianth that exposes the stigma to pollen from external sources (as opposed to cleistogamy).

chiropterophily (chiropterogamy): the floral syndrome involving pollination by bats.

choripetalous (polypetalous): having separate (unfused) petals.

cleistogamy: a condition in which the perianth remains closed and self-pollen fertilizes ovules.

clogging (of stigmas): overcrowding of the stigma by incompatible pollen grains, which might reduce seed set through inhibition of germination or tube growth by compatible grains.

Coleoptera: the order of beetles; insects with four wings, the front pair leathery or hard and covering the membranaceous hind wings; generally with chewing mouthparts; holometabolous; largest insect order.

Appendix 5

Colletidae: a family of bees known as plasterer or yellow-faced bees.

column (of an orchid): the modified style that also bears the pollinia (a gynostemium).

colpus (of pollen grains): an elliptical aperture.

conidia: thin-walled fungal spores, sometimes transferred in a manner similar to pollen by insects, thereby effecting "pollination."

corbicula: pollen basket; scooped-out depression on bee hind leg used for carrying pollen.

corolla: all the petals of a flower.

coronate (of flowers): "tubular or flaring perianth or staminal outgrowth; petaloid appendage" (Radford et al. 1974, p. 103) (as in a daffodil).

corpusculum: the part of the pollinarium, or pair of pollinia of an *Asclepias* (milkweed) flower, that connects the two pollinia. Attached to the two pollinia via translator arms.

corymb: a convex or flat-topped or convex cluster of flowers that open from the outside toward the center; a contracted raceme.

cosexual: only one sexual genotype in the population with individuals capable of functioning as males, females, or both.

crepuscular: active at dusk.

cruciform (of flowers): cross-shaped; with four petals in a cross as in the Brassicaceae.

cryptic dioecy: a condition in which there are functionally male and functionally female plants, but plants do not appear to be dioecious. E.g., **1.** cases of androdioecy where there appear to be male and hermaphroditic plants, but the hermaphroditic plants are male-sterile or **2.** male and female plants appear to be hermaphroditic, differing only in morphological characters like style length (example: dioecious species of *Solanum*).

cryptic self-fertility: the phenomenon in which pollination with loads of pure self-pollen rarely or never results in fruit production, but pollination using mixtures of self- and cross-pollen produces fruits with considerable numbers of selfed seeds.

cryptic self-incompatibility: a condition in which self-pollen tubes grow more slowly than outcross tubes and do not generally fertilize ovules when outcross pollen is present.

cucullus: a floral hood; the nectarial hood on *Asclepias* flowers.

culm: a hollow, jointed stalk as in grasses.

cuticle (of stigma): a detached covering ("stigma membrane") over the surface of the stigma, which must be ruptured before pollination can take place.

cyme: a dichasium or compound dichasium; inflorescence with a terminal flower that opens first, followed by flowers in subtending axillary bracts.

deception: attracting pollinators either by appearing to offer rewards that are not actually provided (e.g., resembling flowers that offer pollen or nectar, resembling ovipositon sites, mates, etc.) (Dafni 1984).

decussate: with alternating pairs placed along an axis at right angles to adjacent pairs.

dehisce (of anthers): open to release pollen.

determinate inflorescence: inflorescence with a terminal flower that opens first, followed by subtending flowers.

dichasium: a simple cyme; an inflorescence with terminal flower and a pair of flowers with equal-length pedicels below.

dichogamy: the temporal separation of male and female functions of a flower, with anthers and stigma(s) of the same flower or on different flowers in monoecious plants ripening at different times; expressed as either protandry or protogyny.

dichotomous: exhibiting bifid branching.

diclinous: bearing male and female flowers.

diclinous breeding systems: gynodioecy, subdioecy, or dioecy.

didynamous: bearing four stamens of two different lengths.

dioecious: a group of plants with both androecious and gynoecious individuals.

diphasy: a sexual system in which individual plants belonging to a single genetic class can vary their sexual mode from year to year depending on circumstances; gender choice.

Diptera: the order of flies; holometabolous insects with one pair of functional wings, the second pair reduced to halteres (balance organs).

disassortative pollination: mating among dissimilar individuals; e.g., in a heterostylous system, pollination between different morphs.

disk florets: tubular, radially symmetrical flowers that make up the central region of composite flower heads (Asteraceae).

dispersal: movement or scatter from a source; as of pollen, seeds, or genes from the parental plant.

dispersion: spatial pattern, as of nectar or pollen resources.

distyly: a floral polymorphism in which two morphs (pin and thrum) are produced, differing in the anatomy of their reproductive parts. Pin flowers have long styles and anthers placed low in the corolla tube; thrum flowers have short styles and anthers placed high in the corolla tube. Successful pollination in such species requires crossing of the two morphs.

ectexine (ektexine): the outer of two subdivisions constituting the exine (part of angiosperm pollen wall; composed of the sexine and nexine 1 layers).

efflux rate: "rate of movement of solutes from plant into nectar" (Búrquez and Corbet 1991, p. 370).

elaiophor: an oil-secreting epithelial floral gland.

emasculate: to remove the male parts of a flower, i.e., removing the anthers.

embryo rescue: removal of a developing embryo to an artificial medium to complete its growth. This technique may be applied when normal development will not proceed after hybrid fertilizations or after overcoming other crossing barriers.

embryo sac: the female gametophyte of angiosperms, including multiple haploid cells in addition to the egg cell.

Empididae: the family of dance flies; flies that are generally predators on small insects and often visit flowers.

enantiostyly: style deflected from the main floral axis.

endexine: the inner of two subdivisions constituting the exine (part of the angiosperm pollen wall).

endintine: an inner cellulosic subdivision of the intine of a pollen grain.

entomophily (entomogamy): the floral syndrome involving pollination by insects.

epichil: one of the three parts of the fleshy lip of some orchid flowers.

epigynous: flowers in which other floral parts are attached to the hypanthium that surrounds the partially inferior ovary.

epipetalous stamens: stamens borne on the petals.

episepalous stamens: stamens borne on the sepals.

ethodynamic pollination: cases in which insects load pollen onto certain parts of their bodies and then deposit it on a stigma; e.g., in some *Yucca* and *Ficus* species (Galil 1973). Contrast with topocentric pollination.

euglossine bees: bees of the subfamily Euglossinae; metallic and brightly colored, long-tongued bees.

euphilic (of flowers): adapted for highly specialized pollinators.

eutherophily: the floral syndrome involving pollination by placental mammals (a new term introduced by Armstrong 1979).

eutropous (uncommonly used): flower-visiting insects that are completely adapted for pollination, with highly developed morphological adaptations and good flower constancy.

exine: the outer of the two walls of a mature pollen grain.

exintine: an outer pectic polysaccharide subdivision of the intine of a pollen grain.

explosive pollination: pollination caused by the violent movement of anthers/stamen and style, alone or together with restraining petals, triggered by the visit of a pollinator.

exserted: (of anthers or stigmas) protruding beyond the corolla (anthers: "thrum" flowers; stigmas: "pin" flowers).

extraction time: the amount of time it takes a flower visitor to remove nectar (not including handling time).

extrorse (of anthers): facing outward; opening toward the petals.

FAA: formalin–acetic acid–alcohol (see Chapter 4, Section K.1.a for recipe).

FCR: fluorochromatic procedure for determining pollen viability (see Chapter 4, Section E.2).

facultative autogamy: the occurence of cross-pollination in normally self-pollinating species.

facultative xenogamy: the breeding system of flowers that can be cross-pollinated if pollinators are present but that will self-pollinate in their absence.

female choice (in plants): "any evolved mechanism in which **1.** pollen is received from some pollen donors but not others in a population or **2.** nonrandom fertilization occurs following pollen deposition" (Stephenson and Bertin 1983, p. 112).

fertilization: fusion of gametes; in angiosperms — double fertilization, fusion of egg and sperm nuclei to produce the zygote, plus fusion of a second male nucleus with the polar bodies to form the triploid endosperm.

filament: a stalk that attaches the anthers to the receptacle.

flagelliflory (penduliflory): the production of flowers or inflorescences at the end of long, ropelike branches dangling down from the crown of a tree, presumably to facilitate access by pollinators.

floral mimicry: 1. Batesian: nonrewarding flower resembles flowers that offer rewards for pollinators; **2.** Müllerian: different species share the same floral characters and both benefit by attracting the same pollinator.

floral syndrome: a suite of characters (color, scent, petal morphology) generally recognized as attractants for a specific class of pollinators.

floret: a small flower; often used to describe flowers of composites, grasses, or clovers.

flower: the reproductive structure of angiosperms, composed of calyx, corolla, androecium, and gynoecium, or some modified combination of these organs.

flower constancy: the tendency of flower visitors to visit a single species of flower during a foraging bout.

flower fly: a fly of the family Syrphidae.

fluorescence microscopy: the use of a microscope equipped with special filters that permit appropriate wavelengths to reach excitable substances on the preparation being viewed. These substances emit light of another, visible frequency, causing them to glow.

fodder stamens: stamens that provide food in a form other than pollen to attract pollinators.

functional gender: operative sex, defined by the relative reproductive success derived from genes contributed through pollen versus ovules.

galeae: the outer lobes of the maxilla of an insect.

Appendix 5

gametophyte (of angiosperms): the pollen grain or embryo sac; the haploid structure carrying the gamete nucleus.

gametophytic incompatibility: a multiallelic, single-locus control system in which cross pollination can only occur between pollen bearing one allele and female plants that do not bear that allele; the incompatibility reaction is based on the male gametophyte genotype.

gamosepalous calyx: with united sepals.

GC: gas chromatography.

geitonogamy: interflower pollination on an individual plant.

gender adjustment: a change in allocation to male or female investment in a system where there is variation along a broad continuum of sex investment ratios (Lloyd and Bawa 1984).

gender choice: a change in sex expression in response to a plant's circumstances in a system where there is a bimodal distribution of sex expression.

genet: a genetically identical unit, often consisting of multiple ramets such as in a clone.

geoflorous: flower heads borne near ground level.

gross secretion rate: "rate of change of solute content in nectar of repeatedly sampled flowers" (see apparent secretion rate). Gross secretion rate should approach efflux rate if sampling of flowers affects only influx and not efflux (Búrquez and Corbet 1991, p. 370).

gullet flower: a flower with the sexual parts on the upper surface resulting in pollen deposition on the back or head of a pollinator.

gymnosperms: vascular plants that produce "naked seeds" that are not surrounded by carpel tissue; includes several plant divisions: Pteridospermophyta, Cycadophyta, Ginkgophyta, Coniferophtya, and Gnetophyta.

gynobasic: "at the base of an invaginated ovary" (Radford et al. 1974, p. 108).

gynodioecious: a group of plants with both hermaphroditic and gynoecious individuals; female morph is unisexual and male morph is ambisexual.

gynoecious: possessing only pistillate flowers.

gynoecium: all the female parts of a flower.

gynomonoecious: cosexual but possessing both hermaphroditic and pistillate flowers.

gynostegium: a combined androecium and gynoecium (e.g., in *Asclepias* flowers).

Halictidae: a family of small- to medium-sized ground-nesting bees.

handling time: the amount of time it takes a flower visitor to land on a flower, determine whether it has nectar and/or pollen, and depart.

haptonasty: movement of a plant in response to being touched.

haustellum: the part of a Dipteran proboscis that lies distal to the maxillary palps.

hawkmoths: moths of the family Sphingidae, including several large species (hummingbird moths) that hover in front of a flower extending their long proboscides for nectar.

heliotropism: the bending or turning of plants (leaves and/or flowers) toward and with the movement of the sun; solar tracking.

hemiphilic (of flowers): flowers that are adapted for pollination by insects with intermediate levels of specialization.

Hemiptera: the order of true bugs; insects characterized by front wings that are leathery at the basal end and membranaceous at the distal end; hind wings are membranaceous; with piercing-sucking mouthparts.

hemitropous (uncommonly used): flower-visiting insects that are partially adapted for pollination, with weakly developed structural adaptations and an intermediate degree of flower constancy.

herkogamy (also hercogamy): the spatial separation of male and female sex organs in a flower, which presumably means that they cannot self-pollinate.

hermaphroditic: 1. a flower with both stamens and pistils; **2.** a plant with only perfect flowers, functionally both staminate and pistillate; **3.** a group of plants containing only hermaphroditic plants.

heteranthery: possession of a dimorphic androecium, with two or more different types of anthers.

heterodichogamy: a breeding system in which there are two mating types differing in their temporal pattern of development, such as both protandrous and protogynous individuals.

heterostyly: a genetic polymorphism in which plant populations have two or three morphs, differing reciprocally in the heights at which the stigmas and anthers are positioned.

homogamy: no difference in the timing of anther dehiscence and stigma receptivity.

homostyle: a flower with anthers and stigmas at the same level; used to refer to the condition in a species with heterostylous ancestors, or to individuals in a normally heterostylous population.

honeybee: *Apis mellifera,* the common domesticated social bee used for honey production and pollination.

hoverfly: a fly of the family Syrphidae.

hydrophily (hydrogamy): pollination by water.

HPLC: high-performance liquid chromatography.

Hymenoptera: the order including the social insects, ants, bees, and wasps; holometabolous insects with membranous front and hind wings, chewing mouthparts, and often modified structures for taking fluids.

hypanthium: a floral tube or cup, formed by the fused bases of the sepals, petals, and stamens.

hypochil: one of the three parts of the fleshy lip of some orchid flowers.

hypogynous: a flower in which the sepals, petals, and stamens are attached at a level below the ovary.

illegitimate pollination: pollination with the "wrong" pollen in a heterostylous pollination system; e.g., pollen from a long stamen deposited on a short style instead of a long one.

inbreeding coefficient: F; a quantitative estimate of the amount of inbreeding expressed in terms of the decrease in heterozygosity compared to a large, randomly mating population with the same allele frequencies.

included style (thrum flowers): a style that does not extend above the perianth.

incompatibility: a genetically based inability for cross-fertilization between two genotypes.

indehiscent anthers: anthers that do not split open.

indeterminate inflorescence: inflorescence with outer or lower flowers opening first.

inferior ovary: stamens, petals, and sepals attach above the ovary.

inflorescence: a cluster of flowers on a plant.

influx rate: "rate of movement of solutes from nectar into plant" (Búrquez and Corbet 1991).

infructescence: a cluster of fruits on a plant.

infundibular (of flowers): shaped like a funnel.

integument: female sporophyte tissue surrounding the nucellus.

interalar: the thoracic area between the wings of an insect.

intine: the inner of the two walls of a mature pollen grain.

involucre: the bracts around a flower cluster or head of a composite.

isantherous (of flowers): with monomorphic anthers.

isostemonic: with a number of stamens equal to the number of sepals or petals.

keel: a central ridge; in leguminous flowers, the two fused petals that enclose the stamens.

labellum (plural labella): 1. lip of a flower; on an orchid, what appears to be the lowermost petal (actually the upper petal morphologically, but the pedicel twists). 2. a fleshy pad at the end of a fly's proboscis.

leaf-cutter bee: a bee of the family Megachilidae (they line nest cells with pieces of leaves).

Lepidoptera: the order of insects that includes butterflies and moths; holometabolous insects with scaled wings and generally long, coiled, sucking mouthparts.

leptokurtic: a statistical distribution with more units at the mean and in the tails than in intermediate regions (with a tall narrow peak compared to a normal distribution).

locule: one cavity of an ovary or anther.

magnet species: a species of flower that is highly attractive to pollinators and can generate an increase in visitation to co-flowering species that are rarely visited by pollinators.

matinal: active in the early morning (e.g., of bees).

mating system: those factors that determine the pattern of gene inheritance between generations; includes such parameters as self-incompatibility system; geitonogamy; herkogamy; dicliny; spatial distribution, etc.; breeding system.

maxilla: one of the paired mouthparts posterior to the mandibles of an insect.

maxillary palp: a feeler on the maxilla of an insect.

meconium: larval feces.

Megachilidae: the family of leaf-cutter bees.

megagametophyte: the female haploid generation or gametophyte.

megaspore: a haploid female meiospore that produces the female gametophyte by mitosis.

megasporophyll: a leaflike structure that bears megasporangia.

melissopalynology: the pollen analysis of honey; commonly used to determine geographical and floral origin of honey.

melittophily (melittogamy): the floral syndrome involving pollination by bees.

mentor pollen: pollen that is mixed with an incompatible pollen type in order to stimulate fruit and seed development and production of some seeds fertilized by the incompatible pollen (see rescue pollination). Mentor pollen is commonly treated to destroy its fertilizing abilities or to destroy the incompatibility mechanism (Brown and Adiwilaga 1991).

mesochil: one of the three parts of the fleshy lip of some orchid flowers.

mesothorax: the middle portion of an insect thorax.

metatherophily: the floral syndrome involving pollination by marsupials (a new term introduced by Armstrong 1979).

micropyle: the opening in the integuments of the ovule through which the pollen tube enters.

microsporangia: structures in which microspores are produced; in angiosperms, the elongated paired pollen sacs that constitute one of the two lobes of a typical anther; thecae.

microspore: a haploid male spore; uninucleate structure released from tetrads after meiosis that produces the male gametophyte (pollen grain) by mitosis.

microsporocytes: pollen mother cells contained within the microsporangia (thecae) that undergo meiosis to produce haploid microspores.

microsporogenesis: the production of microspores by meiosis.

microsporophyll: a leaflike structure that bears microsporangia.

mistake pollination: a pollinator visits a nonrewarding flower "by mistake," such as in plants where male and female flowers look similar but only one sex offers a reward (Dafni 1984).

monocarpic: flowering only once during a lifetime; semelparous.

monoclinous: bearing hermaphrodite flowers.

monoecious: a plant with both staminate and pistillate flowers, or a group of plants with only monoecious plants.

monolectic: insects that visit a single plant species for pollen.

monophilic (of flowers): pollinated by one or a few related species of pollinators.

monotelic inflorescence: one in which the apex of the inflorescence axis ends with a terminal flower.

monotropic: insects that visit a single plant species.

MS: mass spectrometry.

Muscidae: a family of muscoid flies usually similar in appearance to the common housefly.

myiophily (myiogamy): the floral syndrome involving pollination by flies.

myrmecophily (myrmecogamy): the floral syndrome involving pollination by ants.

NAA: neutron activation analysis.

nectar: a secretion produced by flowers to attract pollinators, usually containing sugars, amino acids, and other compounds that are of nutritional importance to flower visitors.

nectar guide: a marking in a contrasting color, not always visible to humans, that serves as an orientation cue to a flower visitor of where to find nectar.

nectar robbing: biting into a flower in order to extract the nectar "illegitimately" and usually without effecting pollination. A behavior exhibited most commonly by some species of bees, but also by ants, hummingbirds, and other animals.

nectar theft: obtaining nectar in a fashion other than that for which a flower has evolved and without effecting pollination, but not by causing mechanical damage to the flower (e.g., bypassing the reproductive parts of a flower with a long, thin proboscis).

nectariferous: producing nectar.

neighborhood size: 1. the genetically effective size of a nonideal population derived from a model of the decay of gene frequency variances in an "ideal," random mating population; **2.** the number of breeding individuals within a circle with a radius equal to twice the standard deviation of gene dispersal.

Nematocera: a suborder of Diptera; generally small flies with more than five antennal segments; the larvae have well-developed heads and mandibles that move laterally, unlike other groups of flies.

NMR: nuclear magnetic resonance.

nototribic: zygomorphic flowers with stamens and style placed so that they come into contact with the dorsal surface of the forager's body.

nucellus: the ovule tissue within the integuments and surrounding the embryo sac.

oligandry: possessing few stamens.

oligolectic (of flower visitors): specializing on a few related plant species for pollen.

oligolege: a flower visitor that specializes on a few related plant species.

oligophilic (of flowers): pollinated by a few related animal taxa.

oligotropic (of flower visitors): visiting only a few related host plants.

ombrogamy (ombrogamie): pollination by rain.

ornithophily (ornithogamy): the floral syndrome involving pollination by birds.

osmophore: scent-producing gland on a flower.

ovary: the part of the pistil that bears the ovules.

ovule: a sporangium surrounded by integuments that contains the female gametophyte or embryo sac; those sporophyte and gametophyte tissues that develop into a seed after fertilization.

paleomorphic (of flowers): lacking symmetry, often bearing multiple floral parts with bracts or leaves below.

panicle: a compound inflorescence with pedicellate flowers.

panmixis: random mating.

papilionaceous: flowers such as many in the Fabaceae that bear a standard, wings, and keel.

pappus: the modified calyx of Composite flowers that functions in seed dispersal.

parthenocarpy: development of fruit without pollination.

pedicel: the stalk of a single flower.

peduncle: the stalk of a single flower or an inflorescence.

pellicle: a protein coating on the stigmatic surface, overlying the cutinised outer layer of cells and functionally important in the capture and hydration of pollen grains.

pentamerous: having five parts.

perianth: the calyx and corolla.

pericarp: the ovary wall in a mature fruit.

perigynous: with sepals, petals, and stamens attached to the floral tube.

petal: a segment of the corolla within the whorl of sepals; often the showy, attractive part of the flower.

petiole: a leaf stalk.

phalaenophily (phalaenogamy): the floral syndrome involving pollination by small moths.

Appendix 5

phyllotaxis: the arrangement of leaves on a stem.

pin flower: in a heteromorphic system, a flower with an exerted stigma and short stamens.

pioneer pollen: pollen that is applied first in "rescue pollination" techniques. This pollen carries the genotype desired in the cross but is normally not cross-compatible with the ovules.

pistil: the ovary, style, and stigma.

pistillate flowers: functionally female, with pistils only.

pistillate sorting: processes occurring in the pistil that result in nonrandom seed parentage with respect to the pollen deposited on the stigma (Bertin et al. 1989).

pleomorphic (of flowers): radially symmetrical with small numbers of floral parts.

P/O ratio: pollen-ovule ratio; number of pollen grains produced by a flower divided by the number of ovules; has been used as an indication of breeding system by some authors.

pollen carryover: the deposition of pollen beyond the first flower after the one where it originated.

pollen flow: the movement of pollen grains to flowers other than those where they originated.

pollen grain: the multicellular male gametophyte.

pollen limitation: production of less than the maximum seed potential due to inadequate numbers of pollen grains to fertilize ovules.

pollen mother cells: the cells that divide to produce microspores that mature into pollen grains.

pollen presenter: a specialized region of the style in the genus *Banksia* (Proteaceae) onto which pollen dehisces from the anthers just before the flower opens.

pollen robbing: biting into a flower in order to remove pollen, usually without effecting pollination. A behavior exhibited by some species of bees and perhaps other animals.

pollen theft: obtaining pollen in a fashion other than that for which a flower has evolved and usually without effecting pollination, but not by causing mechanical damage to the flower (e.g., gathering of pollen from a large flower by small bees that will not effect pollination).

pollen tube: an outgrowth from the pollen grain that allows pollen nuclei to migrate down the style to effect fertilization.

pollenkitt: the adhesive surface of pollen grains; can contain lipoidal or pigmented substances that may add color or odor to the pollen and may cause pollen grains to adhere to each other during dehiscence or to pollinators.

pollinarium (plural pollinaria): the pair of pollen packets (pollinia), connected by a corpusculum and translator arms, that constitute the pollen package of *Asclepias* (milkweed) flowers.

pollination: the transfer of male gametophytes (pollen) to a stigma. Often used incorrectly as a synonym of fertilization, as pollination doesn't necessarily result in fertilization.

pollination intensity: the number of viable pollen grains that are deposited on a stigma.

pollinator abundance: a population characteristic affecting the number of visits a plant receives from a given species of pollinator (Young 1988).

pollinator effectiveness: a measure of the accomplishments of a single visit of an individual animal in terms of pollen deposited, pollen removed, seeds produced, percentage of available florets pollinated, etc. (Spears 1983, Young 1988, Neff and Simpson 1990).

pollinator efficiency: a measure of both the costs (flower damage, pollen eaten, etc.) and benefits (pollen deposited, pollen removed, seeds produced, etc.) of a single visit of an individual animal (Neff and Simpson 1990).

pollinator importance: a measure combining the pollinator abundance and pollinator effectiveness (Young 1988).

pollinium (plural pollinia): a discrete packet containing the pollen produced by a flower, characteristic of Orchidaceae and Asclepiadaceae.

polyad: pollen grains attached in groups of more than four.

polycarpic: flowering more than once during a lifetime; iteroparous.

polylectic (of flower visitors): visiting many plant species.

polylege: a generalist flower visitor.

polypetalous (choripetalous): having separate (unfused) petals.

polyphilic: pollinated by a variety of different taxa of flower visitors.

polytelic inflorescence: one in which there is no terminal flower at the summit of the primary axis but rather a multiflowered "polytelic florescence."

polytropic: visiting many different species of flowers for pollen.

poricidal anthers: anthers that dehisce apically, through opening of a pore at the distal end of the anther. Sometimes anthers with partial longitudinal loculicidal slits are included in this definition.

postzygotic: acting after fertilization (such as seed abortion).

prezygotic: acting before fertilization (such as genetic self-incompatibility mechanisms).

proboscis (plural probosces or proboscides): extended, beaklike mouthparts.

progamic phase: the phase that elapses between pollination to fertilization.

protandry: when a flower's anthers dehisce before its stigma is receptive.

protogyny: when a flower's stigma is receptive before its anthers dehisce.

proximate reason: an immediate or functional explanation.

pseudanthium: a collection of small flowers often surrounded by colorful bracts or leaves that give the appearance of being petals of a single flower.

pseudocopulation: male insects attempt copulation with flowers that resemble females of their species (pollination mechanism for some Orchidaceae) (Dafni 1984).

pseudogamy: pollen is required to trigger development of seeds, but pollen nuclei are not involved in fertilization.

pseudonectaries: structures that look like nectaries but do not secrete nectar (Dafni 1984).

pseudopollen: 1. a nutritive mass of nonpollen cells offered as a floral reward; **2.** cells that resemble pollen and attract pollinators by deception (Dafni 1984).

psychophily (psychogamy): the floral syndrome involving pollination by butterflies.

pyrheliograph: a mechanical instrument for measuring solar radiation.

raceme: a simple, elongate inflorescence with pedicellate flowers.

rachis: the axis of an inflorescence or compound leaf.

ramet: an individual belonging to a genet or clone.

ray florets: the showy ligulate flowers of Composite flower heads (Asteraceae).

reabsorption: "rate of net solute loss from unvisited flowers; negative apparent secretion rate" (Búrquez and Corbet 1991).

receptacle: the enlarged region bearing floral parts or bearing the florets of a Composite.

reciprocal herkogamy: different arrangements of spatial separation of anthers and stigmas; distyly involves two different arrangements, tristyly three.

rendezvous pollination: a flower elicits a sexual response from a pollinator (Dafni 1984).

reproductive status: a plant's absolute (male and female) allocation to sexual reproduction in a given season (from Lloyd and Bawa 1984).

rescue pollination: application of a second pollen load of a different genotype to promote seed formation from the first pollen applied. E.g., pollination of cultivated tetraploid potato flowers with pollen from a triploid hybrid does not result in seed set. However, if this first pollination is followed by pollination with pollen from cultivated tetraploids, fruit development is normal, and some percentage of seeds are derived from the first pollination (Brown and Adiwilaga 1991).

resource limitation: production of less than the potential seed set because of insufficient nutrients, water, or light.

reticulate: netlike.

rostellum: a small beak; a stigma protrusion on some orchid species.

rostrate: beaked.

salverform: with a narrow tube that opens into a flat, expanded circle of petals at the distal end.

sapromyiophily (sapromyiogamy): the floral syndrome involving pollination by carrion- and dung- flies.

scopae: specialized regions on a bee for transporting pollen (pollen baskets or a hairy region on hind legs or abdomen).

scutum: the middle section of the dorsal thorax of insects.

secondary pollen carryover: transport of previously deposited pollen grains from a stigma to another flower by a vector.

secondary pollen presentation: deposition of pollen, typically before anthesis, onto some floral part such as the style, from which it is then transferred to flower visitors (E.g., some Rubiaceae and Proteaceae).

seed: a mature ovule containing an embryo.

self-compatible: capable of self-fertilization.

self-incompatible: incapable of self-fertilization or of being fertilized by pollen expressing the same compatibility allele(s).

SEM: scanning electron microscopy.

sepal: a division of the calyx.

sex allocation: the partitioning of resources between male and female functions in cosexual plants.

sminthophily: the floral syndrome involving pollination by rodents (a new term introduced by Armstrong 1979).

spadix: an inflorescence of the family Araceae.

sphingophily (sphingogamy): the floral syndrome involving pollination by hawk-moths.

sporangia: structures where spores are produced (e.g., anther locules).

sporophyte: the diploid generation that produces meiospores.

sporophytic incompatibility: an incompatibility system in which the genotype of the pollen donor plant determines the expression of pollen grain compatibility type.

sporopollenin: the resistant material constituting the exine; believed to be derived from polymerization of carotenoids.

stamen: the anther and filament.

staminate flower: functionally male, with stamens only.

staminode (plural staminodia): nonfunctional stamen, sometimes with a broadened and even petallike anther filament and usually lacking a well-developed anther.

stereomorphic (of flowers): three-dimensional, radially symmetrical flowers such as *Aquilegia, Narcissus*.

sternotribic: zygomorphic flowers with stamens and style placed so that they come into contact with the ventral part of the forager's body.

stigma: the distal end of the style, which is where pollen is normally deposited before it germinates.

stigma exsertion: the degree to which the stigma protrudes beyond the corolla.

stipe: the basal stalk of a gynoecium.

strobilus: the cone structure of a gymnosperm, which can contain either microsporophylls or megasporophylls.

style: the area of gynoecium between ovary and stigma.

stylopodium: an enlarged, sometimes colorful, style-base that may produce nectar; characteristic of the Apiaceae.

subdioecy: the presence of male and female morphs, one or both of which is often ambisexual.

sweat bee: some species of bees in the family Halictidae in the genus *Lasioglossum*.

sympetalous: with petals that are fused together.

symsepalous calyx: with sepals that are fused together.

syncarpous: with united carpels.

synergids: the nuclei at the micropylar end of the embryo sac.

syngamy: union of gametes to form a zygote.

Syrphidae: the family of Diptera known as flowerflies or hoverflies; common flower-visiting flies, many of which mimic Hymenoptera in appearance and behavior.

Tabanidae: the family of horse flies.

Tachinidae: a family of muscoid flies; similar in appearance to houseflies but usually larger and bristlier, with larvae that are parasitic on other insects.

tapetum: the nutritive tissue of the anther involved in pollen development.

tarsus: the leg segment of insects distal to the tibia.

TEM: transmission electron microscopy.

tepal: a segment of the perianth when sepals and petals are similar in appearance.

tetrad: a group of four pollen grains or pollinia.

thecae: anther lobes; typically each anther comprises two thecae.

therophily: the floral syndrome involving pollination by mammals.

thrips: members of the insect order Thysanoptera; they live in flowers, eat pollen and nectar, and may even pollinate some flowers.

thrum flower: in a heteromorphic system, a flower with a short style and elongate stamen.

thyrse: an inflorescence with determinate terminal axis and indeterminate secondary (tertiary, etc.) axes.

thysanopterophily: pollination by thrips (a word coined by DWI as there is no term to describe this type of pollination).

topocentric pollination: cases in which contact between a foraging insect's body and pollen and stigma is ensured by the topography of the flower; contrast with ethodynamic pollination.

trap flower: flowers (e.g., *Aristolochia*) that effect pollination by temporarily trapping pollinators in the flower.

traplining: following a regular route or sequence in visitation of flowers.

trimonoecious: possessing hermaphroditic, staminate, and pistillate flowers all on the same plant.

trinucleate: a pollen grain containing three nuclei; two are derived from the generative nucleus to produce two sperms.

trioecious: having plants with separate hermaphrodite, male, and female flowers.

tristyly: a heterostylous condition with three floral morphs that differ in anther and style height and compatibility.

ultimate explanation: the fundamental reason that incorporates an evolutionary explanation.

ultraviolet: electromagnetic radiation with wavelengths between about 40 and 400 nm.

umbel: an inflorescence with pedicellate flowers that all originate from a central point.

umbellet: a secondary umbel on a compound umbel.

unguiculate: clawed; with a petal bearing a claw.

unisexual flower: a flower bearing reproductive organs of one gender.

vector: an animal that transports pollen.

vibratile pollination: buzz pollination.

viscidium (plural viscidia): a sticky area on an orchid rostellum where pollinia adhere.

viscin: sticky threads.

xenogamy: cross-pollination between different genets.

zoophilous: pollinated by animals.

zygomorphic (of a flower): bilaterally symmetrical.

zygote: the diploid cell formed by the union of gametes that can develop into an embryo.

Bibliography

Page numbers for each citation of a reference are shown in the brackets at the end of the reference.

Abraham, K., and P. Gopinathan Nair. 1990. Floral biology and artificial pollination in *Dioscorea alata* L. Euphytica 48(1):45–51. [75, 121]

Ackerman, J. D. 1983. Diversity and seasonality of male euglossine bees (Hymenoptera: Apidae) in central Panamá. Ecology 64(2):274–283. [271]

Adams, W. T., and D. S. Birkes. 1991. Estimating mating patterns in forest tree populations. Pages 157–172 *in* S. Fineschi, M. E. Malvolti, F. Cannata, and H. H. Hattemer, editors. Biochemical markers in the population genetics of forest trees. SPB Academic Publishing, the Hague, Netherlands. [230]

Addicott, J. F. 1986. Variation in the costs and benefits of mutualism: the interaction between yuccas and yucca moths. Oecologia 70(4):486–494. [30]

Ågren, J., and D. W. Schemske. 1991. Pollination by deceit in a neotropical monoecious herb, *Begonia involucrata*. Biotropica 23(3):235–341. [408]

Aizen, M. A., K. B. Searcy, and D. L. Mulcahy. 1990. Among- and within-flower variation comparisons of pollen tube growth following self- and cross-pollinations in *Dianthus chinensis* (Caryophyllaceae). American Journal of Botany 77(5):671–676. [124–125, 127]

Akey, D. H. 1991. A review of marking techniques in arthropods and an introduction to elemental marking. Southwestern Naturalist, supplement 14:1–8, with Volume 16(2), June 1991. [321]

Alcorn, S. M., S. E. McGregor, and G. Olin. 1961. Pollination of saguaro cactus by doves, nectar-feeding bats, and honey bees. Science 133(3464):1594–1595. [391]

Alexander, M. P. 1969. Differential staining of aborted and non-aborted pollen. Stain Technology 44(3):117–122. [110]

Alexander, M. P. 1980. A versatile stain for pollen, fungi, yeast and bacteria. Stain Technology 55(1):13–18. [110, 447, 454–456]

Alexander, M. P. 1987. A method for staining pollen tubes in pistil. Stain Technology 62(2):107–112. [131]

Allan, S. A.; J. G. Stoffolano, Jr.; and R. R. Bennett. 1991. Spectral sensitivity of the horse fly *Tabanus nigrovittatus* (Diptera: Tabanidae). Canadian Journal of Zoology 69(2):369–374. [315]

Allison, T. D. 1990. Pollen production and plant density affect pollination and seed production in *Taxus canadensis*. Ecology 71(2):516–622. [33]

Alm, J., T. E. Ohnmeiss, J. Lanza, and L. Vriesenga. 1990. Preference of cabbage white butterflies and honey bees for nectar that contains amino acids. Oecologia 84(1):53–57. [355–356]

Alspach, P. A., N. B. Pyke, C.G.T. Morgan, and J. E. Ruth. 1991. Influence of application rates of bee-collected pollen on the fruit size of kiwifruit. New Zealand Journal of Crop and Horticultural Science 19:19–24. [121]

Bibliography

Altenburger, R., and P. Matile. 1990. Further observations on rhythmic emission of fragrance in flowers. Planta 180(2):194–197. [47]

Aluja, M., R. J. Prokopy, J. S. Elkinton, and F. Laurence. 1989. Novel approach for tracking and quantifying the movement patterns of insects in three dimensions under seminatural conditions. Environmental Entomology 18(1):1–7. [345]

Aluri, R.J.S. 1990. The explosive pollination mechanism and mating system of the weedy *Hyptis suaveolens* (Lamiaceae). Plant Species Biology 5(2):235–241. [336–337]

Alves, L. M., A. E. Middleton, and D. J. Morre. 1968. Localization of callose deposits in pollen tubes of *Lilium longiflorum* Thunb. by fluorescence microscopy. Proceedings of the Indiana Academy of Science 77:144–147. [129, 447]

Andersen, A. N. 1989. How important is seed predation to recruitment in stable populations of long-lived perennials? Oecologia 81(3):310–315. [29]

Anderson, G. J., and D. A. Levine. 1982. Three taxa constitute the sexes of a single dioecious species of *Solanum*. Taxon 31(4):667–672. [240]

Anderson, G. J., and D. E. Symon. 1989. Functional dioecy and andromonoecy in *Solanum*. Evolution 43(1):204– 219. [240]

Anderson, J. M., and S.C.H. Barrett. 1986. Pollen tube growth in tristylous *Pontederia cordata* (Pontederiaceae). Canadian Journal of Botany 64(11):2602–2607. [125–128]

Andersson, S. 1988. Size-dependent pollination efficiency in *Anchusa officinalis* (Boraginaceae): causes and consequences. Oecologia 76(1):125–130. [33]

Antonovics, J., and J. Schmitt. 1986. Paternal and maternal effects on propagule size in *Anthoxanthum odoratum*. Oecologia 69(2):277–282. [14]

Arditti, J., N. M. Hogan, and A. V. Chadwick. 1973. Post-pollination phenomena in orchid flowers. IV. Effects of ethylene. American Journal of Botany 60(9):883–888. [338]

Argus, G. W. 1974. An experimental study of hybridization and pollination in *Salix*. Canadian Journal of Botany 52(7):1613–1619. [20]

Armbruster, W. S. 1984. The role of resin in angiosperm pollination: ecological and chemical considerations. American Journal of Botany 71(8):1149–1160. [215]

Armbruster, W. S., S. Keller, M. Matsuki, and T. P. Clausen. 1989. Pollination of *Dalechampia magnoliifolia* (Euphorbiaceae) by male euglossine bees. American Journal of Botany 76(9):1279–1285. [49, 52]

Armbruster, W. S., and W. R. Mziray. 1987. Pollination and herbivore ecology of an African *Dalechampia* (Euphorbiaceae): comparisons with New World species. Biotropica 19(1):64–73. [215]

Armbruster, W. S., and G. L. Webster. 1979. Pollination of two species of *Dalechampia* (Euphorbiaceae) in Mexico by euglossine bees. Biotropica 11(4):278–283. [51, 215]

Armstrong, D. P. 1991. Nectar depletion and its implications for honeyeaters in heathland near Sydney. Australian Journal of Ecology 16:99–109. [160]

Armstrong, D. P., and D. C. Paton. 1990. Methods for measuring amounts of energy available from *Banksia* inflorescences. Australian Journal of Ecology 15(3):291–297. [163]

Armstrong, J. A. 1979. Biotic pollination mechanisms in the Australian flora — a review. New Zealand Journal of Botany 17:467-508. [479, 484, 490]

Arnold, R. M. 1982. Floral biology of *Chaenorrhinum minus* (Scrophulariaceae) a self-compatible annual. American Midland Naturalist 108(2):317–324. [68]

Bibliography

Arroyo, M.T.K., J. J. Armesto, and R. B. Primack. 1985. Community studies in pollination ecology in high temperate Andes of central Chile. II. Effect of temperature on visitation rates and pollination possibilities. Plant Systematics and Evolution 149(3-4):187–203. [342, 393]

Arroyo, M.T.K., R. Primack, and J. Armesto. 1982. Community studies in pollination ecology in the high temperate Andes of central Chile. I. Pollination mechanisms and altitudinal variation. American Journal of Botany 69(1):82–97. [342]

Arroyo, M.T.K., and F. A. Squeo. 1987. Experimental detection of anemophily in *Pernettya mucronata* (Ericaceae) in western Patagonia, Chile. Botanische Jahrbücher für Systematik, Pflanzengeschichte und Pflanzengeographie 108:537–546. [20]

Ashman, T.-L., and I. Baker. 1992. Variation in floral sex allocation with time of season and currency. Ecology 73(4):1237-1243. [47]

Ashman, T.-L., and M. Stanton. 1991. Seasonal variation in pollination dynamics of sexually dimorphic *Sidalcea oregana* ssp. *spicata* (Malvaceae). Ecology 72(3):993–1003. [163, 361, 381, 391]

Augspurger, C. K. 1980. Mass-flowering of a tropical shrub (*Hybanthus prunifolius*): influence on pollinator attraction and movement. Evolution 34(3):475–488. [28, 85, 401]

Bach, C. E. 1990. Plant successional stage and insect herbivory: flea beetles on sand-dune willow. Ecology 71(2):598–609. [331]

Baker, H. G. 1977. Non-sugar chemical constituents of nectar. Apidologie 8:349–356. [200]

Baker, H. G. 1983. An outline of the history of anthecology, or pollination biology. Pages 7–28 *in* L. Real, editor. Pollination biology. Academic Press, Orlando, Florida, USA. [1]

Baker, H. G., and I. Baker. 1973. Some anthecological aspects of the evolution of nectar-producing flowers, particularly amino acid production in nectar. Pages 243–264 *in* V. H. Heywood, editor. Taxonomy and ecology. Academic Press, London, England. [194–196, 205, 450, 455, 457–458, 474]

Baker, H. G., and I. Baker. 1975. Studies of nectar-constitution and pollinator-plant coevolution. Pages 100–140 *in* L. E. Gilbert and P. H. Raven, editors. Animal and plant coevolution. University of Texas Press, Austin, Texas, USA. [195, 200, 203, 207–209, 211, 449, 453, 457, 460, 462]

Baker, H. G., and I. Baker. 1977. Intraspecific constancy of floral nectar amino acid complements. Botanical Gazette 138(2):183–191. [196, 454]

Baker, H. G., and I. Baker. 1979. Sugar ratios in nectars. Phytochemical Bulletin 12(1):43–45. [168]

Baker, H. G., and I. Baker. 1982. Chemical constituents of nectar in relation to pollination mechanisms and phylogeny. Pages 131–171 *in* M. H. Nitecki, editor. Biochemical aspects of evolutionary biology. Proceedings of the 4th annual spring systematics symposium. University of Chicago Press, Chicago, IL, USA. [153–154, 169, 456]

Baker, H. G., I. Baker and P. A. Opler. 1973. Stigmatic exudates and pollination. Pages 47–60 *in* N.B.M. Brantjes and H. F. Linskens, editors. Pollination and dispersal. Proceedings of a symposium, published by the Department of Botany, University of Nijmegen, Netherlands. [65–66, 452]

Baker, H. G., and P. D. Hurd, Jr. 1968. Intrafloral ecology. Annual Review of Entomology 13:385–414. [4]

Bibliography

Baker, I. 1979. Methods for the determination of volumes and sugar concentrations from nectar spots on paper. Phytochemical Bulletin 12(1):40–42. [169, 171, 174, 447–448, 451, 458]

Baker, I., and H. G. Baker. 1976a. Analyses of amino acids in flower nectars of hybrids and their parents, with phylogenetic implications. New Phytologist 76(1):87–98. [194, 196, 452]

Baker, I., and H. G. Baker. 1976b. Analysis of amino acids in nectar. Phytochemical Bulletin 9(1):4–6. [194, 196]

Baker, J. D., and R. W. Cruden. 1991. Thrips-mediated self-pollination of two facultatively xenogamous wetland species. American Journal of Botany 78(7):959–963. [18–19]

Baltosser, W. H. 1978. New and modified methods for color-marking hummingbirds. Bird-Banding 49(1):47–49. [333]

Bamberg, J. B., and R. E. Hanneman, Jr. 1991. An effective method for culturing pollen tubes of potato. American Potato Journal 68:373–380. [105]

Bancroft, J. D. 1975. Histochemical techniques. 2nd edition. Butterworths, London, England. [69, 455, 461]

Barrett, S.C.H. 1985. Floral trimorphism and monomorphism in continental and island populations of *Eichhornia paniculata* (Spreng.) Solms. (Pontederiaceae). Biological Journal of the Linnean Society 25(1):41–60. [43, 96, 239]

Barrett, S.C.H. 1990. The evolution and adaptive significance of heterostyly. Trends in Ecology and Evolution 5(5):144–148. [239, 474]

Barrett, S.C.H., and C. G. Eckert. 1990. Variation and evolution of mating systems in seed plants. Pages 229–254 *in* S. Kawano, editor. Biological approaches and evolutionary trends in plants. Academic Press, London, England. [218]

Barrett, S.C.H., and D. E. Glover. 1985. On the Darwinian hypothesis of the adaptive significance of tristyly. Evolution 39(4):766–774. [89, 239]

Barrett, S.C.H., and K. Helenurm. 1987. The reproductive biology of boreal forest herbs. I. Breeding systems and pollination. Canadian Journal of Botany 65(10):2036–2046. [15]

Barrett, S.C.H., and L. M. Wolfe. 1986. Pollen heteromorphism as a tool in studies of the pollination process in *Pontederia cordata* L. Pages 435–442 *in* D. L. Mulcahy, G. Bergamini Mulcahy, and G. Ottaviano, editors. Biotechnology and ecology of pollen. Springer Verlag, Berlin, Germany. [94, 239]

Barrow, D. A., and R. S. Pickard. 1985. Estimating corolla length in the study of bumble bees and their food plants. Journal of Apicultural Research 24(1):3–6. [41]

Barth, F. G. 1985. Insects and flowers. The biology of a partnership. Princeton University Press, Princeton, New Jersey, USA. [7]

Baskin, S. I., and C. A. Bliss. 1969. Sugar occurring in the extrafloral exudates of the Orchidaceae. Phytochemistry 8(7):1139–1145. [184, 192, 448, 453, 455, 462]

Bateman, A. J. 1951. The taxonomic discrimination of bees. Heredity 5(2):271–278. [375]

Bawa, K. S. 1974. Breeding systems of tree species of a lowland tropical community. Evolution 28(1):85–92. [260]

Bawa, K. S., and J. H. Beach. 1981. Evolution of sexual systems in flowering plants. Annals of the Missouri Botanical Garden 68:254–274. [245]

Bazzaz, F. A., R. W. Carlson, and J. L. Harper. 1979. Contribution to reproductive effort by photosynthesis of flowers and fruits. Nature 279:554–555. [167]

Bibliography

Beals, E. W. 1984. Bray-Curtis ordination: an effective strategy for analysis of multivariate ecological data. Pages 1–55 *in* A. Macfadyen and E. D. Ford, editors. Advances in ecological research, Volume 14. Academic Press, London, England. [29]

Beaman, R. S., P. J. Decker, and J. H. Beaman. 1988. Pollination of *Rafflesia* (Rafflesiaceae). American Journal of Botany 75(8):1148–1162. [379]

Beare, M. H., and W. E. Perkins. 1982. Effects of variation in floral morphology on pollination mechanisms in *Asclepias tuberosa* L. butterfly weed (Asclepiadaceae). American Journal of Botany 69(4):579–584. [43]

Beattie, A. J. 1971a. A technique for the study of insect-borne pollen. Pan-Pacific Entomologist 47:82. [115, 117, 289, 449]

Beattie, A. J. 1971b. Pollination mechanisms in *Viola*. New Phytologist 70(2):343–360. [293, 296]

Beattie, A. J. 1971c. Itinerant pollinators in a forest. Madroño 21:120–124. [391]

Beattie, A. J. 1976. Plant dispersion, pollination and gene flow in *Viola*. Oecologia 25(4):291–300. [34, 253, 403]

Beattie, A. J., D. E. Breedlove, and P. R. Ehrlich. 1973. The ecology of the pollinators and predators of *Frasera speciosa*. Ecology 54(1):81–91. [293–294]

Beattie, A. J., C. Turnbull, R. B. Knox, and E. G. Williams. 1984. Ant inhibition of pollen function: a possible reason why ant pollination is rare. American Journal of Botany 71(3):421–426. [103, 114]

Beckmann, H. 1974. Beeinflussung des Gedächtnisses der Honigbiene durch Narkose, Kühlung und Stress. Journal of comparative Physiology 94:249–266. [268]

Beker, R., A. Dafni, D. Eisikowitch, and U. Ravid. 1989. Volatiles of two chemotypes of *Majorana syriaca* L. (Labiatae) as olfactory cues for the honeybee. Oecologia 79(4):446–451. [365]

Bell, M. R. 1988. *Heliothis virescens* and H. *zea* (Lepidoptera: Noctuidae) feasibility of using oil-soluble dye to mark populations developing on early-season host plants. Journal of Entomological Science 23(3):223–228. [322, 450]

Bené, F. 1941. Experiments on the color preference of black-chinned hummingbirds. Condor 43(5):237–242. [368]

Benedict, J. H.; D. A. Wolfenbarger; V. M. Bryant, Jr.; and D. M. George. 1991. Pollens ingested by boll weevils (Coleoptera: Curculionidae) in southern Texas and northeastern Mexico. Journal of Economic Entomology 84(1):126–131. [80, 299]

Bentley, B., and T. Elias, editors. 1983. The biology of nectaries. Columbia University Press, New York, New York, USA. [7]

Bergström, G., M. Appelgren, A.-K. Borg-Karlson, I. Groth, S. Strömberg, and S. Strömberg. 1980. Studies on natural odoriferous compounds. Chemica Scripta 16:173–180. [47, 49, 52, 450, 452, 455–456, 458]

Berlyn, G. P., and J. P. Miksche. 1976. Botanical microtechnique and cytochemistry. Iowa State University Press, Ames, Iowa, USA. [73]

Bernhardt, P. 1983. Dimorphic *Amyema melaleucae*: a shift towards obligate autogamy. Bulletin of the Torrey Botanical Club 110(2):195–202. [69]

Bernhardt, P., and D. M. Calder. 1981. Hybridization between *Amyema pendulum* and *Amyema quandang* (Loranthaceae). Bulletin of the Torrey Botanical Club 108(4):456–466. [130]

Bibliography

Bernhardt, P., R. B. Knox, and D. M. Calder. 1980. Floral biology and self-incompatibility in some Australian mistletoes of the genus *Amyema* (Loranthaceae). Australian Journal of Botany 28(4):437–451. [118, 127–128]

Berry, P. E., and R. N. Calvo. 1989. Wind pollination, self-incompatibility, and altitudinal shifts in pollination systems in the high Andean genus *Espeletia* (Asteraceae). American Journal of Botany 76(11):1602–1614. [342, 393]

Bertin, R. I. 1982. Floral biology, hummingbird pollination and fruit production of trumpet creeper (*Campsis radicans*, Bignoniaceae). American Journal of Botany 69(1):122–134. [67, 115, 118]

Bertin, R. I. 1985. Nonrandom fruit production in *Campsis radicans*: between-year consistency and effects of prior pollination. American Naturalist 126(6):750–759. [76, 261]

Bertin, R. I. 1988. Paternity in plants. Pages 30–59 *in* J. Lovett Doust and L. Lovett Doust, editors. Plant Reproductive Ecology. Oxford University Press, New York, New York, USA. [248]

Bertin, R. I. 1990a. Effects of pollination intensity in *Campsis radicans*. American Journal of Botany 77(2):178–187. [117]

Bertin, R. I. 1990b. Paternal success following mixed pollinations of *Campsis radicans*. American Midland Naturalist 124:153–163. [230]

Bertin, R., I. C. Barnes, and S. I. Guttman, 1989. Self-sterility and cryptic self-fertility in *Campsis radicans* (Bignoniaceae). Botanical Gazette 150:397–403. [487]

Bertin, R. I., and P. Peters. 1991. Paternal effects on offspring quality in *Campsis radicans*. Abstract. Supplement to American Journal of Botany 78(6):49. [101]

Bertin, R. I., and M. Sullivan. 1988. Pollen interference and cryptic self-fertility in *Campsis radicans*. American Journal of Botany 75(8):1140–1147. [239]

Bertin, R. I., and M. F. Willson. 1980. Effectiveness of diurnal and noctural pollination of two milkweeds. Canadian Journal of Botany 58(16):1744–1746. [13, 390]

Bertsch, A. 1984. Foraging in male bumblebees (*Bombus lucorum* L.): maximizing energy or minimizing water load? Oecologia 62(3):325–336. [304, 311, 350]

Best, L. S., and P. Bierzychudek. 1982. Pollinator foraging on foxglove (*Digitalis purpurea*): a test of a new model. Evolution 36(1):70–79. [346]

Bierzychudek, P. 1981. Pollinator limitation of plant reproductive effort. American Naturalist 117(5):838–840. [255]

Bierzychudek, P. 1987. Pollinators increase the cost of sex by avoiding female flowers. Ecology 68(2):444–447. [37]

Block, R. J., E. L. Durrum, and G. Zweig. 1958. A manual of paper chromatography and paper electrophoresis. Academic Press, New York, New York, USA. [187, 188, 449–450, 459]

Boggs, C. L. 1986. Ecology of nectar and pollen feeding in Lepidoptera. Pages 369–391 *in* F. Slansky, Jr., and J. G. Rodriguez, editors. Nutritional ecology of insects, mites, spiders, and related invertebrates. John Wiley & Sons, New York, New York, USA. [316]

Boggs, C. L. 1988. Rates of nectar feeding in butterflies: effects of sex, size, age and nectar concentration. Functional Ecology 2:289–295. [358, 373]

Bolten, A. B., P. Feinsinger, H. G. Baker, and I. Baker. 1979. On the calculation of sugar concentration in flower nectar. Oecologia 41(3):301–304. [171]

Bibliography

Borgia, G. 1985. Bower quality, number of decorations, and mating success of male satin bowerbirds (*Ptilonorhynchus violaceus*): an experimental analysis. Animal Behaviour 33:266–271. [339]

Borror, D. J., D. M. DeLong, and C. A. Triplehorn. 1981. Introduction to the study of insects. Saunders College Publishing, Philadelphia, Pennsylvania, USA. [272–273]

Borthwick, H. A. 1931. Development of the macrogametophyte and embryo of *Daucus carota*. Botanical Gazette 92(1):23–44. [122]

Bosi, G., and M. Battaglini. 1978. Gas chromatographic analysis of free and protein amino acids in some unifloral honeys. Journal of Apicultural Research 17(3):152–166. [197–198]

Bowers, M. D. 1986. Density dynamics of bumblebees in subalpine meadows: competition and resource limitation. Holarctic Ecology 9:175–184. [382]

Bradford, M. M. 1976. A rapid and sensitive method for the quantitation of microgram quantities of protein utilizing the principle of protein-dye binding. Analytical Biochemistry 72:248–254. [206, 451, 459]

Brantjes, N.B.M. 1973. Sphingophilous flowers, function of their scent. Pages 27–46 *in* N.B.M. Brantjes and H. F. Linskens, editors. Pollination and dispersal. Proceedings of a symposium, published by the Department of Botany, University of Nijmegen, Netherlands. [351]

Brantjes, N.B.M. 1978. Sensory responses to flowers in night-flying moths. Pages 13–19 *in* A. J. Richards, editor. The pollination of flowers by insects. Linnean Society Symposium Series No. 6. Academic Press, London, England. [56, 455]

Brewbaker, J. L., and B. H. Kwack. 1963. The essential role of calcium ion in pollen germination and pollen tube growth. American Journal of Botany 50(9):747–858. [102, 450, 456, 459]

Brian, A. D. 1951. The pollen collected by bumble-bees. Journal of Animal Ecology 20(2):191–194. [300]

Brian, A. D. 1957. Differences in the flowers visited by four species of bumble-bees and their causes. Journal of Animal Ecology 26(1):71–98. [350, 367]

Brink, D. E. 1980. Reproduction and variation in *Aconitum columbianum* (Ranunculaceae), with emphasis on California populations. American Journal of Botany 67(3):263–273. [42]

Brink, D., and J.M.J. deWet. 1980. Interpopulation variation in nectar production in *Aconitum columbianum* (Ranunculaceae). Oecologia 47(2):160–163. [42, 158, 431]

Broom, D. M. 1976. Duration of feeding bouts and responses to salt solutions by hummingbirds at artificial feeders. Condor 78(1):135–138. [365]

Brown, A.H.D. 1990. Genetic characterization of plant mating systems. Pages 145–162 *in* A.H.D. Brown, M. T. Clegg, A. L. Kahler, and B. S. Weir, editors. Plant population genetics, breeding, and genetic resources. Sinauer Associates, Sunderland, Massachusetts, USA. [217–218, 225, 228–230]

Brown, A.H.D., S.C.H. Barrett, and G. F. Moran. 1985. Mating system estimation in forest trees: models, methods and meanings. Pages 32–49 *in* H. R. Gregorius, editor. Population genetics in forestry. Springer Verlag, New York, New York, USA. [218, 226–227, 229]

Brown, A.H.D., J. J. Burdon, and A. M. Jarosz. 1989. Isozyme analysis of plant mating systems. Pages 73–86 *in* D. E. Soltis and P. S. Soltis, editors. Isozymes in plant biology. Dioscorides Press, Portland, Oregon, USA. [222–223]

499

Bibliography

Brown, A.H.D., M. T. Clegg, A. L. Kahler, and B. S. Weir, editors. 1990. Plant population genetics, breeding, and genetic resources. Sinauer Associates, Sunderland, Massachusetts, USA. [222]

Brown, B. A., and M. T. Clegg. 1984. Influence of flower color polymorphism on genetic transmission in a natural population of the common morning glory *Ipomoea purpurea*. Evolution 38(4):796–803. [218]

Brown, C. R., and K. D. Adiwilaga. 1991. Use of rescue pollination to make a complex interspecific cross in potato. American Potato Journal 68:813–820. [240, 484, 489]

Brown, J. H., and A. Kodric-Brown. 1979. Convergence, competition, and mimicry in a temperate community of hummingbird-pollinated flowers. Ecology 60(5):1022–1035. [118, 296]

Broyles, S. B., and R. Wyatt. 1990a. Paternity analysis in a natural population of *Asclepias exaltata*: multiple paternity, functional gender, and the "pollen-donation hypothesis." Evolution 44(6):1454–1468. [229–230, 242]

Broyles, S. B., and R. Wyatt. 1990b. Plant parenthood in milkweeds: a direct test of the pollen donation hypothesis. Plant Species Biology 5:131–142. [230, 244]

Brussard, P. F. 1971. Field techniques for investigations of population structure in a "ubiquitous" butterfly. Journal of the Lepidopterists' Society. 25(1):22–29. [328–330]

Bryant, V. M., M. Pendleton, R. E. Murry, P. D. Lingren, and J. R. Raulston. 1991. Techniques for studying pollen adhering to nectar-feeding corn earworm (Lepidoptera: Noctuidae) moths using scanning electron microscopy. Journal of Economic Entomology 84(1):237–240. [88, 294]

Buchmann, S. L. 1983. Buzz pollination in angiosperms. Chapter 4, pages 73–113, *in* C. E. Jones and R. J. Little, editors. Handbook of experimental pollination biology. Scientific and Academic Editions, Van Nostrand Reinhold Company, New York, New York, USA. [48]

Buchmann, S. L. 1992. Buzzing is necessary for tomato flower pollination. Bumblebeequest 2(2):1–3. [340]

Buchmann, S. L., and M. D. Buchmann. 1981. Anthecology of *Mouriri myrtilloides* (Melastomataceae: Memecyleae), an oil flower in Panama. Biotropica (Reproductive Botany Supplement) 13:7–24. [57, 86, 90, 208, 210, 214, 449–450, 452, 455–458]

Buchmann, S. L., and J. H. Cane. 1989. Bees assess pollen returns while sonicating *Solanum* flowers. Oecologia 81(3):289–294. [93, 371]

Buchmann, S. L., and C. W. Shipman. 1990. Pollen harvesting rate for *Apis mellifera* L. on *Gossypium* (Malvaceae) flowers. Journal of the Kansas Entomological Society 63(1):92–100. [94, 290]

Buck, P., and E. Levetin. 1985. Airborne pollen and mold spores in a subalpine environment. Annals of Allergy 55:794–801. [92]

Burg, S. P., and M. J. Dijkman. 1967. Ethylene and auxin participation in pollen-induced fading of Vanda orchid blossoms. Plant Physiology 42:1648–1650. [338]

Burley, N., G. Krantzberg, and P. Radman. 1982. Influence of colour-banding on the conspecific preferences of Zebra Finches. Animal Behaviour 30:444–455. [334]

Búrquez, A. 1987. Leaf thickness and water deficit in plants: a tool for field studies. Journal of Experimental Botany 38(186):109–114. [402]

Búrquez, A. 1989. Blue tits, *Parus caeruleus,* as pollinators of the crown imperial, *Fritillaria imperialis,* in Britain. Oikos 55(3):335–340. [312]

Bibliography

Búrquez, A., and S. A. Corbet. 1991. Do flowers reabsorb nectar? Functional Ecology 5(3):369–379. [157–159, 215, 474, 478, 481, 483, 489]

Byrne, D. N., S. L. Buchmann, and H. G. Spangler. 1988. Relationship between wing loading, wingbeat frequency and body mass in homopterous insects. Journal of Experimental Biology 135:9–23. [314]

Calder, W. A. 1971. Temperature relationships and nesting of the Calliope Hummingbird. Condor 73(3):314– 321. [316]

Calder, W. A., and J. Booser. 1973. Hypothermia of Broad-tailed hummingbirds during incubation in nature with ecological correlations. Science 180(4087):751–753. [316]

Calder, W. A., L. L. Calder, and T. D. Fraizer. 1990. The hummingbird's restraint: a natural model for weight control. Experientia 46:999–1002. [311–312]

Calder, W. A., III; and S. M. Hiebert. 1983. Nectar feeding, diuresis, and electrolyte replacement of hummingbirds. Physiological Zoology 56(3):325–334. [320]

Calder, W. A., S. M. Hiebert, N. M. Waser, D. W. Inouye, and S. J. Miller. 1983. Site fidelity, longevity, and population dynamics of Broad-tailed Hummingbirds: a ten year study. Oecologia 56:689–700. [381]

Cameron, S. A. 1981. Chemical signals in bumble bee foraging. Behavioral Ecology and Sociobiology 9:257–260. [348–349]

Campbell, D. R. 1985. Pollinator sharing and seed set of *Stellaria pubera*: competition for pollination. Ecology 66(2):544–553. [34, 135, 138–139, 134–235, 345]

Campbell, D. R. 1989. Measurements of selection in a hermaphroditic plant: variation in male and female pollination success. Evolution 43(2):318–334. [3, 40, 43, 137, 259, 408]

Campbell, D. R. 1991a. Effects of floral traits on sequential components of fitness in *Ipomopsis aggregata*. American Naturalist 137(6):713–737. [3, 75]

Campbell, D. R. 1991b. Comparing pollen dispersal and gene flow in a natural population. Evolution 45(8):1965–1968. [135, 140]

Campbell, D. R., and A. F. Motten. 1985. The mechanism of competition for pollination between two forest herbs. Ecology 66(2):554–563. [371]

Campbell, D. R., and N. M. Waser. 1989. Variation in pollen flow within and among populations of *Ipomopsis aggregata*. Evolution 43(7):1444–1455. [137, 139, 140]

Campbell, D. R., N. M. Waser, M. V. Price, E. A. Lynch, and R. J. Mitchell. 1991. Components of phenotypic selection: pollen export and flower corolla width in *Ipomopsis aggregata*. Evolution 45(6):1458–1467. [40, 42, 259]

Cane, J. H. 1991. An ideal hang tag for flowers, fruits and petioles. Plant Science Bulletin 37:10. [36]

Cane, J. H., and S. L. Buchmann. 1989. Novel pollen-harvesting behavior by the bee *Protandrena mexicanorum* (Hymenoptera: Andrenidae). Journal of Insect Behavior 2(3):431–436. [340]

Cane, J. H., and J. A. Payne. 1988. Foraging ecology of the bee *Habropoda laboriosa* (Hymenoptera: Anthophoridae), an oligolege of blueberries (Ericaceae: *Vaccinium*) in the southeastern United States. Annals of the Entomological Society of America 81(3):419–427. [340, 376]

Carpenter, F. L. 1983. Pollination energetics in avian communities: simple concepts and complex realities. Chapter 10, pages 215–234 *in* C. E. Jones and R. J. Little, editors.

Bibliography

Handbook of experimental pollination biology. Scientific and Academic Editions, Van Nostrand Reinhold Company, New York, New York, USA. [162]

Carpenter, F. L., M. A. Hixon, A. Hunt, and R. W. Russell. 1991. Why hummingbirds have such large crops. Evolutionary Ecology 5(4):405–414. [333]

Carpenter, F. L., D. C. Paton, and M. A. Hixon. 1983. Weight gain and adjustment of feeding territory size in migrant hummingbirds. Proceedings of the National Academy of Sciences (USA) 80:7259–7263. [311–312]

Carr, D. 1990. The reproductive ecology of a dioecious understory tree, *Ilex opaca* Ait. Ph.D. dissertation, University of Maryland, College Park, Maryland, USA. [236]

Cartar, R. V. 1991. A test of risk-sensitive foraging in wild bumble bees. Ecology 72(3):888–895. [166, 341]

Cartar, R. V., and L. M. Dill. 1990. Why are bumble bees risk-sensitive foragers? Behavioral Ecology and Sociobiology 26:121–127. [348]

Carter, A. L., and T. McNeilly. 1975. Effects of increased humidity on pollen tube growth and seed set following self pollination in Brussels sprout (*Brassica oleracea* var. *gemmifera*). Euphytica 24:805–813. [239]

Casper, B. B., and T. R. La Pine. 1984. Changes in corolla color and other floral characteristics in *Cryptantha humilis* (Boraginaceae): cues to discourage pollinators. Evolution 38(1):128–141. [64, 182]

Chaplin, S. J., and J. L. Walker. 1982. Energetic constraints and adaptive significance of the floral display of a forest milkweed. Ecology 63(6):1857–1870. [274]

Charlesworth, D. 1988. A method for estimating outcrossing rates in natural populations of plants. Heredity 61:469–471. [222]

Charnov, E. L. 1979. Simultaneous hermaphroditism and sexual selection. Proceedings of the National Academy of Sciences (USA) 76:2480–2482. [248]

Charnov, E. L. 1982. The theory of sex allocation. Princeton University Press, Princeton, New Jersey, USA. [243]

Chen, D., J. S. Collins, and T. H. Goldsmith. 1984. The ultraviolet receptor of bird retinas. Science 225(4659):337–340. [316]

Chrambach, A., R. A. Reisfeld, M. Wyckoff, and F. Zaccari. 1967. A procedure for rapid and sensitive staining of protein fractionated by polyacrylamide gel electrophoresis. Analytical Biochemistry 20:150–154. [205, 462]

Ciampolini, F., K. R. Shivanna, and M. Cresti. 1990. The structure and cytochemistry of the pistil of *Sternbergia lutea* (Amaryllidaceae). Annals of Botany 66:703–712. [66–67, 448–449, 454, 456–458, 460–463]

Cibula, D. A., and M. A. Zimmerman. 1984. The effect of plant density on departure decisions: testing the marginal value theorem using bumblebees and *Delphinium nelsonii*. Oikos 43(2):154–158. [33]

Cibula, D. A., and M. Zimmerman. 1987. Bumblebee foraging behavior: changes in departure decisions as a function of experimental nectar manipulations. American Midland Naturalist 117(2):386–394. [370]

Clark, C. 1979. Ultraviolet absorption by flowers of the Eschscholzioideae (Papaveraceae). Madroño 26:22–25. [65]

Clark, J. M. 1964. Experimental biochemistry. W. H. Freeman and Co., San Francisco, California, USA. [179]

Classenbockhoff, R. 1991. Untersuchungen zur Konstruktion des Bestäubungsapparates von *Thalia geniculata* (Marantaceen). Botanica Acta 104(3):183–193. [337]

Clegg, M. T. 1980. Measuring plant mating systems. BioScience 30(12):814–818. [225–227]

Clegg, M. T. 1990. Molecular diversity in plant populations. Pages 98–115 *in* A.H.D. Brown, M. T. Clegg, A. L. Kahler, and B. S. Weir, editors. Plant population genetics, breeding, and genetic resources. Sinauer Associates, Sunderland, Massachusetts, USA. [224]

Clegg, M. T., and B. K. Epperson. 1985. Recent developments in population genetics. Advances in Genetics 23:235–269. [218]

Clements, F. E., and F. L. Long. 1923. Experimental pollination. An outline of the ecology of flowers and insects. Publication No. 336. Carnegie Institution of Washington. Washington, D.C., USA. [1, 38, 347, 362]

Cock, M.J.W. 1978. The assessment of preference. Journal of Animal Ecology 47(3):805–816. [367, 376]

Coen, E. S., and E. M. Meyerowitz. 1991. The war of the whorls: genetic interactions controlling flower developments. Nature 353(6339):31–37. [2]

Cohen, S. A., and D. J. Strydom. 1988. Amino acid analysis utilizing phenylisothiocyanate derivatives. Analytical Biochemistry 174(1):1–16. [199]

Collias, N. E., and E. C. Collias. 1968. Anna's hummingbirds trained to select different colors in feeding. Condor 70(3):273–274. [368]

Collins, B. G., and D. C. Paton. 1989. Consequences of differences in body size, wing length and leg morphology for nectar-feeding birds. Australian Journal of Ecology 14(3):269–289. [312]

Colwell, R. N. 1951. The use of radioactive isotopes in determining spore distribution patterns. American Journal of Botany 38(7):511–523. [142]

Cook, S. A., and R. G. Stanley. 1960. Tetrazolium chloride as an indicator of pine pollen germinability. Silvae Genetica 9:134–136. [109]

Copland, B. J., and R. J. Whelan. 1989. Seasonal variation in flowering intensity and pollination limitation of fruit set in four co-occurring *Banksia* species. Journal of Ecology 77(2):509–523. [126–128]

Corbet, S. A. 1978a. Bee visits and the nectar of *Echium vulgare* L. and *Sinapsis alba* L. Ecological Entomology 3:25–37. [157, 160, 166, 169, 403]

Corbet, S. A. 1978b. Bee visits and the nectar of *Echium vulgare*. Pages 21–30 *in* A. J. Richards, editor. The pollination of flowers by insects. Linnean Society Symposium Series No. 6. Academic Press, London, England. [158–159, 169]

Corbet, S. A. 1990. Pollination and the weather. Israel Journal of Botany 39(1-2):13–30. [151]

Corbet, S. A. 1991. Applied pollination ecology. Trends in Ecology and Evolution 6(1):3–4. [3]

Corbet, S. A., J. Beament, and D. Eisikowitch. 1982. Are electrostatic forces involved in pollen transfer? Plant, Cell and Environment 5:125–129. [150]

Corbet, S. A., H. Chapman, and N. Saville. 1988. Vibratory pollen collection and flower form: bumble-bees on *Actinidia, Symphytum, Borago* and *Polygonatum*. Functional Ecology 2(2):147–155. [93]

Corbet, S. A., and E. S. Delfosse. 1984. Honeybees and the nectar of *Echium plantagineum* L. in southeastern Australia. Australian Journal of Ecology 9(2):125–139. [17, 36, 393, 398]

Bibliography

Corbet, S. A., and J. R. Plumridge. 1985. Hydrodynamics and the germination of oil-seed rape pollen. Journal of Agricultural Science (Cambridge) 104:445–451. [98, 151]

Corbet, S. A., D. M. Unwin, and O. E. Prŷs-Jones. 1979. Humidity, nectar and insect visits to flowers, with special reference to *Crataegus*, *Tilia* and *Echium*. Ecological Entomology 4:9–22. [169, 171, 398]

Corbet, S. A., I. H. Williams, and J. L. Osborne. 1991. Bees and the pollination of crops and wild flowers in the European Community. Bee World 72(2):47–51. [253, 256–257]

Corbet, S. A., and P. G. Willmer. 1981. The nectar of *Justicia* and *Columnea*: composition and concentration in a humid tropical climate. Oecologia 51(3):412–418. [17, 159, 181–182, 194, 399, 450]

Corbet, S. A., P. G. Willmer, J.W.L. Beament, D. M. Unwin, and O. E. Prŷs-Jones. 1979. Post-secretory determinants of sugar concentration in nectar. Plant, Cell and Environment 2:293–308. [169]

Costin, A. B., M. Gray, C. J. Totterdell, and D. J. Wimbush. 1979. Kosciusko Alpine Flora. CSIRO and Collins (jointly published), East Melbourne and Sydney, Australia. [337]

Cox, P. A. 1988. Hydrophilous pollination. Annual Review of Ecology and Systematics 19:261–280. [20]

Cox, P. A. 1991. Hydrophilous pollination of a dioecious seagrass, *Thalassodendron ciliatum* (Cymodoceaceae) in Kenya. Biotropica 23(2):159–165. [20]

Cox, P. A., T. Elmqvist, and P. B. Tomlinson. 1990. Submarine pollination and reproductive morphology in *Syringodium filiforme* (Cymodoceaceae). Biotropica 22(3):259–265. [20]

Craig, J. L. 1989. Seed set in *Phormium*: interactive effects of pollinator behaviour, pollen carryover and pollen source. Oecologia 81(1):1–5. [138, 338]

Crawford, T. J. 1984. The estimation of neighbourhood parameters for plant populations. Heredity 52(2):273– 283. [232]

Cresswell, J. E. 1989. Optimal foraging theory applied to bumblebees gathering nectar from wild bergamot. Ph.D. dessertation, University of Michigan, Ann Arbor.

Cresswell, J. E., and C. Galen. 1991. Frequency-dependent selection and adaptive surfaces for floral character combinations: the pollination of *Polemonium viscosum*. American Naturalist 138(6):1342–1353. [165]

Cruden, R. W. 1977. Pollen-ovule ratios: a conservative indicator of breeding systems in flowering plants. Evolution 31(1):32–46. [91, 94, 247]

Cruden, R. W., K. K. Baker, T. E. Cullinan, K. A. Disbrow, K. L. Douglas, J. D. Erb, K. J. Kirsten, M. L. Malik, E. A. Turner, J. A. Weier, and S. R. Wilmot. 1990. The mating system and pollination biology of three species of *Verbena* (Verbenaceae). Journal of the Iowa Academy of Science 97(4):178–183. [18, 118, 249, 425]

Cruden, R. W., and S. M. Hermann. 1983. Studying nectar? Some observations on the art. Pages 223–241 *in* B. Bentley and T. Elias, editors. The biology of nectaries. Columbia University Press, New York, New York, USA. [158, 160–161, 170–171]

Cruden, R. W., S. M. Hermann, and S. Peterson. 1983. Patterns of nectar production and plant-pollinator coevolution. Pages 80–125 *in* B. Bentley and T. Elias, editors. The biology of nectaries. Columbia University Press, New York, New York, USA. [17]

Cruden, R. W., and D. L. Lyon. 1989. Facultative xenogamy: examination of a mixed mating system. Pages 171– 207 *in* J. H. Bock and Y. B. Linhart, editors. The evolutionary ecology of plants. Westview Press, Boulder, Colorado, USA. [220]

Bibliography

Crumpacker, D. W. 1974. The use of micronized fluorescent dusts to mark adult *Drosophila pseudobscura*. American Midland Naturalist 91:118–129. [323–324]

Cruzan, M. B. 1986. Pollen tube distribution in *Nicotiana glauca*: Evidence for density dependent growth. American Journal of Botany 73(6):902–907. [124]

Cruzan, M. B. 1989. Pollen tube attrition in *Erythronium grandiflorum*. American Journal of Botany 76(4):562– 570. [38, 448, 451]

Cruzan, M. B. 1990. Pollen-pollen and pollen-style interactions during pollen tube growth in *Erythronium grandiflorum* (Liliaceae). American Journal of Botany 77(1):116–122. [122]

Cumming, W. C., and F. I. Righter. 1948. Methods used to control pollination of pines in the Sierra Nevada of California. USDA Circular No. 792. Government Printing Office, Washington, D.C., USA. [17, 93, 121]

Cunningham, S. A. 1991. Experimental evidence for pollination of *Banksia* spp. by non-flying mammals. Oecologia 87(1):86–90. [339]

Currier, H. B. 1957. Callose substance in plant cells. American Journal of Botany 44(6):478– 488. [127, 460]

Dafni, A. 1984. Mimicry and deception in pollination. Annual Review of Ecology and Systematics 15:259–278. [475, 478, 485, 489].

Dafni, A. 1991. Advertisement, flower longevity, reward and nectar protection in Labiatae. Acta Horticulturae 288:340–346. [43–44]

Dafni, A., P. Bernhardt, A. Shmida, Y. Ivri, S. Greenbau, Ch. O'Toole, and L. Losito. 1990. Red bowl-shaped flowers: convergence for beetle pollination in the Mediterranean region. Israel Journal of Botany 39(1-2):81– 92. [368]

Dafni, A., and R. Dukas. 1986. Insect and wind pollination in *Urginea maritima* (Liliaceae). Plant Systematics and Evolution 154(1):1–10. [92, 261]

Dafni, A., D. Eisikowitch, and Y. Ivri. 1987. Nectar flow and pollinators' efficiency in two co-occurring species of *Capparis* (Capparaceae) in Israel. Plant Systematics and Evolution 157(3-4):181–186. [389]

Daly, H. V., and C. L. Jordan. 1989. Computer generated collection labels: they are fast, but will they last? Insect Collection News 2(2):26–27. [281]

Daniel, T. L., J. G. Kingsolver, and E. Meyhöfer. 1989. Mechanical determinants of nectar-feeding energetics in butterflies: muscle mechanics, feeding geometry, and functional equivalence. Oecologia 79(1):66–75. [373]

D'Arcy, W. G., N. S. D'Arcy, and R. C. Keating. 1990. Scented anthers in the Solanaceae. Rhodora 92(870):50–53. [49]

Darwin, C. 1876. The effects of cross and self fertilization in the vegetable kingdom. John Murray, London, England. [7]

Darwin, C. 1877. The various contrivances by which orchids are fertilised by insects. 2nd edition. D. Appleton and Company, New York, New York, USA. [7]

Daumer, K. 1958. Blumenfarben: wie sie die Bienen sehen. Zeitschrift für vergleichende Physiologie 38:413– 478. [59]

Davey, J. F., and R. S. Ersser. 1990. Amino acid analysis of physiological fluids by high-performance liquid chromatography with phenylisothiocyanate derivatization and comparison with ion-exchange chromatography. Journal of Chromatography 528(1):9–23. [199]

Bibliography

DeFoliart, G. R., and C. D. Morris. 1967. A dry ice-baited trap for the collection and field storage of hematophagous Diptera. Journal of Medical Entomology 4(3):360–362. [272]

Deinzer, M. L., P. A. Thomson, D. M. Burgett, and D. L. Isaacson. 1977. Pyrrolizidine alkaloids: their occurrence in honey from tansy ragwort (*Senecio jacobaea* L.). Science 195(4277):497–499. [200, 448, 451]

Delph, L. F. 1986. Factors regulating fruit and seed production in the desert annual *Lesquerella gordonii*. Oecologia 69(3):471–476. [255]

Delph, L. F., and C. M. Lively. 1989. The evolution of floral color change: pollinator attraction versus physiological constraints in *Fuchsia excorticata*. Evolution 43(6):1252–1262. [339]

del Rio, C. M. 1990. Nectar preferences in hummingbirds: the influence of subtle chemical differences on food choice. Condor 92(4):1022–1030. [365, 388]

del Rio, C., and A. Búrquez. 1986. Nectar production and temperature dependent pollination in *Mirabilis jalapa* L. Biotropica 18(1):28–31. [394]

Dempster, J. P., K. H. Lakhani, and P. A. Coward. 1986. The use of chemical composition as a population marker in insects: a study of the Brimstone butterfly. Ecological Entomology 11:51–65. [322]

de Nettancourt, D. 1977. Incompatibility in angiosperms. Springer Verlag, New York, New York, USA. [236]

DesGranges, J.-L. 1979. Organization of a tropical nectar feeding bird guild in a variable environment. Living Bird 1978:199–236. [164]

Devlin, B., K. Roeder, and N. C. Ellstrand. 1988. Fractional paternity assignment: theoretical development and comparison to other methods. Theoretical and Applied Genetics 76:369–380. [230]

Devlin, B., and A. G. Stephenson. 1985. Sex differential floral duration, nectar secretion and pollinator foraging in a protandrous species. American Journal of Botany 72(2):303–310. [27]

DeVries, P. J., and F. G. Stiles. 1990. Attraction of pyrrolizidine alkaloid seeking Lepidoptera to *Epidendrum paniculatum* orchids. Biotropica 22(3):290–297. [273]

Difco Laboratories. 1984. Difco manual. 10th edition. Difco Laboratories, Detroit, Michigan, USA. [204]

Di-Giovanni, F., and P. G. Kevan. 1991. Factors affecting pollen dynamics and its importance to pollen contamination: a review. Canadian Journal of Forestry Research 21:1155–1170. [134]

Digonnet-Kerhoas, C., and G. Gay. 1990. Qualité du pollen: définition et estimation. Bulletin de la Société Botanique de France 137:97–100. [98, 151]

Dobson, H.E.M. 1987. Role of flower and pollen aromas in host-plant recognition by solitary bees. Oecologia 72(4):618–623. [47, 367]

Dobson, H.E.M. 1988. Survey of pollen and pollenkitt lipids — chemical cues to flower visitors? American Journal of Botany 75(2):170–182. [56, 148]

Dobson, H.E.M. 1991. Analysis of flower and pollen volatiles. Pages 231–251 *in* H. F. Linskens and J. F. Jackson, editors. Essential oils and waxes. New Series Volume 12, Modern Methods of Plant Analysis. Springer Verlag, Berlin, Germany. [48–49, 53]

Bibliography

Dobson, H.E.M., G. Bergström, and I. Groth. 1990. Differences in fragrance chemistry between flower parts of *Rosa rugosa* Thunb. (Rosaceae). Israel Journal of Botany 39(1-2):143–156. [47, 53, 56]

Dobson, H.E.M., J. Bergström, G. Bergström, and I. Groth. 1987. Pollen and flower volatiles in two *Rosa* species. Phytochemistry 26(12):3171–3173. [56]

Dodson, G., and D. Yeates. 1990. The mating system of a bee fly (Diptera: Bombyliidae). II. Factors affecting male territorial and mating success. Journal of Insect Behavior 3(5):619–636. [314]

Dole, J. A. 1990. Role of corolla abscission in delayed self-pollination of *Mimulus guttatus* (Scrophulariaceae). American Journal of Botany 77(11):1505–1507. [75]

Dreisig, H. 1989. Nectar distribution assessment by bumblebees foraging at vertical inflorescences. Oikos 55(2):239–249. [348]

Drost, Y. C., W. J. Lewis, P. O. Zanen, and M. A. Keller. 1986. Beneficial arthropod behavior mediated by airborne semiochemicals. I. Flight behavior and influence of preflight handling of *Microplitis croceipes* (Cresson). Journal of Chemical Ecology 12(6):1247–1262. [367]

D'Sousa, L. 1972. Staining pollen tubes in the styles of cereals with cotton blue: fixation by ethanol-lactic acid for an enhanced differentiation. Stain Technology 47:107–108. [131]

Dubois, M., K. A. Gilles, J. K. Hamilton, P. A. Rebers, and F. Smith. 1956. Colorimetric method for determination of sugars and related substances. Analytical Chemistry 28(3):350–356. [177, 186]

Ducker, S. C., and R. B. Knox. 1976. Submarine pollination in seagrasses. Nature 263(5579):705–706. [20, 69]

Dudash, M. R. 1991. Plant size effects on female and male function in hermaphroditic *Sabatia angularis* (Gentianaceae). Ecology 72(3):1004–1012. [33, 91, 96, 138–139]

Dumas, C., A. Clarke, and B. Knox. 1985. Pollination and cellular recognition. Outlook on Agriculture 14(2):68–78. [112–113]

Dumas, C., and R. B. Knox. 1983. Callose and determination of pistil viability and incompatibility. Theoretical and Applied Genetics 67:1–10. [70, 75, 128, 453]

East, E. M. 1940. The distribution of self-sterility in the flowering plants. Proceedings of the American Philosophical Society 82:449–518. [219]

Ebadi, R., N. E. Gary, and K. Lorenzen. 1980. Effects of carbon dioxide and low temperature narcosis on honey bees, *Apis mellifera*. Environmental Entomology 9(1):144–147. [268]

Eckblad, J. W. 1991. How many samples should be taken? BioScience 41(5):346–348. [406–407]

Edmonds, R. L. 1972. Collection efficiency of Rotorod samplers for sampling fungus spores in the atmosphere. Plant Disease Reporter 56(8):704–708. [92]

Eguiarte, L. E., and A. Búrquez. 1988. Reducción en la fecundidad en *Manfreda brachystachya* (Cav.) Rose, una agavácea polinizada por murciélagos: los resgos de la especialización en la polinización. Boletín de la Sociedad Botánica de México 48:147–149. [13]

Ehrlich, P. R., and S. E. Davidson. 1960. Techniques for capture-recapture studies of Lepidoptera populations. Journal of the Lepidopterists' Society 14:227–229. [328]

Ehrlich, P. R., and P. H. Raven. 1969. Differentiation of populations. Science 165 (3899):1228–1232. [233]

Bibliography

Eijnde, J., van den. 1990. Ganzjährige Züchtung von Hummelvölkern für die Bestäubung in Gewächshäusern: eine rasche Entwicklung. ADIZ 1990 (6):12–14. [3]

Eisner, T., M. Eisner, P. A. Hyypio, D. Aneshansley, and R. E. Silberglied. 1973. Plant taxonomy: ultraviolet patterns of flowers visible as fluorescent patterns in pressed herbarium specimens. Science 179(4072):486–487. [64]

Eisner, T., R. E. Silberglied, D. Aneshansley, J. E. Carrel, and H. C. Howland. 1969. Ultraviolet video-viewing: the television camera as an insect eye. Science 166(3909):1172–1174. [64]

Elger, R. 1969. Freilandstudien zur Biologie und Ökologie von *Panaxia quadripunctaria* (Lepidoptera, Arctiidae) auf der Insel Rhodos. Oecologia 2(2):162–198. [330]

El-Kassaby, Y. A., and R. Davidson. 1991. Impact of pollination environment manipulation on the apparent outcrossing rate in a Douglas-fir seed orchard. Heredity 66(1):55–59. [218]

Ellstrand, N. C. 1984. Multiple paternity within the fruits of the wild radish, *Raphanus sativus*. American Naturalist 123(6):819–828. [230]

Ellstrand, N. C., A. M. Tores, and D. A. Levin. 1978. Density and the rate of apparent outcrossing in *Helianthus* (Asteraceae). Systematic Botany 3:403–407. [34, 218]

Elmqvist, T., J. Ågren, and A. Tunlid. 1988. Sexual dimorphism and between-year variation in flowering, fruit set and pollinator behaviour in a boreal willow. Oikos 53(1):58–66. [168, 198, 460]

Endress, P. K. 1982. Syncarpy and alternative modes of escaping disadvantages of apocarpy in primitive angiosperms. Taxon 31:48–52. [122]

English-Loeb, G. M., and R. Karban. 1992. Consequences of variation in flowering phenology for seed head herbivory and reproductive success in *Erigeron glaucus* (Compositae). Oecologia 89:588-595. [22]

Ennos, R. A., and M. T. Clegg. 1982. Effect of population substructuring on estimates of outcrossing rate in plant populations. Heredity 48(2):283–292. [221]

Erdtman, G. 1945. Pollen morphology and plant taxonomy. II. *Morina* L. with an addition on pollen morphological terminology. Svensk Botanisk Tidskrift 39:187–191. [89]

Erdtman, G. 1954. An introduction to pollen analysis. Chronica Botanica, Waltham, Massachusetts, USA. [79, 81]

Erdtman, G. 1960. The acetolysis method. A revised description. Svensk Botanisk Tidskrift 54(4):561–564. [79, 83, 461]

Erdtman, G. 1969. Handbook of Palynology. Munksgaard, Copenhagen, Denmark. [79–83, 459, 461–462]

Erhardt, A. 1991a. Nectar sugar and amino acid preferences of *Battus philenor* (Lepidoptera, Papilionidae). Ecological Entomology 16:425–434. [356–357]

Erhardt, A. 1991b. Pollination of *Dianthus superbus* L. Flora 185:99–106. [389]

Erickson, E. H., R. W. Thorp, D. L. Briggs, J. R. Estes, R. J. Daun, M. Marks, and C. H. Schroeder. 1979. Characterization of floral nectars by high-performance liquid chromatography. Journal of Apicultural Research 18(2):148–152. [209]

Estabrook, G. F., J. A. Winsor, A. G. Stephenson, and H. F. Howe. 1982. When are two phenological patterns different? Botanical Gazette 143(3):374–378. [29]

Ewald, P. W., and F. L. Carpenter. 1978. Territorial responses to energy manipulations in the Anna Hummingbird. Oecologia 31(3):277–292. [334–335, 360]

508

Bibliography

Ewald, P. W., and W. A. Williams. 1982. Function of the bill and tongue in nectar uptake by hummingbirds. Auk 99(3):573–576. [313]

Faegri, K., and L. van der Pijl. 1979. The principles of pollination ecology. 3rd edition. Pergamon Press, Oxford, England. [6, 7, 19]

Farnsworth, P. B., and P. A. Cox. 1988. A laser illuminator designed for pollination studies with a night vision device. Biotropica 20(4):334–335. [388]

Farris, M. A., and J. B. Mitton. 1984. Population density, outcrossing rate, and heterozygote superiority in ponderosa pine. Evolution 38(5):1151–1154. [218]

Fasman, G. D., editor. 1975. Viscosity and density tables. Pages 415–418 *in* Handbook of biochemistry and molecular biology, Volume 1: Physical and chemical data. 3rd edition. CRC Press, Cleveland, Ohio, USA. [211]

Feder, W. A., and R. Shrier. 1990. Combination of u.v.-B and ozone reduces pollen tube growth more than either stress alone. Environmental and Experimental Botany 30(4):451–454. [113, 125]

Feinsinger, P. 1983. Variable nectar secretion in a *Heliconia* species pollinated by hermit hummingbirds. Biotropica 15(1):48–52. [158]

Feinsinger, P., J. H. Beach, Y. B. Linhart, W. H. Busby, and K. G. Murray. 1987. Disturbance, pollinator predictability, and pollination success among Costa Rican cloud forest plants. Ecology 68(5):1294–1305. [297]

Feinsinger, P., and W. H. Busby. 1987. Pollen carryover: experimental comparisons between morphs of *Palicourea lasiorrachis* (Rubiaceae), a distylous, bird-pollinated, tropical treelet. Oecologia 73(3):231–235. [126–127, 135, 234]

Feinsinger, P., R. K. Colwell, J. Terborgh, and S. B. Chaplin. 1979. Elevation and the morphology, flight energetics, and foraging ecology of tropical hummingbirds. American Naturalist 113(4):481–497. [305]

Feinsinger, P., K. G. Murray, S. Kinsman, and W. H. Busby. 1986. Floral neighborhood and pollination success in four hummingbird-pollinated cloud forest plant species. Ecology 67(2):449–464. [118–119]

Feinsinger, P.; H. M. Tiebout, III; and B. E. Young. 1991. Do tropical bird-pollinated plants exhibit density- dependent interactions? Field experiments. Ecology 72(6):1953–1963. [361]

Fendrick, I., and H. Glubrecht. 1967. Investigation of the propagation of plant pollen by an indicator activation method. *In* Nuclear activation techniques in the life sciences. Proceedings of a symposium, Amsterdam, 1967. International Atomic Energy Agency, Vienna, Austria. [142]

Fenster, C. B. 1991a. Effect of male pollen donor and female seed parent on allocation of resources to developing seeds and fruit in *Chamaecrista fasciculata* (Leguminosae). American Journal of Botany 78(1):13– 23. [34, 120]

Fenster, C. B. 1991b. Selection on floral morphology by hummingbirds. Biotropica 23(1):98–101. [39]

Fenster, C. B. 1991c. Gene flow in *Chamaecrista fasciculata* (Leguminosae). I. Gene dispersal. Evolution 45(2):398–409. [232–233, 381]

Fenster, C. B. 1991d. Gene flow in *Chamaecrista fasciculata* (Leguminosae). II. Gene establishment. Evolution 45(2):410–422. [233]

Bibliography

Fenster, C. B., and V. L. Sork. 1988. Effect of crossing distance and male parent on in vivo pollen tube growth in *Chamaecrista fasciculata*. American Journal of Botany 75(12):1898–1903. [126–127]

Feuer, S. 1987. Combined cryo- and thin sectioning (TEM) in the elucidation of mimosoid tetrad/polyad ultrastructure. Review of Paleobotany and Palynology 52:367–374. [88]

Feuer, S. 1990. Pollen aperture evolution among the families Ersoonioideae, Sphalmioideae, and Carnarvonioideae (Proteaceae). American Journal of Botany 77(6):783–794. [88–89, 449, 456]

Finch, S., and R. H. Collier. 1989. Effects of the angle of inclination of traps on the numbers of large Diptera caught on sticky boards in certain vegetable crops. Entomologica Experimentale et Applicata 52:23–27. [268]

Firmage, D. H., and F. R. Cole. 1988. Reproductive success and inflorescence size of *Calopogon tuberosus* (Orchidaceae). American Journal of Botany 75(9):1371–1377. [33]

Fitzwater, W. D. 1943. Color marking of mammals with special reference to squirrels. Journal of Wildlife Management 7(2):190–192. [335]

Fleischer, S. J., F. W. Ravlin, R. Delorme, R. J. Stipes, and M. L. McManus. 1990. Marking gypsy moth (Lepidoptera: Lymantriidae) life stages and products with low doses of rubidium injected or implanted into pin oak. Journal of Economic Entomology 83(6):2343–2348. [321]

Fleming, T. H., and B. L. Partridge. 1984. On the analysis of phenological overlap. Oecologia 62(3):344–350. [29]

Flores, R. 1978. A rapid and reproducible assay for quantitative estimation of proteins using bromphenol blue. Analytical Biochemistry 88:605–611. [207]

Ford, H. A. 1979. Interspecific competition in Australian honeyeaters — depletion of common resources. Australian Journal of Ecology 4(2):145–164. [14]

Ford, H. A., and D. C. Paton. 1976. Resource partitioning and competition in honeyeaters of the genus *Meliphaga*. Australian Journal of Ecology 1(4):281–287. [409]

Ford, H. A., and D. C. Paton. 1977. The comparative ecology of ten species of honeyeaters in South Australia. Australian Journal of Ecology 2(4):399–407. [312]

Frankie, G. W. 1973. A simple field technique for marking bees with fluorescent powders. Annals of the Entomological Society of America 66:690–691. [324]

Franks, F. 1988. Characterization of proteins. Humana Press, Clifton, New Jersey, USA. [194]

Free, J. B. 1970a. Insect pollination of crops. Academic Press, London, England. [3, 326, 336–337, 387]

Free, J. B. 1970b. Effect of flower shapes and nectar guides on the behaviour of foraging honeybees. Behaviour 37:269–285. [350]

Free, J. B., and C. G. Butler. 1959. Bumblebees. The Macmillan Company, New York, New York, USA. [385]

Freeland, P. W. 1976. Tests for the viability of seeds. Journal of Biological Education 10:57–64. [30]

Freeman, C. E., W. H. Reid, J. E. Becvar, and R. Scogin. 1984. Similarity and apparent convergence in the nectar-sugar composition of some hummingbird-pollinated flowers. Botanical Gazette 145(1):132–135. [193]

Bibliography

Freeman, C. E., and D. H. Wilken. 1987. Variation in nectar sugar composition at the intraplant level in *Ipomopsis longiflora* (Polemoniaceae). American Journal of Botany 74(11):1681–1689. [193]

Freilich, J. E. 1989. A method for tagging individual benthic macroinvertebrates. Journal of the North American Benthological Society 8:351–354. [327]

Fricke, J. M. 1991. Trap-nest design for small trap-nesting Hymenoptera. Great Lakes Entomologist 24(2):121–122. [384]

Friend, D.T.C. 1961. A simple method of measuring integrated light values in the field. Ecology 42(3):577–580. [397]

Fryer, J. C., and C. L. Meek. 1989. Further studies on marking an adult mosquito, *Psorophora columbiae,* in situ using fluorescent pigments. Southwestern Entomologist 14:409–418. [325]

Fryxell, P. A. 1957. Mode of reproduction of higher plants. Botanical Review 23:135–233. [219]

Furgala, B., E. C. Mussen, D. M. Noetzel, and R. G. Robinson. 1976. Observations on nectar secretion in sunflowers. Proceedings of the First Sunflower Forum 1:11–12. [162]

Galen, C. 1985. Regulation of seed-set in *Polemonium viscosum*: floral scents, pollination, and resources. Ecology 66(3):792–797. [19, 254]

Galen, C. 1989. Measuring pollinator-mediated selection on morphometric floral traits: bumblebees and the alpine sky pilot, *Polemonium viscosum*. Evolution 43(4):882–890. [43]

Galen, C., and T. Gregory. 1989. Interspecific pollen transfer as a mechanism of competition: consequences of foreign pollen contamination for seed set in the alpine wildflower, *Polemonium viscosum*. Oecologia 81(1):120–123. [146]

Galen, C., T. Gregory, and L. F. Galloway. 1989. Costs of self-pollination in a self-incompatible plant, *Polemonium viscosum*. American Journal of Botany 76(11):1675–1680. [127]

Galen, C., and M.E.A. Newport. 1987. Bumble bee behavior and selection on flower size in the sky pilot, *Polemonium viscosum*. Oecologia 74(1):20–23. [250, 364]

Galen, C., and R. C. Plowright. 1985. The effects of nectar level and flower development on pollen carry-over in inflorescences of fireweed (*Epilobium angustifolium*) (Onagraceae). Canadian Journal of Botany 63(3):488–451. [234]

Galen, C., and R. C. Plowright. 1987. Testing the accuracy of using peroxidase activity to indicate stigma receptivity. Canadian Journal of Botany 65(1):107–111. [68–69]

Galen, C., R. C. Plowright, and J. D. Thomson. 1985. Floral biology and regulation of seed set and seed size in the lily, *Clintonia borealis*. American Journal of Botany 72(10):1544–1552. [68, 261, 449, 455]

Galen, C., and M. L. Stanton. 1989. Bumble bee pollination and floral morphology: factors influencing pollen dispersal in the alpine sky pilot, *Polemonium viscosum* (Polemoniaceae). American Journal of Botany 76:419–426. [260]

Galen, C., K. A. Zimmer, and M. E. Newport. 1987. Pollination in floral scent morphs of *Polemonium viscosum*: a mechanism for disruptive selection on flower size. Evolution 41(3):599–606. [43, 123]

Galil, J. 1973. Topocentric and Ethodynamic Pollination. Pages 85–100 *in* N.B.M. Brantjes and H. F. Linskens, editors. Pollination and dispersal. Proceedings of a symposium, published by the Department of Botany, University of Nijmegen, Netherlands. [479]

Bibliography

Gambino, P. 1990. Mark-recapture studies on *Vespula pensylvanica* (Saussure) queens (Hymenoptera: Vespidae). Pan-Pacific Entomologist 66(3):227–231. [267]

Gambino, P., A. C. Medeiros, and L. L. Loope. 1990. Invasion and colonization of upper elevations on East Maui (Hawaii) by *Vespula pensylvanica* (Hymenoptera: Vespidae). Annals of the Entomological Society of America 83(6):1088–1095. [273]

Gamboa, G. J., R. L. Foster, and K. W. Richards. 1987. Intraspecific nest and brood recognition by queens of the bumble bee, *Bombus occidentalis* (Hymenoptera: Apidae). Canadian Journal of Zoology 65(12):2893–2897. [384]

Ganders, F. R. 1974. Disassortative pollination in the distylous plant *Jepsonia heterandra*. Canadian Journal of Botany 52(11):2401–2406. [80]

Gary, N. E. 1967. A method for evaluating honey bee flight activity at the hive entrance. Journal of Economic Entomology 60(1):102–105. [382]

Gary, N. E. 1971. Magnetic retrieval of ferrous labels in a capture-recapture system for honey bees and other insects. Journal of Economic Entomology 64(4):961–965. [327]

Gary, N. E., and K. Lorenzen. 1976. A method for collecting the honey-sac contents from honeybees. Journal of Apicultural Research 15(2):73–79. [301, 303]

Gary, N. E., and K. Lorenzen. 1987. Vacuum device for collecting and dispensing honey bees (Hymenoptera: Apidae) and other insects into small cages. Annals of the Entomological Society of America 80:664–666. [271]

Gary, N. E., and J. M. Marston. 1976. A vacuum apparatus for collecting honey bees and other insects in trees. Annals of the Entomological Society of America 69(2):287–289. [270]

Gary, N. E., P. C. Witherell, K. Lorenzen, and J. M. Marston. 1977. Area fidelity and intra-field distribution of honey bees during the pollination of onions. Environmental Entomology 6(2):303–310. [327]

Gary, N. E., P. C. Witherell, and J. M. Marston. 1976. The inter- and intra-orchard distribution of honeybees during almond pollination. Journal of Apicultural Research 15(1):43–50. [327]

Gass, C. L., and G. D. Sutherland. 1985. Specialization by territorial hummingbirds on experimentally enriched patches of flowers: energetic profitability and learning. Canadian Journal of Zoology 63(9):2125–2133. [370]

Gaudreau, M. M., and J. W. Hardin. 1974. The use of neutron activation analysis in pollination ecology. Brittonia 26:316–320. [142–143, 453, 460]

Geber, M. A. 1985. The relationship of plant size to self-pollination in *Mertensia ciliata*. Ecology 66(3):762–772. [234]

Gerlach, G., and R. Schill. 1989. Fragrance analyses, an aid to taxonomic relationships of the genus *Coryanthes* (Orchidaceae). Plant Systematics and Evolution 168(3-4):159–165. [53]

Giesen, Th.G., and G. Van der Velde. 1983. Ultraviolet reflectance and absorption patterns in flowers of *Nymphaea alba* L., *Nymphaea candida* Presl and *Nuphar lutea* (L.) Sm. (Nymphaeaceae). Aquatic Botany 16:369–376. [63]

Gilbert, F. S. 1981. Foraging ecology of hoverflies: morphology of the mouthparts in relation to feeding on nectar and pollen in some common urban species. Ecological Entomology 6:245–262. [310]

Gilbert, J., M. J. Shepherd, M. A. Wallwork, and R. G. Harris. 1981. Determination of the geographical origin of honeys by multivariate analysis of gas chromatographic data on

Bibliography

their free amino acid content. Journal of Apicultural Research 20(2):125–135. [197–199, 448–449, 455, 459]

Gilissen, L.J.W. 1977. Style-controlled wilting of the flower. Planta 133(3):275–280. [338]

Gill, D. E. 1986. Individual plants as genetic mosaics: ecological organisms versus evolutionary individuals. Pages 321–343 *in* M. J. Crawley, editor. Plant ecology. Blackwell Scientific Publications, Oxford, England. [24]

Gill, F. B. 1988. Effects of nectar removal on nectar accumulation in flowers of *Heliconia imbricata* (Heliconiaceae). Biotropica 20(2):169–171. [159, 215]

Gill, F. B., A. L. Mack, and R. T. Ray. 1982. Competition between hermit hummingbirds Phaethorninae and insects for nectar in a Costa Rican rain forest. Ibis 124:44–49. [38]

Gill, F. B., and L. L. Wolf. 1975. Foraging strategies and energetics of East African sunbirds at mistletoe flowers. American Naturalist 109(969):491–510. [319, 337, 341]

Gilliam, M., J. O. Moffett, and N. M. Kauffeld. 1983. Examination of floral nectar of citrus, cotton, and Arizona desert plants for microbes. Apidologie 14:299–302. [209]

Ginsberg, H. 1986. Honey bee orientation behaviour and the influence of flower distribution on foraging movements. Ecological Entomology 11:173–179. [361]

Gleason, H. A., and A. Cronquist. 1963. Manual of vascular plants of northeastern United States and adjacent Canada. Van Nostrand Reinhold, New York, New York, USA. [337]

Gleeson, S. K. 1982. Character displacement in flowering phenologies. Oecologia 51(2):294–295. [29]

Goldblatt, P., and P. Bernhardt. 1990. Pollination biology of *Nivenia* (Iridaceae) and the presence of heterostylous self-compatibility. Israel Journal of Botany 39(1-2):93–111. [36, 457]

Goldblatt, P., P. Bernhardt, and J. Manning. 1989. Notes on the pollination mechanisms of *Moraea inclinata* and *M. brevistyla* (Iridaceae). Plant Systematics and Evolution 163(3-4):201–209. [292]

Goldingay, R. L. 1990. The foraging behaviour of a nectar feeding marsupial, *Petaurus australis*. Oecologia 85(2):191–199. [389]

Goldingay, R. L., S. M. Carthew, and R. J. Whelan. 1991. The importance of non-flying mammals in pollination. Oikos 61(1):79–87. [13, 16, 288, 297, 299, 336–337, 339, 389]

Goldsmith, K. M., and T. H. Goldsmith. 1982. Sense of smell in the Black-chinned Hummingbird. Condor 84(2):237–238. [39]

Goldsmith, T. H., and G. D. Bernard. 1974. The visual system of insects. Pages 165–272 *in* M. Rockstein, editor. The physiology of insecta, Volume 2. Academic Press, New York, New York, USA. [315]

Goldsmith, T. H., and K. M. Goldsmith. 1979. Discrimination of colors by the black-chinned hummingbird, *Archilochus alexandri*. Journal of comparative Physiology 130:209–220. [369]

Goldwasser, L., G. E. Schatz, and H. J. Young. 1993. A new method for marking Scarabaeidae and other Coleoptera. Coleopterists' Bulletin, 47(1):21–26. [330]

Goot, V. S. van der, and R.A.J. Grabandt. 1970. Some species of the genera *Melanostoma*, *Platycheirus* and *Pyrophaena* (Diptera, Syrphidae) and their relation to flowers. Entomologische Berichten 30:135–143. [299]

Bibliography

Gorchov, D. L. 1988. Effects of pollen and resources on seed number and other fitness components in *Amelanchier arborea* (Rosaceae: Maloideae). American Journal of Botany 75(9):1275–1285. [122]

Gori, D. F. 1983. Post-pollination phenomena and adaptive floral changes. Chapter 2, pages 31–49, *in* C. E. Jones and R. J. Little, editors. Handbook of experimental pollination biology. Scientific and Academic Editions, Van Nostrand Reinhold Company, New York, New York, USA. [58, 339]

Gori, D. F. 1989. Floral color change in *Lupinus argenteus* (Fabaceae): Why should plants advertise the location of unrewarding flowers to pollinators? Evolution 43(4):870–881. [58, 71, 95, 338]

Gottsberger, G. 1977. Some aspects of beetle pollination in the evolution of flowering plants. Plant Systematics and Evolution, Supplement 1:211–226. [30]

Gottsberger, G., T. Arnold, and H. F. Linskens. 1990. Variation in floral nectar amino acids with aging of flowers, pollen contamination, and flower damage. Israel Journal of Botany 39(1-2):167–176. [194, 200]

Gottsberger, G., J. Schrauwen, and H. F. Linskens. 1984. Amino acids and sugars in nectar, and their putative evolutionary significance. Plant Systematics and Evolution 145(1-2):55–77. [200]

Grabe, D. F., editor. 1970. Tetrazolium testing handbook for agricultural seeds. Contribution No. 29 to the Handbook on Seed Testing, prepared by the Tetrazolium Testing Committee of the Association of Official Seed Analysts. Published by the Association. [30, 32]

Grace, J. 1989. Measurement of wind speed near vegetation. Chapter 4, pages 57–73 *in* R. W. Pearcy, J. R. Ehleringer, H. A. Mooney, and P. W. Rundel, editors. Plant physiological ecology. Field methods and instrumentation. Chapman and Hall, New York, New York, USA. [400]

Grant, V. 1950. The flower constancy of bees. Botanical Review 16:379–398. [315]

Grant, V., and K. A. Grant. 1965. Flower pollination in the phlox family. Columbia University Press, New York, New York, USA. [7]

Grant, W. D., and J. R. Beggs. 1989. Carbohydrate analysis of beech honeydew. New Zealand Journal of Zoology 16:283–288. [186, 448, 450, 453, 457, 460]

Green, T. W., and G. E. Bohart. 1975. The pollination ecology of *Astragalus cibarius* and *Astragalus utahensis* (Leguminosae). American Journal of Botany 62(4):379–386. [92, 293]

Gregg, K. B. 1991. Reproductive strategy of *Cleistes divaricata* (Orchidaceae). American Journal of Botany 78(3):350–360. [120]

Gregory, P. H. 1973. The microbiology of the atmosphere. 2nd edition. John Wiley & Sons, New York, New York, USA. [91]

Grogan, D. E., and J. H. Hunt. 1979. Pollen proteases: their potential role in insect digestion. Insect Biochemistry 9:309–313. [148]

Gross, R. S., and P. A. Werner. 1983. Relationships among flowering phenology, insect visits, and seed-set of individuals: experimental studies on four co-occurring species of goldenrod (*Solidago*: Compositae). Ecological Monographs 53(1):95–117. [75]

Groth, I., G. Bergström, and O. Pellmyr. 1987. Floral fragrances in *Cimicifuga*: chemical polymorphism and incipient speciation in *Cimicifuga simplex*. Biochemical Systematics and Ecology 15(4):441–444. [53]

Bibliography

Guldberg, L. D., and P. R. Atsatt. 1975. Frequency of reflection and absorption of ultraviolet light in flowering plants. American Midland Naturalist 93:35–43. [61, 64]

Guo, Y.-H., R. Sperry, C.D.K. Cook, and P. A. Cox. 1990. The pollination ecology of *Zannichellia palustris* L. (Zannichelliaceae). Aquatic Botany 38:341–356. [20]

Guth, C. J., and S. G. Weller. 1986. Pollination, fertilization and ovule abortion in *Oxalis magnifica*. American Journal of Botany 73(2):246–253. [72]

Hagler, J. R., A. C. Cohen, D. Bradley-Dunlop, and F. J. Enriquez. 1992. New approach to mark insects for feeding and dispersal studies. Environmental Entomology 21(1):20–25. [323]

Haig, D., and M. Westoby. 1988. On limits to seed production. American Naturalist 131(5):757–759. [254]

Hainsworth, F. R. 1973. On the tongue of a hummingbird: its role in the rate and energetics of feeding. Comparative Biochemistry and Physiology 46A:65–78. [360, 377]

Hainsworth, F. R., and L. L. Wolf. 1972a. Energetics of nectar extraction in a small, high altitude, tropical hummingbird, *Selasphorus flammula*. Journal of comparative Physiology 80:377–387. [169, 314]

Hainsworth, F. R., and L. L. Wolf. 1972b. Crop volume, nectar concentration and hummingbird energetics. Comparative Biochemistry and Physiology 42A:359–366. [377]

Hainsworth, F. R., and L. L. Wolf. 1976. Nectar characteristics and food selection by hummingbirds. Oecologia 25(2):101–114. [360]

Hainsworth, F. R., L. L. Wolf, and T. Mercier. 1984. Pollination and pre-dispersal seed predation: net effects on reproduction and inflorescence characteristics. Oecologia 63:405–409. [33]

Hamilton-Kemp, T. R., J. H. Loughrin, D. D. Archbold, R. A. Andersen, and D. F. Hildebrand. 1991. Inhibition of pollen germination by volatile compounds including 2-hexenal and 3-hexenal. Journal of Agriculture and Food Chemistry 39:952–956. [99]

Hamrick, J. L. 1982. Plant population genetics and evolution. American Journal of Botany 69(10):1685–1693. [218]

Hamrick, J. L. 1987. Gene flow and distribution of genetic variation in plant populations. Pages 53–67 *in* K. M. Urbanska, editor. Differentiation patterns in higher plants. Academic Press, New York, New York, USA. [233]

Hamrick, J. L., and D. A. Murawski. 1990. The breeding structure of tropical tree populations. Plant Species Biology 5:157–165. [230]

Hamrick, J. L., and A. Schnabel 1985. Understanding the genetic structure of plant populations: some old problems and a new approach. Pages 50–70 *in* S. Levin, editor. Population genetics in forestry. (Lecture notes in biomathematics, Volume 60.) Springer Verlag, New York, New York, USA. [230]

Handel, S. N. 1976. Restricted pollen flow of two woodland herbs determined by neutron-activation analysis. Nature 260:422–423. [142–143, 460, 463]

Handel, S. N. 1982. Dynamics of gene flow in an experimental population of *Cucumis melo* (Cucurbitaceae). American Journal of Botany 69(10):1538–1546. [144, 407]

Handel, S. N. 1983a. Pollination ecology, plant population structure, and gene flow. Chapter 8, pages 163–211 *in* L. Real, editor. Pollination biology. Academic Press, Orlando, Florida, USA. [138–139, 144, 347]

Handel, S. N. 1983b. Contrasting gene flow patterns and genetic subdivision in adjacent populations of *Cucumis sativus* (Cucurbitaceae). Evolution 37(4):760–771. [144]

Bibliography

Hanny, B., and C. D. Elmore. 1974. Amino acid composition of cotton nectar. Agricultural and Food Chemistry 22:476–478. [197–199]

Harder, L. D. 1982. Measurement and estimation of functional proboscis length in bumble-bees (Hymenoptera: Apidae). Canadian Journal of Zoology 60(5):1073–1079. [308–309]

Harder, L. D. 1983. Flower handling efficiency of bumble bees: morphological aspects of probing time. Oecologia 57(2):274–280. [336, 377]

Harder, L. D. 1986. Effects of nectar concentration and flower depth on flower handling efficiency of bumble bees. Oecologia 69(2):309–315. [353]

Harder, L. D. 1988. Choice of individual flowers by bumble bees: interaction of morphology, time and energy. Behaviour 104:60–77. [347]

Harder, L. D. 1990. Pollen removal by bumble bees and its implications for pollen dispersal. Ecology 71(3):1110–1125. [97, 377]

Harder, L. D., and S.C.H. Barrett. 1992. The energy cost of bee pollination for *Pontederia cordata* (Pontederiaceae). Functional Ecology 6:226–233. [46, 167]

Harder, L. D., and M. B. Cruzan. 1990. An evaluation of the physiological and evolutionary influences of inflorescence size and flower depth on nectar production. Functional Ecology 4:559–572. [42, 163]

Harder, L. D., and J. D. Thomson. 1989. Evolutionary options for maximizing pollen dispersal of animal-pollinated plants. American Naturalist 133(3):323–344. [97, 117, 149]

Harder, L. D., J. D. Thomson, M. B. Cruzan, and R. S. Unnasch. 1985. Sexual reproduction and variation in floral morphology in an ephemeral vernal lilly, *Erythronium americanum*. Oecologia 67(2):286–291. [44, 90, 97]

Harper, J. L., and H. L. Wallace. 1987. Control of fecundity through abortion in *Epilobium montanum* L. Oecologia 74(1):31–38. [72]

Harrison, S., J. F. Quinn, J. F. Baughman, D. D. Murphy, and P. R. Ehrlich. 1991. Estimating the effects of scientific study on two butterfly populations. American Naturalist 137(2):227–243. [321]

Harriss, F.C.L., and A. J. Beattie. 1991. Viability of pollen carried by *Apis mellifera* L., *Trigona carbonaria* Smith and *Vespula germanica* (F.) (Hymenoptera: Apidae, Vespidae). Journal of the Australian Entomological Society 30(1):40–47. [114–115]

Hartling, L. K., and R. C. Plowright. 1979. Foraging by bumble bees on patches of artificial flowers: a laboratory study. Canadian Journal of Zoology 57(10):1866–1970. [351–352]

Haslett, J. R. 1983. A photographic account of pollen digestion by adult hoverflies. Physiological Ecology 8:167– 171. [297–298]

Haslett, J. R. 1989. Interpreting patterns of resource utilization: randomness and selectivity in pollen feeding by adult hoverflies. Oecologia 78(4):433–442. [60, 267, 297, 368]

Hatton, T. J. 1989. Spatial patterning of sweet briar (*Rosa rubiginosa*) by two vertebrate species. Australian Journal of Ecology 14(2):199–205. [30]

Hauser, E.J.P., and J. H. Morrison. 1964. The cytochemical reduction of nitro blue tetrazolium as an index of pollen viability. American Journal of Botany 51(7):748–752. [98, 108–109, 457, 461]

Hawkins, R. P. 1969. Length of tongue in a honey bee in relation to the pollination of red clover. Journal of Agricultural Science (Cambridge) 73:489–493. [309]

Bibliography

Hawkins, R. P. 1971. Selection for height of nectar in the corolla tube of English singlecut Red Clover. Journal of Agricultural Science (Cambridge) 77:347–350. [159]

Hayes, J. L. 1989. Detection of single and multiple trace element labels in individual eggs of diet-reared *Heliothis virescens* (Lepidoptera: Noctuidae). Annals of the Entomological Society of America 82(3):340–345. [321]

Hayes, J. L., and K. G. Reed. 1989. Using rubidium-treated artificial nectar to label adults and eggs of *Heliothis virescens* (Lepidoptera: Noctuidae). Environmental Entomology 18(5):807–810. [321]

Heemert, C. van, A. de Ruijter, J. van den Eijnde, and J. van der Steen. 1990. Year-round production of bumble bee colonies for crop pollination. Bee World 71(2):54–56. [385–386]

Heideman, P. D. 1989. Temporal and spatial variation in the phenology of flowering and fruiting in a tropical rainforest. Journal of Ecology 77(4):1059–1079. [23–24]

Heinrich, B. 1972. Energetics of temperature regulation and foraging in a bumblebee, *Bombus terricola* Kirby. Journal of comparative Physiology 77(1):49–64. [302, 317]

Heinrich, B. 1976. The foraging specializations of individual bumblebees. Ecological Monographs 46(1):105–128. [374]

Heinrich, B. 1979a. Keeping a cool head: honeybee thermoregulation. Science 205(4412):1269–1271. [302]

Heinrich, B. 1979b. Resource heterogeneity and patterns of movement in foraging bumblebees. Oecologia 40(3):235–245. [362, 379]

Heinrich, B. 1979c. "Majoring" and "minoring" by foraging bumblebees, *Bombus vagans*: an experimental analysis. Ecology 60(2):245–255. [385]

Heinrich, B. 1981. Insect foraging energetics. Chapter 9, pages 187–214 *in* C. E. Jones and R. J. Little, editors. Handbook of experimental pollination biology. Scientific and Academic Editions, Van Nostrand Reinhold Company, New York, New York, USA. [324, 328]

Heinrich, B, and T. M. Casey. 1973. Metabolic rate and endothermy in sphinx moths. Journal of comparative Physiology 82(2):195–206. [317]

Heinrich, B., P. R. Mudge, and P. G. Deringis. 1977. Laboratory analysis of flower constancy in foraging bumblebees: *Bombus ternarius* and *B. terricola*. Behavioral Ecology and Sociobiology 2:247–265. [348]

Heinrich, B., and C. Pantle. 1975. Thermoregulation in small flies (*Syrphus* sp.): basking and shivering. Journal of Experimental Biology 62:599–610. [317]

Herr, J. M., Jr. 1971. A new clearing-squash technique for the study of ovule development in Angiosperms. American Journal of Botany 55(8):785–790. [71]

Herrera, C. M. 1987. Components of pollinator "quality": comparative analysis of a diverse insect assemblage. Oikos 50(1):79–90. [68]

Heslop-Harrison, J., and Y. Heslop-Harrison. 1970. Evaluation of pollen viability by enzymatically induced fluorescence; intra-cellular hydrolysis of fluorescein diacetate. Stain Technology 45: 115–122. [106]

Heslop-Harrison, J., Y. Heslop-Harrison, and K. R. Shivanna. 1984. The evaluation of pollen quality, and a further appraisal of the fluorochromatic (FCR) test procedure. Theoretical and Applied Genetics 67:367–375. [30, 98, 100–102, 106, 108–110]

Bibliography

Heslop-Harrison, J., R. B. Knox, and Y. Heslop-Harrison. 1974. Pollen-wall proteins: exine-held fractions associated with the incompatibility response in Cruciferae. Theoretical and Applied Genetics 44:133–137. [66, 448]

Heslop-Harrison, Y., and K. R. Shivanna. 1977. The receptive surface of the angiosperm stigma. Annals of Botany 41:1233–1258. [65]

Hessing, M. B. 1988. Geitonogamous pollination and its consequences in *Geranium caespitosum*. American Journal of Botany 75(9):1324–1333. [234]

Hewitt, F. R., T. Hough, P. O'Neill, J. M. Sasse, E. G. Williams, and K. S. Rowan. 1985. Effect of brassinolide and other growth regulators on the germination and growth of pollen tubes of *Prunus avium* using a multiple hanging-drop assay. Australian Journal of Plant Physiology 12:201–211. [105, 448]

Heyneman, A. J. 1983. Optimal sugar concentrations of floral nectars - dependence on sugar intake efficiency and foraging costs. Oecologia 60(2):198–213. [211]

Hiebert, S. M., and W. A. Calder. 1983. Sodium, potassium, and chloride in floral nectars: energy-free contributions to refractive index and salt balance. Ecology 64(2):399–402. [170, 320]

Hill, R. J. 1977. Technical note: ultraviolet reflectance-absorbance photography; an easy, inexpensive research tool. Brittonia 29:382–390. [63, 65]

Hills, H. G., and B. Schutzman. 1990. Considerations for sampling floral fragrances. Phytochemical Bulletin 22(1-2):2–9. [48]

Hilsenbeck, R. A. 1990. Pollen morphology and systematics of *Siphonoglossa* sensu lato (Acanthaceae). American Journal of Botany 77(1):27–40. [90]

Hobbs, G. A. 1967. Obtaining and protecting red-clover pollinating species of *Bombus* (Hymenoptera: Apidae). Canadian Entomologist 99:943–951. [383]

Hobbs, G. A., J. F. Virostek, and W. O. Nummi. 1960. Establishment of *Bombus* spp. (Hymenoptera: Apidae) in artificial domiciles in southern Alberta. Canadian Entomologist 92:868–872. [383]

Hocking, B. 1968. Insect-flower associations in the high Arctic with special reference to nectar. Oikos 19(2):359–388. [14–15]

Hodges, C. M., and R. B. Miller. 1981. Pollinator flight directionality and the assessment of pollen returns. Oecologia 50(3):376–379. [345]

Hodges, C. M., and L. L. Wolf. 1981. Optimal foraging in bumblebees: why is nectar left behind in flowers? Behavioral Ecology and Sociobiology 9:41–44. [167]

Holm, S. N. 1966. The utilization and management of bumble bees for red clover and alfalfa seed production. Annual Review of Entomology 11:155–182. [385]

Holman, R. T., and W. H. Heimermann. 1973. Identification of components of orchid fragrances by gas chromatography–mass spectrometry. American Orchid Society Bulletin 42:678–682. [50]

Holman, R. T., and W. H. Heimermann. 1976. The chemical composition of fragrances of some orchids. Pages 75–89 *in* H. H. Szmant and J. Wemple, editors. First symposium on the scientific aspects of orchids. Chemistry Department, University of Detroit, Detroit, Michigan, USA. [50]

Holsinger, K. E. 1991. Inbreeding depression and the evolution of plant mating systems. Trends in Ecology and Evolution 6:307–308. [218]

Bibliography

Holtsford, T. P., and N. C. Ellstrand. 1990. Inbreeding effects in *Clarkia tembloriensis* (Onagraceae) populations with different natural outcrossing rates. Evolution 44(8):2031–2046. [218]

Hoopingarner, R., and M. Adey. 1985. Proposal to standardize the times used when recording bee-foraging data. Pollination Research Newsletter, 1:2. December 1985. International Bee Research Association and Royal Botanic Gardens (Kew). [403]

Horovitz, A., and Y. Cohen. 1972. Ultraviolet reflectance characteristics in flowers of crucifers. American Journal of Botany 59:706–713. [65]

Horovitz, A., and J. Harding. 1972. Genetics of *Lupinus*. V. Intraspecific variability for reproductive traits in *Lupinus nanus*. Botanical Gazette 133(2):155–165. [59, 218]

Horvitz, C. C., and D. W. Schemske. 1988. A test of the pollinator limitation hypothesis for a neotropical herb. Ecology 69(2):200–206. [256]

Hough, L., J.K.N. Jones, and W. H. Wadman. 1950. Quantitative analysis of mixtures of sugars by the method of partition chromatography. V. Improved methods for the separation and detection of their methylated derivatives on the paper chromatogram. Journal of the Chemical Society 1950 Part II:1702–1706. [184, 450]

Hough, T., P. Bernhardt, R. B. Knox, and E. G. Williams. 1985. Applications of fluorochromes to pollen biology. II. The DNA probes ethidium bromide and Hoechst 33258 in conjunction with the callose-specific aniline blue fluorochrome. Stain Technology 60(3):155–162. [128–129, 449, 453]

House, S. M. 1989. Pollen movement to flowering canopies of pistillate individuals of three rain forest tree species in tropical Australia. Australian Journal of Ecology 14(1):77–93. [24, 268, 292]

Hribar, L. J., D. J. Leprince, and L. D. Foil. 1991. Increasing horse fly (Diptera: Tabanidae) catch in canopy traps by reducing ultraviolet light reflectance. Journal of Medical Entomology 28(6):874–877. [273–274]

Hutchings, M. J. 1986. Plant population biology. Pages 377–435 *in* P. D. Moore and S. B. Chapman, editors. Methods in plant ecology. Blackwell Scientific Publications, Oxford, England. [32–33]

Immers, J. 1964. A microchromatograph for quantitative estimation of sugars using a paper strip as partition support. Journal of Chromatography 15:252–256. [187]

Inouye, D. W. 1975. Flight temperatures of male euglossine bees (Hymenoptera: Apidae: Euglossini). Journal of the Kansas Entomological Society 48(3):366–370. [271, 317]

Inouye, D. W. 1977. Species structure of bumblebee communities in North America and Europe. Pages 35–40 *in* W. J. Mattson, editor. The role of arthropods in forest ecosystems. Springer Verlag, New York, New York, USA. [305]

Inouye, D. W. 1978. Resource partitioning in bumblebee guilds: experimental studies of foraging behavior. Ecology 59(4):672–678. [3, 305, 342, 371, 390]

Inouye, D. W. 1980a. The effect of proboscis and corolla tube lengths on patterns and rates of flower visitation by bumblebees. Oecologia 45(2):197–201. [6, 338]

Inouye, D. W. 1980b. The terminology of floral larceny. Ecology 61(5):1251–1253. [305, 308, 390, 409]

Inouye, D. W. 1986. Long-term preformation of leaves and inflorescences by a long-lived perennial monocarp, *Frasera speciosa*, Gentianaceae. American Journal of Botany 73(11):1535–1540. [28]

Inouye, D. W. 1991. Quick and easy insect labels. Journal of the Kansas Entomological Society 64(2):242–243. [280]

Inouye, D. W., W. A. Calder, and N. M. Waser. 1991. The effect of floral abundance on feeder censuses of hummingbird populations. Condor 93(2):279–285. [381]

Inouye, D. W., N. D. Favre, J. A. Lanum, D. M. Levine, J. B. Meyers, F. C. Roberts, F. C. Tsao, and Y. Wang. 1980. The effects of nonsugar nectar constituents on nectar energy content. Ecology 61(4):992–996. [170]

Inouye, D. W., D. E. Gill, M. R. Dudash, and C. B. Fenster. 1994. A model and lexicon for pollen fate. American Journal of Botany, in press. [250]

Inouye, D. W., and A. D. McGuire. 1991. Effects of snowpack on timing and abundance of flowering in *Delphinium nelsonii* (Ranunculaceae): implications for climate change. American Journal of Botany 78(7):997–1001. [25]

Inouye, D. W., and G. H. Pyke. 1988. Pollination biology in the Snowy Mountains of Australia: comparisons with montane Colorado, USA. Australian Journal of Ecology 13(2):191–210. [65, 308, 342, 393, 408]

Inouye, D. W., and O. R. Taylor, Jr. 1979. A temperate region plant-ant-seed predator system: consequences of extrafloral nectar secretion by *Helianthella quinquenervis*. Ecology 60(1):1–7. [29]

Inouye, D. W., and G. D. Waller. 1984. Responses of honeybees (*Apis mellifera*) to amino acid solutions mimicking nectars. Ecology 65(2):618–625. [353–354]

Jacobs, G. H., J. Neitz, and J. F. Deegan, II. 1991. Retinal receptors in rodents maximally sensitive to ultraviolet light. Nature 353:655–656. [316]

Jain, S. K. 1984. Breeding systems and the dynamics of plant populations. Proceedings of the International Congress of Genetics (New Delhi) 4:291–316. [217]

Jain, S. K., and D. R. Marshall. 1967. Population studies in predominantly self-pollinating species. X. Variation in natural populations of *Avena fatua* and *A. barbata*. American Naturalist 101(1):15–33. [227]

Janick, J., and J. N. Moore, editors. 1975. Advances in fruit breeding. Purdue University Press, West Lafayette, Indiana, USA. [4]

Jefferies, C. J., and A. R. Belcher. 1974. A fluorescent brightener used for pollen tube identification *in vivo*. Stain Technology 49(4):199–202. [127, 129, 454, 460–461]

Jeffrey, D. C., J. Arditti, and R. Ernst. 1969. Determination of di-, tri-, and tetrasaccharides in mixtures with their component moieties by thin layer chromatography. Journal of Chromatography 41:475–580. [185, 190–191, 449, 456]

Jennersten, O. 1983. Butterfly visitors as vectors of *Ustilago violacea* spores between caryophyllaceous plants. Oikos 40(1):125–130. [76]

Jennersten, O. 1988. Pollination of *Viscaria vulgaris* (Caryophyllaceae): the contribution of diurnal and noctural insects to seed set and seed predation. Oikos 52(3):319–327. [390]

Jennersten, O., L. Berg, and C. Lehman. 1988. Phenological differences in pollinator visitation, pollen deposition and seed set in the sticky catchfly, *Viscaria vulgaris*. Journal of Ecology 76(4):1111–1132. [119, 261, 391]

Jennersten, O., and M. M. Kwak. 1991. Competition for bumblebee visitation between *Melampyrum pratense* and *Viscaria vulgaris* with healthy and *Ustilago*-infected flowers. Oecologia 86(1):88–98. [76]

Bibliography

Jennersten, O., and D. H. Morse. 1991. The quality of pollination by diurnal and nocturnal insects visiting common milkweed, *Asclepias syriaca*. American Midland Naturalist 125:18–28. [13, 390]

Jensen, W. A. 1962. Botanical histochemistry: principles and practice. W. H. Freeman, San Francisco, California, USA. [208]

Joel, D. M., B. E. Juniper, and A. Dafni. 1985. Ultraviolet patterns in the traps of carnivorous plants. New Phytologist 101(4):585–593. [65]

Johnston, M. O. 1991a. Pollen limitation of female reproduction in *Lobelia cardinalis* and *L. siphilitica*. Ecology 72(4):1500–1503. [254–255, 261]

Johnston, M. O. 1991b. Natural selection on floral traits in two species of *Lobelia* with different pollinators. Evolution 45(6):1468–1479. [408]

Jones, B. N. 1986. Amino acid analysis by o-pthaldialdehyde precolumn derivatization and reverse-phase HPLC. Pages 121–151 *in* J. E. Shively, editor. Methods of protein microcharacterization: a practical handbook. Humana Press, Clifton, New Jersey, USA. [194]

Jones, C. E., and S. L. Buchmann. 1974. Ultraviolet floral patterns as functional orientation cues in hymenopterous pollination systems. Animal Behaviour 22:481–485. [46, 63–64, 379]

Jones, C. E., and R. J. Little, editors. 1983. Handbook of experimental pollination biology. Scientific and Academic Editions, Van Nostrand Reinhold Company, New York, New York, USA. [7]

Kahn, A. P., and D. H. Morse. 1991. Pollinium germination and putative ovule penetration in self- and cross-pollinated common milkweed *Asclepias syriaca*. American Midland Naturalist 126:61–67. [236]

Kahn, T. L., and D. A. DeMason. 1988. *Citrus* pollen tube development in cross-compatible gynoecia, self-incompatible gynoecia, and *in vitro*. Canadian Journal of Botany 66(12):2527–2532. [125]

Kaiser, R. 1991. Trapping, investigation and reconstitution of flower scents. Pages 213–250 *in* P. M. Müller and D. Lamparsky, editors. Perfumes: art, science and technology. Elsevier Applied Science, London, England. [48–49]

Kambal, A. E., D. A. Bond, and G. Toynbee-Clarke. 1976. A study on the pollination mechanism in field beans (*Vicia faba* L.). Journal of Agricultural Science 87:519–526. [131, 463]

Kang, H., and R. B. Primack. 1991. Temporal variation of flower and fruit size in relation to seed yield in celandine poppy (*Chelidonium majus*; Papaveraceae). American Journal of Botany 78(5):711–722. [76]

Kaplan, S. M., and D. L. Mulcahy. 1971. Mode of pollination and floral sexuality in *Thalictrum*. Evolution 25(4):659–668. [92]

Käpylä, M. 1978. Amount and type of nectar sugar in some wild flowers in Finland. Annales Botanici Fennici 15:85–88. [179, 452, 461]

Kato, M., T. Itino, M. Hotta, and T. Inoue. 1991. Pollination of four Sumatran *Impatiens* species by hawkmoths and bees. Tropics 1:59–73. [166]

Kay, Q. O. N. 1978. The role of preferential and assortative pollination in the maintenance of flower colour polymorphisms. Pages 175–190 *in* A. J. Richards, editor. The pollination of flowers by insects. Linnean Society Symposium Series No. 6. Academic Press, London, England. [315]

Bibliography

Kearns, C. A. 1990. The role of fly pollination in montane habitats. Ph.D. dissertation, University of Maryland, College Park, Maryland, USA. [116, 220, 273, 298, 342, 382, 391, 393]

Keijzer, C. J., M. C. Reinders, J. Janson, and J. M. van Tuyl. 1988. Tracing sperm cells in styles, ovaries, and ovules of *Lilium longiflorum* after pollination with DAPI-stained pollen. Pages 149–152 *in* H. J. Wilms and C. J. Keijzer, editors. Plant sperm cells as tools for biotechnology. Pudoc, Wageningen, Netherlands. [240]

Kendall, D. A., and B. D. Smith. 1975. The pollinating efficiency of honeybee and bumblebee visits to field bean flowers (*Vicia faba* L.). Journal of Applied Ecology 12(3):709–717. [35, 252]

Kendall, D. A., and B. D. Smith. 1976. The pollinating efficiency of honeybee and bumblebee visits to flowers of the runner bean *Phaseolus coccineus* L. Journal of Applied Ecology 13(3):749–752. [35, 252]

Kendall, D. M., and P. G. Kevan. 1981. Nocturnal flight activity of moths (Lepidoptera) in alpine tundra. Canadian Entomologist 113:607–614. [274]

Kephart, S. R. 1990. Starch gel electrophoresis of plant isozymes: a comparative analysis of techniques. American Journal of Botany 77(5):693–712. [224]

Kerner von Marilaun, A. 1902. The natural history of plants. Translated by F. W. Oliver. Blackie, London, England. [7]

Kevan, P. G. 1972a. Insect pollination of high arctic flowers. Journal of Ecology 60(3):831–847. [14, 20]

Kevan, P. G. 1972b. Floral colors in the high arctic with reference to insect-flower relations and pollination. Canadian Journal of Botany 50(11):2289–2316. [65]

Kevan, P. G. 1975. Suntracking solar furnaces in high arctic flowers: Significance for pollination and insects. Science 189(4204):723–726. [44]

Kevan, P. G. 1978. Floral coloration, its colorimetric analysis and significance in anthecology. Pages 51–78 *in* A. J. Richards, editor. The pollination of flowers by insects. Linnean Society Symposium Series No. 6. Academic Press, London, England. [62]

Kevan, P. G. 1979. Vegetation and floral colors using ultraviolet light: interpretational difficulties for functional significance. American Journal of Botany 66(6):749–751. [62, 64]

Kevan, P. G. 1983. Floral colors through the insect eye: what they are and what they mean. Chapter 1, pages 3–30 *in* C. E. Jones and R. J. Little, editors. Scientific and Academic Editions, Van Nostrand Reinhold Company, New York, New York, USA. [58–89, 61]

Kevan, P. G., E. A. Clark, and V. G. Thomas. 1990. Insect pollinators and sustainable agriculture. American Journal of Alternative Agriculture 5(1):13–22. [3]

Kevan, P. G., D. Eisikowitch, S. Fowle, and K. Thomas. 1988. Yeast-contaminated nectar and its effects on bee foraging. Journal of Apicultural Research 27(1):26–29. [209]

Kevan, P. G., N. D. Grainger, G. A. Mulligan, and A. R. Robertson. 1973. A gray-scale for measuring reflectance and color in the insect and human visual spectra. Ecology 54(4):924–926. [61]

Kevan, P. G., S. D. St. Helens, and I. Baker. 1983. Hummingbirds feeding from exudates on diseased scrub oak. Condor 85(2):251–252. [207, 462]

Keyes, B. E., and C. E. Grue. 1982. Capturing birds with mist nets: a review. North American Bird Bander 7(1):2–14. [285]

Kho, Y. O., and J. Baër. 1968. Observing pollen tubes by means of fluorescence. Euphytica 17:298–302. [127–128]

King, J. R. 1960. The peroxidase reaction as an indicator of pollen viability. Stain Technology 36:225–227. [109]

Kingsolver, J. G., and T. L. Daniel. 1983. Mechanical determinants of nectar feeding strategy in hummingbirds: energetics, tongue morphology, and licking behavior. Oecologia 60(2):214–226. [211, 373]

Kipp, L. R. 1987. The flight directionality of honeybees foraging on real and artificial inflorescences. Canadian Journal of Zoology 65(3):587–593. [325, 351]

Kirk, W.D.J. 1984. Ecologically selective coloured traps. Ecological Entomology 9(1):35–41. [269, 314]

Kirk, W.D.J. 1985. Effect of some floral scents on host finding by thrips (Insecta: Thysanoptera). Journal of Chemical Ecology 11(1):35–43. [49]

Kislev, M. E., Z. Kraviz, and J. Lorch. 1972. A study of hawkmoth pollination by a palynological analysis of the proboscis. Israel Journal of Botany 21:57–75. [252, 274, 293, 310, 459]

Knoll, F. 1921-1926. Insekten und Blumen. Abhandlungen der Zoologisch-Botanische Gesellschaft in Wien. 12:1–645. [351, 378]

Knoll, F. 1925. Lichtsinn und Blütenbesuch des Falters von *Deilephila livornica*. Zeitschrift für Vergleichende Physiologie 2:329–380. [315, 368, 378]

Knudsen, J. T. and J. M. Olesen. 1992. Sexual trends and origin of buzz-pollination in North European Pyrolaceae. Unpublished manuscript. [57]

Knudsen, J. T., and L. Tollsten. 1991. Floral scent and intrafloral scent differentiation in *Moneses* and *Pyrola* (Pyrolaceae). Plant Systematics and Evolution 177(1):81–91. [53, 56–57, 457]

Knuth, P. 1906–1909. Handbook of flower pollination. Translated by J. R. Ainsworth Davis (3 volumes, 1906, 1908, 1909). Oxford University Press, Oxford, England. [7]

Kochmer, J. P., and S. N. Handel. 1986. Constraints and competition in the evolution of flowering phenology. Ecological Monographs 56:303–325. [29]

Kodric-Brown, A., and J. H. Brown. 1978. Influence of economics, interspecific competition, and sexual dimorphism on territoriality of migrant Rufous Hummingbirds. Ecology 59(2):285–296. [369–370]

Kohn, J. R., and N. M. Waser. 1985. The effect of *Delphinium nelsonii* pollen on seed set in *Ipomopsis aggregata*, a competitor for hummingbird pollination. American Journal of Botany 72(7): 1144–1148. [34]

Koptur, S. 1984. Outcrossing and pollinator limitation of fruit set: breeding systems of neotropical *Inga* trees (Fabaceae: Mimosoideae). Evolution 38(5):1130–1143. [95, 247]

Kowalski, R., S. E. Davies, and C. Hawkes. 1989. Metal composition as a natural marker in anthomyiid fly Delia radicum (L.). Journal of Chemical Ecology 15:1231–1239. [323]

Kramer, C. L., M. G. Eversmeyer, and T. I. Collins. 1976. A new 7-day spore sampler. Phytopathology 66:60–61. [91]

Krannitz, P. G., and M. A. Maun. 1991. An experimental study of floral display size and reproductive success in *Viburnum opulus*: importance of grouping. Canadian Journal of Botany 69:394–399. [362]

Bibliography

Krebs, C. J. 1989. Ecological methodology. Harper and Row, New York, New York, USA. [407]

Kugler, H. 1951. Blütenökologische Untersuchungen mit Goldfliegen (Lucilien). Berichte der Deutschen Botanischen Gesellschaft 64:327–341. [315]

Kugler, H. 1956. Über die optische Wirkung von Fliegenblumen auf Fliegen. Berichte der Deutschen Botanischen Gesellschaft 69:387–398. [358]

Kuno, E. 1991. Sampling and analysis of insect populations. Annual Review of Entomology 36:285–304. [382]

Kunz, T. H., editor. 1988. Ecological and behavioral methods for the study of bats. Smithsonian Institution Press, Washington, D.C., USA. [390]

Ladyman, J.A.R., and R. E. Taylor. 1988. ^{31}P and ^{1}H NMR as a non-destructive method for measuring pollen viability. Pages 69–74 in M. Cresti, P. Gori, and E. Pacini, editors. Sexual reproduction in higher plants. Springer Verlag, Berlin, Germany. [112]

Lamont, B. B., and B. G. Collins. 1988. Flower colour change in Banksia ilicifolia: A signal for pollinators. Australian Journal of Ecology 13(2):129–135. [58]

Lance, D. R., and N. C. Elliott. 1990. Marking western corn rootworm beetles (Coleoptera: Chrysomelidae): effects on survival and a blind evaluation for estimating bias in mark-recapture data. Journal of the Kansas Entomological Society 63(1):1–8. [331, 450]

Lande, R., and D. W. Schemske. 1985. The evolution of self-fertilization and inbreeding depression in plants. I. Genetic models. Evolution 39(1):24–40. [217]

Langford, M., G. E. Taylor, and J. R. Flenley. 1990. Computerized identification of pollen grains by texture analysis. Review of Palaeobotany and Palynology 64:197–203. [87]

Lanza, J. 1988. Ant preferences for Passiflora nectar mimics that contain amino acids. Biotropica 20(4):341– 344. [358]

La Porta, N., and G. Roselli. 1991. Relationship between pollen germination in vitro and fluorochromatic reaction in cherry clone F12/1 (Prunus avium L.) and some of its mutants. Journal of Horticultural Science 66(2):171–175. [102]

Larson, K. S., and R. J. Larson. 1990. Lure of the locks: showiest ladies-tresses orchids, Spiranthes romanzoffiana, affect bumblebee, Bombus spp., foraging behavior. Canadian Field-Naturalist 104(4):519–525. [33]

Lasiewski, R. C. 1962. The capture and maintenance of hummingbirds for experimental purposes. Avicultural Magazine 68:59–64. [287–388]

Laverty, T. M. 1980. The flower-visiting behaviour of bumble bees: floral complexity and learning. Canadian Journal of Zoology 58(7):1324–1335. [377, 385]

Laverty, T. M. 1992. Plant interactions for pollinator visits: a test of the magnet species effect. Oecologia 89:502–508. [392]

Lawrence, W.J.C. 1968. Plant breeding. Edward Arnold, Ltd., London, England. [4, 14, 240]

Leduc, N., G. C. Douglas, M. Monnier, and V. Connolly. 1990. Pollination in vitro: effects on the growth of pollen tubes, seed set and gametophytic self-incompatibility in Trifolium pratense L. and T. repens L. Theoretical and Applied Genetics 80:657–664. [72, 151]

Leduc, N., M. Monnier, and G. C. Douglas. 1990. Germination of trinucleated pollen: formulation of a new medium for Capsella bursa-pastoris. Sexual Plant Reproduction 3:228–235. [99, 102–104, 448, 450–451, 453–454, 456, 459–461, 463]

Bibliography

Lee, T. D. 1988. Patterns of fruit and seed production. Pages 179–202 *in* J. Lovett Doust and L. Lovett Doust, editors. Plant reproductive ecology. Oxford University Press, New York, New York, USA. [256]

Leereveld, H., A.D.J. Meeuse, and P. Stelleman. 1981. Anthecological relations between reputedly anemophilous flowers and syrphid flies. IV. A note on the anthecology of *Scirpus maritimus* L. Acta Botanica Neerlandica 30(5/6):465–473. [376]

Leereveld, H. 1982. Anthecological relations between reputedly anemophilous flowers and syrphid flies. III. Worldwide survey of crop and intestine contents of certain anthophilous syrphid flies. Tijdschrift voor Entomologie 125(2):25–35. [299]

Leffingwell, H. A., and N. Hodgkin. 1971. Techniques for preparing fossil palynomorphs for study with the scanning and transmission electron microscopes. Review of Palaeobotany and Palynology 11:177–199. [87]

Lertzman, K. P., and C. L. Gass. 1983. Alternative models of pollen transfer. Chapter 24, pages 474–489 *in* C. E. Jones and R. J. Little, editors. Scientific and Academic Editions, Van Nostrand Reinhold Company, New York, New York, USA. [234]

Levin, D. A. 1981. Dispersal versus gene flow in plants. Annals of the Missouri Botanical Garden 68:233–253. [233]

Levin, D. A. 1983. Plant parentage: an alternative view of the breeding structure of populations. Pages 171– 188 *in* C. E. King and P. S. Dawson, editors. Population biology: retrospect and prospect. Columbia University Press, New York, New York, USA. [233]

Levin, D. A. 1990. Sizes of natural microgametophyte populations in pistils of *Phlox drummondii*. American Journal of Botany 77(3):356–363. [130, 453]

Levin, D. A., and H. W. Kerster. 1969. Density-dependent gene dispersal in *Liatris*. American Naturalist 103:61–74. [218]

Levin, D. A., and H. W. Kerster. 1971. Neighborhood structure in plants under diverse reproductive methods. American Naturalist 105(944):345–354. [217]

Levin, D. A., and H. W. Kerster. 1974. Gene flow in seed plants. Evolutionary Biology 7:139–220. [22, 231, 233–234]

Levin, D. A., H. W. Kerster, and M. Niedzlek. 1971. Pollinator flight directionality and its effect on pollen flow. Evolution 25(1):113–118. [344]

Lewis, A. C. 1986. Memory constraints and flower choice in *Pieris rapae*. Science 232(4752):863–865. [375]

Lewis, A. C. 1989. Flower visit consistency in *Pieris rapae,* the cabbage butterfly. Journal of Animal Ecology 58(1):1–13. [344, 374–375]

Lewis, A. C., and G. A. Lipani. 1990. Learning and flower use in butterflies: hypotheses from honey bees. Pages 95–110 *in* E. A. Bernays, editor. Insect-plant interactions, Volume 2. CRC Press, Boca Raton, Florida, USA. [375]

Lewis, W. J., and K. Takasu. 1990. Use of learned odours by a parasitic wasp in accordance with host and food needs. Nature 348:635–636. [367]

Lindsey, A. H. 1982. Floral phenology patterns and breeding systems in *Thaspium* and *Zizia* (Apiaceae). Systematic Botany 7(1):1–12. [27]

Lindsey, A. H. 1984. Reproductive biology of Apiaceae. I. Floral visitors to *Thaspium* and *Zizia* and their importance in pollination. American Journal of Botany 71(3):375–387. [252, 290]

Bibliography

Lindsey, A. H., and C. R. Bell. 1985. Reproductive biology of Apiaceae. II. Cryptic specialization and floral evolution in *Thaspium* and *Zizia*. American Journal of Botany 72(2):231–237. [45, 61]

Linhart, Y. B. 1973. Ecological and behavioral determinants of pollen dispersal in hummingbird-pollinated *Heliconia*. American Naturalist 107(956):511–523. [135–136, 457]

Linhart, Y. B., and P. Feinsinger. 1980. Plant-hummingbird interaction: effects of island size and degree of specialization on pollination. Journal of Ecology 68(3):745–760. [135]

Linskens, H. F., and J. F. Jackson, editors. 1987. High performance liquid chromatography in plant sciences. Springer Verlag, New York, New York, USA. [193–194]

Linskens, H. F., and J. Schrauwen. 1969. The release of free amino acids from germinating pollen. Acta Botanica Neerlandica 18:605–614. [200]

Liu, C., J. J. Leonard, and J. J. Feddes. 1990. Automated monitoring of flight activity at a beehive entrance using infrared light sensors. Journal of Apicultural Research 29(1):20–27. [383]

Lloyd, D. G. 1979. Parental strategies of angiosperms. New Zealand Journal of Botany 17:595–606. [241]

Lloyd, D. G. 1980. The distributions of gender in four Angiosperm species illustrating two evolutionary pathways to dioecy. Evolution 34(1):123–134. [241]

Lloyd, D. G. 1984. Gender allocations in outcrossing cosexual plants. Pages 277–300 *in* R. Dirzo and J. Sarukhán, editors. Perspectives in plant population ecology. Sinauer Associates, Sunderland, Massachusetts, USA. [241, 243]

Lloyd, D. G., and K. S. Bawa. 1984. Modification of the gender of seed plants in varying conditions. Evolutionary Biology 17:255–338. [242, 245–246, 481, 489]

Loper, G. M., G. D. Waller, and R. L. Berdel. 1976. Effect of flower age on sucrose content in nectar of citrus. HortScience 11:416–417. [193, 460]

Lord, E. M., and K. J. Eckard. 1984. Incompatibility between the dimorphic flowers of *Collomia grandiflora*, a cleistogamous species. Science 223(4637):695–696. [239]

Lord, E. M., and L. U. Kohorn. 1986. Gynoecial development, pollination, and the path of pollen tube growth in the tepary bean, *Phaseolus acutifolius*. American Journal of Botany 73(1):70–78. [66, 69, 449, 451–452, 454, 458–460]

Louda, S. 1982. Distribution ecology: variation in plant recruitment over a gradient in relation to insect seed predation. Ecological Monographs 52(1):25–41. [29]

Lovett Doust, J., and L. Lovett Doust, editors. 1988. Plant reproductive ecology. Oxford University Press, New York, New York, USA. [222, 248]

Lowry, O. H., N. J. Rosebrough, A. L. Farr, and R. N. Randall. 1951. Protein measurement with the Folin phenol reagent. Journal of Biological Chemistry 193:265–275. [206, 454]

Lubbers, A. E., and N. L. Christensen. 1986. Intraseasonal variation in seed production among flowers and plants of *Thalictrum thalictroides* (Ranunculaceae). American Journal of Botany 73(2):190–203. [397]

Lunau, K. 1991. Innate flower recognition in bumblebees (*Bombus terrestris, B. lucorum*; Apidae): optical signals from stamens as landing reaction releasers. Ethology 88(3):203–214. [380]

Luo, D., E. S. Coen, S. Doyle, and R. Carpenter. 1991. Pigmentation mutants produced by transposon mutagenesis in *Antirrhinum majus*. Plant Journal 1(1):59–69. [2, 369]

Bibliography

Lüttge, U. 1962. Über den Vitamin-C-Gehalt von Bienenhonig. Zeitschrift für Untersuchung der Lebensmittel 117:289. [204, 452]

Lynch, S. P., and G. L. Webster. 1975. A new technique of preparing pollen for scanning electron microscopy. Grana 15:127–136. [80, 86, 447–448, 457]

Lyons, E. E., and J. Antonovics. 1991. Breeding system evolution in *Leavenworthia*: breeding system variation and reproductive success in natural populations of *Leavenworthia crassa* (Cruciferae). American Journal of Botany 78(2):270–287. [218]

Lyons, E. E., N. M. Waser, M. V. Price, J. Antonovics, and A. F. Motten. 1989. Sources of variation in plant reproductive success and implications for concepts of sexual selection. American Naturalist 134(3):409–433. [1, 231]

Mabry, T. J., K. Markham, and M. B. Thomas. 1970. Systematic identification of the flavonoids. Springer Verlag, New York, New York, USA. [210]

Macior, L. W. 1964. An experimental study of the floral ecology of *Dodecatheon meadia*. American Journal of Botany 51(1):96–108. [377]

Macior, L. W. 1968. Pollination adaptation in *Pedicularis groenlandica*. American Journal of Botany 55(8):927–932. [377]

Macior, L. W. 1978. Pollination interactions in sympatric *Dicentra* species. American Journal of Botany 65(1):57–62. [60]

Macior, L. W. 1983. The pollination dynamics of sympatric species of *Pedicularis* (Scrophulariaceae). American Journal of Botany 70(6):844–853. [59, 290, 310, 457]

Macior, L. W. 1986. Floral resource sharing by bumblebees and hummingbirds in *Pedicularis* (Scrophulariaceae) pollination. Bulletin of the Torrey Botanical Club 113(2):101–109. [63, 65, 68]

Madden, G. D., and H. L. Malstrom. 1975. Pecans and hickories. Pages 420–438 *in* J. Janick and J. N. Moore, editors. Advances in fruit breeding. Purdue University Press, West Lafayette, Indiana, USA. [37, 71]

Maeta, Y., S. F. Sakagami, and C. D. Michener. 1985. Laboratory studies on the life cycle and nesting biology of *Braunsapis sauteriella*, a social xylocopine bee (Hymenoptera: Apidae). Sociobiology 10(1):17–41. [384]

Magnarelli, L. A. 1979. Diurnal nectar-feeding of *Aedes cantator* and A. *sollicitans* (Diptera: Culicidae). Environmental Entomology 8:949–955. [212]

Magnarelli, L. A., and J. F. Anderson. 1977. Follicular development in salt marsh Tabanidae (Diptera) and incidence of nectar feeding with relation to gonotrophic activity. Annals of the Entomological Society of America 70(4):529–533. [213, 448, 450, 453, 456]

Magnarelli, L. A., J. F. Anderson, and J. H. Thorne. 1979. Diurnal nectar-feeding of salt marsh Tabanidae (Diptera). Environmental Entomology 8:544–548. [212, 447–448, 450]

Maneval, W. E. 1936. Lacto-phenol preparations. Stain Technology 11 (1):9–11. [111]

Manning, A. 1956a. The effect of honey-guides. Behaviour 9(2):114–139. [58]

Manning, A. 1956b. Some aspects of the foraging behaviour of bumble-bees. Behaviour 9:164–201. [361]

Manning, A. 1957. Some evolutionary aspects of the flower constancy of bees. Proceedings of the Royal Philosophical Society 25:67–71. [348, 368]

Marden, J. H. 1984a. Intrapopulation variation in nectar secretion in *Impatiens capensis*. Oecologia 63:418–422. [160]

Bibliography

Marden, J. H. 1984b. Remote perception of floral nectar by bumblebees. Oecologia 64:232–240. [352]

Marden, J. H., and K. D. Waddington. 1981. Floral choices by honeybees in relation to the relative distances to flowers. Physiological Entomology 6:431–435. [352, 363]

Marshall, D. L. 1991. Nonrandom mating in wild radish: variation in pollen donor success and effects of multiple paternity among one- to six-donor pollinations. American Journal of Botany 78(10):1404–1418. [34, 409]

Marshall, D. L., and N. C. Ellstrand. 1985. Proximal causes of multiple paternity in wild radish, *Raphanus sativus*. American Naturalist 126(5):596–605. [120]

Marshall, D. L., and N. C. Ellstrand. 1988. Effective mate choice in wild radish: evidence for selective seed abortion and its mechanism. American Naturalist 131(5):739–756. [33]

Martin, F. W. 1959. Staining and observing pollen tubes in the style by means of fluorescence. Stain Techniques 34:125–128. [126, 128]

Masters, A. R. 1990. Pyrrolizidine alkaloids in artificial nectar protect adult ithomiine butterflies from a spider predator. Biotropica 22(3):298–304. [200]

Masters, A. R. 1991. Dual role of pyrrolizidine alkaloids in nectar. Journal of Chemical Ecology 17(1):195–205. [357–358]

Mathur, G., and H. Y. Mohan Ram. 1978. Significance of petal colour in thrips-pollinated *Lantana camara* L. Annals of Botany 42:1473–1476. [338]

Mattocks, A. R. 1967. Spectrophotometric determination of unsaturated pyrrolizidine alkaloids. Analytical Chemistry 39:443–447. [201, 452, 455–456, 461]

Mattsson, O., R. B. Knox, J. Heslop-Harrison, and Y. Heslop-Harrison. 1974. Protein pellicle of stigmatic papillae as a probable recognition site in incompatibility reactions. Nature 247:298–300. [69, 453]

Maurer, H. R. 1971. Disc electrophoresis and related techniques of polyacrylamide gel electrophoresis. Walter de Gruyter, New York, New York, USA. [205, 224]

May, P. G. 1985. A simple method for measuring nectar extraction rates in butterflies. Journal of the Lepidopterists' Society 39:53–55. [373]

Mayer, S. S. 1991. Artificial hybridization in Hawaiian *Wikstroemia* (Thymelaeaceae). American Journal of Botany 78(1):122–130. [111, 457]

Mayer, S. S., and D. Charlesworth. 1991. Cryptic dioecy in flowering plants. Trends in Ecology and Evolution 6:320–325. [240]

McAlpine, J. F., editor. 1987. Manual of Nearctic Diptera. Biosystematics Research Centre, Agriculture Canada, Ottawa, Ontario, Canada. [307]

McCall, C., and R. B. Primack. 1992. Influence of flower characteristics, weather, time of day, and season on insect visitation rates in three plant communities. American Journal of Botany, 79(4):434–442. [342, 393]

McCrea, K. D., and M. Levy. 1983. Photographic visualization of floral colors as perceived by honeybee pollinators. American Journal of Botany 70(3):369–375. [61, 63, 316]

McCrea, K. D., W. G. Abrahamson, and A. E. Weis. 1985. Goldenrod ball gall effects on *Solidago altissima*: ^{14}C translocation and growth. Ecology 66(6):1902–1907. [141]

McDade, L. A., and P. Davidar. 1984. Determinants of fruit and seed set in *Pavonia dasypetala* (Malvaceae). Oecologia 64(1):61–67. [342]

McElwee, R. L. 1970. Radioactive tracer techniques for pine pollen flight studies and an analysis of short-range pollen behavior. Unpublished Ph.D. dissertation. Department of Forestry, North Carolina State University, Raleigh, North Carolina, USA. [142]

McGuire, A. D., and W. S. Armbruster. 1991. An experimental test for reproductive interactions between two sequentially blooming *Saxifraga* species (Saxifragaceae). American Journal of Botany 78(2):214–219. [35]

McKenna, M., and J. D. Thomson. 1988. A technique for sampling and measuring small amounts of floral nectar. Ecology 69(4):1306–1307. [161, 163, 169, 176, 448]

McKone, M. J. 1990. Characteristics of pollen production in a population of New Zealand snow-tussock grass (*Chionochloa pallens* Zotov). New Phytologist 116(3):555–562. [97]

McKone, M. J., and C. J. Webb. 1988. A difference in pollen size between the male and hermaphrodite flowers of two species of Apiaceae. Australian Journal of Botany 36(3):331–337. [90]

Meagher, T. R. 1986. Analysis of paternity within a natural population of *Chamaelirium luteum*: I. Identification of most-likely male parents. American Naturalist 128(1):199–215. [230]

Meagher, T. R. 1988. Sex determination in plants. Pages 125–138 *in* J. Lovett Doust and L. Lovett Doust, editors. Plant reproductive ecology. Oxford University Press, New York, New York, USA. [241]

Medler, J. T. 1962a. Measurements of the labium and radial cell of *Psithyrus* (Hymenoptera: Apidae). Canadian Entomologist 94:444–447. [309]

Medler, J. T. 1962b. Morphometric studies on bumble bees. Annals of the Entomological Society of America 55:212–218. [409]

Meeuse, B.J.D. 1973. Films of liquid crystals as an aid in pollination studies. Pages 18–20 *in* N.B.M. Brantjes and H. F. Linskens, editors. Pollination and dispersal. Proceedings of a symposium, published by the Department of Botany, University of Nijmegen, Netherlands. [395, 451]

Meeuse, B.J.D. 1982. Reproductive biology of flowering plants. Laboratory manual for Botany 475. Published by Lecture Notes, 113 HUB, FK-10, University of Washington, Seattle, WA 98195. [155, 187–188, 209, 380]

Meeuse, B., and S. Morris. 1984. The sex life of flowers. Facts on File Publications, New York, New York, USA. [7]

Menzel, R. 1979. Spectral sensitivity and color vision in invertebrates. Pages 503–580 *in* H. Autrum, editor. Handbook of sensory physiology, Volume VII/6A. Springer Verlag, Berlin, Germany. [315]

Menzel, R. 1990. Color vision in flower visiting insects. Jahresbericht der KFA Jülich. A 16-page booklet published by the Institut für Neurobiologie der Freien Universität Berlin, Königin-Luise Strasse 28-30, 1000 Berlin 33. [63, 65, 315]

Menzel, R., and W. Backhaus. 1991. Colour vision in insects. Pages 262–293 *in* P. Gouras, editor. The perception of colour. CRC Press, Boca Raton, Florida, USA. [65, 615–316]

Menzel, R., and A. W. Snyder. 1974. Polarised light detection in the bee, *Apis mellifera*. Journal of comparative Physiology 88:247–270. [316]

Michener, C. D., and D. J. Brothers. 1971. A simplified observation nest for burrowing bees. Journal of the Kansas Entomological Society 44(2):236–239. [387]

Bibliography

Miller, J. S., and J. W. Nowicke. 1990. Dioecy and a reevaluation of *Lepidocordia* and *Antrophora* (Boraginaceae: Ehretioideae). American Journal of Botany 77(4):543–551. [89, 457]

Miller, R. S., and R. E. Miller. 1971. Feeding activity and color preference of Ruby-throated Hummingbirds. Condor 73(3):309–313. [368]

Miller, R. S., S. Tamm, G. D. Sutherland, and C. L. Gass. 1985. Cues for orientation in hummingbird foraging: color and position. Canadian Journal of Zoology 63(1):18–21. [369]

Miller, S. J., and D. W. Inouye. 1983. Roles of the wing whistle in the territorial behaviour of male Broad-tailed hummingbirds (*Selasphorus platycercus*). Animal Behaviour 31:689–700. [332–333]

Mitchell, T. B. 1960. Bees of the Eastern United States. 2 volumes. North Carolina Agricultural Experiment Station, Technical Bulletin 141. [265, 276, 307]

Mitton, J. 1992. The dynamic mating systems of conifers. New Forests 6:197–216. [218, 225–230]

Mjelde, A. 1983. The foraging strategy of *Bombus consobrinus* (Hymenoptera, Apidae). Acta Entomologica Fennica 42:51–56. [385]

Mokrasch, L. C. 1954. Analysis of hexose phosphates and sugar mixtures with the anthrone reagent. Journal of Biological Chemistry 208:55–59. [177, 186]

Montalvo, A. 1992. Relative success of self and outcross pollen comparing mixed- and single-donor pollinations in *Aquilegia caerulea*. Evolution 46(4):1181-1198. [122]

Montgomerie, R. D. 1984. Nectar extraction by hummingbirds: response to different floral characters. Oecologia 63:229–236. [312–313]

Montgomerie, R. D., and C. L. Gass. 1981. Energy limitation of hummingbird populations in tropical and temperate communities. Oecologia 50(2):162–165. [319]

Montgomerie, R. D., M. McA. Eadie, and L. D. Harder. 1984. What do foraging hummingbirds maximize? Oecologia 63:357–363. [360, 388]

Moore, P. D., J. A. Webb, and M. E. Collinson. 1991. Pollen analysis. 2nd edition. Blackwell Scientific Publications. Oxford, England. [79]

Moore, R. P. 1962. Tetrazolium as a universally acceptable quality test of viable seed. Proceedings of the International Seed Test Association 27: 795–805. [30, 32]

Morrow, C. A., and W. M. Sandstrom. 1935. Biochemical laboratory methods for students in the biological sciences. John Wiley, New York, New York, USA. [155, 458, 461]

Morse, D. H. 1977a. Estimating proboscis length from wing length in bumblebees (*Bombus* spp.). Annals of the Entomological Society of America 70(3):311–315. [308]

Morse, D. H. 1977b. Resource partitioning in bumble bees: the role of behavioral factors. Science 197(4304):678–680. [390]

Morse, D. H. 1978. Size-related foraging differences of bumble bee workers. Ecological Entomology 3:189–192. [390]

Morse, D. H. 1981. Modification of bumblebee foraging: the effect of milkweed pollinia. Ecology 62(1):89–97. [327]

Morse, D. H. 1982a. The turnover of milkweed pollinia on bumble bees, and implications for outcrossing. Oecologia 53(2):187–196. [327]

Bibliography

Morse, D. H. 1982b. Behavior and ecology of bumble bees. Chapter 2, pages 245–322 *in* H. R. Hermann, editor. Social insects, Volume 3. Academic Press, New York, New York, USA. [385]

Morse, D. H. 1982c. Foraging relationships within a guild of bumble bees. Insectes Sociaux 29(3):445–454. [390]

Morse, D. H. 1987. Roles of pollen and ovary age in follicle production of the common milkweed *Asclepias syriaca*. American Journal of Botany 74(6):851–856. [35, 67, 244]

Morse, D. H., and R. S. Fritz. 1985. Variation in the pollinaria, anthers, and alar fissures of common milkweed (*Asclepias syriaca* L.). American Journal of Botany 72(7):1032–1038. [43, 408]

Motten, A. F. 1982. Autogamy and competition for pollinators in *Hepatica americana* (Ranunculaceae). American Journal of Botany 69(8):1296–1305. [69]

Motten, A. F. 1986. Pollination ecology of the spring wildflower community of a temperate deciduous forest. Ecological Monographs 56:21–42. [249, 290, 447]

Muirhead-Thomson, R. C. 1991. Trap responses of flying insects. Academic Press, London, England. [263]

Mulcahy, D. 1979. The rise of angiosperms: A genecological factor. Science 206(4414):20–23. [65, 124, 248]

Mulcahy, D. L., P. S. Curtis, and A. A. Snow. 1983. Pollen competition in a natural population. Chapter 16, pages 330–337 *in* C. E. Jones and R. J. Little, editors. Handbook of experimental pollination biology. Scientific and Academic Editions, Van Nostrand Reinhold Company, New York, New York, USA. [65, 116, 124]

Mulcahy, D. L., G. B. Mulcahy, and E. M. Ottaviano. 1978. Further evidence that gametophytic selection modified the genetic quality of the sporophyte. Société Botanique de France. Actualités Botaniques 1:57–60. [249]

Mulcahy, G. B., and D. L. Mulcahy. 1982. The two phases of growth of *Petunia hybrida* (*Hort. Vilm-Andz.*) pollen tubes through compatible styles. Journal of Palynology 18(1-2):61–64. [127]

Müller, H. 1883. The fertilisation of flowers. Macmillan, London, England. Translated by D'Arcy W. Thompson. [7]

Mulligan, G. A., and P. G. Kevan. 1973. Color, brightness, and other floral characteristics attracting insects to the blossoms of some Canadian weeds. Canadian Journal of Botany 51(10):1939–1952. [63, 65]

Murawski, D. A. 1987. Floral resource variation, pollinator response, and potential pollen flow in *Psiguria warscewiczii*. Ecology 68(5):1273–1282. [135]

Murawski, D. A., and L. E. Gilbert. 1986. Pollen flow in *Psiguria warscewiczii*: a comparison of *Heliconius* butterflies and hummingbirds. Oecologia 68(2):161–167. [135]

Murcia, C. 1990. Effect of floral morphology and temperature on pollen receipt and removal in *Ipomoea trichocarpa*. Ecology 71(3):1098–1109. [118, 126, 291]

Murphy, D. D., A. E. Launer, and P. R. Ehrlich. 1983. The role of adult feeding in egg production and population dynamics of the checkerspot butterfly *Euphydryas editha*. Oecologia 56:257–263. [364]

Murphy, S. D. 1992. The determination of the allelopathic potential of pollen and nectar. Pages 333–357 *in* H. F. Linskens and J. F. Jackson, editors. Modern methods of plant analysis (New Series), Volume 13: Plant toxin analysis. Springer Verlag, New York, New York, USA. [146]

Bibliography

Murphy, S. D., and L. W. Aarssen. 1989. Pollen allelopathy among sympatric grassland species: *in vitro* evidence in *Phleum pratense* L. New Phytologist 112(2):295–305. [146–147]

Murray, K. G., P. Feinsinger, W. H. Busby, Y. B. Linhart, J. H. Beach, and S. Kinsman. 1987. Evaluation of character displacement among plants in two tropical pollination guilds. Ecology 68(5):1283–1293. [29, 39]

Nakamura, R. R., M. L. Stanton, and S. J. Mazer. 1989. Effects of mate size and mate number on male reproductive success in plants. Ecology 70(1):71–76. [244]

Neff, J. L., and B. B. Simpson. 1990. The roles of phenology and reward structure in the pollination biology of wild sunflower (*Helianthus annuus* L., Asteraceae). Israel Journal of Botany 39(1-2):197–216. [71, 193, 251, 338, 488]

Neill, D. A. 1987. Trapliners in the trees: hummingbird pollination of *Erythrina* sect. Erythrina (Leguminosae: Papilionoideae). Annals of the Missouri Botanical Garden 74:27–41. [432]

Nelson, N. 1944. A photometric adaptation of the Somogyi method for the determination of glucose. Journal of Biological Chemistry 153:375–380. [180]

Nichols, R. 1977. Sites of ethylene production in the pollinated and unpollinated senescing carnation (*Dianthus caryophyllus*) inflorescence. Planta 135(2):155–159. [338]

Nicholson, S. W. 1990. Osmoregulation in a nectar-feeding insect, the carpenter bee *Xylocopa capitata*: water excess and ion concentration. Physiological Entomology 15:433–440. [320]

Niesenbaum, R. A. 1992. Sex ratio, components of reproduction, and pollen deposition in *Lindera benzoin* (Lauraceae). American Journal of Botany 79(5):495-500. [95, 261]

Niklas, K. J. 1984. The motion of windborne pollen grains around conifer ovulate cones: implications on wind pollination. American Journal of Botany 71(3):356–374. [22]

Niklas, K. J. 1985a. The aerodynamics of wind pollination. Botanical Review 51(3):328–386. [21]

Niklas, K. J. 1985b. Wind pollination— a study in controlled chaos. American Scientist 73:462–470. [22]

Niklas, K. J. 1985c. Wind pollination of *Taxus cuspidata*. American Journal of Botany 72(1):1–13. [22]

Niklas, K. J., and S. L. Buchmann. 1988. Aerobiology and pollen capture of orchard-grown *Pistacia vera* (Anacardiaceae). American Journal of Botany 75(12):1813–1829. [22]

Nilsson, L. A. 1978. Pollination ecology and adaptation in *Platanthera chlorantha* (Orchidaceae). Botaniska Notiser 131:35–51. [49, 53]

Nilsson, L. A. 1979. The pollination ecology of *Herminium monorchis* (Orchidaceae). Botaniska Notiser 132:537– 549. [182, 340, 378]

Nilsson, L. A. 1981. The pollination ecology of *Listera ovata* (Orchidaceae). Nordic Journal of Botany 1:461–480. [49, 53]

Nilsson, L. A. 1983. Anthecology of *Orchis mascula* (Orchidaceae). Nordic Journal of Botany 3:157–179. [60, 62]

Nilsson, L. A. 1984. Anthecology of *Orchis morio* (Orchidaceae) at its outpost in the north. Nova Acta regiae Societatis Scientiarum Upsalensis, Serie V:C,3:167–179. [60]

Nilsson, L. A. 1988. The evolution of flowers with deep corolla tubes. Nature 334:147–149. [362]

Bibliography

Nilsson, L. A., L. Jonsson, L. Rason, and E. Randrianjohany. 1985. Monophily and pollination mechanisms in *Angraecum arachnites* Schltr. (Orchidaceae) in a guild of long-tongued hawk-moths (Sphingidae) in Madagascar. Biological Journal of the Linnean Society 26:1–19. [53]

Nilsson, L. A., L. Jonsson, L. Rason, and E. Randrianjohany. 1986. The pollination of *Cymbidiella flabellata* (Orchidaceae) in Madagascar: a system operated by sphecid wasps. Nordic Journal of Botany 6(4):411–422. [63]

Nilsson, L. A., and E. Rabakonandrianina. 1988. Hawk-moth scale analysis and pollination specialization in the epilitic Malagasy endemic *Aerangis ellisii* (Reichenb. fil.) Schltr. (Orchidaceae). Botanical Journal of the Linnean Society 97:49–61. [337]

Núñez, J. 1977. Nectar flow by melliferous flora and gathering flow by *Apis mellifera ligustica*. Journal of Insect Physiology 23:265–275. [164, 165]

O'Brien, T. P., and M. E. McCully. 1981. The study of plant structure. Principles and selected methods. Termarcarphi Pty. Ltd., Melbourne, Australia. [208]

Olesen, J. M., and E. Warncke. 1989a. Flowering and seasonal changes in flower sex ratio and frequency of flower visitors in a population of *Saxifraga hirculus*. Holarctic Ecology 12(1):21–30. [150]

Olesen, J. M., and E. Warncke. 1989b. Predation and potential transfer of pollen in a population of *Saxifraga hirculus*. Holarctic Ecology 12(1):87–95. [291, 298]

Olesen, J. M., and E. Warncke. 1989c. Temporal changes in pollen flow and neighbourhood structure in a population of *Saxifraga hirculus* L. Oecologia 79(2):205–211. [345]

O'Neill, K. M., W. P. Kemp, and K. A. Johnson. 1990. Behavioural thermoregulation in three species of robber flies (Diptera, Asilidae: *Efferia*). Animal Behaviour 39:181–191. [317–318]

O'Neill, P., M. B. Singh, and R. B. Knox. 1988. Cell biology of the stigma of *Brassica campestris* in relation to CO_2 effects on self-pollination. Journal of Cell Science 89:541–549. [240]

O'Neill, P. M., M. B. Singh, and R. B. Knox. 1989. Biosynthesis of S-associated proteins following self- and cross-pollinations in *Brassica campestris* L. var. 'T. 15'. Sexual Plant Reproduction 2:103–108. [240]

Oni, O. 1990. Between-tree and floral variations in pollen viability and pollen tube growth in obeche (*Triplochiton scleroxylon*). Forest Ecology and Management 37:259–265. [98, 151]

Ordway, E., S. L. Buchmann, R. O. Kuehl, and C. W. Shipman. 1987. Pollen dispersal in *Cucurbita foetidissima* (Cucurbitaceae) by bees of the genera *Apis, Peponapis* and *Xenoglossa* (Hymenoptera: Apidae, Anthophoridae). Journal of the Kansas Entomological Society 60(4):489–503. [137]

Ornduff, R. 1975a. Complementary roles of halictids and syrphids in the pollination of *Jepsonia heterandra* (Saxifragaceae). Evolution 29(2):371–373. [19, 80]

Ornduff, R. 1975b. Pollen flow in *Lythrum junceum*, a tristylous species. New Phytologist 75(1):161–166. [89, 118]

Ornduff, R. 1980. Heterostyly, population composition, and pollen flow in *Hedyotis caerulea*. American Journal of Botany 67(1):95–103. [89]

Osborn, M. M., P. G. Kevan, and M. A. Lane. 1988. Pollination biology of *Opuntia polyacantha* and *Opuntia phaeacantha* (Cactaceae) in southern Colorado. Plant Systematics and Evolution 159(1-2):85–94. [68, 97]

Bibliography

Palmer, M., J. Travis, and J. Antonovics. 1988. Seasonal pollen flow and progeny diversity in *Amianthium muscaetoxicum*: ecological potential for multiple mating in a self-incompatible, hermaphroditic perennial. Oecologia 77(1):19–24. [138]

Palser, B. F., J. L. Rouse, and E. G. Williams. 1989. Coordinated timetables for megagametophyte development and pollen tube growth in *Rhododendron nuttallii* from anthesis to early postfertilization. American Journal of Botany 76(8):1167–1202. [71–73, 126–128, 451, 461]

Pandey, K. K., L. Przywara, and P. M. Sanders. 1990. Induced parthenogenesis in kiwifruit (*Actinidia deliciosa*) through the use of lethally irradiated pollen. Euphytica 51:1–9. [100, 151]

Paoletti, E., and L. M. Bellani. 1990. The in-vitro response of pollen germination and tube length to different types of acidity. Environmental Pollution 67:279–286. [126]

Park, O. W. 1933. Studies on the changes in nectar concentration produced by the honeybee, *Apis mellifera*. Part I. Changes which occur between the flower and the hive. Iowa Agriculture and Home Economics Experiment Station Research Bulletin 151:211–243. [169, 302]

Parker, F. D. 1981. How efficient are bees in pollinating sunflowers? Journal of the Kansas Entomological Society 54(1):61–67. [252, 291]

Parker, F. D. 1982. Efficiency of bees pollinating onion flowers. Journal of the Kansas Entomological Society 55(1):171–176. [252]

Partridge, S. M. 1949. Aniline hydrogen phthalate as a spraying reagent for chromatography of sugars. Nature 164:443. [187, 459]

Partridge, S. M. 1951. Partition chromatography and its application to carbohydrate studies. Biochemical Society Symposium 3:52. [187, 448]

Partridge, S. M., and R. G. Westhall. 1948. Filter paper partition chromatography of sugars. Biochemical Journal 42:238. [188, 457, 462]

Paton, D. C., and B. G. Collins. 1989. Bills and tongues of nectar-feeding birds: a review of morphology, function and performance, with intercontinental comparisons. Australian Journal of Ecology 14(4):473–506. [312–313, 373]

Paton, D. C., and H. A. Ford. 1977. Pollination by birds of native plants in South Australia. Emu 77:73–85. [117, 294, 451]

Patt, J. M., J. C. French, C. Schal, and T. G. Hartman. 1991. Pollination biology of *Peltandra virginica* (Araceae). I. Chemical composition and emission pattern of floral odor. Supplement to American Journal of Botany 78(6):67. [47]

Patt, J. M., T. G. Hartman, R. W. Creekmore, J. J. Elliott, C. Schal, J. Lech, and R. T. Rosen. 1992. The floral odor of *Peltandra virginica* contains novel trimethyl-2,5-dioxabicyclo[3.2.1.]nonanes. Phytochemistry 31 (2):487–491. [76]

Patt, J. M., M. W. Merchant, D.R.E. Williams, and B.J.D. Meeuse. 1989. Pollination biology of *Platanthera stricta* (Orchidaceae) in Olympic National Park, Washington. American Journal of Botany 76(8):1097–1106. [155, 180, 182]

Patt, J. M., D. F. Rhoades, and J. A. Corkill. 1988. Analysis of the floral fragrance of *Platanthera stricta*. Phytochemistry 27(1):91–95. [47, 55]

Paulson, G. S., and R. D. Akre. 1991. Role of predaceous ants in pear psylla (Homoptera: Psyllidae) management: estimating colony size and foraging range of *Formica neoclara* (Hymenoptera: Formicidae) through a mark-recapture technique. Journal of Economic Entomology 84(5):1437–1440. [328]

Bibliography

Pavlovic, N. B., M. DeMauro, and M. Bowles. 1992. Perspectives on plant competition - plant collection rate should be positively correlated with plant population size. Plant Science Bulletin 38(1):8. [11]

Peakall, R. 1989. A new technique for monitoring pollen flow in orchids. Oecologia 79(3):361–365. [135, 449, 451, 457–458, 460, 462]

Peakall, R. 1990. Responses of male *Zaspilothynnus trilobatus* Turner wasps to females and the sexually deceptive orchid it pollinates. Functional Ecology 4:159–167. [114, 325]

Peakall, R., C. J. Angus, and A. J. Beattie. 1990. The significance of ant and plant traits for ant pollination in *Leporella fimbriata*. Oecologia 84(4):457–460. [114]

Peakall, R., and A. J. Beattie. 1989. Pollination of the orchid *Microtis parviflora* R. Br. by flightless worker ants. Functional Ecology 3(5):515–522. [101, 103, 455]

Peakall, R., S. N. Handel, and A. J. Beattie. 1991. The evidence for, and importance of, ant pollination. Pages 421–429 *in* C. R. Huxley and D. F. Cutler, editors. Ant-plant interactions. Oxford University Press, Oxford, England. [6]

Pearse, A.G.E. 1980. Histochemistry, theoretical and applied, Volume 1: Preparative and optical technology. Churchill Livingstone, Edinburgh, Scotland. [69, 107]

Pearson, A. C., G. R. Ballmer, V. Sevacherian, and P. V. Vail. 1989. Interpretation of rubidium marking levels in beet armyworm eggs (Lepidoptera: Noctuidae). Environmental Entomology 18(5):844–848. [321, 460]

Pedersen, M. W. 1953. Seed production in alfalfa as related to nectar production and honeybee visitation. Botanical Gazette 115(2):129–138. [159]

Pellmyr, O. 1984. The pollination ecology of *Actaea spicata* (Ranunculaceae). Nordic Journal of Botany 4:443– 456. [119]

Pellmyr, O. 1985. Pollination ecology of *Cimicifuga arizonica* (Ranunculaceae). Botanical Gazette 146(3):404–412. [294, 298]

Pellmyr, O. 1986a. Pollination of two nectariferous *Cimicifuga* sp. (Ranunculaceae) and the evolution of andromonoecy. Nordic Journal of Botany 6(2):129–138. [20]

Pellmyr, O. 1986b. Three pollination morphs in *Cimicifuga simplex;* incipient speciation due to inferiority in competition. Oecologia 68(2):304–307. [378, 392]

Pellmyr, O. 1987. Temporal patterns of ovule allocation, fruit set, and seed predation in *Anemonopsis macrophylla* (Ranunculaceae). Botanical Magazine 100:175–183. [76]

Pellmyr, O. 1989. The cost of mutualism: interactions between *Trollius europaeus* and its pollinating parasites. Oecologia 78(1):53–59. [13–14, 30, 269]

Pellmyr, O., G. Bergström, and I. Groth. 1987. Floral fragrances in *Actaea,* using differential chromatograms to discern between floral and vegetative volatiles. Phytochemistry 26(6):1603–1606. [53]

Pellmyr, O., and J. M. Patt. 1986. Function of olfactory and visual stimuli in pollination of *Lysichiton americanum* (Araceae) by a staphylinid beetle. Madroño 33(1):47–54. [57, 297, 378]

Pellmyr, O., W. Tang, I. Groth, G. Bergström, and L. B. Thien. 1991. Cycad cone and angiosperm floral volatiles: inferences for the evolution of insect pollination. Biochemical Systematics and Ecology 19(8):623– 628. [52]

Pellmyr, O., L. B. Thien, G. Bergström, and I. Groth. 1990. Pollination of new Caledonian Winteraceae: opportunistic shifts or parallel radiation with their pollinators? Plant Systematics and Evolution 173(3- 4):143–157. [56]

Bibliography

Penny, J.H.J. 1983. Nectar guide colour contrast: a possible relationship with pollination strategy. New Phytologist 95(4):707–721. [61]

Percival, M. S. 1955. The presentation of pollen in certain Angiosperms and its collection by *Apis mellifera*. New Phytologist 54(3):353–368. [148–149]

Percival, M. S. 1961. Types of nectar in Angiosperms. New Phytologist 60(3):235–281. [154, 184, 448, 453]

Percival, M. S. 1965. Floral biology. Pergamon Press, Oxford, England. [99, 336–337]

Perdew, P. E., and C. L. Meek. 1990. An improved model of a battery-powered aspirator. Journal of the American Mosquito Control Association 6(4):716–719. [271]

Petanidou, T., and D. Vokou. 1990. Pollination and pollen energetics in Mediterranean ecosystems. American Journal of Botany 77(8):986–992. [147]

Petersen, J. H. 1990. A BASIC program for recording behavioral events on a personal computer. Newsletter of the Animal Behavior Society 35(1):11–12. [464]

Peterson, C. A., and R. A. Fletcher. 1973. Lactic acid clearing and fluorescent staining for demonstration of sieve tubes. Stain Technology 48(1):23–27. [74]

Pettersson, M. W. 1991. Flower herbivory and seed predation in *Silene vulgaris* (Caryophyllaceae): effects of pollination and phenology. Holarctic Ecology 14:45–50. [30, 381]

Pflumm, W. 1977. Welche grössen Beeinflussen die Menge der von Bienen und Wespen an der Futterquelle aufgenommenen Zuckerlösung. Apidologie 401–411. [304]

Philbrick, C. T., and A. L. Bogle. 1988. A survey of floral variation in five populations of *Podostemum ceratophyllum* Michx. (Podostemaceae). Rhodora 90(862):113–121. [198]

Phillipson, J. 1964. A miniature bomb calorimeter for small biological samples. Oikos 15:130–139. [46]

Pleasants, J. M. 1980. Competition for bumblebee pollinators in Rocky Mountain plant communities. Ecology 61(6):1446–1459. [35]

Pleasants, J. M. 1981. Bumblebee response to variation in nectar availability. Ecology 62(6):1648–1661. [341]

Pleasants, J. M. 1983. Nectar production patterns in *Ipomopsis aggregata* (Polemoniaceae). American Journal of Botany 70(10):1468–1475. [17, 158–160, 396]

Pleasants, J. M. 1990. Null-model tests for competitive displacement: the fallacy of not focusing on the whole community. Ecology 71(3):1078–1084. [29]

Pleasants, J. M. 1991. Evidence for short-distance dispersal of pollinia in *Asclepias syriaca* L. Functional Ecology 5(1):75–82. [141]

Pleasants, J. M., and S. J. Chaplin. 1983. Nectar production rates of *Asclepias quadrifolia*: causes and consequences of individual variation. Oecologia 59(2-3):232–238. [37, 161, 167, 170, 432]

Pleasants, J. M., H. T. Horner, and G. Ng. 1990. A labelling technique to track dispersal of milkweek pollinia. Functional Ecology 4:823–827. [141, 449]

Pleasants, J. M., and M. Zimmerman. 1979. Patchiness in the dispersion of nectar resources: evidence for hot and cold spots. Oecologia 41(3):283–288. [157]

Pleasants, J. M., and M. Zimmerman. 1983. The distribution of standing crop of nectar: what does it really tell us? Oecologia 57(3):412–414. [155–156]

Bibliography

Pleasants, J. M., and M. Zimmerman. 1990. The effect of inflorescence size on pollinator visitation of *Delphinium nelsonii* and *Aconitum columbianum*. Collectanea Botanica (Barcelona) 19:21–39. [33]

Pliske, T. E. 1975a. Courtship behavior and use of chemical communication by males of certain species of Ithomiine butterflies (Nymphalidae: Lepidoptera). Annals of the Entomological Society of America 68(6):935– 942. [273]

Pliske, T. E. 1975b. Pollination of pyrrolizidine alkaloid-containing plants by male Lepidoptera. Environmental Entomology 4:474–479. [273]

Pliske, T. E. 1975c. Attraction of Lepidoptera to plants containing pyrrolizidine alkaloids. Environmental Entomology 4(3):455–473. [273]

Plowright, R. C. 1987. Corolla depth and nectar concentration: an experimental study. Canadian Journal of Botany 65(5):1011–1013. [41, 168–169]

Plowright, R. C., and S. C. Jay. 1966. Rearing bumble bee colonies in captivity. Journal of Apicultural Research 5(3):155–165. [385–386]

Poole, R. W., and B. J. Rathcke. 1979. Regularity, randomness, and aggregation in flowering phenologies. Science 203(4379):470–471. [29]

Possingham, H. P. 1989. The distribution and abundance of resources encountered by a forager. American Naturalist 133(1):42–60. [155, 157]

Poulsen, M. H. 1973. The frequency and foraging behaviour of honeybees and bumble bees on field beans in Denmark. Journal of Apicultural Research 12(2):75–80. [382]

Prakash, N. 1986. Methods in plant microtechnique. University of New England Department of Botany, Armidale, N. S. W., Australia. [131]

Preston, R. E. 1991. The intrafloral phenology of *Streptanthus tortuosus* (Brassicaceae). American Journal of Botany 78(8):1044–1053. [68, 120]

Price, M. V., and N. M. Waser. 1979. Pollen dispersal and optimal outcrossing in *Delphinium nelsonii*. Nature 277:294–296. [120, 234]

Price, M. V., and N. M. Waser. 1982. Experimental studies of pollen carryover: hummingbirds and *Ipomopsis aggregata*. Oecologia 54(3):353–358. [135]

Primack, R. B. 1982. Ultraviolet patterns in flowers, or flowers as viewed by insects. Arnoldia 42:139–146. [63]

Primack, R. B., and P. Hall. 1990. Costs of reproduction in the pink lady's slipper orchid: a four-year experimental study. American Naturalist 136(5):638–656. [3, 255]

Primack, R. B., and J. A. Silander. 1975. Measuring the relative importance of different pollinators to plants. Nature 255:143–144. [291]

Pritham, G. H. 1968. Anderson's essentials of biochemistry. C. V. Mosby Company, Saint Louis, Missouri, USA. [184, 190]

Proctor, M., and P. Yeo. 1973. The pollination of flowers. Taplinger Publishing Company, New York, New York, USA. [1, 7, 19]

Prŷs-Jones, O. E. 1982. Ecological studies of foraging and life history in bumble bees. Ph. D. dissertation, Cambridge University, U.K. (cited in Barrow and Pickard 1985). [42]

Pyke, G. H. 1978a. Optimal foraging in bumblebees and coevolution with their plants. Oecologia 36(3):281–293. [157–158, 344]

Pyke, G. H. 1978b. Optimal foraging: movement patterns of bumblebees between inflorescences. Theoretical Population Biology 13(1):72–98. [346]

Bibliography

Pyke, G. H. 1979. Optimal foraging in bumblebees: rule of movement between flowers within inflorescences. Animal Behaviour 27:1167–1181. [46, 319, 346]

Pyke, G. H. 1980. Optimal foraging in bumblebees: calculation of net rate of energy intake and optimal patch choice. Theoretical Population Biology 17(2):232–246. [319–320]

Pyke, G. H. 1981a. Hummingbird foraging on artificial inflorescences. Behaviour Analysis Letters 1:11–15. [46, 345, 359, 363]

Pyke, G. H. 1981b. Optimal foraging in hummingbirds: rule of movement between inflorescences. Animal Behaviour 29:889–896. [346]

Pyke, G. H. 1981c. Optimal nectar production in a hummingbird pollinated plant. Theoretical Population Biology 20(3):326–343. [359]

Pyke, G. H. 1982a. Foraging in bumblebees: rule of departure from an inflorescence. Canadian Journal of Zoology 60(3):417–428. [120]

Pyke, G. H. 1982b. Local geographic distributions of bumblebees near Crested Butte, Colorado: competition and community structure. Ecology 63(2):555–573. [342, 382]

Pyke, G. H. 1988. Yearly variation in seasonal patterns of honeyeater abundance, flower density and nectar production in heathland near Sydney. Australian Journal of Ecology 13(1):1–10. [381]

Pyke, G. H. 1991. What does it cost a plant to produce floral nectar? Nature 350:58–59. [167–168, 215]

Pyke, G. H., L. P. Day, and K. A. Wale. 1988. Pollination ecology of Christmas Bells (*Blanfordia nobilis* Sm.): Effects of adding artificial nectar on pollen removal and seed-set. Australian Journal of Ecology 13(3):279– 284. [96, 370]

Pyke, G. H., and N. M. Waser. 1981. The production of dilute nectars by hummingbird and honeyeater flowers. Biotropica 13(4):260–270. [154]

Queller, D. C. 1984. Pollen-ovule ratios and hermaphroditic sexual allocation strategies. Evolution 38(5):1148– 1151. [248]

Radford, A. E., W. C. Dickison, J. R. Massey, and C. R. Bell. 1974. Vascular plant systematics. Harper and Row, New York, New York, USA. [11, 84, 89–90, 111, 448, 451, 453–454, 458, 477, 481]

Raff, J. W., and R. B. Knox. 1982. Pollen tube growth in *Prunus avium*. Pages 123–134 *in* E. G. Williams, R. B. Knox, J. H. Gilbert, and P. Bernhardt, editors. Pollination '82. Plant Cell Biology Research Center, School of Botany, University of Melbourne, Australia. [94, 104, 125]

Ramanna, M. S. 1973. Euparal as a mounting medium for preserving fluorescence of aniline blue in plant material. Stain Technology 48(3):103–105. [129, 421]

Ramming, D. W., and H. A. Hinrichs. 1973. Sequential staining of callose by aniline blue and lacmoid for fluorescence and regular microscopy on a durable preparation of the same specimen. Stain Technology 48(3):133–134. [448, 455]

Ramsey, M., and G. Vaughton. 1991. Self-incompatibility, protandry, pollen production and pollen longevity in *Banksia menziesii*. Australian Journal of Botany 39:497–504. [69]

Ranta, E. 1983. Proboscis length and the coexistence of bumblebee species. Oikos 43(2):189–196. [305]

Ranta, E., and H. Lundberg. 1980. Resource partitioning in bumblebees: the significance of differences in proboscis length. Oikos 35(3):298–302. [305]

538

Bibliography

Ranta, E., and H. Lundberg. 1981. Food niche analyses of bumblebees: a comparison of three data collecting methods. Oikos 36(1):12–16. [302]

Ranta, E., H. Lundberg, and I. Teräs. 1981. Patterns of resource utilization in two Fennoscandian bumblebee communities. Oikos 36(1):1–11. [382]

Rao, K. S., and U. Kristen. 1990. The influence of the detergent Triton X-100 on the growth and ultrastructure of tobacco pollen tubes. Canadian Journal of Botany 68(5):1131–1138. [132, 450, 456, 460, 462]

Rathcke, B. 1988. Flowering phenologies in a shrub community: competition and constraints. Journal of Ecology 76(4):975–994. [29]

Raw, A. 1974. Pollen preferences of three *Osmia* species (Hymenoptera). Oikos 25(1):54–60. [300]

Reader, R. J. 1977. Bog ericad flowers: self-compatibility and relative attractiveness to bees. Canadian Journal of Botany 55(17):2279–2287. [169, 432]

Reader, R. J. 1983. Heatsum models to account for geographic variation in the floral phenology of two ericaceous shrubs. Journal of Biogeography 10:47–64. [22]

Real, L. A. 1981. Uncertainty and pollinator-plant interactions: the foraging behavior of bees and wasps on artificial flowers. Ecology 62(1):20–26. [348]

Real, L., editor. 1983. Pollination biology. Academic Press, Orlando, Florida, USA. [7]

Real, L. A., and B. J. Rathcke. 1991. Individual variation in nectar production and its effect on fitness in *Kalmia latifolia*. Ecology 72(1):149–155. [215, 216]

Redmond, A. M., L. E. Robbins, and J. Travis. 1989. The effects of pollination distance on seed production in three populations of *Amianthium muscaetoxicum* (Liliaceae). Oecologia 79(2):260–264. [34]

Reese, C.S.L., and E. M. Barrows. 1980. Co-evolution of *Claytonia virginica* (Portulacaceae) and its main native pollinator, *Andrena erigeniae* (Andrenidae). Proceedings of the Entomological Society of Washington 82:685–694. [44, 58]

Reinke, D. C., and W. L. Bloom. 1979. Pollen dispersal in natural populations: a method for tracking individual pollen grains. Systematic Botany 4(3):223–229. [144]

Ribbands, C. R. 1950. Changes in the behavior of honey bees following their recovery from anaesthesia. Journal of Experimental Biology 27:302–310. [267]

Richards, A. J., editor. 1978. The pollination of flowers by insects. Linnean Society Symposium Series #6. Academic Press, London, England. [7, 384]

Richards, A. J. 1986. Plant breeding systems. George Allen and Unwin, London, England. [77, 99, 101–102, 236–238]

Richards, A. J., and H. Ibrahim. 1978. Estimation of neighbourhood size in two populations of *Primula veris*. Pages 165–174 *in* A. J. Richards, editor. The pollination of flowers by insects. Linnean Society Symposium Series #6. Academic Press, London, England. [133, 232]

Richards, K. W. 1978. Nest site selection by bumble bees (Hymenoptera: Apidae) in southern Alberta. Canadian Entomologist 110:841–846. [384]

Richards, K. W. 1987. Diversity, density, efficiency, and effectiveness of pollinators of cicer milkvetch, *Astragalus cicer* L. Canadian Journal of Zoology 65(9):2168–2176. [384]

Richardson, T. E., and A. G. Stephenson. 1991. Effects of parentage, prior fruit set and pollen load on fruit and seed production in *Campanula americana* L. Oecologia 87(1):80–85. [38, 123]

Bibliography

Rieseberg, L. H., and D. E. Soltis. 1987. Allozymic differentiation between *Tolmiea menziesii* and *Tellima grandiflora* (Saxifragaceae). Systematic Botany 12:154–161. [224]

Rinderer, T. E., B. G. Harville, J. J. Lackett, and J. R. Baxter. 1981. Variation in honey bee morphology, behavior, and seed set in white clover. Annals of the Entomological Society of America 74:459–461. [409]

Ritland, K. 1983. Estimation of mating systems. Pages 289–302 *in* S. D. Tanksley and T. J. Orton, editors. Isozymes in plant genetics and breeding. Elsevier Science Publishers, Amsterdam, Netherlands. [225–227, 229]

Ritland, K. 1984. The effective proportion of self-fertilization with consanguineous matings in inbred populations. Genetics 106:139–152. [228]

Ritland, K. 1986. Joint maximum likelihood estimation of genetic and mating structure using open-pollinated progenies. Biometrics 42:25–43. [225, 227–228]

Ritland, K. 1989. Correlated matings in the partial selfer *Mimulus guttatus*. Evolution 43(4):848–859. [228]

Ritland, K. 1990a. A series of FORTRAN computer programs for estimating plant mating systems. Journal of Heredity 81:235–237. [225–226]

Ritland, K. 1990b. Inferences about inbreeding depression based on changes of the inbreeding coefficient. Evolution 44(5):1230–1241. [225, 229]

Ritland, K., and S. Jain. 1981. A model for the estimation of outcrossing rate and gene frequency using n-independent loci. Heredity 47(1):35–52. [229]

Ritland, C., and K. Ritland. 1989. Variation of sex allocation among eight taxa of the *Mimulus guttatus* species complex (Scrophulariaceae). American Journal of Botany 76(12):1731–1739. [95, 337, 408]

Robbins, L., and J. Travis. 1986. Examining the relationship between functional gender and gender specialization in hermaphroditic plants. American Naturalist 128(3):409–415. [243]

Roberds, J. H., S. T. Friedman, and Y. A. El-Kassaby. 1991. Effective number of pollen parents in clonal seed orchards. Theoretical and Applied Genetics 82:313–320. [134, 233]

Roberts, E. H. 1972. Viability of seeds. Syracuse University Press, Syracuse, New York, USA. [30]

Roberts, R. B. 1979. Spectrophotometric analysis of sugars produced by plants and harvested by insects. Journal of Apicultural Research 18(3):191–195. [161, 164, 166, 177–178, 187, 304, 462]

Rocha, O. J., and A. G. Stephenson. 1990. Effect of ovule position on seed production, seed weight, and progeny performance in *Phaseolus coccineus*. American Journal of Botany 77(10):1320–1329. [74, 76]

Roland, J. 1978. Variation in spectral reflectance of alpine and arctic *Colias* (Lepidoptera: Pieridae). Canadian Journal of Zoology 56(6):1447–1453. [59]

Rooum, D. 1989. Karl von Frisch and the 'spot codes' for marking insects. Bee World 70:120–126. [329]

Röseler, P.-F. 1985. A technique for year-round rearing of *Bombus terrestris* (Apidae, Bombini) colonies in captivity. Apidologie 16(2):165–170. [386]

Rotenberry, J. T. 1990. Variable floral phenology: temporal resource heterogeneity and its implication for flower visitors. Holarctic Ecology 13:1–10. [23–24, 29]

Bibliography

Roubik, D. W. 1989. Ecology and natural history of tropical bees. Cambridge University Press, New York, New York, USA. [90, 325]

Roubik, D. W., and J. D. Ackerman. 1987. Long-term ecology of euglossine orchid-bees (Apidae: Euglossini) in Panama. Oecologia 73(3):321–333. [271]

Roubik, D. W., J. D. Ackerman, C. Copenhaver, and B. H. Smith. 1982. Stratum, tree, and flower selection by tropical bees: implications for the reproductive biology of outcrossing *Cochlospermum vitifolium* in Panama. Ecology 63(3):712–720. [33]

Roubik, D. W., and M. Aluja. 1983. Flight ranges of *Melipona* and *Trigona* in tropical forest. Journal of the Kansas Entomological Society 56(2):217–222. [327]

Rouse, J. L. 1985. Pollen storage in *Rhododendron* breeding. Pages 185–186 *in* E. G. Williams and R. B. Knox, editors. Pollination '84. Proceedings of a symposium held at the Plant Cell Biology Research Centre, School of Botany, University of Melbourne, Parkville, Victoria, 3052, Australia. The School of Botany, University of Melbourne, Melbourne, Australia. [113, 450]

Rundel, P. W., and W. M. Jarrell. 1989. Water in the environment. Chapter 3, pages 29–56 *in* R. W. Pearcy, J. R. Ehleringer, H. A. Mooney, and P. W. Rundel, editors. Plant physiological ecology. Field methods and instrumentation. Chapman and Hall, New York, New York, USA. [402]

Rust, R. W. 1977. Pollination in *Impatiens capensis* and *Impatiens pallida* (Balsaminaceae). Bulletin of the Torrey Botanical Club 104(4):361–367. [200]

Sacchi, C. F., and P. W. Price. 1988. Pollination of the arroyo willow, *Salix lasiolepis*: role of insects and wind. American Journal of Botany 75(9):1387–1393. [21]

Sado, M. 1990. Study of atmospheric pollen by volumetric methods. Review of Palaeobotany and Palynology 64:61–69. [92]

Sage, T. L., S. B. Broyles, and R. Wyatt. 1990. The relationship between the five stigmatic chambers and two ovaries of milkweed (*Asclepias amplexicaulis* Sm.) flowers: a three-dimensional assessment. Israel Journal of Botany 39(1-2):187–196. [74]

Saini, A. S. 1966. Some technical improvements in the paper chromatography of sugars. A method of sample desalting and a sensitive staining reagent. Journal of Chromatography 24:484–486. [186, 188, 448, 450, 459]

Sanders, L. C., and E. M. Lord. 1989. Directed movement of latex particles in the gynoecia of three species of flowering plants. Science 243(4898):1606–1608. [100, 151]

Sanford, M. T. 1991. Comments on blueberry pollination. Florida Extension Beekeeping Newsletter 9(2) (distributed via electronic mail on Bee-L bulletin board). [3]

Scagel, R. F., R. J. Bandoni, G. E. Rouse, W. B. Schofield, J. R. Stein, and T.M.C. Taylor. 1969. Plant diversity: an evolutionary approach. Wadsworth Publishing, Belmont, California, USA. [77]

Schaal, B. A. 1978. Density dependent foraging on *Liatris pycnostachya*. Evolution 32(2):452–454. [361]

Schaal, B. A. 1980. Measurement of gene flow in *Lupinus texensis*. Nature 284:450–451. [233]

Scheithauer, W. 1967. Hummingbirds. Thomas Y. Crowell, New York, New York, USA. [388]

Schemske, D. W. 1980a. Floral ecology and hummingbird pollination of *Combretum farinosum* in Costa Rica. Biotropica 12(3):169–181. [37, 166]

Schemske, D. W. 1980b. Evolution of floral display in the orchid *Brassavola nodosa*. Evolution 34(3):489–493. [45]

Bibliography

Schemske, D. W., and C. C. Horvitz. 1984. Variation among floral visitors in pollination ability: a precondition for mutualism specialization. Science 225(4661):519–521. [251, 337]

Schemske, D. W., and C. C. Horvitz. 1988. Plant-animal interactions and fruit production in a neotropical herb: a path analysis. Ecology 69(4):1128–1137. [258–259]

Schemske, D. W., and C. C. Horvitz. 1989. Temporal variation in selection on a floral character. Evolution 43(2):461–465. [259]

Schemske, D. W., M. F. Willson, M. N. Melampy, L. J. Miller, L. Verner, K. M. Schemske, and L. B. Best. 1978. Flowering ecology of some spring woodland herbs. Ecology 59(2):351–366. [37, 182, 462]

Schlamowitz, R., F. R. Hainsworth, and L. L. Wolf. 1976. On the tongues of sunbirds. Condor 78(1):104–107. [360, 377]

Schlessman, M. A. 1986a. Floral protogyny, self-compatibility and the pollination of Ourisia macrocarpa (Scrophulariaceae). New Zealand Journal of Botany 24:651–656. [45, 448, 408]

Schlessman, M. A. 1986b. Interpretation of evidence for gender choice in plants. American Naturalist 128(3):416–420. [240, 246]

Schlessman, M. A. 1987. Gender modification in North American ginseng: Dichotomous choice versus adjustment. BioScience 37(7):469–475. [240]

Schlessman, M. A. 1988. Gender diphasy ("sex choice"). Pages 139–153 in J. Lovett Doust and L. Lovett Doust, editors. Plant reproductive ecology. Oxford University Press, New York, New York, USA. [245–246]

Schlessman, M. A. 1991. Size, gender, and sex change in dwarf ginseng, Panax trifolium (Araliaceae). Oecologia 87(4):588–595. [245–246]

Schlessman, M. A.; P. P. Lowry, II; and D. G. Lloyd. 1990. Functional dioecism in the New Caledonian endemic Polyscias pancheri (Araliaceae). Biotropica 22:133–139. [240, 260]

Schlichting, C. D., A. G. Stephenson, and L. E. Small. 1990. Pollen loads and progeny vigor in Cucurbita pepo: the next generation. Evolution 44(5):1358–1372. [249]

Schlising, R. A., and R. A. Turpin. 1971. Hummingbird dispersal of Delphinium cardinale pollen treated with radioactive iodine. American Journal of Botany 58(5):401–406. [142]

Schmid-Hempel, P., and B. Speiser. 1988. Effects of inflorescence size on pollination in Epilobium angustifolium. Oikos 53(1):98–104. [116, 372]

Schmid-Hempel, R., and C. B. Müller. 1991. Do parasitized bumblebees forage for their colony? Animal Behaviour 41:910–912. [335]

Schmid-Hempel, P., and R. Schmid-Hempel. 1990. Endoparasitic larvae of conopid flies alter pollination behavior of bumblebees. Naturwissenschaften 77:450–452. [335]

Schmid-Hempel, R., and P. Schmid-Hempel. 1991. Endoparasitic flies, pollen-collection by bumblebees and a potential host-parasite conflict. Oecologia 87(2):227–232. [335]

Schmidt, J. O. 1985. Phagostimulants in pollen. Journal of Apicultural Research 24:107–114. [148]

Schmidt, J. O., S. C. Thoenes, and M. D. Levin. 1987. Survival of honey bees, Apis mellifera (Hymenoptera: Apidae), fed various pollen sources. Annals of the Entomological Society of America 80:176–183. [148]

Bibliography

Schmitt, J. 1980. Pollinator foraging behavior and gene dispersal in *Senecio* (Compositae). Evolution 34(5):934– 943. [233, 391]

Schmitt, J. 1983a. Individual flowering phenology, plant size, and reproductive success in *Linanthus androsaceus*, a California annual. Oecologia 59(1):135–140. [22, 391]

Schmitt, J. 1983b. Density-dependent pollinator foraging, flowering phenology, and temporal pollen dispersal patterns in *Linanthus bicolor*. Evolution 37(6):1247–1257. [33]

Schmitt, J. 1983c. Flowering plant density and pollinator visitation in *Senecio*. Oecologia 60:97–102. [362]

Schmitt, J., J. Eccleston, and D. W. Ehrhardt. 1987. Density-dependent flowering phenology, outcrossing, and reproduction in *Impatiens capensis*. Oecologia 72(3):341–347. [33–34]

Schneider, E. L., and L. A. Moore. 1977. Morphological studies of the Nymphaeaceae. VII. The floral biology of *Nuphar lutea* ssp. *macrophylla*. Brittonia 29:88–99. [30]

Schoen, D. J. 1982. The breeding system of *Gilia achilleifolia*: variation in floral characteristics and outcrossing rate. Evolution 36(2):352–360. [218]

Schoen, D. J., and M. T. Clegg. 1984. Estimation of mating system parameters when outcrossing events are correlated. Proceedings of the National Academy of Sciences (USA) 81:5258–5262. [228]

Schrauwen, J., and H. F. Linskens. 1974. Influence of the extraction conditions on the recovery of free amino acids in plant material. Acta Botanica Neerlandica 23:42–47. [200]

Scogin, R. 1979. Nectar constituents in the genus *Fremontia* (Sterculiaceae): sugars, flavonoids, and proteins. Botanical Gazette 140(1):29–31. [205, 451]

Scogin, R. 1980. Anthocyanins of the Bignoniaceae. Biochemical Systematics and Ecology 8:273–276. [58, 206]

Scogin, R. 1983. Visible floral pigments and pollinators. Chapter 7, pages 160–172, *in* C. E. Jones and R. J. Little, editors. Handbook of experimental pollination biology. Scientific and Academic Editions, Van Nostrand Reinhold Company, New York, New York, USA. [58]

Scora, R. W. 1964. Dependency of pollination on patterns in *Monarda* (Labiatae). Nature 204(4962):1011–1012. [380]

Sedgley, M., and A. R. Griffin. 1989. Sexual reproduction of tree crops. Academic Press, London, England. [4]

Seigler, D., B. B. Simpson, C. Martin, and J. L. Neff. 1978. Free 3-acetoxyfatty acids in floral glands of *Krameria* species. Phytochemistry 17:995–996. [215]

Shaw, C. R., and R. Prasad. 1970. Starch gel electrophoresis of enzymes — a compilation of recipes. Biochemical Genetics 4:297–320. [205–206, 224]

Shaw, D. V., and R. W. Allard. 1982. Estimation of outcrossing rates in Douglas-fir using isozyme markers. Theoretical and Applied Genetics 62:113–120. [218]

Shaw, D. V., A. L. Kahler, and R. W. Allard. 1981. A multilocus estimator of mating system parameters in plant populations. Proceedings of the National Academy of Sciences (USA) 78:1298–1302. [229]

Shelly, T. E. 1984. Comparative foraging behavior of Neotropical robber flies (Diptera: Asilidae). Oecologia 62(2):188–195. [317–318]

Shivanna, K. R., and J. Heslop-Harrison. 1981. Membrane state and pollen viability. Annals of Botany 467:759– 770. [107, 454]

Shivanna, K. R., and B. M. Johri. 1985. The angiosperm pollen. John Wiley & Sons, New York, New York, USA. [98–99, 102, 112–113, 125, 151]

Shivanna, K. R., H. F. Linskens, and M. Cresti. 1991a. Responses of tobacco pollen to high humidity and heat stress: viability and germinability in vitro and in vivo. Sexual Plant Reproduction 4:104–109. [99–100, 106, 452]

Shivanna, K. R., H. F. Linskens, and M. Cresti. 1991b. Pollen viability and pollen vigor. Theoretical and Applied Genetics 81:38–42. [100, 106–107]

Shivanna, K. R., and N. S. Rangaswamy. 1992. Pollen biology: a laboratory manual. Springer Verlag, Heidelberg, Germany. [5, 79]

Shively, J. E., editor. 1986. Methods of protein microcharacterization: a practical handbook. Humana Press, Clifton, New Jersey, USA. [194]

Shmida, A., and R. Dukas. 1990. Progressive reduction in the mean body sizes of solitary bees active during the flowering season and its correlation with the sizes of bee flowers of the mint family (Lamiaceae). Israel Journal of Botany 39(1-2):133–141. [309]

Shmida, A., and R. Kadmon. 1991. Within-plant patchiness in nectar standing crop in *Anchusa strigosa*. Vegetatio 94:95–99. [156]

Shore, J. S., and S.C.H. Barrett. 1984. The effect of pollination intensity and incompatible pollen on seed set in *Turnera ulmifolia* (Turneraceae). Canadian Journal of Botany 62(6):1298–1303. [96, 121]

Showers, W. B., F. Whitford, R. B. Smelser, A. J. Keaster, J. F. Robinson, J. D. Lopez, and S. E. Taylor. 1989. Direct evidence for meteorologically driven long-range dispersal of an economically important moth. Ecology 70(4):987–992. [322, 450]

Silander, J. A., and R. B. Primack. 1978. Pollination intensity and seed set in the evening primrose (*Oenothera fruiticosa*). American Midland Naturalist 100:213–216. [116]

Silberglied, R. E. 1979. Communication in the ultraviolet. Annual Review of Ecology and Systematics 10:373–398. [65]

Simpson, B. B., and J. L. Neff. 1983. Evolution and diversity of floral rewards. Chapter 6, pages 142–159 *in* C. E. Jones and R. J. Little, editors. Handbook of experimental pollination biology. Scientific and Academic Editions, Van Nostrand Reinhold Company, New York, New York, USA. [90, 213]

Simpson, B. B., J. L. Neff, and D. Seigler. 1977. *Krameria,* free fatty acids and oil-collecting bees. Nature 267:150–151. [214]

Simpson, B. B., D. S. Seigler, and J. L. Neff. 1979. Lipids from the floral glands of *Krameria.* Biochemical Systematics and Ecology 7:193–194. [214]

Simpson, J. 1954. Effects of some anaesthetics on honey bees: nitrous oxide, carbon dioxide, ammonium nitrate smoker fumes. Bee World 35:149–155. [268]

Sims, L. L. 1989. How permanent are they? Insect Collection News 2(2):26. [281]

Singer, M. C., and P. Wedlake. 1981. Capture does affect probability of recapture in a butterfly species. Ecological Entomology 6:215–216. [330]

Slater, A. T., and D. M. Calder. 1988. The pollination biology of *Dendrobium speciosum* Smith: a case of false advertising? Australian Journal of Botany 36(2):145–158. [57, 69, 447, 457–458]

Smith, A. P. 1975. Insect pollination and heliotropism in *Oritrophium limnophilum* Compositae of the Andean paramo. Biotropica 7(4):284–286. [44]

Bibliography

Smith, C. C., J. L. Hamrick, and C. L. Kramer. 1988. The effects of stand density on frequency of filled seeds and fecundity in lodgepole pine (*Pinus contorta* Dougl.). Canadian Journal of Forestry Research 18:453–460. [91]

Smith, G. F. 1991. Studies on the reproductive biology and palynology of *Chortolirion* Berger (Asphodelaceae: Alooideae) in southern Africa. Taxon 40(1):61–73. [127]

Smith, G. F., and L. R. Tiedt. 1991. A rapid, non-destructive osmium tetroxide technique for preparing pollen for scanning electron microscopy. Taxon 40(2):195–200. [87]

Smith, I. 1969. Chromatographic and electrophoretic techniques. Volume I. Chromatography. 3rd edition. Heinemann, London, England. [203, 459]

Smith-Huerta, N. L., and F. C. Vasek. 1984. Pollen longevity and stigma pre-emption in *Clarkia*. American Journal of Botany 71:1183–1191. [99]

Snow, A. A. 1982. Pollination intensity and potential seed set in *Passiflora vitifolia*. Oecologia 55:231–237. [117, 122, 249]

Snow, A. A. 1986. Pollination dynamics in *Epilobium canum* (Onagraceae): consequences for gametophytic selection. American Journal of Botany 73(1):139–151. [116, 249]

Snow, A. A. 1990. Effects of pollen-load size and number of donors on sporophyte fitness in wild radish (*Raphanus raphanistrum*). American Naturalist 136(6):742–758. [151]

Snow, A. A., and S. J. Mazer. 1988. Gametophytic selection in *Raphanus raphanistrum*: a test for heritable variation in pollen competitive ability. Evolution 42(5):1065–1075. [58]

Snow, A. A., and D. W. Roubik. 1987. Pollen deposition and removal by bees visiting two tree species in Panama. Biotropica 19(1):57–63. [97, 372]

Snow, A. A., and T. P. Spira. 1991a. Differential pollen-tube growth rates and nonrandom fertilization in *Hibiscus moscheutos* (Malvaceae). American Journal of Botany 78(10):1419–1426. [101, 122, 125, 248–249]

Snow, A. A., and T. P. Spira. 1991b. Pollen vigour and the potential for sexual selection in plants. Nature 352:796–797. [248, 409]

Sobrevila, C. 1988. Effects of distance between pollen donor and pollen recipient on fitness components in *Espeletia schultzii*. American Journal of Botany 75(5):701–724. [235]

Sobrevila, C. 1989. Effects of pollen donors on seed formation in *Espeletia schultzii* (Compositae) populations at different elevations. Plant Systematics and Evolution 166(1-2):45–67. [30, 120]

Sokal, R. R., and N. L. Oden. 1978. Spatial autocorrelation in biology. I. Methodology. Biological Journal of the Linnean Society 10:199–228. [157]

Sol, R. 1966. The occurrence of aphidovorous syrphids and their larvae on different crops, with the help of coloured water traps. Pages 181–184 *in* I. Hodek, editor. Ecology of aphidophagous insects. Dr. W. Junk, The Hague, Netherlands. [269]

Soltis, D. E., C. H. Haugler, D. C. Darrow, and G. J. Gastony. 1983. Starch gel electrophoresis of ferns: a compilation of grinding buffers, gel and electrode buffers, and staining schedules. American Fern Journal 73:9–26. [224]

Soltis, D. E, and P. S. Soltis, editors. 1989. Isozymes in plant biology. Dioscorides Press, Portland, Oregon, USA. [222, 224]

Soltz, R. L. 1986. Foraging path selection in bumblebees: hindsight or foresight? Behaviour 99:1–21. [345, 362]

Bibliography

Soltz, R. L. 1987. Interspecific competition and resource utilization between bumblebees. Southwestern Naturalist 32(1):39–52. [363]

Somogyi, M. 1945. A new reagent for the determination of sugars. Journal of Biological Chemistry 160:61–68. [180]

Southwick, E. E. 1982a. "Lucky hit" nectar rewards and energetics of plant and pollinators. Comparative Physiology and Ecology 7:51–55. [156]

Southwick, E. E. 1982b. Nectar biology and pollinator attraction in the north temperate climate. Pages 19–23 *in* M. D. Breed, C. D. Michener, and H. E. Evans, editors. The biology of social insects. Westview Press, Boulder, Colorado, USA. [168]

Southwick, E. E. 1983. Nectar biology and nectar feeders of common milkweed, *Asclepias syriaca* L. Bulletin of the Torrey Botanical Club 110:324–334. [17, 156]

Southwick, E. E. 1984. Photosynthate allocation to floral nectar: a neglected energy investment. Ecology 65(6):1775–1779. [167]

Southwick, E. E. 1990. Floral nectar. American Bee Journal 130(8):517–519. [154, 168]

Southwick, E. E., G. M. Loper, and S. E. Sadwick. 1981. Nectar production, composition, energetics and pollinator attractiveness in spring flowers of Western New York. American Journal of Botany 68(7):994– 1002. [17, 156, 161, 193, 432, 447, 459]

Southwick, E. E., and A. K. Southwick. 1983. Aging effect on nectar production in two clones of *Asclepias syriaca*. Oecologia 56:121–125. [157–159, 166]

Southwood, T.R.E. 1978. Ecological methods. 2nd edition. Chapman and Hall, London, England. [273–274, 321, 368, 382]

Sowig, P. 1989. Effects of flowering plant's patch size on species composition of pollinator communities, foraging strategies, and resource partitioning in bumblebees (Hymenoptera: Apidae). Oecologia 78(4):550–558. [182]

Spangler, H. G. 1969. Photoelectrical counting of outgoing and incoming honey bees. Journal of Economic Entomology 62(5):1183–1184. [383]

Spears, E. E., Jr. 1983. A direct measure of pollinator effectiveness. Oecologia 57(1-2):196–199. [488]

Spencer, F.G.T., P.A.J. Gorin, G. A. Hobbs, and D. A. Cooke. 1970. Yeasts isolated from bumblebee honey from Western Canada: identification with the aid of proton magnetic resonance spectra of their mannose-containing polysaccharides. Canadian Journal of Microbiology 16:117–119. [210]

Spira, T. P., A. A. Snow, D. F. Whigham, and J. Leak. 1992. Flower visitation, pollen deposition, and pollen tube competition in *Hibiscus moscheutos* (Malvaceae). American Journal of Botany 79(4):428–433. [119]

Stanley, R. G., and H. F. Linskens. 1974. Pollen: biology, biochemistry, and management. Springer Verlag, New York, New York, USA. [91, 98–99, 101, 105, 109–110, 120–121, 151, 458–459, 462]

Stanton, M. L. 1987a. Reproductive biology of petal color variants in wild populations of *Raphanus sativus*: I. Pollinator response to color morphs. American Journal of Botany 74(2):178–187. [39]

Stanton, M. L. 1987b. Reproductive biology of petal color variants in wild populations of *Raphanus sativus*: II. Factors limiting seed production. American Journal of Botany 74(2):188–196. [39]

Stanton, M. L., and C. Galen. 1989. Consequences of flower heliotropism for reproduction in an alpine buttercup (*Ranunculus adoneus*). Oecologia 78(4):477–485. [44–45]

Bibliography

Stanton, M. L., A. A. Snow, and S. N. Handel. 1986. Floral evolution: attractiveness to pollinators increases male fitness. Science 232(4758):1625–1627. [244, 248, 253, 407]

Stanton, M. L., H. J. Young, N. C. Ellstrand, and J. M. Clegg. 1991. Consequences of floral variation for male and female reproduction in experimental populations of wild radish, *Raphanus sativus* L. Evolution 45(2):268–280. [43]

Stead, A. D., and K. G. Moore. 1977. Flower development and senescence in *Digitalis purpurea* L., cv. Foxy. Annals of Botany 41:283–292. [338]

Stead, A. D., and M. S. Reid. 1990. The effect of pollination and ethylene on the colour change of the banner spot of *Lupinus albifrons* (Bentham) flowers. Annals of Botany 66:655–663. [338]

Stead, A. D., I. N. Roberts, and H. G. Dickinson. 1979. Pollen-pistil interaction in *Brassica oleracea*. Planta 146(2):211–216. [78, 151]

Stebbings, R. E. 1978. Marking bats. Pages 81–94 *in* B. Stonehouse, editor. Animal marking: recognition marking of animals in research. University Park Press, Baltimore, Maryland, USA. [335]

Steiner, J. J., P. R. Beuselinck, R. N. Peaden, W. P. Kojis, and E. T. Bingham. 1992. Pollinator effects on crossing and genetic shift in a three-flower-color alfalfa population. Crop Science 32:73–77. [369]

Steiner, K. E., and V. B. Whitehead. 1990. Pollinator adaptation to oil-secreting flowers *Rediviva* and *Diascia*. Evolution 44(6):1701–1707. [42, 308]

Steiner, K. E., and V. B. Whitehead. 1991a. Resin collection and the pollination of *Dalechampia capensis* (Euphorbiaceae) by *Pachyanthidium cordatum* (Hymenoptera: Megachilidae) in South Africa. Journal of the Entomological Society of South Africa 54(1):67–72. [213]

Steiner, K. E., and V. B. Whitehead. 1991b. Oil flowers and oil bees: further evidence for pollinator adaptation. Evolution 45(6):1493–1501. [308, 408]

Stelleman, P. 1978. The possible role of insect visits in pollination of reputedly anemophilous plants, exemplified by *Plantago lanceolata,* and syrphid flies. Pages 41–46 *in* A. J. Richards, editor. The pollination of flowers by insects. Linnean Society Symposium Series, No. 6. Academic Press, London, England. [457]

Stelleman, P., and A.D.J. Meeuse. 1976. Anthecological relationships between reputedly anemophilous flowers and syrphid flies. I. The possible role of syrphid flies as pollinators of *Plantago*. Tijdschrift voor Entomologie 119:15–31. [136, 294]

Stelly, D. M., S. J. Peloquin, R. G. Palmer, and C. F. Crane. 1984. Mayer's hemalum-methyl salicylate: a stain- clearing technique for observations within whole ovules. Stain Technology 59:155–161. [72–73, 455]

Stephens, S. G., and M. D. Finkner. 1953. Natural crossing in cotton. Economic Botany 7(3):257–269. [135–136, 456]

Stephenson, A. G. 1979. An evolutionary examination of the floral display of *Catalpa speciosa* (Bignoniaceae). Evolution 33(4):1200–1209. [362]

Stephenson, A. G. 1981. Flower and fruit abortion: proximate causes and ultimate functions. Annual Review of Ecology and Systematics 12:253–279. [254, 256]

Stephenson, A. G. 1982. Iridoid glycosides in the nectar of *Catalpa speciosa* are unpalatable to nectar thieves. Journal of Chemical Ecology 8: 1025–1034. [210]

Stephenson, A. G. 1984. The regulation of maternal investments in an indeterminate flowering plant (*Lotus corniculatus*). Ecology 65(1):113–121. [255]

Bibliography

Stephenson, A. G., and R. I. Bertin. 1983. Male competition, female choice, and sexual selection in plants. Chapter 6, pages 109–149 *in* L. Real, editor. Pollination biology. Academic Press, Orlando, Florida, USA. [248, 480]

Sterling, C. 1963. Structure of the male gametophyte in Gymnosperms. Biological Reviews 38:167–203. [77]

Stern, W. L., K. J. Curry, and W. M. Whitten. 1986. Staining fragrance glands in orchid flowers. Bulletin of the Torrey Botanical Club 113(3):288–297. [57, 450, 458, 462]

Steyskal, G. C., W. L. Murphy, and E. M. Hoover, editors. 1986. Insects and mites: Techniques for collection and preservation. U. S. Department of Agriculture, Miscellaneous Publication No. 1443. [263–265, 268, 270, 272, 274, 276, 278, 280]

Stiles, F. G. 1976. Taste preferences, color preferences, and flower choice in hummingbirds. Condor 78(1):10–26. [181, 359, 368, 387]

Stiles, F. G., and L. L. Wolf. 1973. Techniques for color-marking hummingbirds. Condor 75(2):244–245. [332–334]

Stockhouse, R. E., II. 1976. A new method for studying pollen dispersal using micronized fluorescent dusts. American Midland Naturalist 96:241–245. [137–138, 430]

Stonehouse, B., editor. 1978. Animal marking: recognition marking of animals in research. University Park Press, Baltimore, Maryland, USA. [321]

Strand, M. R., B. D. Roitberg, and D. R. Papaj. 1990. Acridine orange: a potentially useful internal marker of Hymenoptera and Diptera. Journal of the Kansas Entomological Society 63(4):634–637. [322, 447]

Strickler, K. 1979. Specialization and foraging efficiency of solitary bees. Ecology 60(5):998–1009. [341, 372]

Stuessy, T. F., D. M. Spooner, and K. A. Evans. 1986. Adaptive significance of ray corollas in *Helianthus grosseserratus* (Compositae). American Midland Naturalist 115:191–197. [379]

Sukhada, K., and Jayachandra. 1980. Pollen allelopathy — a new phenomenon. New Phytologist 84(4):739–746. [146]

Sutherland, S. 1986. Patterns of fruit-set: what controls fruit-flower ratios in plants? Evolution 40(1):117–128. [253]

Sutherland, S., and L. F. Delph. 1984. On the importance of male fitness in plants: patterns of fruit-set. Ecology 65(4):1093–1104. [253]

Svensson, L. 1986. Secondary pollen carryover by ants in a natural population of *Scleranthus perennis* (Caryophyllaceae). Oecologia 70(4):631–632. [139]

Swanson, C. A., and R. W. Shuel. 1950. The centrifuge method for measuring nectar yield. Plant Physiology 25:513–520. [162]

Sweeley, C. C., R. Bentley, M. Makita, and W. W. Wells. 1963. Gas-liquid chromatography of trimethylsilyl derivates of sugars and related substances. Journal of the American Chemical Society 85:2497–2507. [192, 455, 460, 462]

Taylor, O. R., Jr., and D. W. Inouye. 1985. Synchrony and periodicity of flowering in *Frasera speciosa* (Gentianaceae). Ecology 66(2):521–527. [28]

Tepedino, V. J. 1981. The pollination efficiency of the squash bee (*Peponapis pruinosa*) and the honey bee (*Apis mellifera*) on summer squash (*Cucurbita pepo*). Journal of the Kansas Entomological Society 54(2):359–377. [252, 382]

Bibliography

Theunis, C. H., E. S. Pierson, and M. Cresti. 1991. Isolation of male and female gametes in higher plants. Sexual Plant Reproduction 4:145–154. [260]

Thien, L. B. 1969. Mosquito pollination of *Habenaria obtusata* (Orchidaceae). American Journal of Botany 56(2):232–237. [272]

Thien, L. B., P. Bernhardt, G. W. Gibbs, O. Pellmyr, G. Bergström, I. Groth, and G. McPherson. 1985. The pollination of *Zygogynum* (Winteraceae) by a moth, *Sabatinca* (Micropterigidae): An ancient association? Science 227(4686):540–543. [53]

Thien, L. B., W. H. Heimermann, and R. T. Holman. 1975. Floral odors and quantitative taxonomy of *Magnolia* and *Liriodendron*. Taxon 24(5/6):557–568. [50]

Thomson, J. D. 1978. Effects of stand composition on insect visitation in two-species mixtures of *Hieracium*. American Midland Naturalist 100:431–440. [340]

Thomson, J. D. 1981a. Spatial and temporal components of resource assessment by flower-feeding insects. Journal of Animal Ecology 50(1):49–59. [33]

Thomson, J. D. 1981b. Field measures of flower constancy in bumblebees. American Midland Naturalist 105(2):377–380. [116, 376]

Thomson, J. D. 1982. Patterns of visitation by animal pollinators. Oikos 39(2):241–250. [138, 338]

Thomson, J. D. 1985. Pollination and seed set in *Diervilla lonicera* (Caprifoliaceae): temporal patterns of flower and ovule development. American Journal of Botany 72(5):737–740. [76]

Thomson, J. D. 1986. Pollen transport and deposition by bumble bees in *Erythronium*: influences of floral nectar and bee grooming. Journal of Ecology 74(2):329–341. [117, 134, 163]

Thomson, J. D. 1988. Effects of variation in inflorescence size and floral rewards on the visitation rates of traplining pollinators of *Aralia hispida*. Evolutionary Ecology 2:65–76. [33, 361, 364]

Thomson, J. D. 1989a. Germination schedules of pollen grains: implications for pollen selection. Evolution 43(1):220–223. [134]

Thomson, J. D. 1989b. Reversal of apparent preferences of bumble bees by aggression from *Vespula* wasps. Canadian Journal of Zoology 67(10):2588–2591. [361, 371]

Thomson, J. D., B. J. Andrews, and R. C. Plowright. 1981. The effect of a foreign pollen on ovule development in *Diervilla lonicera* (Caprifoliaceae). New Phytologist 90(4):777–783. [118, 146]

Thomson, J. D., and S.C.H. Barrett. 1981. Temporal variation of gender in *Aralia hispida* Vent. (Araliaceae). Evolution 35(6):1094–1107. [35, 67]

Thomson, J. D., W. P. Maddison, and R. C. Plowright. 1982. Behavior of bumble bee pollinators of *Aralia hispida* Vent. (Araliaceae). Oecologia 54(3):326–336. [234, 326, 364]

Thomson, J. D., M. A. McKenna, and M. B. Cruzan. 1989. Temporal patterns of nectar and pollen production in *Aralia hispida*: implications for reproductive success. Ecology 70(4):1061–1068. [90, 93, 97, 127–128, 163, 174, 176]

Thomson, J. D., S. C. Peterson, and L. D. Harder. 1987. Response of traplining bumble bees to competition experiments: shifts in feeding location and efficiency. Oecologia 71(2):295–300. [301, 344]

Bibliography

Thomson, J. D., and R. C. Plowright. 1980. Pollen carryover, nectar rewards, and pollinator behavior with special reference to *Diervilla lonicera*. Oecologia 46(1):68–74. [116–118, 134–135, 234, 407]

Thomson, J. D., M. V. Price, N. M. Waser, and D. A. Stratton. 1986. Comparative studies of pollen and fluorescent dye transport by bumble bees visiting *Erythronium grandiflorum*. Oecologia 69(4):561–566. [116, 134, 137–139, 409]

Thomson, J. D., K. R. Shivanna, J. Kenrick, and R. B. Knox. 1989. Sex expression, breeding system, and pollen biology of *Ricinocarpos pinifolius*: a case of androdioecy? American Journal of Botany 76(7):1048–1059. [38]

Thomson, J. D., and D. A. Stratton. 1985. Floral morphology and cross-pollination in *Erythronium grandiflorum* (Liliaceae). American Journal of Botany 72(3):433–437. [134, 218]

Thomson, J. D., and B. A. Thomson. 1989. Dispersal of *Erythronium grandiflorum* pollen by bumblebees: implications for gene flow and reproductive success. Evolution 43(3):657–661. [97, 134, 364]

Thomson, J. D., and B. A. Thomson. 1992. Pollen presentation and viability schedules and their consequences for reproductive success through animal pollination.Pages 1–24 *in* R. Wyatt, editor. Ecology and evolution of plant reproduction: new approaches. Chapman and Hall, New York, New York, USA. [149, 151, 253, 391]

Thorp, R. W.; D. L. Briggs, Jr.; R. Estes; and E. H. Erickson. 1975. Nectar fluorescence under ultraviolet irradiation. Science 189(4201):476–478. [304]

Tiebout, H. M., III. 1991. Daytime energy management by tropical hummingbirds: responses to foraging constraint. Ecology 72(3):839–851. [388]

Toledo, V. M., and H. M. Hernández. 1979. *Erythrina oliviac*: a new case of oriole pollination in Mexico. Annals of the Missouri Botanical Garden 66:503–511. [432]

Tomkins, S. P., and P. H. Williams. 1990. Fast plants for finer science — an introduction to the biology of rapid-cycling *Brassica campestris* (*rapa*) L. Journal of Biological Education 24(4):239–250. [120, 409, 431]

Tomlinson, P. B., R. B. Primack, and J. S. Blunt. 1979. Preliminary observations on floral biology in mangrove Rhizophoraceae. Biotropica 11(4):256–277. [93, 97]

Torchio, P. F. 1972. *Sapyga pumila* Cresson, a parasite of *Megachile rotundata* (F.) (Hymenoptera: Sapygidae; Megachilidae). Melanderia 10:1–22. [384]

Torchio, P. F. 1976. Use of *Osmia lignaria* Say (Hymenoptera: Apoidea, Megachilidae) as a pollinator in an apple and prune orchard. Journal of the Kansas Entomological Society 49(4):475–482. [384]

Torchio, P. F. 1981. Field experiments with *Osmia lignaria propinqua* Cresson as a pollinator in almond orchards: I, 1975 studies (Hymenoptera: Megachilidae). Journal of the Kansas Entomological Society 54(4):815–823. [384]

Torchio, P. F. 1982a. Field experiments with the pollinator species, *Osmia lignaria propinqua* Cresson, in apple orchards: II, 1976 studies (Hymenoptera: Megachilidae). Journal of the Kansas Entomological Society 55(4):759–778. [384]

Torchio, P. F. 1982b. Field experiments with *Osmia lignaria propinqua* Cresson as a pollinator in almond orchards: III, 1977 studies (Hymenoptera: Megachilidae). Journal of the Kansas Entomological Society 55(1):101–116. [384]

Torchio, P. F. 1984a. Field experiments with the pollinator species, *Osmia lignaria propinqua* Cresson, in apple orchards: IV, 1978 studies (Hymenoptera: Megachilidae). Journal of the Kansas Entomological Society 57(4):689–694. [384]

Bibliography

Torchio, P. F. 1984b. The nesting biology of *Hylaeus bisinuatus* Forster and development of its immature forms (Hymenoptera: Colletidae). Journal of the Kansas Entomological Society 57(2):276–297. [384]

Torchio, P. F. 1987. Use of non-honey bee species as pollinators of crops. Proceedings of the Entomological Society of Ontario 118:111–124. [3]

Torchio, P. F. 1989a. Biology, immature development, and adaptive behavior of *Stelis montana*, a cleptoparasite of *Osmia* (Hymenoptera: Megachilidae). Annals of the Entomological Society of America 82(5):616–632. [384]

Torchio, P. F. 1989b. In-nest biologies and development of immature stages of three *Osmia* species (Hymenoptera: Megachilidae). Annals of the Entomological Society of America 82(5):599–615. [384]

Torchio, P. F. 1990. Diversification of pollination strategies for U.S. crops. Environmental Entomology 19(6):1649–1656. [3]

Tremblay, R. L. 1991. Ecological significance of pollen diversity for the orchid, *Cypripedium calceolus* L. Abstract. Bulletin of the Ecological Society of America 72(2):270. [136]

Trevelyan, W. E., D. P. Procter, and J. S. Harrison. 1950. Detection of sugars on paper chromatograms. Nature 166(4219):444–445. [189, 455, 460–461]

Tucker, A. O., M. J. Maciarello, and S. S. Tucker. 1991. A survey of color charts for biological descriptions. Taxon 40(2):201–214. [58]

Tupý, J., E. Hrabetová, and V. Balatková. 1977. Evidence for ribosomal RNA synthesis in pollen tubes in culture. Biologia Plantarum (Praha) 19:226–230. [125]

Turchin, P., F. J. Odendaal, and M. D. Rausher. 1991. Quantifying insect movement in the field. Environmental Entomology 20(4):955–963. [344–345, 464]

Turner, J.R.G. 1971. Experiments on the demography of tropical butterflies. II. Longevity and home-range behaviour in *Heliconius erato*. Biotropica 3(1):21–31. [328]

Turner, V. 1982. Marsupials as pollinators in Australia. Pages 55–66 *in* J. A. Armstrong, J. M. Powell, and A. J. Richards, editors. Pollination and evolution. Royal Botanic Gardens, Sydney, Australia. [388]

Turnock, W. J., J. Chong, and B. Luit. 1978. Scanning electron microscopy: a direct method of identifying pollen grains on moths (Noctuidae: Lepidoptera). Canadian Journal of Zoology 56(9):2050–2054. [88, 293–294]

Turpin, R. A., and R. A. Schlising. 1971. A new method for studying pollen dispersal using Iodine-131. Radiation Botany 11:75–78. [142]

Turrell, F. M. 1946. Tables of surfaces and volumes of spheres and prolate and oblate spheroids and spheroidal coefficients. University of California Press, Berkeley, California, USA. [90]

Tuttle, M. 1991. Bats. The cactus connection. National Geographic 179(6):130–140. [288]

Udovic, D. 1981. Determinants of fruit set in *Yucca whipplei*: reproductive expenditure vs. pollinator availability. Oecologia 48(3):389–399. [343]

Uenoyama, M. K. 1986. Inbreeding and the cost of meiosis: the evolution of selfing in populations practicing biparental inbreeding. Evolution 40(2):388–404. [228]

Umbeck, P. F., K. A. Barton, E. V. Nordheim, J. C. McCarty, W. L. Parrott, and J. N. Jenkins. 1991. Degree of pollen dispersal by insects from a field test of genetically engineered cotton. Journal of Economic Entomology 84(6):1943–1950. [4]

Bibliography

Unwin, D. M. 1980. Microclimate measurement for ecologists. Academic Press, London, England. [311, 394, 396, 398–399, 459]

Unwin, D. M., and S. A. Corbet. 1991. Insects, plants and microclimate. Richmond Publishing Co. Ltd., Slough, England. [This book is not distributed in the United States but is available from D. W. Inouye.] [394, 400]

Unwin, D. M., and C. P. Ellington. 1979. An optical tachometer for measurement of the wing-beat frequency of free-flying insects. Journal of Experimental Biology 82:377–378. [314]

Unwin, D. M., and P. G. Willmer. 1978. A simple field cryoscope-osmometer for freezing-point determination with small fluid samples. Physiological Entomology 3:341–345. [182, 421]

Utech, F. H., and S. Kawano. 1975. Spectral polymorphisms in Angiosperm flowers determined by differential ultraviolet reflectance. Botanical Magazine (Tokyo) 88:9–30. [65]

Utrio, P. 1983. Sugaring for moths: why are noctuids attracted more than geometrids? Ecological Entomology 8:437–445. [272]

Vail, S. G. 1983. Density effects in pollination and herbivory of *Potentilla gracilis*. Unpublished M.S. thesis, University of Maryland, College Park, Maryland, USA. [362]

Vandemark, F. L., and G. J. Schmidt. 1981. Determination of water soluble vitamins in pharmaceutical preparations using liquid chromatography. Journal of Liquid Chromatography 4(7):1157–1171. [205]

Van Handel, E. 1967. Determination of fructose and fructose-yielding carbohydrates with cold anthrone. Analytical Biochemistry 19:193–194. [175, 462]

Van Handel, E. 1968. Direct microdetermination of sucrose. Analytical Biochemistry 22:280–283. [175]

Van Handel, E. 1972. The detection of nectar in mosquitoes. Mosquito News 32:458. [212, 451, 456]

Vanstone, V. A., and D. C. Paton. 1988. Extrafloral nectaries and pollination of *Acacia pycnantha* Benth. by birds. Australian Journal of Botany 36(5):519–531. [14]

Vasil, I. K. 1964. Effect of boron on pollen germination and pollen tube growth. Pages 107–119 *in* Linskens, H. F., editor. Pollen physiology and fertilization. North-Holland, Amersterdam, The Netherlands. [102]

Vaughton, G. 1990. Predation by insects limits seed production in *Banksia spinulosa* var. *neoanglica* (Proteaceae). Australian Journal of Botany 38(4):335–340. [29]

Vaughton, G., and M. Ramsey. 1991. Floral biology and inefficient pollen removal in *Banksia spinulosa* var. *Ineoangelica* (Proreaceae). Australian Journal of Botany 39:167–177. [19, 29, 68]

Vithanage, V. 1984. Pollination techniques in pistachio breeding. Pages 167–175 *in* E. G. Williams and R. B. Knox, editors. Pollination '84. Proceedings of a symposium held at the Plant Cell Biology Research Centre, School of Botany, University of Melbourne, Parkville, Victoria, 3052, Australia. The School of Botany, University of Melbourne, Melbourne, Australia. [112–113]

Vogel, S. 1954. Blütenbiologische Typen als Elemente der Sippengliederung, dargestellt anhand der Flora Südafrikas. Botanische Studien, No. 1, G. Fischer Verlag, Jena, Germany. [306]

Bibliography

Vogel, S. 1963. Duftdrüsen im Dienste der Bestäubung. Über Bau und Funktion der Osmophoren. Akademie der Wissenschaften und der Literatur in Mainz. Abhandlungen der Mathematisch-Naturwissenschaftlichen Klasse 10:598–763. [57]

Vogel, S., and I. C. Machado. 1991. Pollination of 4 sympatric species of *Angelonia* (Scrophulariaceae) by oil- collecting bees in NE Brazil. Plant Systematics and Evolution 178(3-4):153–178. [213]

Vogel, S., and C. D. Michener. 1985. Long bee legs and oil-producing floral spurs, and a new *Rediviva* (Hymenoptera, Melittidae; Scrophulariaceae). Journal of the Kansas Entomological Society 58(2):359–364. [308]

Waddington, K. D. 1979a. Flight patterns of three species of sweat bees (Halictidae) foraging at *Convolvulus arvensis*. Journal of the Kansas Entomological Society 52(4):751–758. [344–345, 361]

Waddington, K. D. 1979b. Quantification of the movement patterns of bees: a novel method. American Midland Naturalist 101(2):278–285. [345, 348]

Waddington, K. D. 1980. Flight patterns of foraging bees relative to density of artificial flowers and distribution of nectar. Oecologia 44(2):199–204. [345, 348]

Waddington, K. D. 1981. Factors influencing pollen flow in bumblebee-pollinated *Delphinium virescens*. Oikos 37(2):153–159. [120, 135–136, 234–235]

Waddington, K. D. 1982. Honey bee foraging profitability and round dance correlates. Journal of comparative Physiology 148:297–301. [349–350]

Waddington, K. D. 1983a. Floral-visitation-sequences by bees: models and experiments. Chapter 23, pages 461–473 *in* C. E. Jones and R. J. Little, editors. Handbook of experimental pollination biology. Scientific and Academic Editions, Van Nostrand Reinhold Company, New York, New York, USA. [347, 374]

Waddington, K. D. 1983b. Foraging behavior of pollinators. Chapter 9, pages 213–239 *in* L. Real, editor. Pollination biology. Academic Press, Orlando, Florida, USA. [347]

Waddington, K. D. 1985. Cost-intake information used in foraging. Journal of Insect Physiology 31(11):891–897. [353, 363]

Waddington, K. D., T. Allen, and B. Heinrich. 1981. Floral preferences of bumblebees (*Bombus edwardsii*) in relation to intermittent versus continuous rewards. Animal Behaviour 29:779–784. [348, 350]

Waddington, K. D., and B. Heinrich. 1979. The foraging movements of bumblebees on vertical "inflorescences": an experimental analysis. Journal of comparative Physiology 134:113–117. [348, 350, 355, 363]

Wagner, D. H. 1991. The "1-in-20 rule" for plant collectors. Plant Science Bulletin 37(2):11. [11]

Wagner, H., S. Bladt, and E. M. Zgainski. 1984. Plant drug analysis: a thin layer chromatography atlas. Springer Verlag, Berlin, Germany. [200]

Wainwright, C. M. 1978. The floral biology and pollination ecology of two desert lupines. Bulletin of the Torrey Botanical Club 105:24–38. [338]

Waller, D. M., and S. E. Knight. 1989. Genetic consequences of outcrossing in the cleistogamous annual *Impatiens capensis*. II. Outcrossing rates and genotypic correlations. Evolution 43(4):860–869. [228]

Waller, G. D. 1972. Evaluating responses of honeybees to sugar solutions using an artificial-flower feeder. Annals of the Entomological Society of America 65:857–862. [353–354]

Waller, G. D., E. W. Carpenter, and O. A. Ziehl. 1972. Potassium in onion nectar and its probable effect on attractiveness of onion flowers to honey bees. Journal of the American Society for Horticultural Science 97:535–539. [210, 301, 320, 353–354]

Waller, G. D., G. M. Loper, and R. L. Berdel. 1973. A bioassay for determining honey bee responses to flower volatiles. Environmental Entomology 2(2):255–259. [365–366]

Waller, G. D., G. M. Loper, and R. L. Berdel. 1974. Olfactory discrimination by honeybees of terpenes identified from volatiles of alfalfa flowers. Journal of Apicultural Research 13(3):191–197. [365]

Waller, G. D., and J. H. Martin. 1978. Fluorescence for identification of onion nectar in foraging honey bees. Environmental Entomology 7(5):766–768. [304]

Waller, G. D., N. D. Waters, E. H. Erickson, and J. H. Martin. 1976. The use of potassium to identify onion nectar-collecting honey bees. Environmental Entomology 5(4):780–782. [304]

Wartenberg, D., S. Ferson, and F. J. Rohlf. 1987. Putting things in order: a critique of detrended correspondence analysis. American Naturalist 129(3):434–448. [29]

Waser, N. M. 1978. Competition for hummingbird pollination and sequential flowering in two Colorado wildflowers. Ecology 59(5):934–944. [14, 28, 146]

Waser, N. M. 1979. Pollinator availability as a determinant of flowering time in ocotillo (*Fouquieria spendens*). Oecologia 39(1):107–121. [14, 28, 382]

Waser, N. M. 1983. Competition for pollination and floral character differences among sympatric plant species: a review of evidence. Chapter 13, pages 277–293 *in* C. E. Jones and R. J. Little, editors. Handbook of experimental pollination biology. Scientific and Academic Editions, Van Nostrand Reinhold Company, New York, New York, USA. [29]

Waser, N. M. 1986. Flower constancy: definition, cause, and measurement. The American Naturalist 127(5):593–603. [374–375]

Waser, N. M. 1987. Spatial genetic heterogeneity in a population of the montane perennial plant *Delphinium nelsonii*. Heredity 58:249–256. [125, 235]

Waser, N. M. 1988. Comparative pollen and dye transfer by pollinators of *Delphinium nelsonii*. Functional Ecology 2:41–48. [137–140, 285]

Waser, N. M., and W. A. Calder. 1975. Possible impairment of nest-building of hummingbirds by acetate leg tags. Condor 77(3):361. [333]

Waser, N. M., and M. L. Fugate. 1986. Pollen precedence and stigma closure: a mechanism of competition for pollination between *Delphinium nelsonii* and *Ipomopsis aggregata*. Oecologia 70(4):573–577. [146, 453]

Waser, N. M., and R. J. Mitchell. 1990. Nectar standing crops in *Delphinium nelsonii* flowers: spatial autocorrelation among plants? Ecology 71(1):116–123. [157]

Waser, N. M., and M. V. Price. 1981. Pollinator choice and stabilizing selection for flower color in *Delphinium nelsonii*. Evolution 35(2):376–390. [57, 361]

Waser, N. M., and M. V. Price. 1982. A comparison of pollen and fluorescent dye carry-over by natural pollinators of *Ipomopsis aggregata* (Polemoniaceae). Ecology 63(4):1168–1172. [133, 135, 138–140]

Waser, N. M., and M. V. Price. 1983. Optimal and actual outcrossing in plants, and the nature of plant-pollinator interaction. Chapter 17, pages 341–359 *in* C. E. Jones and R. J. Little, editors. Handbook of experimental pollination biology. Scientific and Academic Editions, Van Nostrand Reinhold Company, New York, New York, USA. [34, 135, 235]

Bibliography

Waser, N. M., and M. V. Price. 1984. Experimental studies of pollen carryover: effects of floral variability in *Ipomopsis aggregata*. Oecologia 62(2):262–268. [234, 363, 408]

Waser, N. M., and M. V. Price. 1985a. The effect of nectar guides on pollinator preference: experimental studies with a montane herb. Oecologia 67(1):121–126. [64, 235, 341, 367–377, 379]

Waser, N. M., and M. V. Price. 1985b. Reciprocal transplant experiments with *Delphinium nelsonii* (Ranunculaceae): evidence for local adaptation. American Journal of Botany 72(11):1726–1732. [235]

Waser, N. M., and M. V. Price. 1989. Optimal outcrossing in *Ipomopsis aggregata*: seed set and offspring fitness. Evolution 43(5):1097–1109. [34, 235]

Waser, N. M., and M. V. Price. 1990. Pollination efficiency and effectiveness of bumble bees and hummingbirds visiting *Delphinium nelsonii*. Collectanea Botanica (Barcelona) 19:9–20. [249]

Waser, N. M., and M. V. Price. 1991a. Outcrossing distance effects in *Delphinium nelsonii*: pollen loads, pollen tubes, and seed set. Ecology 72(1):171–179. [34, 117, 120, 127, 235]

Waser, N. M., and M. V. Price. 1991b. Reproductive costs of self-pollination in *Ipomopsis aggregata* (Polemoniaceae): are ovules usurped? American Journal of Botany 78(8):1036–1043. [122–123, 146]

Waser, N. M., M. V. Price, A. M. Montalvo, and R. N. Gray. 1987. Female mate choice in a perennial herbaceous wildflower, *Delphinium nelsonii*. Evolutionary Trends in Plants 1(1):29–33. [128, 235, 453]

Washitani, I., H. Namai, R. Osawa, and M. Niwa. 1991. Species biology of *Primula sieboldii* for the conservation of its lowland-habitat population: I. inter-clonal variations in the flowering phenology, pollen load and female fertility components. Plant Species Biology 6:27–37. [339]

Watt, W. B., F. S. Chew, L. R. G. Snyder, A. G. Watt, and D. E. Rothschild. 1977. Population structure of pierid butterflies. I. Numbers and movements of some montane *Colias* species. Oecologia 27(1):1–22. [328, 330]

Watt, W. B., P. C. Hoch, and S. G. Mills. 1974. Nectar resource use by *Colias* butterflies. Oecologia 14(4):353–374. [177, 186, 448, 450, 453]

Weast, R. C. 1972. Handbook of chemistry and physics. 53rd edition. Chemical Rubber Company, Cleveland, Ohio, USA. [401]

Weast, R. C., editor. 1978. CRC handbook of chemistry and physics. CRC Press, West Palm Beach, Florida, USA. [172]

Webb, C. J., and K. S. Bawa. 1983. Pollen dispersal by hummingbirds and butterflies: a comparative study of two lowland tropical plants. Evolution 37(6):1258–1270. [135]

Weberling, F. 1989. Morphology of flowers and inflorescences. Cambridge University Press, Cambridge, England. [45, 337]

Weir, B. S. 1990. Genetic data analysis. Sinauer, Sunderland, Massachusetts, USA. [224]

Weiss, M. R. 1991. Floral colour changes as cues for pollinators. Nature 354:227–229. [58]

Weller, S. G. 1981. Pollination biology of heteromorphic populations of *Oxalis alpina* (Rose) Knuth (Oxalidaceae) in south-eastern Arizona. Botanical Journal of the Linnean Society 83(3):189–198. [89, 95]

Bibliography

Weller, S. G., and R. Ornduff. 1989. Incompatibility in *Amsinckia grandiflora* (Boraginaceae): distribution of callose plugs and pollen tubes following inter- and intramorph crosses. American Journal of Botany 76(2):277–282. [128, 460]

Welte, E., E. Przemeck, and M. C. Nuh. 1971. Über Ursprung und Ausmaß des Analysenfehlers bei der quantitativen Bestimmung des Gehaltes freier Aminosäuren in Pflanzensubstanzen. Zeitschrift für Pflanzenernährung und Bodenkunde 128:243–256. [200]

White, J. W., Jr., and O. N. Rudyj. 1978. Proline content of United States honeys. Journal of Apicultural Research 17(2):89–93. [455]

White, R. H. 1985. Insect visual pigments and color vision. Pages 431–493 *in* G. A. Kerkut and L. I. Gilbert, editors. Comprehensive insect physiology, biochemistry and pharmacology, Volume 6: Nervous systems: sensory. Pergamon Press, New York, New York, USA. [316]

Whitham, T. G., and C. N. Slobodchikoff. 1981. Evolution by individuals, plant-herbivore interactions, and mosaics of genetic variability: the adaptive significance of somatic mutations in plants. Oecologia 49(3):287– 292. [24]

Wickerham, L. J. 1951. Taxonomy of yeasts. U. S. Department of Agriculture Bulletin 1029. [209]

Wiebes, J. T. 1979. Coevolution of figs and their insect pollinators. Annual Review of Ecology and Systematics 10:1–12. [30]

Wilkins, C. K., and B. A. Bohm. 1976. Chemotaxonomic studies in the Saxifragaceae s.l. 4. The flavonoids of *Heuchera micrantha* var. *diversifolia*. Canadian Journal of Botany 54(18):2133–2140. [210]

Willemstein, S. C. 1987. An evolutionary basis for pollination biology. Leiden Botanical Series, Volume 10. E. J. Brill, Leiden, Netherlands. [7]

Williams, E. G., and R. B. Knox. 1982. Quantitative analysis of pollen tube growth in *Lycopersicon peruvianum*. Journal of Palynology 19(1-2):65–74. [124]

Williams, E. G., S. Ramm-Anderson, C. Dumas, S. L. Mau, and A. E. Clarke. 1982. The effect of isolated components of *Prunus avium* L. styles on in vitro growth of pollen tubes. Planta 156(6):517–519. [103–104, 114, 126–127, 129]

Williams, N. H. 1983. Floral fragrances as cues in animal behavior. Chapter 3, pages 50–72 *in* C. E. Jones and R. J. Little, editors. Handbook of experimental pollination biology. Scientific and Academic Editions, Van Nostrand Reinhold Company, New York, New York, USA. [47, 49–50, 56, 451–452, 459, 462]

Williams, P. H., and C. B. Hill. 1986. Rapid-cycling populations of *Brassica*. Science 232(4756):1385–1389. [410]

Willmer, P. G. 1980. The effects of insect visitors on nectar constituents in temperate plants. Oecologia 47(2):270–277. [197, 207, 430, 447, 452]

Willmer, P. G., and S. A. Corbet. 1981. Temporal and microclimatic partitioning of the floral resources of *Justicia aurea* amongst a concourse of pollen vectors and nectar robbers. Oecologia 51(1):67–78. [65]

Willmer, P. G., and D. M. Unwin. 1981. Field analyses of insect heat budgets: reflectance, size and heating rates. Oecologia 50(2):250–255. [314, 318]

Willson, M. F., and R. I. Bertin. 1979. Flower-visitors, nectar production, and inflorescence size of *Asclepias syriaca*. Canadian Journal of Botany 57(12):1380–1388. [33, 161, 389]

Willson, M. F., and N. Burley. 1983. Mate choice in plants. Princeton University Press, Princeton, New Jersey, USA. [248, 254, 256]

Bibliography

Willson, M. F., and B. J. Rathcke. 1974. Adaptive design of the floral display in *Asclepias syriaca* L. American Midland Naturalist 92(1):47–57. [46]

Wilson, P., and J. D. Thomson. 1991. Heterogeneity among floral visitors leads to discordance between removal and deposition of pollen. Ecology 72(4):1503–1507. [253, 371–372, 391]

Wineriter, S. A., and T. J. Walker. 1984. Insect marking techniques: durability of materials. Entomological News 95(3):117–123. [325]

Wittmann, D., and E. Scholz. 1989. Nectar dehydration by male carpenter bees as preparation for mating flights. Behavioral Ecology and Sociobiology 25:387–391. [301, 304]

Wolf, L. L., and F. R. Hainsworth. 1971. Time and energy budgets of territorial hummingbirds. Ecology 52(6):980–988. [319]

Wolf, L. L., F. R. Hainsworth, and F. B. Gill. 1975. Foraging efficiencies and time budgets in nectar feeding birds. Ecology 56(1):117–128. [319]

Wolf, L. L., F. G. Stiles, and F. R. Hainsworth. 1976. Ecological organization of a tropical highland hummingbird community. Journal of Animal Ecology 45(2):349–379. [319]

Wolfe, A. D.; J. R. Estes; and W. F. Chissoe, III. 1991. Tracking pollen flow of *Solanum rostratum* (Solanaceae) using backscatter scanning electron microscopy and x-ray microanalysis. American Journal of Botany 78(11):1503–1507. [145, 462–463]

Wolfe, L. M., and S.C.H. Barrett. 1987. Pollinator foraging behavior and pollen collection on the floral morphs of tristylous *Pontederia cordata* L. Oecologia 74(3):347–351. [362]

Wolfe, L. M., and S.C.H. Barrett. 1988. Temporal changes in the pollinator fauna of tristylous *Pontederia cordata,* an aquatic plant. Canadian Journal of Zoology 66(6):1421–1424. [391]

Wolfe, L. M., and S.C.H. Barrett. 1989. Patterns of pollen removal and deposition in tristylous *Pontederia cordata* L. (Pontederiaceae). Biological Journal of the Linnean Society 36(4):317–329. [80, 96, 372]

Woodell, S.R.J. 1978. Directionality in bumblebees in relation to environmental factors. Pages 31–39 *in* A. J. Richards, editor. The pollination of flowers by insects. Linnean Society Symposium Series No. 6. Academic Press, London, England. [344, 346]

Wooller, R. D., K. C. Richardson, and C. M. Pagendham. 1988. The digestion of pollen by some Australian birds. Australian Journal of Zoology 36(4):357–362. [299, 387]

Wooller, R. D., E. M. Russell, and M. B. Renfree. 1983. A technique for sampling pollen carried by vertebrates. Australian Wildlife Research 10:433–434. [297]

Wright, D. H. 1985. Patch dynamics of a foraging assembly of bees. Occologia 65:558–65. [432]

Wright, S. 1943. Isolation by distance. Genetics 28:114–138. [231]

Wright, S. 1946. Isolation by distance under diverse mating systems. Genetics 31:39–59. [231]

Wyatt, R. 1982. Inflorescence architecture: how flower number, arrangement, and phenology affect pollination and fruit-set. American Journal of Botany 69(4):585–594. [45]

Wyatt, R. 1983. Pollinator-plant interactions and the evolution of breeding systems. Chapter 4, pages 51–95 *in* L. Real, editor. Pollination biology. Academic Press, Orlando, Florida, USA. [228, 475]

Wyatt, R., S. B. Broyles, and G. S. Derda. 1992. Environmental influences on nectar production in milkweeds (*Asclepias syriaca* and *A. exaltata.* American Journal of Botany 79(6):636-642. [18, 160]

Bibliography

Yeates, D., and G. Dodson. 1990. The mating system of a bee fly (Diptera: Bombyliidae). I. Non-resource-based hilltop territoriality and a resource-based alternative. Journal of Insect Behavior 3(5):603–617. [325]

Young, A. M. 1989. Pollination biology of *Theobroma* and *Herrania* (Sterculiaceae). IV. Major volatile constituents of steam-distilled floral oils as field attractants to cacao-associated midges (Diptera: Cecidomyiidae and Ceratopogonidae) in Costa Rica. Turrialba 39(4):454–458. [271]

Young, D. R., and W. K. Smith. 1980. Influence of sunlight on photosynthesis, water relations, and leaf structure in the understory species *Arnica cordifolia*. Ecology 61(6):1380–1390. [396, 400]

Young, H. J. 1988. Differential importance of beetle species pollinating *Dieffenbachia longispatha* (Araceae). Ecology 69(3):832–844. [291, 488]

Young, H. J., and M. L. Stanton. 1990a. Influence of environmental quality on pollen competitive ability in wild radish. Science 248:1631–1633. [33, 151]

Young, H. J., and M. L. Stanton. 1990b. Influences of floral variation on pollen removal and seed production in wild radish. Ecology 71(2):536–547. [91, 97, 364]

Young, H. J., and T. P. Young. 1992. Alternative outcomes of natural and experimental high pollen loads. Ecology 73(2):639–647. [151, 254, 256]

Zeisler, M. 1938. Über die Abgrenzung der eigentlichen Narbenfläche mit Hilfe von Reaktionen. Beihefte zum Botanisches Zentralblatt A 58:308–318. [68]

Ziegler, H., U. Lüttge, and U. Lüttge. 1964. Die wasserlöslichen Vitamine des Nektars. Flora 154:215–229. [204]

Ziegler, H., and I. Ziegler. 1962. Die wasserlöslichen Vitamine in den Siebröhrensäften einiger Bäume. Flora 152:257–278. [204]

Zimmerman, M. 1981. Optimal foraging, plant density and the marginal value theorem. Oecologia 49(2):148– 153. [12]

Zimmerman, M. 1983a. Calculating nectar production rates: residual nectar and optimal foraging. Oecologia 58(2):258–259. [167]

Zimmerman, M. 1983b. Plant reproduction and optimal foraging: experimental nectar manipulations in *Delphinium nelsonii*. Oikos 41(1):57–63. [261, 370]

Zimmerman, M. 1987. Reproduction in *Polemonium*: factors influencing outbreeding potential. Oecologia 72(4):624–632. [12]

Zimmerman, M. 1988a. Pollination biology of montane plants: relationship between rate of nectar production and standing crop. American Midland Naturalist 120(1):50–57. [37]

Zimmerman, M. 1988b. Nectar production, flowering phenology, and strategies for pollination. Pages 157–178 *in* J. Lovett Doust and L. Lovett Doust, editors. Plant reproductive ecology. Oxford University Press, New York, New York, USA. [234, 254]

Zimmerman, M., and G. H. Pyke. 1988a. Reproduction in *Polemonium*: Assessing the factors limiting seed set. American Naturalist 131(5):723–738. [35, 255]

Zimmerman, M., and G. H. Pyke. 1988b. Pollination ecology of Christmas Bells (*Blandfordia nobilis*): effects of pollen quality and source on seed set. Australian Journal of Ecology 13(1):93–99. [120]

Zimmerman, M., and G. H. Pyke. 1988c. Pollination ecology of Christmas Bells (*Blandfordia nobilis*): patterns of standing crop of nectar. Australian Journal of Ecology 13:301–309. [161]

Bibliography

Zimmerman, M., and G. H. Pyke. 1988d. Experimental manipulations of *Polemonium foliosissimum*: effects on subsequent nectar production, seed production and growth. Journal of Ecology 76(3):777–789. [370]

Index of Subjects

Index of Subjects

Index of Subjects

Flowers: age, 157–158, 194; albino, 361; bagging, 75, 96, 160, 166, 251, 256, 390; clearing, 74; covering, 13; energy content, 46–47; excess, 253–254; hermaphroditic, 240; longevity, 44; manipulating, 362, 379; marking, 35–37; orientation, 44–45; pistillate, 240–241, 245, 260; presenting to pollinators, 364–365, 376; size, 43–44; staminate, 241, 245, 260; trip, 251. *See also* Artificial flowers; Bags
Flower visit consistency, 374–375
Flower volatiles. *See* Floral fragrance
Fluorescence microscopy, 2, 66, 114, 126, 133. *See also* Fluorochromatic procedure
Fluorescent powdered dyes: for marking pollinators, 323–326, 338; as a pollen analogue, 136–140, 259; reliability as a pollen analogue, 139
Fluorochromatic procedure, 98, 100, 106–108, 114
Fodder anthers, 145
Folin reagent, 206
Footprints, 339
Footprint substance, 366–367
Foraging behavior, 46, 154, 156, 209, 231, 252, 311, 314, 333, 335–380, 371, 374–376; effects of CO_2, 267, 308. *See also* Flower constancy; Optimal foraging
Foraging efficiency, 319
Forceps, 40, 120, 147, 164, 279, 290, 299, 302, 309–311, 326
Freezer, 267, 277, 310
Fruit set, 215, 251; resource limitation of, 244, 261
Functional gender, 94, 228, 230, 240–244
Fungus, 76
Fur dye, 335

G-tests, 246
Gametes: isolation of, 260
Gametic selection, 226
Garden sprayer, 165
Gas chromatography. *See* Chromatography
Geitonogamy, 33, 146, 220, 228, 244
Gelatin capsules, 17, 37
Gender: adjustment, 244–247; diphasy, 244–247; specialization, 243. *See also* Functional gender
Gene dispersal, 231–232
Gene flow, 3, 133, 140, 144, 231, 233–235; and flower constancy, 374

Generative cell (nucleus), 77, 99, 106
Genetic markers, 3, 58, 139, 144–145, 219, 221
Genital armature, 276
Glass chimney, 14
Glassine bags (envelopes), 14, 17, 38, 138, 239. *See also* Pollination bags
Glossa length, 308–310, 390. *See also* Proboscis length
Glossometer, 309
Gloves: handball, 318, 326
Glue, 162, 277, 285, 287, 326–327, 333, 371, 379
Glycerin (glycerol) jelly, 289–290, 310; slides, 20, 82–84, 90, 117, 289–291, 297
Glycosides, 154
Graduated cylinders: for making hummingbird feeders, 359
Gram stain, 209
Graphics tablet, 125
Gray scale, 61, 64. *See also* Reflectance scale
Greenhouse, 13, 14, 28, 338, 385
Grooming behavior, 288, 291, 324, 326, 332, 377
Growth chamber, 13, 14, 160
Gut contents, 273
Gypsum blocks, 402

Hand lens, 116, 136
Handling time, 336, 340–341, 360, 377
Hand-pollination, 67, 75, 93, 106, 119–123, 132, 222, 255–256, 258
Hanging-drop cultures, 104–105
Hardy-Weinberg equilibrium, 225
Heat budgets, 318
Heliotropism, 44
Hemacytometer, 95, 96, 291, 372
Herbarium specimens, 11
Heteromorphic systems. *See* Heterostyly
Heterostyly, 94, 117, 238–239
Highlighter pen, 35
High-performance liquid chromatography (HPLC). *See* Chromatography
Histidine scale, 195–196
Hoffman Modulation Contrast Optics, 119, 297
Honey, 197, 200–201, 204, 210; for artificial flowers, 352, 363; as a bait, 288
Honey pot, 301
Honeysac, 212, 302
Honey stomach, 301–302, 304
Hoyer's medium, 84–85, 90
HPLC. *See* Chromatography

Index of Subjects

Index of Plants

Genus and Species Names

Index of Plants

Common Names

Index of Plants

Index of Animals

Genus and Species Names

Agrostis ipsilon, 322
Apis, 59. *See also* Honeybees

Bombus, 304, 308. *See also* Bumblebees
Bombylius fuliginosus, 378. *See also* Bombyliidae, Bee flies

Comptosia, 314

Donacia, 30
Drosophila, 324, 388

Hadena, 30
Heriades, 215
Hypanthidium, 215

Lygus, 323

Macroglossum stellatarum, 378
Megistorhynchus longirostris, 306
Melipona, 85
Myrmecia, 114

Nomada, 235

Osmia, 300, 384

Petaurus australis, 389
Pieris rapae, 355
Psithyrus, 309

Selasphorus platycercus, 25. *See also* Broad-tailed Hummingbird

Tegeticula, 30
Trigona, 39; *carbonaria,* 115

Vespa germanica, 115
Vespula pensylvanica, 273

Xylocopa, 301, 304, 320. *See also* Carpenter bees

Common Names

Anthophorine bees, 213
Anthomyiid flies, 323
Ants: artificial flowers for, 354, 358; effects on pollen viability, 6, 114; excluding, 19, 38–39, 348; marking, 328
Aphids, 18–19, 221, 274
Arctiid moths, 330

Bats, 14, 288; marking, 334–335; viewing, 388
Bee flies, 139, 168, 235, 310, 345. *See also* Bombyliidae
Bees, 148, 288, 380; and artificial flowers, 345; collecting, 265, 271, 279; collecting pollen from, 291; marking, 325–326, 329, 383; measuring proboscis lengths, 310; oil-collecting, 308; pollen in feces, 297; population size, 381; recording sounds from, 340; seed set from, 252; trap nests for, 383–384; weighing, 301. *See also* Bee flies; Bumblebees; Hymenoptera; Solitary bees
Beetles, 265, 291, 297, 330–331, 378; color preferences, 368. *See also* Coleoptera
Birds, 299; banding, 332; bill length, 305, 312–313, 409; netting and banding, 14, 284–288; permits for banding, 284–334. *See also* Honeycreepers; Honeyeaters; Hummingbirds; Sugarbirds; Sunbirds
Blowflies, 387. *See also* Calliphoridae
Boll weevils, 299
Bombyliid flies, 378. *See also* Bee flies
Bowerbirds, 339
Broad-tailed Hummingbird, 25, 333, 363, 382. *See also* Hummingbirds
Brimstone butterflies, 322
Bumblebees, 116, 117, 135, 167, 244, 269, 291, 304, 338; captive colonies, 385; collecting nectar from, 301–302; community structure, 305; counting at nests, 383; energy budgets, 319–320; flight temperatures, 317; following,

Index of Chemicals

Index of Chemicals

Index of Chemicals

Index of Chemicals

Phosphoric acid, 187–188, 193, 206–207, 380
Phosphorus pentoxide, 198
O-phthaldialdehyde, 194
Phthalic acid, 187
Platinic chloride, 203
Polyethylene glycol (PEG), 104
Polyoxyethylene [20] sorbitan monooleate, 143
Polysaccharides, 133, 187, 210
Polyvinyl alcohol, 85, 95, 113
Porapak Q, 50–53
Potassium: in nectar, 210, 301, 304, 320
Potassium acetate, 399–400
Potassium biphthalate, 168
Potassium carbonate, 80
Potassium chloride, 103
Potassium cyanide, 264–265
Potassium dichromate, 73
Potassium hydroxide, 80, 82, 141, 175–176, 192, 293, 298–299, 309–310
Potassium iodide, 104, 113, 150, 201, 203
Potassium nitrate, 102–103
Potassium permanganate, 203; reagent, 203
Potassium phosphate, 103, 127; tribasic, 128
Proline, 195
Propanol, 186, 198
Proprionic acid, 71
Pyridine, 184, 186, 192
Pyridoxine, 154, 204
Pyrrolizidine alkaloids, 154, 200, 273

Raffinose, 154
Reducing sugars, 189
Reynold's lead citrate, 89
Rhodamine, 135
Rhodium, 142
Riboflavin, 154, 204
Rubidium, 321
Ruthenium red, 67, 289

Safranine, 298
Samarium, 142–143; oxide, 144; sequioxide, 143
Silica gel, 190
Silver nitrate, 189, 361; reagent, 189
Silver oxide, 189
Skatol, 272
Sodium: in nectar, 210, 320
Sodium acetate, 133, 155
Sodium amytal, 109
Sodium bicarbonate, 72, 74, 168, 196
Sodium carbonate, 80

Sodium chloride, 97, 123, 143
Sodium cyanide, 264–265
Sodium hydroxide, 70, 104, 106, 119, 127, 180, 189–190, 206
Sodium molybdate, dihydrate, 104
Sodium nitrite, 70
Sodium potassium tartrate, 179–180
Sodium pyrophosphate, 201–202
Sodium succinate, 109
Sodium sulfite, 70, 127, 129, 380
Somogyi's copper–phosphate reagent, 180
Sorensen's phosphate buffer, 108
Spurr's medium, 89
Spurr's resin, 133
Sterols, 208
Stockwell's solution, 73–74
Succinic acid, 154
Sucrose, 101–104, 107, 109, 114, 130, 133, 153–154, 168, 174–176, 179, 190, 193, 209–213, 304, 314, 347; optimal concentrations, 373; solution for artificial flowers or feeders, 350, 353, 358–360, 363, 367, 369–370, 399–400
Sudan black B, 57, 66, 136, 208
Sudan III, 208
Sudan IV, 208
Sulfur dioxide, 380
Sulfuric acid, 80–81, 118, 175–178, 181–182, 186, 212, 400

Tartaric acid, 154
Tenax, 50–52, 55
Terpenes: bait for euglossine bees, 271–272
Terpenoids, 57, 146
Tetrasaccharides, 190
Tetrazolium, 31
Thiamin, 154, 204
Thioglycollate medium, 209
Threonine, 197
Tin, 145
Toluidine blue, 72, 128, 136
Tribasic potassium phosphate, 129
Trichloroacetic acid, 188, 205
Triketohydrindene hydrate (ninhydrin), 195
Trimethylchlorosilane, 192
Trimethylsilyation procedure, 192–193
Triphenyl tetrazolium, 109; chloride, 189; chloride enzyme test, 109
Triphenyltetrazolium chloride reagent, 189–190
Trisaccharides, 184, 190
Tris-citrate buffer, 205

Index of Chemicals

Tris-HCl buffer, 109, 150
Triterpenes, 215
Trypan blue, 131
Trypan red, 135
Tween 20, 143

Uranyl acetate, 67, 89, 133

Xylene, 71, 73, 84, 268

Zinc, 145
Zinc sulfate, heptahydrate, 104